Preface

The technology of industrial hydraulics has been developing at a high rate for some time. With the beginning of World War II, hydraulics filled a need for an energy transmission and control system with muscle which could be easily adapted to automated machinery.

During this period, Parker Hannifin was also growing and today has evolved into a full-line manufacturer of motion control components. The material in this text is a by-product of Parker Hannifin's experience in the areas of design, manufacture, application, and servicing of hydraulic components and systems over the years. As a result, we feel the text material is pertinent and accurately describes industrial hydraulics as it is presently seen.

The organization of the text material is designed for the beginning student. Starting with "The Physical World of a Machine," the student is led through topics ranging from fluids and basic physical concepts to component operation and its typical system application.

In some instances, methods of verifying component performance and maintenance procedures are illustrated. The intent of these items is to indicate how individual components as well as system performance can be determined and/or improved. They are not intended to show how components are disassembled and serviced; that procedure is the concern of maintenance manuals.

Also, in some instances we have taken metric conversions and rounded them up to the nearest whole number. This was done for ease of calculation.

Exercises are placed at the end of the chapters. These exercises are designed to be a summary of the text high points and at the same time, to be interesting and self-checking.

We hope that the student will find the course of study logical and easily understood. If we have failed in this regard, your comments will be greatly appreciated.

M & C Training Department W3MC01
Hydraulics Group
Parker Hannifin Corporation
6035 Parkland Blvd.
Cleveland, OH 44124-4141
Phone 216/896-2495
Fax 216/514-6738
E-mail mctrain@parker.com

Table of Contents

Chapter 1

The Physical World of a Machine

Machinery was invented to perform work. Machines were invented to be servants, yet most individuals are afraid of or puzzled by machines because they don't understand how they operate.

Machines are surrounded by physical elements which hinder the performance of work. To understand, then, how a machine operates, we will look at and attempt to define some of these elements, and then determine how a machine contends with them.

NOTE: The elements with which a machine deals are defined in this and the following sections. These definitions are not intended to be all inclusive, but only show in what sense these terms will be used throughout this text.

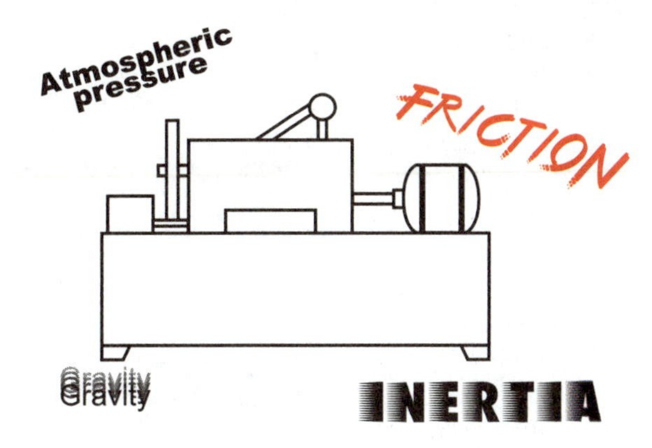

Figure 1-1

Force

A force is any influence capable of producing a change in the motion of a body.

Force

Figure 1-2

Pound

One unit for measuring force is the pound (Newton).

Changes Motion

Force, as we will deal with it, can change the motion of a body in basically three ways:

1. It can cause a body to move.
2. It can retard or stop a body which is moving.
3. It can change the direction of motion.

Resistance

Figure 1-3

Resistance

Any force which can stop or retard the movement of a body is a resistance. Examples of resistances are friction and inertia during acceleration.

Friction as Resistance

Frictional resistance is always present between the contacting surfaces of two objects when they are moving across one another.

Figure 1-4

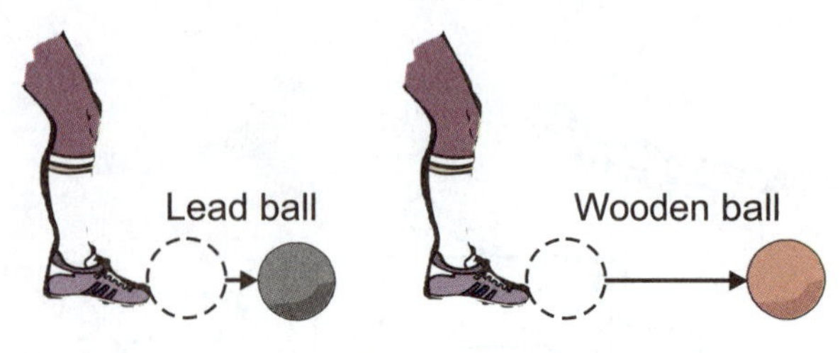

Figure 1-5

Inertia as Resistance

Inertia is the reluctance of a body to change in its motion.

Inertia is directly related to the quantity of matter in a body. The more mass or matter an object has, the heavier it is, and consequently the harder it is to move.

The inertia of a lead ball is greater than a wooden ball. If both balls are kicked with the same force, the wooden ball will move faster and farther than the lead ball. The lead ball has more of a resistance to being moved.

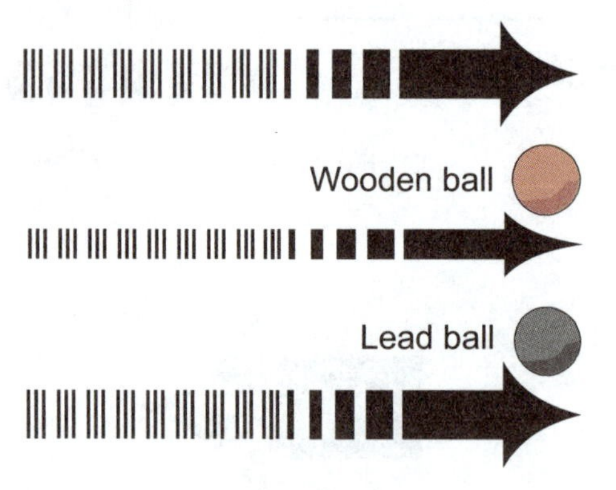

Figure 1-6

Energy

A force which can cause a body to move is energy.

Inertia as Energy

Inertia, being the reluctance of a body to a change in its motion, can also be energy. A moving body exhibits a reluctance to be stopped, and can, therefore, strike another body and cause its motion.

With a wooden ball and lead ball moving at the same speeds, the lead ball exhibits more inertia since it is more difficult to stop. The lead ball has more kinetic energy than the wooden ball.

Figure 1-7

Some Forms of Energy

Some forms of energy are mechanical energy, heat energy, electrical energy, light energy, chemical energy, and sound energy.

Law of Conservation of Energy

The principle of conservation of energy says that energy can neither be created nor destroyed, although it may change from one form to another.

Energies Change Form

Energy exists under various forms, and has the ability to change from one form to another. For instance, electrical energy may be changed to several other forms. Depending upon what device or appliance is plugged into the outlet, electrical energy changes to light energy, heat energy, mechanical energy, or sound energy.

Another example of energy changing form is a person sliding down a rope. When it comes time to stop or slow down, the rope is squeezed and some mechanical energy of the falling body is changed into heat energy, as most people are well aware.

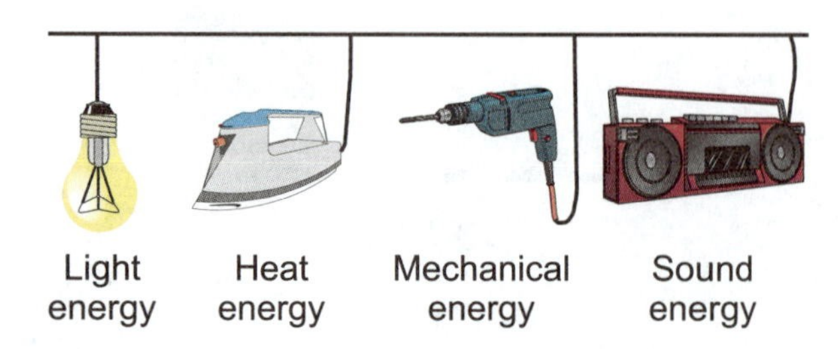

Figure 1-8

Energy States

An important consideration when dealing with energy is the state or condition in which it is found.

Kinetic State of Energy

Energy in the kinetic state is moving. It is an indication of the amount of work done on, or the amount of work an object can do.

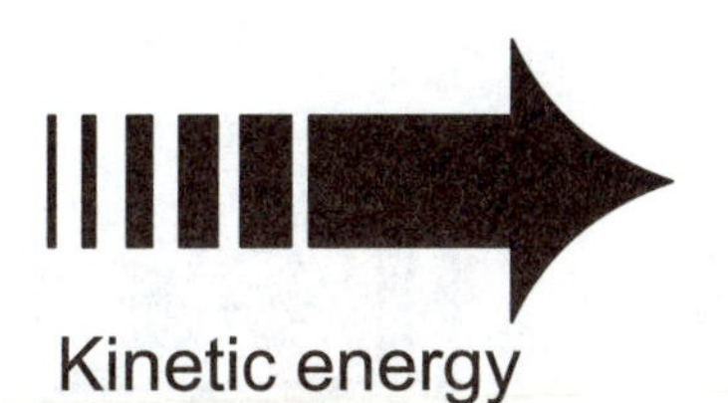

Figure 1-9

Potential State of Energy

When in a potential state, energy is stored. It is waiting to spring into action, to change to a kinetic state as soon as an opportunity arises. Potential energy has the ability to become kinetic because of its physical makeup or its position above a reference point.

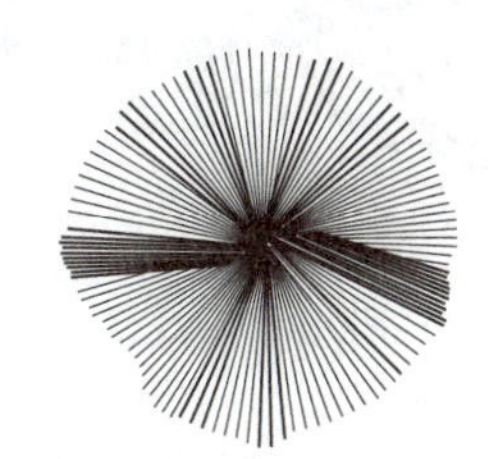

Potential energy

Figure 1-10

Because of its elevation, the water contained in a water tower is potential energy. It has the ability to be drawn off at a household water tap at a lower level.

Figure 1-11

A storage battery, when not connected in a circuit, is in a state of potential energy. Because of their physical makeup, the chemicals in a battery have the ability to change to electrical kinetic energy.

Energies Change States

As has been discussed, potential energy has the ability to change to kinetic. But, kinetic energy can change to potential as well. The water in a water tower is potential energy which changes to hydraulic kinetic energy at a water tap. This kinetic energy changes to a potential state while it fills a glass.

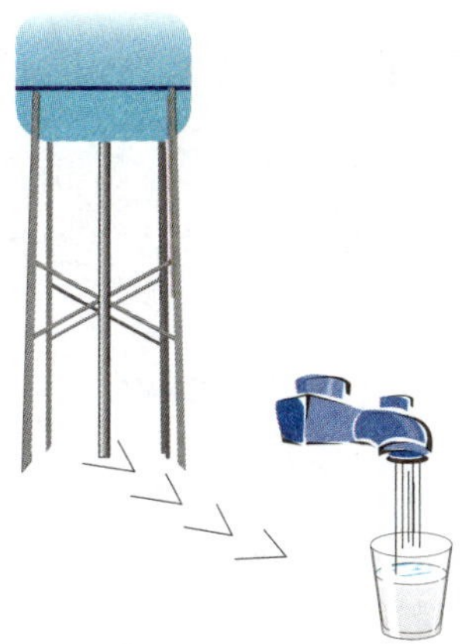

Figure 1-12

It is not only convenient that energy can change from one state to another, but most forms of energy must be in the kinetic state before any work can be done.

Work

Work, as we will deal with it, is the application of a force to cause movement of an object through a distance.

“Work” is getting things done. Machinery exists to perform work.

Figure 1-13

Foot—Pound

The unit for measuring work is the ft. lb. (Joule = Newton-meter).

Description of Work

The expression which describes work is:

work	=	distance moved	x	force exerted
ft. lbs.		ft.		lbs.
Joule		meter		Newton

An example of doing work would be a forklift loading a truck. If the forklift exerted a force of 2000 lbs. (8880 Newton) over a vertical distance of 5 ft. (1.524 m) to load each pallet, then 10,000 ft. lbs. (13533.1 N•m or 13533.1 J) of work would be done per pallet.

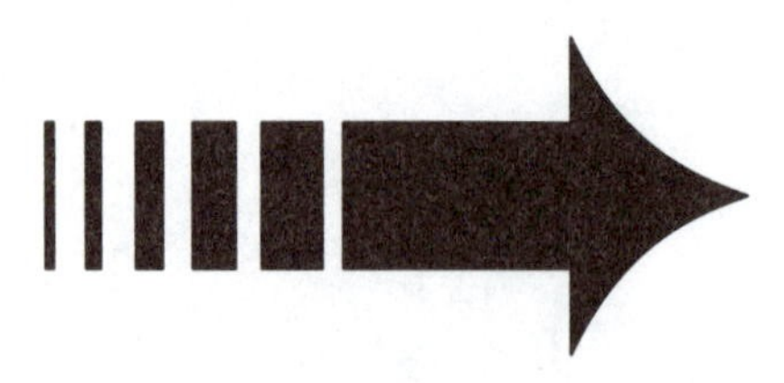

Power

Figure 1-14

Power

Most work is done within a certain time. Power is the speed or rate at which work is done.

Description of Power

The expression which describes power is:

$$\text{power} = \frac{\underset{\text{ft. (meter)}}{\text{distance moved}} \times \underset{\text{lbs. (Newton)}}{\text{force exerted}}}{\text{time (seconds)}}$$

In our example of loading the truck, if the 10,000 ft. lbs. were done in 5 seconds, the rate of doing work would be:

$$\frac{10{,}000 \text{ ft. lbs.}}{5 \text{ seconds}} = 2000 \text{ ft. lbs./sec OR}$$

$$\frac{13{,}533.1 \text{ J}}{5 \text{ seconds}} = 2706.6 \text{ J/sec}$$

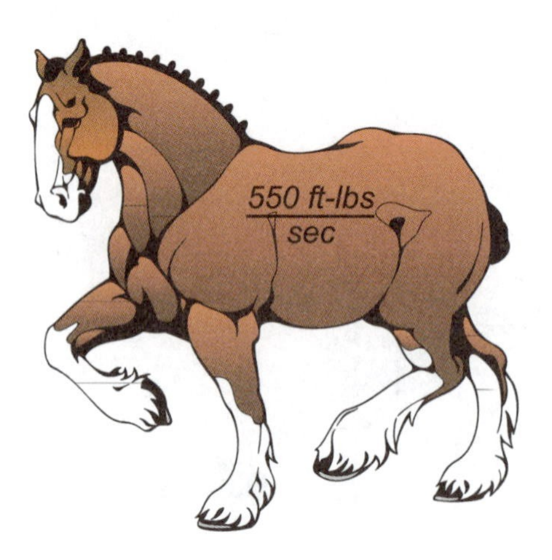

Figure 1-15

Horsepower

The unit for measuring power is horsepower. James Watt, the inventor of the steam engine, wanted to compare the amount of power his engine could produce against the power produced by a horse. From experimentation, Watt discovered that a horse could lift a 550 lb (2446 N•m) load one foot (.3048 m) in one second. He thus said that one horsepower is equal to 550 ft. lbs. per second (746 Watts).

One horsepower is equal to approximately 746 watts or .746 kW.

Description of Horsepower

The expression that describes horsepower is:

$$hp = \frac{\text{distance moved (ft.)} \times \text{force exerted (lbs.)}}{\text{time (seconds)}} \div 550$$

Kilowatts = hp x .746 Using the example of loading the truck once again, the 2000 ft. lbs./sec. (2706.6 J/s) rate of doing work would translate into 3.6 horsepower (2706.6 J/s).

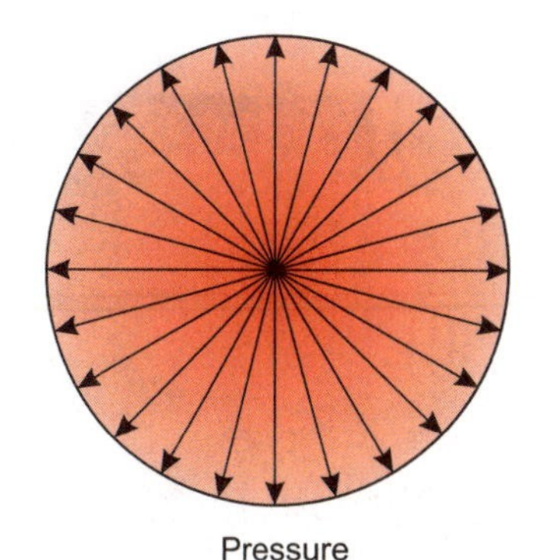

Figure 1-16

Pressure

Pressure is a measure of a force's intensity.

Many times, the intensity of a force is of more interest and of greater concern than the actual force itself. To determine pressure—the intensity of a force—the total force is divided by the area (square inches or square centimeters) on which it is acting. The result is the pressure (amount of force per square inch or square centimeter).

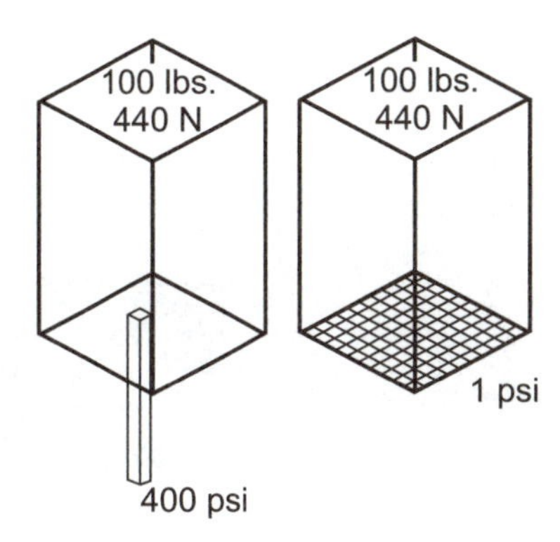

Figure 1-17

Description of Pressure

The expression used to describe pressure is:

$$\text{pressure} = \frac{\text{total force (lbs.)}}{\text{area on which total force acts (in}^2\text{)}}$$

To date, the accepted SI (meter) unit for pressure is the bar. $1 \text{ bar} = \frac{10^5 \text{ N}}{\text{m}^2} = 10^5 \text{ Pa}$

A 100 pound (444N) weight with a base area of 100 sq. in. (0.065 m^2) exerts a pressure of one pound per square inch (0.07 bar) on a stationary surface upon which it is laid.

The same 100 pound (444 N) weight placed on a square steel rod with a base area of 1/4 sq. in. (0.0001612 m^2) exerts a pressure of 400 psi (27.6 bar). The total forces are equal, but the intensities differ greatly.

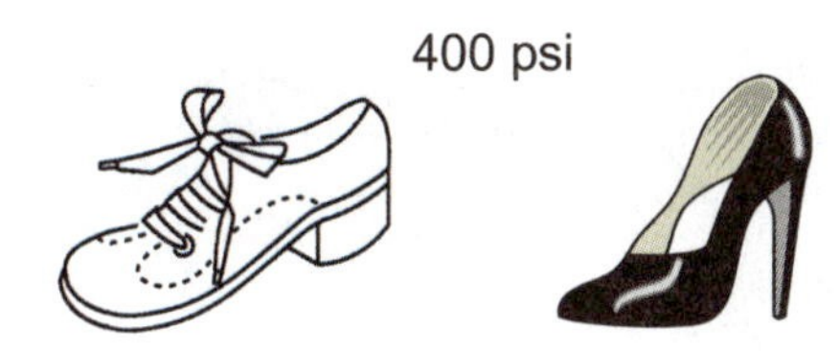

Figure 1-18

An example of this from everyday life would be the difference in pressures generated by heels of various styles of shoes. Anyone who has had his foot stepped on by a woman wearing high-heeled shoes with a tiny heel area knows how painful it can be. The same woman wearing low, large heel area shoes would probably cause less pain and discomfort.

Figure 1-19

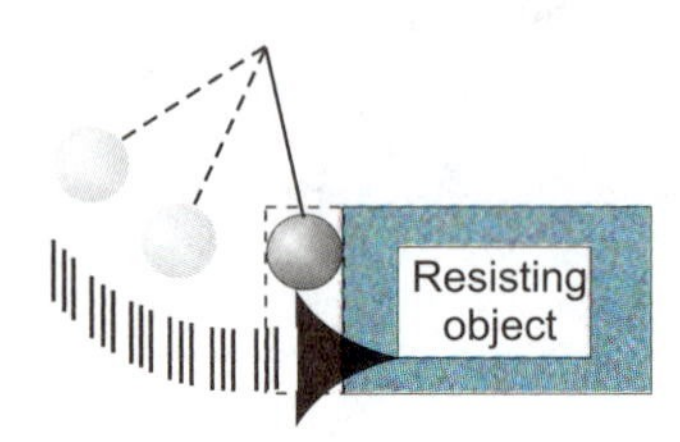

Figure 1-20

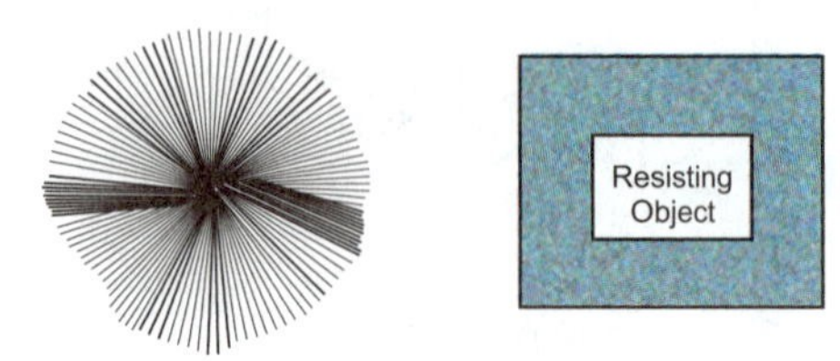

Figure 1-21

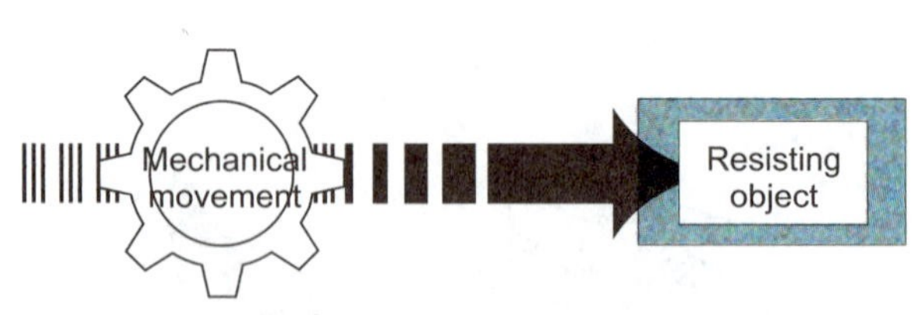

Figure 1-22

Working Energy

Kinetic energy used by a machine is usually in a working form which is characterized by an accompanying pressure. The pressure is the result of the force of kinetic energy being applied to a resistant object.

Working energy is the combination of kinetic energy and pressure.

Working Energy Changes Form

Common sense tells us that we can't get something for nothing. Something must happen to the energy in moving an object.

A swinging wrecking ball is an example of kinetic energy. If it is applied to a resisting object like a stone block, it will result in a movement of the block through a distance (assuming that the ball has sufficient inertia and will not rebound.) After this work is done, both the ball and object will stop.

What happens is that working energy is transformed. If the object moves along the same level, working energy changes to heat because of friction at the sliding surface of the object. If the object is raised to a higher level, as in the case of a forklift raising a pallet, working energy changes to potential energy.

In all instances of machines doing work, working energy is not destroyed, but changes form. Machines effect energy transformation in the process of doing work.

Transmission of Energy

Usually the source of energy for a machine is not at the point where it is to do work. Energy must be transmitted to the resisting object. This is usually done mechanically, electrically, pneumatically, or hydraulically.

Mechanical Transmission of Energy

In mechanical transmission, energy in the form of mechanical movement is transmitted and controlled through levers, chains, pulleys, belts, cams, and gears to the point where the work is to be done.

Electrical Transmission of Energy

In electrical transmission, energy in the form of electricity is transmitted and controlled through wire, to an electrical actuator where the work is to be done.

Pneumatic Transmission of Energy

In pneumatic transmission, energy in the form of compressed air flow is transmitted and controlled through a conduit system to a pneumatic actuator where the work is to be done.

Figure 1-23

Hydraulic Transmission of Energy

In hydraulic transmission, energy in the form of pressurized liquid flow is transmitted and controlled through a conduit system to a hydraulic actuator to the point of work.

Figure 1-24

For almost all machines, the energy that does the ultimate work is mechanical energy. Even other forms of energy transmission generally result in mechanical energy. For this reason, they require an actuator before the point of work. Actuators transform electrical, pneumatic, and hydraulic energy into mechanical energy.

Each method of energy transmission has its own advantages and disadvantages. For this reason, a machine may be equipped with a combination of mechanical, electrical, pneumatic, and hydraulic systems.

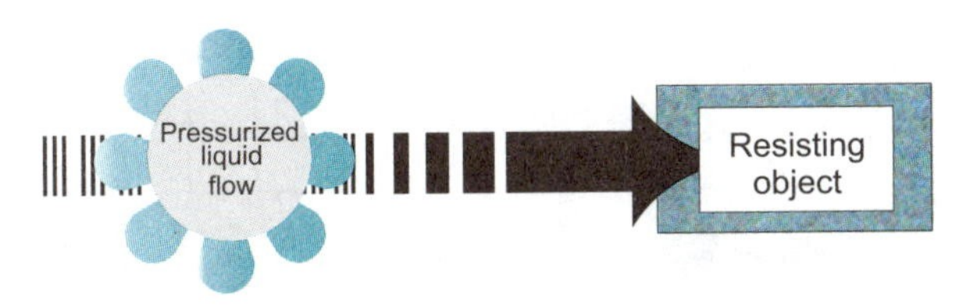

Figure 1-25

Inefficiency

The objective of the various transmission systems is to perform useful work; that is, move a resisting object through a distance. Useful work is performed by the application of kinetic energy to a surface of the resisting object. This we defined as working energy.

Energy transmitted through the various systems is also working energy. The conductors of energy in each system are physical objects with surfaces which also act as a resistance. The kinetic energy applies a pressure to the conductor's surface. This is working energy, but energy that performs non-useful work, since no resisting object is moved.

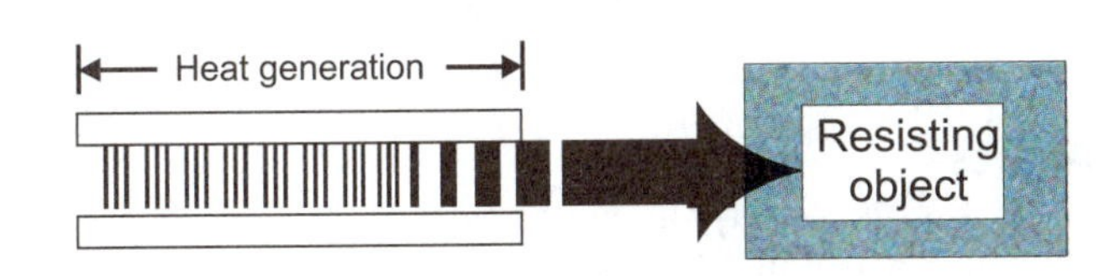

Figure 1-26

Traveling through the system, the pressure of the working energy becomes less and less as it gets to the point of work. This pressure is not destroyed, but changes to the form of heat energy because of friction. The degree to which this happens is a measure of a system's inefficency.

Exercise
The Physical World of a Machine
50 Points

Instructions: Complete the crossword puzzle.

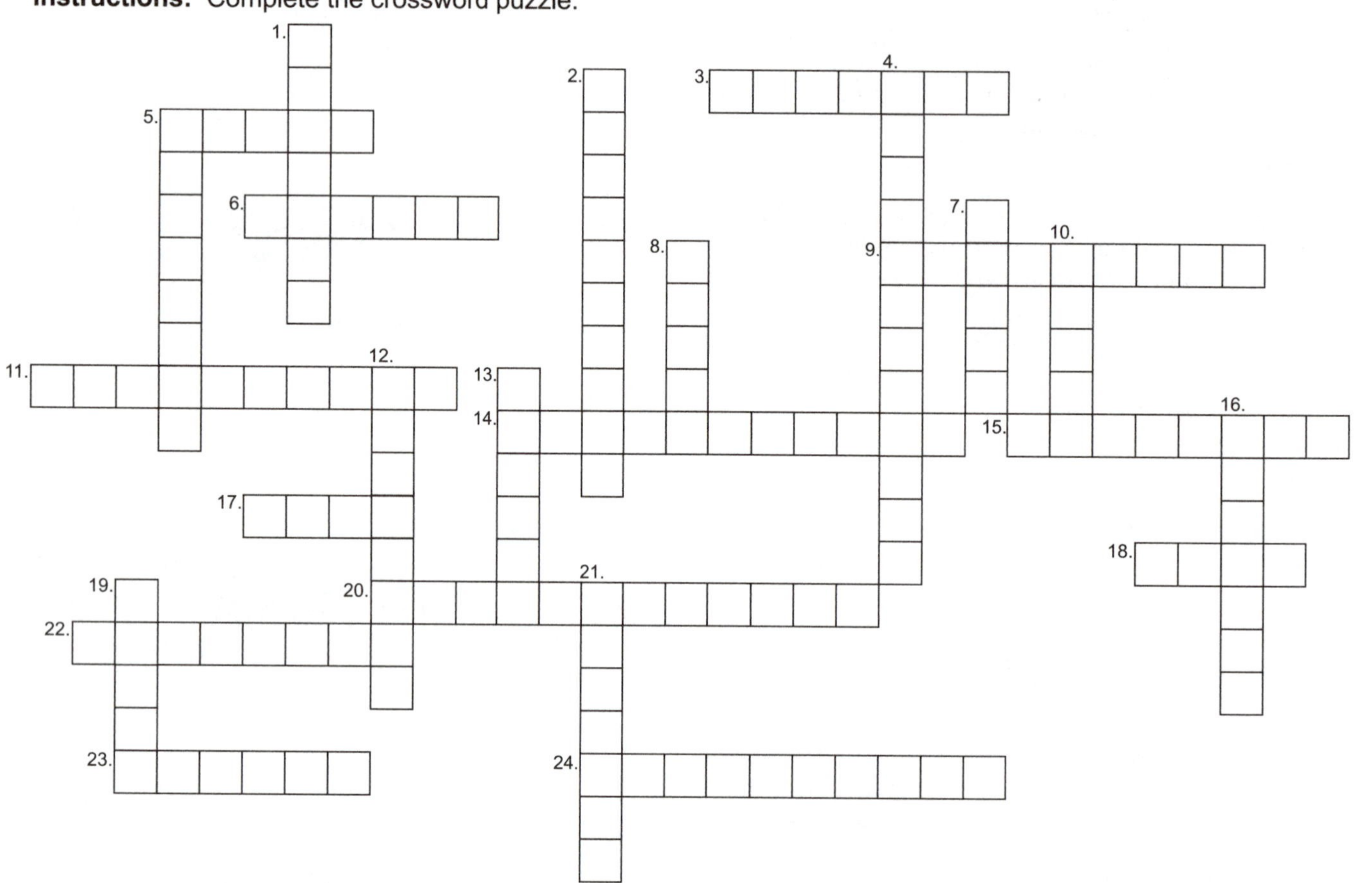

Across

3. Kinetic energy accompanied by pressure is _____ energy.
5. A force measure
6. Potential energy is _____ energy
9. A work measure
11. A means of energy transmission
14. Since the source of kinetic energy for a machine is not at the point of work, it must be _____.
15. Kinetic energy changes to heat when this resistance is overcome.
17. An energy form
18. Moving an object through a distance
20. Energy cannot be created or destroyed, so says this law
22. What is present when work is done?
23. Force which causes movement
24. Friction is a _____

Down

1. A state of energy which can do work
2. One type of energy transmission system
4. The degree to which a system's working energy decreases is a measure of a system's _____
5. Force intensity
7. A form of kinetic energy
8. Energy has many of these
10. Term indicating how long it takes to do work
12. Kinetic energy does work as it _____ an object's surface
13. Condition of various energies
16. Can be energy or resistance
19. It influences motion

21 Resistance _____ motion

CHAPTER 2

Hydraulic Transmission of Force and Energy

Before dealing with energy transmission through a liquid, it will help our understanding of hydraulics to first concentrate on some characteristics of a liquid, and then how a force is transmitted through a liquid.

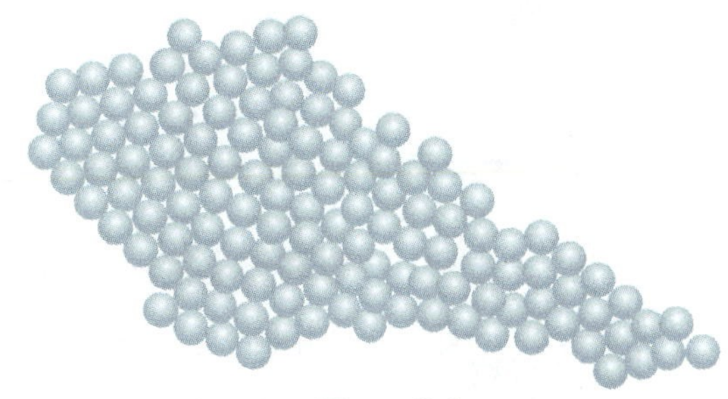

Figure 2-1

Fluid

A fluid is a substance that can flow and possess no definite shape. A fluid can be a liquid or gas.

Liquid

A liquid, like a gas, is a substance made up of molecules. Unlike a gas, however, these molecules are closely attracted to one another. Also, unlike a solid, these molecules are not so attracted to each other that they are in a rigid position.

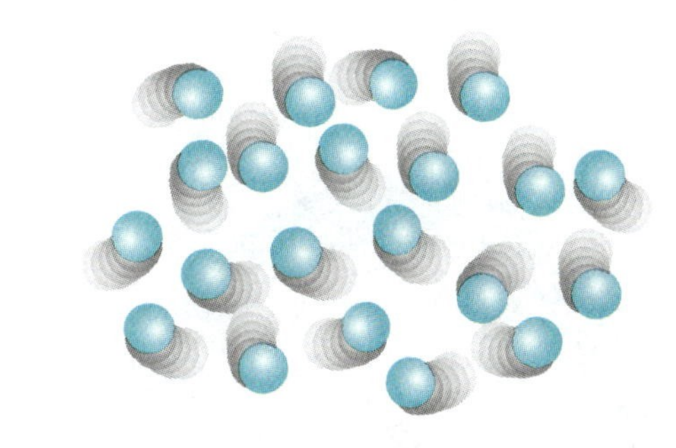

Figure 2-2

Molecular Energy

Liquid molecules are continuously moving. They slip and slide past one another even when the liquid is apparently at rest. This movement of molecules is molecular energy.

Liquids Take Any Shape

Since this slipping and sliding action is continuously taking place, a liquid is able to take the shape of any container. It may or may not fill the container depending upon the quantity of liquid. The ability of a liquid to accomplish this is directly related to its viscosity. Viscosity is discussed in more detail in later chapters.

Figure 2-3

Liquids are Relatively Incompressible

With molecules in close contact with one another, liquids exhibit a characteristic of solids. Liquids are relatively unable to be compressed.

This is the reason a diver tries to “knife” his way into the water and avoid a belly smacker.

Since liquids are relatively incompressible and can take the shape of any container, they possess certain advantages for transmitting a force.

Figure 2-4

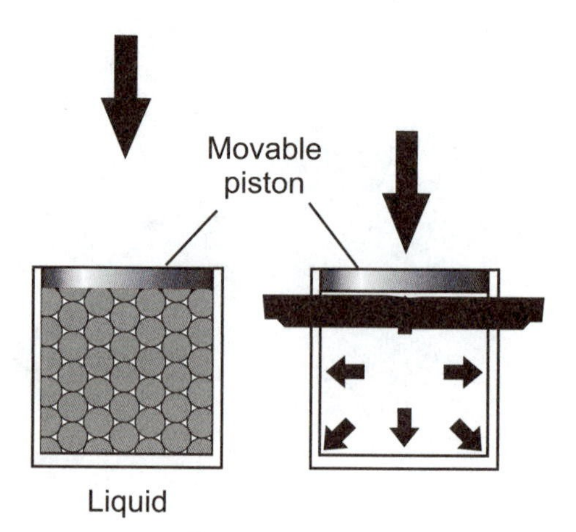

Figure 2-5

Figure 2-6

Force Transmission

The four methods of energy transmission (mechanical, electrical, hydraulic, pneumatic) are capable of transmitting a static force (potential energy) as well as kinetic energy. When a static force is transmitted in a liquid, it happens in a special way. To illustrate, we will compare how it is transmitted through a solid and through a confined liquid.

Force Transmitted Through a Liquid

Unlike a solid, a force applied to a confined liquid is transmitted equally throughout the liquid in the form of hydraulic pressure.

If we pushed on a container filled with liquid, the pressure of the applied force would be transmitted equally throughout the liquid.

A confined liquid will transmit pressure in the same manner regardless of how it is generated. As far as the liquid is concerned, an applied force results in pressure whether the application of force comes from a hammer, by hand, weight, fixed or adjustable spring, compressed air, or any combination of forces.

Since fluid can take the shape of any container, pressure will also be transmitted regardless of the shape of the container.

Pascal's Law

The property of a liquid to transmit pressure equally throughout itself is known as "Pascal's Law," in honor of Blaise Pascal who defined it.

The mathematical expression which describes Pascal's Law is the same as that used to describe any pressure:

$$\text{Pressure (psi)} = \frac{\text{force (lbs.)}}{\text{area (in2)}}$$

$$\text{Bar} = \frac{\text{force (Newton)}}{\text{area (meter2)}}$$

Pressure Gage

A pressure gage is a device which measures the intensity of a force applied to a liquid. Two types of pressure gages that are commonly used in hydraulic systems are the Bourdon tube gage and the plunger gage.

Bourdon Tube Pressure Gage

A Bourdon tube gage basically consists of a dial face calibrated in units of psi, bar, PA and needle pointer attached through a linkage to a flexible metal coiled tube, called a Bourdon tube. The Bourdon tube is connected to system pressure.

How a Bourdon Tube Gage Works

As pressure in a system rises the Bourdon tube tends to straighten out because of the difference in areas between its inside and outside diameters. This action causes the pointer to move and indicate the appropriate pressure on the dial face.

Bourdon tube gages are generally precision instruments with accuracies ranging from 0.1% to 3.0% of full scale. They are frequently used for laboratory purposes and on systems where pressure determination is important.

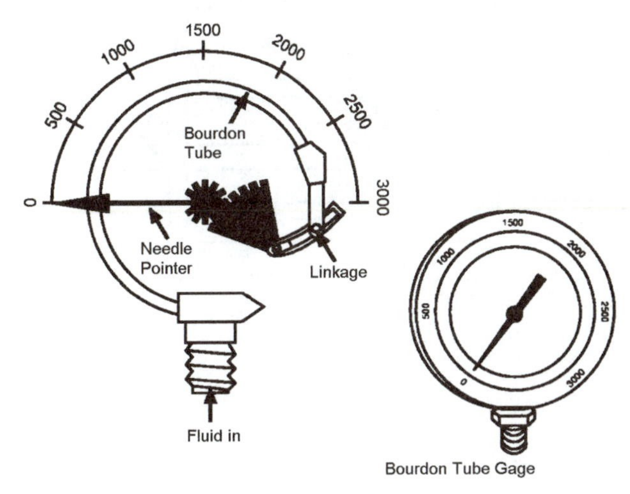

Figure 2-7

Plunger Pressure Gage

A plunger gage consists of a plunger connected to system pressure, a bias spring, pointer, and a calibrated scale calibrated in appropriate units of psi (bar).

How a Plunger Gage Works

As pressure in a system rises, the plunger is moved by the pressure acting against the force of the bias spring. This movement causes the pointer attached to the plunger to indicate the appropriate pressure on the scale.

Plunger gages are a durable, economical means of measuring system pressure.

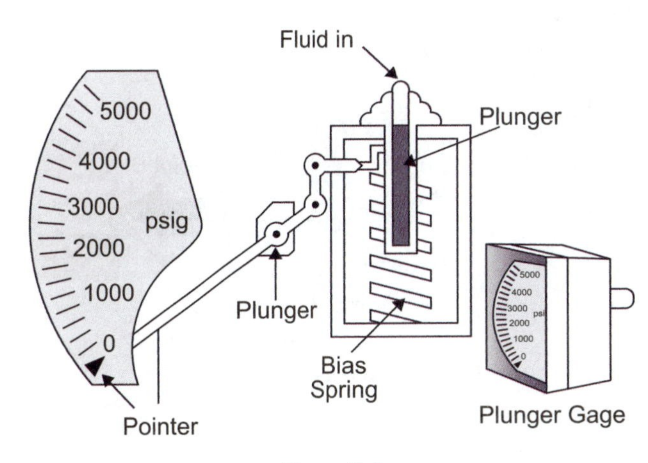

Figure 2-8

Hydraulic Pressure to Mechanical Force

Applying a force to a liquid and transmitting the resulting pressure throughout a liquid in various shaped containers does very little good for its own sake. Hydraulic pressure must be converted into mechanical force before any work can be done. This is the function of a hydraulic actuator - to accept hydraulic pressure and convert it into a mechanical force.

One type of actuator is the hydraulic cylinder.

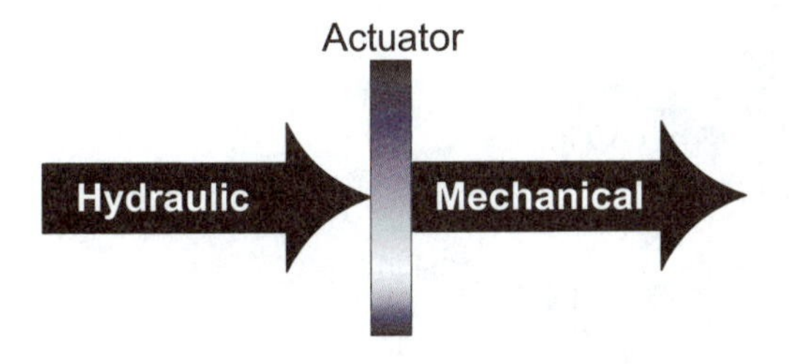

Figure 2-9

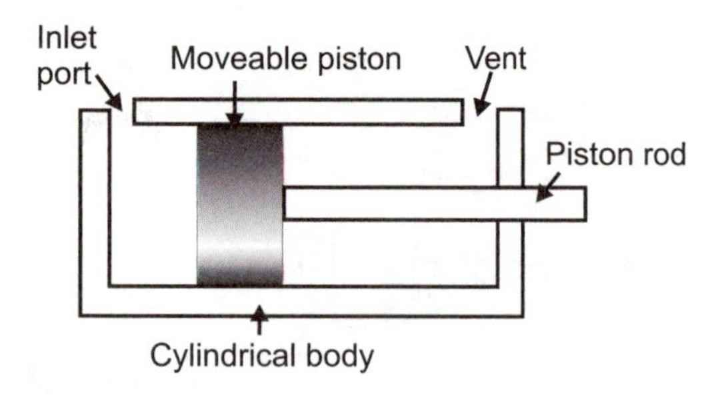

Figure 2-10

Hydraulic Cylinder

A hydraulic cylinder accepts hydraulic pressure and converts it into a straight-line or linear mechanical force and with the appropriate mechanical connections can convert to rotary motion.

What Cylinders Consist Of

Hydraulic cylinders basically consist of a cylinder body, a closure at each end, a movable piston, and attached to the piston a piston rod. At one end, the cylinder body has an inlet port by which the fluid enters the body. The other end is vented.

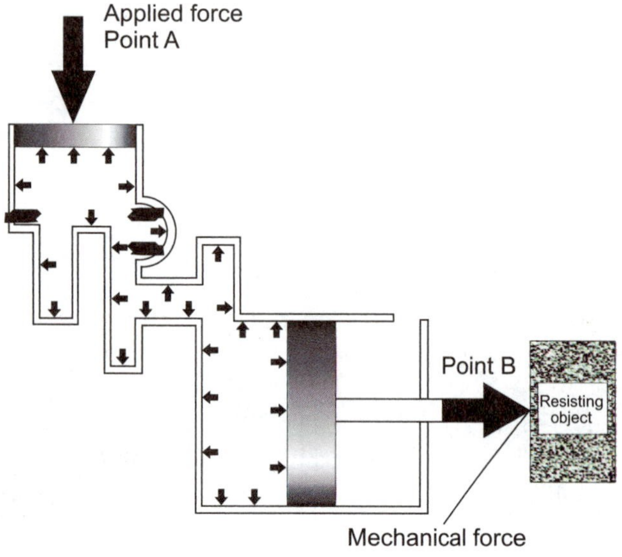

Figure 2-11

How Cylinders Work

With the inlet port of the cylinder connected to the system, the cylinder becomes part of the system. In our illustration, when a force is applied at point A, the resulting pressure is transmitted throughout the system and to the piston in the cylinder. The pressure acts on the piston resulting in a mechanical force at point B.

Applying Pressure

In transmitting pressure through a confined liquid, some sort of movable member has been used to apply the pressure. In the examples used so far, the movable member has been a piston.

To determine the intensity of the force or pressure being applied to the system, the force is divided by the area of the movable member. (P = F/A).

Mechanical Force Multiplication

Mechanical forces can be multiplied using hydraulics. The determining factor for force multiplication is the square inch area (cm^2) on which hydraulic pressure is applied. Since pressure is transmitted equally throughout a confined liquid, if the piston in a cylinder has more area than the movable member on which the force is added, the output force will be greater than the input force. With a smaller cylinder area the force would be less for the same system pressure (500 psi/34.5 bar).

In our example, assume that the resisting object is stationary and will not move. A 5000 lb. (22200 N) force acting on the 10 in^2 (65.52 cm^2) area piston results in a pressure of 500 psi (34.5 bar) being transmitted throughout the system. The 500 psi (34.5 bar) acts on the cylinder piston with an area of 15 in^2 (96.78 cm^2) and results in a mechanical force of 7500 lbs. (33360 N).

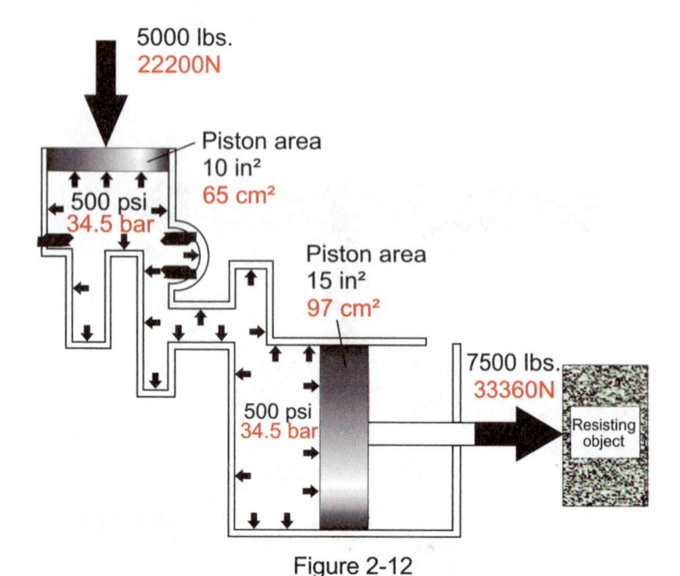

Figure 2-12

force = Pressure x area (lbs.)
(PSI) (in^2)

(Newton) = *(bar) (cm^2) x 10,000
*In Newton / m^2

Intensifier

An intensifier multiplies hydraulic pressure.

What Intensifiers Consist Of

The common intensifier basically consists of a housing with inlet and outlet ports and a large area piston connected by a rod to a small area piston. The area between the two pistons is vented.

How Intensifiers Work

An intensifier multiplies, or intensifies an existing hydraulic pressure by accepting hydraulic pressure at the large area piston and applying the resultant force to the small area piston. Fluid pressure is therefore intensified or multiplied at the actuator.

In our example, assume that the resisting object is stationary and will not move. A 5000 lb. (22200 N) force at point A ultimately results in a 30,000 lb. (133200 N) mechanical force at point B. This is accomplished by applying hydraulic pressures to various piston areas.

A common application for an intensifier is a clamping device.

Hydraulic Transmission of Energy

The reason for using hydraulics or any other type of energy transmission on a machine, is to perform useful productive work.

As illustrated previously, accomplishment of work requires the application of energy to a resisting object resulting in the object moving through a distance. To do any work then, a hydraulic system requires something which can apply energy.

Hydraulic Accumulator

Up to this point, the device used to apply a pressure to a confined liquid has consisted of a movable piston on which the force is applied and a cylinder body to confine the liquid. This device is called an accumulator.

F = P x A

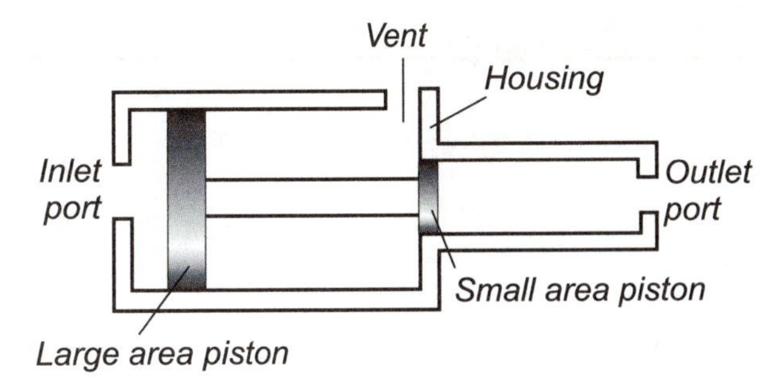

Figure 2-13

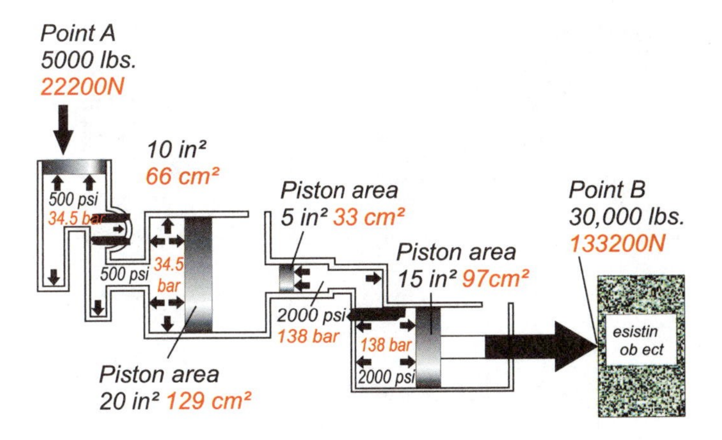

Figure 2-14

Figure 2-15

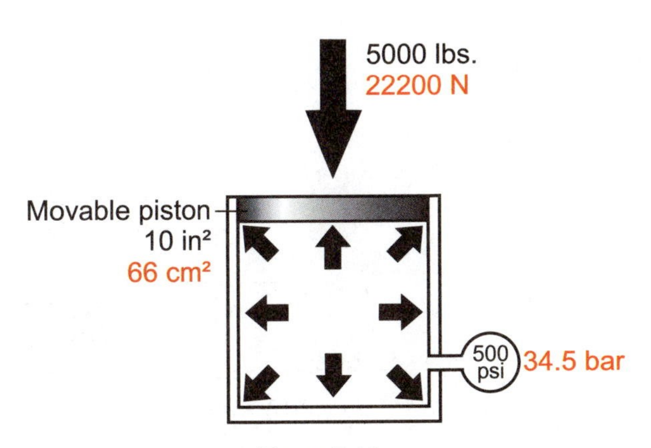

Figure 2-16

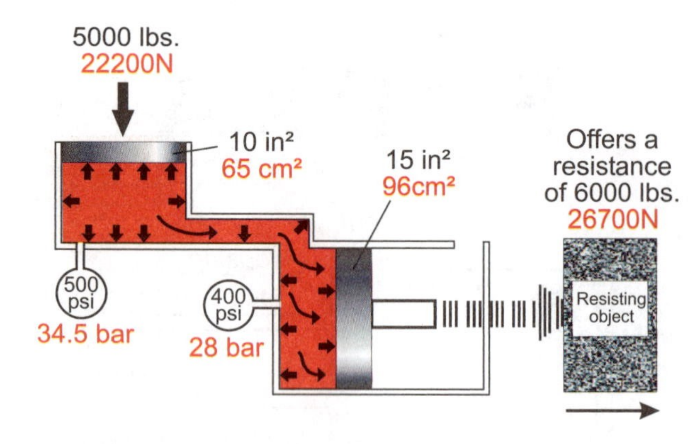

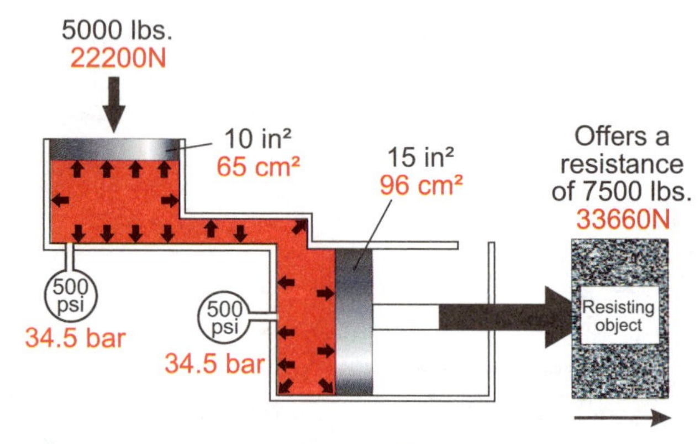

Figure 2-17

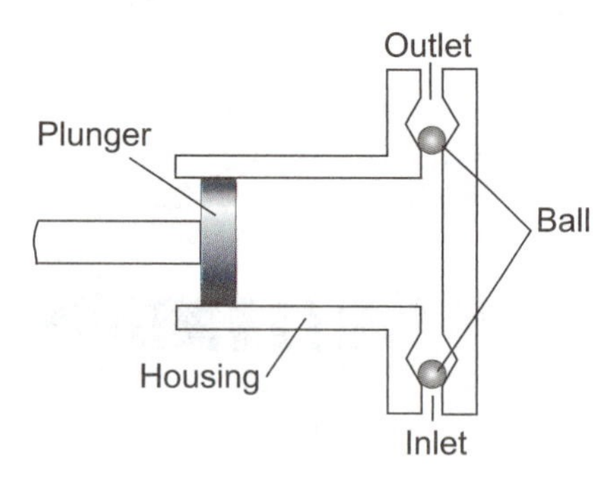

Figure 2-18

An accumulator may store hydraulic pressure which is potential energy. This pressure can turn to working energy (flow and pressure).

In our illustration, the pressure stored in the accumulator is used to move the resisting object. 400 psi (27.6 bar) of the 500 psi (34.5 bar) accumulator pressure is used to overcome the resistance offered by the load. The remaining pressure is transformed into a liquid flow.

An accumulator has its disadvantages when applying energy to a resisting object. If the resistance is large enough, there will be no movement and therefore no work. In addition, an accumulator has a definite volume. Once it is discharged, flow ceases.

A device is needed which can apply the necessary pressure to overcome a resistance and whose flow is continuous. This device is a positive displacement pump.

Positive Displacement Pump

With constant cycling of its movable member, a positive displacement pump will deliver a constant flow of liquid, and will apply (within limits) whatever pressure is required.

In other words, a positive displacement pump develops working energy - kinetic energy (flow) and pressure.

What a Positive Displacement Pump Consists Of

A positive displacement pump basically consists of a movable member inside a housing. The pump in the illustration has a plunger for a movable member. The shaft of the piston is connected to a prime mover (an engine or electric motor) which is modified to produce a constant reciprocating motion of the plunger. The inlet port is connected to fluid reservoir. At inlet and outlet ports, a ball allows liquid to flow only one way through the housing.

How a Positive Displacement Pump Works

As the piston is pulled out, an increasing volume is formed within the housing. The ball at the inlet port allows fluid to enter filling the void.

After the housing is filled, the plunger is pushed in. The ball at the inlet closes and the ball at the outlet port unseats. A decreasing volume is formed within the housing. The plunger applies the pressure necessary to get the liquid out into the system.

The constant cycling of the piston results in a pulsating rate of flow at whatever pressure is required.

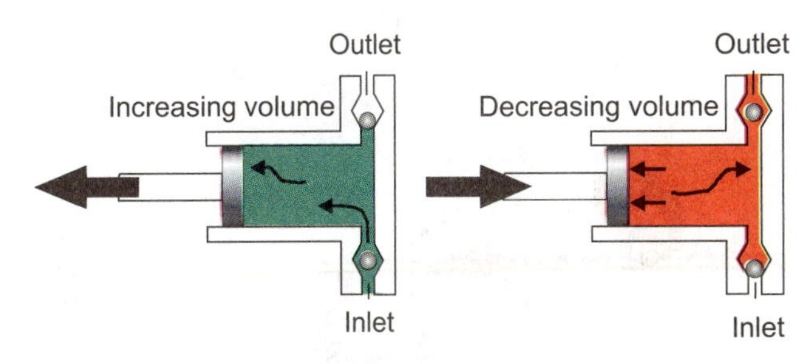

Figure 2-19

Rotary Positive Displacement Pump

The most commonly used positive displacement pump in a machine's hydraulic system is not a reciprocating plunger, but a rotary type. A rotary positive displacement pump develops a relatively smooth, pressurized flow of liquid, and it can be easily driven by an electric motor or internal combustion engine. For each revolution the pump is rotated, a definite volume of liquid is displaced.

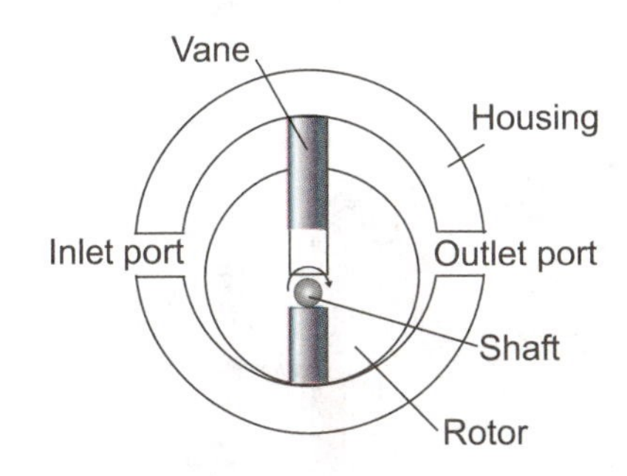

Figure 2-20

What Rotary Positive Displacement Pumps Consist Of

A rotary positive displacement pump basically consists of a housing with inlet and outlet ports, and a rotating group which develops fluid flow and pressure. One type of rotating group in the illustration consists of a rotor and vanes which are free to move in and out.

How the Pump Operates

The rotating group of the pump is positioned off-center to the housing. The rotor is connected to a prime mover by means of a shaft. As the rotor is turned, the vanes are initially thrown out by centrifugal force and contact the housing forming a positive seal. The increasing volume at the inlet side allows liquid to fill the pump. The decreasing volume at the outlet forcefully pushes the liquid out into the system. The amount of pressure applied to the liquid by the pump will only be as great as the least resistance to flow in the system.

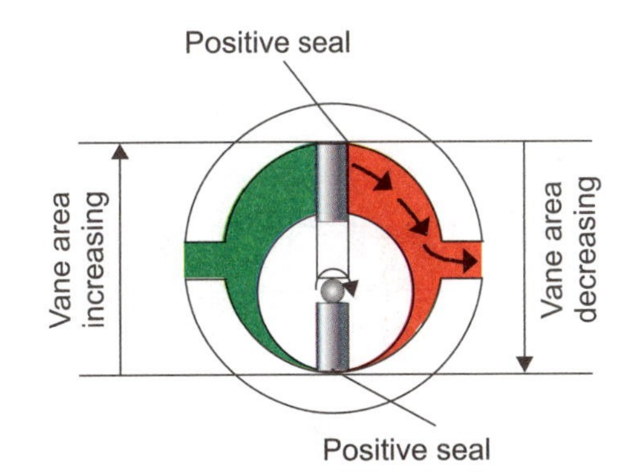

Figure 2-21

Resistance vs. Pressure

In a hydraulic system, there is a direct relationship between pressure and resistance. Pressure is applied by the pump in forcing fluid out into the system. The amount of pressure is determined by the degree of resistance. If the resistance is high, the pump must apply an equally high pressure. If the resistance is low, the pump applies a low pressure.

Resistance to flow determines how much pressure is exerted.

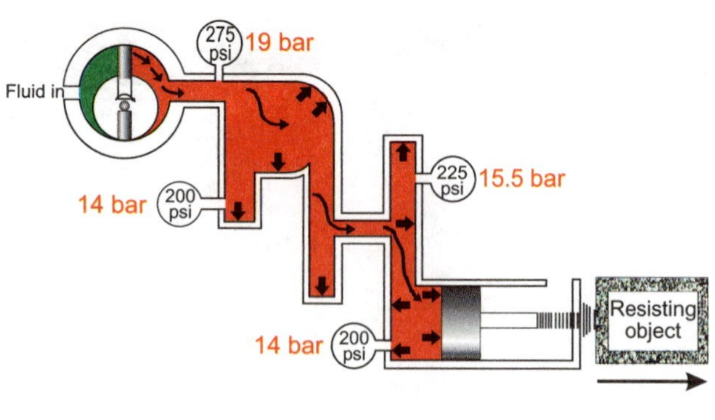

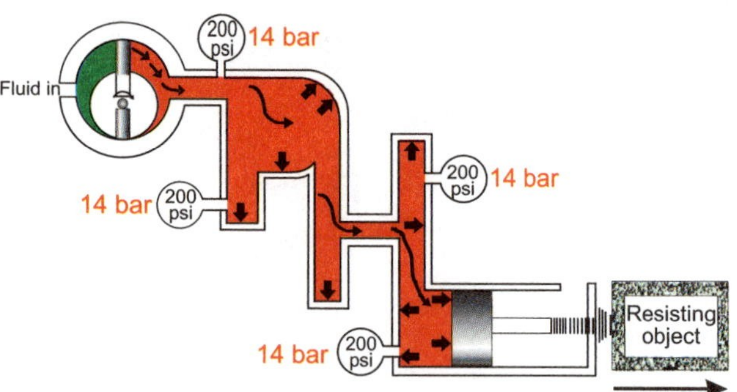

Figure 2-22

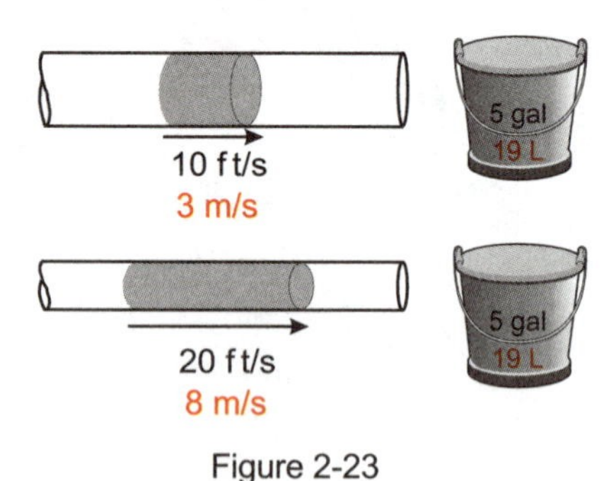

Figure 2-23

Resistance to the Flow of a Positive Displacement Pump

Resistance to the flow of a positive displacement pump comes basically from two sources - the object resisting being moved, and the flow forces needed to move the liquid itself.

If the resistance of the liquid could be eliminated from the system illustrated, the only resistance to pump flow would be the resisting object, which can be overcome with a pressure of 200 psi (13.8 bar) at the actuator. The pump would apply the required 200 psi (13.8 bar) to get the liquid out into the system. This hydraulic working energy would be applied to the actuator and the object would move.

Resistance of a liquid to the pump's flow is always present. To discharge its flow against the liquid's resistance, the pump absorbs more energy from its prime mover, and applies an additional pressure to the liquid.

Extra Energy Changes Form

The additional energy applied by the pump does not result in additional hydraulic working energy at the actuator, because it is used up by a liquid's resistance. The working energy is "used up"—not in the sense of being destroyed, but it changes to heat energy resulting in a rise in the temperature of the fluid. As was pointed out previously, this is the inefficiency in a system.

Velocity vs. Flow Rate

In a dynamic system, fluid flowing through a pipe is traveling at a certain speed. This is the fluid's velocity which is usually measured in feet per second (f/s) or meters per second (m/s).

The volume of fluid flowing through tubing in a period of time is a rate of flow. Flow rate in a typical hydraulic system is usually measured in gallons per minute (gpm) or litres per minute (lpm).

The relationship between velocity and flow rate can be seen from the illustration. In order to fill the 5 gallon (18.95 L) container in one minute, a 5 gallon (18.95 L) volume of fluid in the large tube must travel at a speed of 10 feet per second (3.048 m/s). In the small tube a 5 gallon volume (18.95 L) must travel at a speed of 20 feet per second (6.096 m/s) in order to fill the container in one minute.

In both cases the rate of flow is 5 gpm (18.95 lpm). The fluid velocities are different.

Friction Generates Heat

In a hydraulic system, friction, and therefore heat, is generated because of the liquid flowing through the pipe. The faster a liquid travels, the more heat is generated.

In industrial applications, the recommended maximum fluid velocity between pump and actuator is usually 15 fps (4.572 m/s).

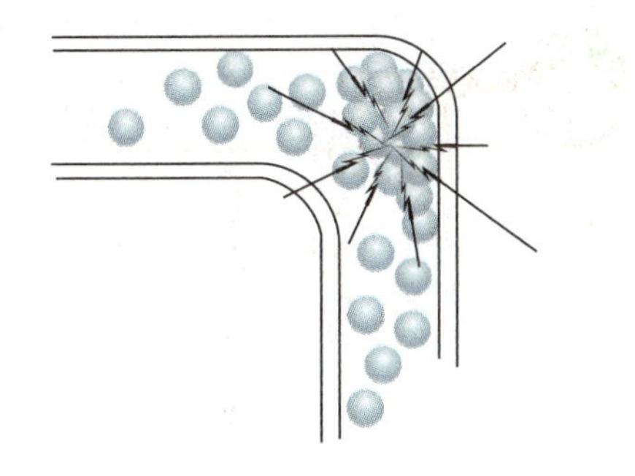

Figure 2-24

Changing Fluid Direction Generates Heat

A mainstream of liquid flowing along in a straight line generates heat when it crashes into other liquid molecules while forced to change direction because of a pipe bend or elbow. Depending upon the conduit size, one 90° elbow could generate as much heat as several feet of the flow conduit.

Pressure Differential

Pressure differential is simply the difference in pressure between any two points in a system.

It is a symptom of what is happening in a system :

1. By indicating that working energy in the form of a moving, pressurized liquid is present in the system.
2. By measuring the amount of working energy that changes to heat energy between the two points.

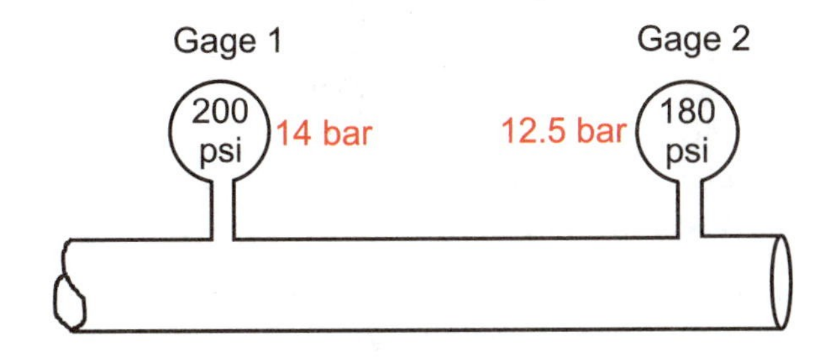

Figure 2-25

In our illustration, the pressure differential between the two gage points is 20 psi (1.38 bar). This indicates that:

1. Working energy is moving from gage 1 toward gage 2.
2. While moving between the two gage points, 20 psi (1.38 bar) of the working energy is transformed into heat energy because of liquid resistance.

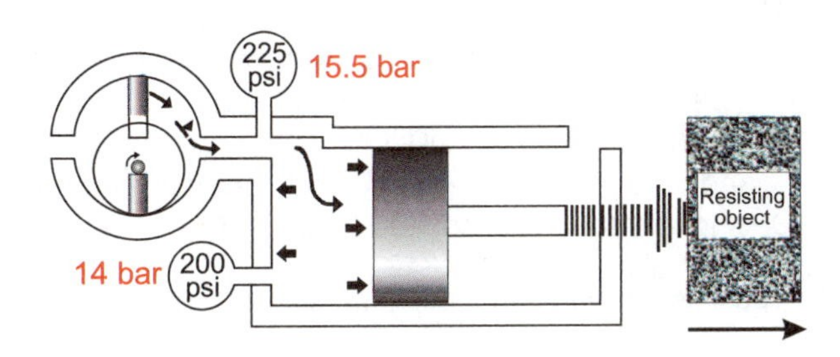

Figure 2-26

Hydraulic System Designed to Avoid Generation of Heat

The generation of heat by working energy traveling through a system to a point of work is system inefficiency. To make a hydraulic system more efficient, a designer chooses oil with the appropriate viscosity, uses piping of a proper size, and keeps the number of bends to a minimum.

Exercise
Hydraulic Transmission of Energy
35 Points

INSTRUCTIONS: For each incomplete sentence, one choice will correctly complete the statement. After reading the sentence and the possible choices, circle the letter next to the most correct answer. After all six statements have been completed, place the letter for each answer in the appropriately labeled box. The letter combination should form a word. Total points possible 35; 5 points per answer, 5 points for correct word.

1. With a constant cycling of its movable member, a positive displacement pump develops a constant flow of liquid, and at the same time will apply whatever _____ is needed to overcome any _____ .
   B. force, actuator
   R. flow, resistance
   M. speed, load
   E. pressure, resistance

2. Liquid enters and is forced out of a pump by _____ .
   L. generating an increasing and decreasing volume within its housing.
   R. centrifugal force forming a positive seal between vane and housing.
   A. absorbing additional energy from its prime mover.
   E. overcoming any resistance to flow.

3. The resistance to pump flow comes from the _____ and _____ .
   I. viscosity, friction
   S. load, liquid
   N. viscosity, changing direction
   F. velocity, liquid

4. A flowing liquid's viscosity, friction, and change of direction all result in _____ .
   U. an increase in efficiency.
   O. a low pressure differential.
   P. the generation of heat.
   N. working energy increasing in velocity.

5. Pressure differential in a hydraulic system (as illustrated)
   A. is a symptom that something is wrong.
   R. illustrates that Pascal's Law is not applicable to a dynamic system.
   I. indicates that working energy is present in the form of a moving, pressurized liquid.
   G. shows that a pressurized liquid is flowing upstream to a point of work.

6. A positive displacement pump applies a higher pressure to the fluid than required at the actuator because of _____ .
   N. viscosity
   L. fluid changing direction
   S. fluid friction
   M. all of above

Answer 3	Answer 5	Answer 6	Answer 4	Answer 2	Answer 1

Exercise
Hydraulic Transmission of Force

1. What does the gage read?

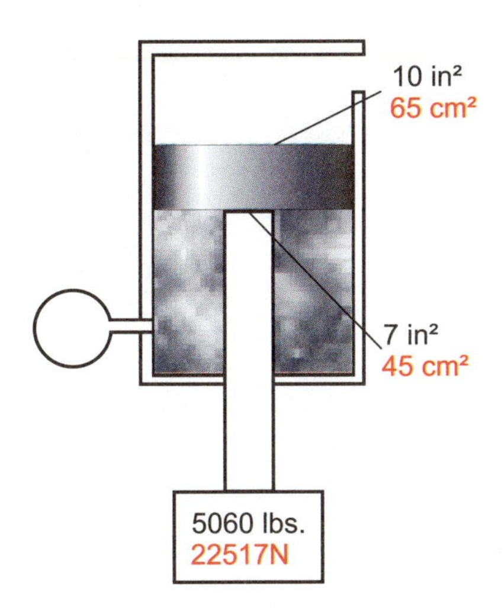

2. What does the gage read?

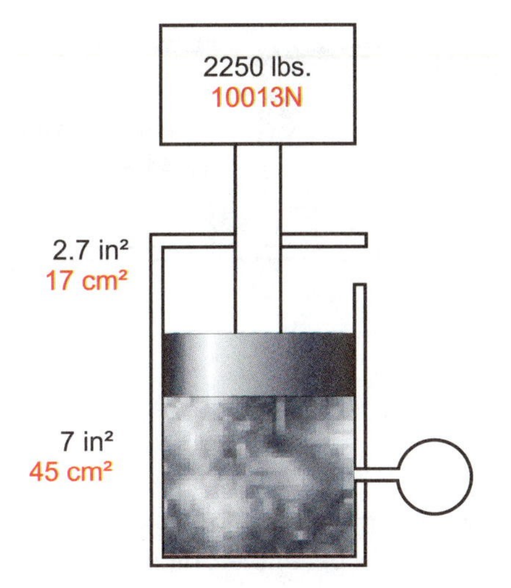

3. What pressure is needed to suspend the 1980 lb. load?

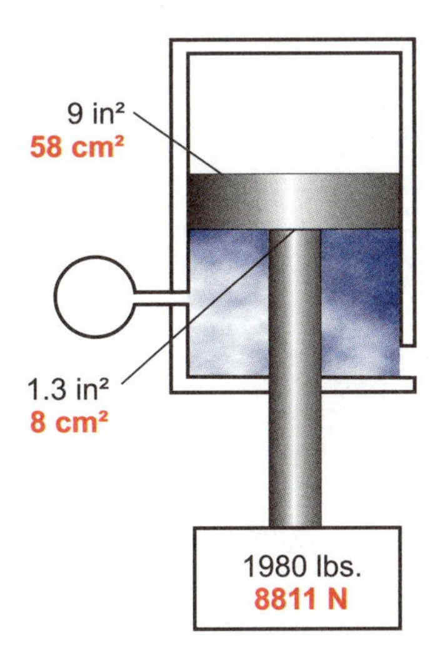

4. What pressure is needed to raise the 3130 lb. load?

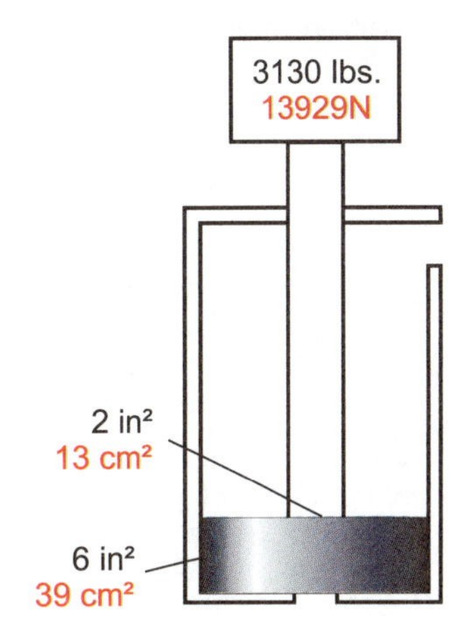

Exercise
Hydraulic Transmission of Force

5. What pressure is needed to move out the load?

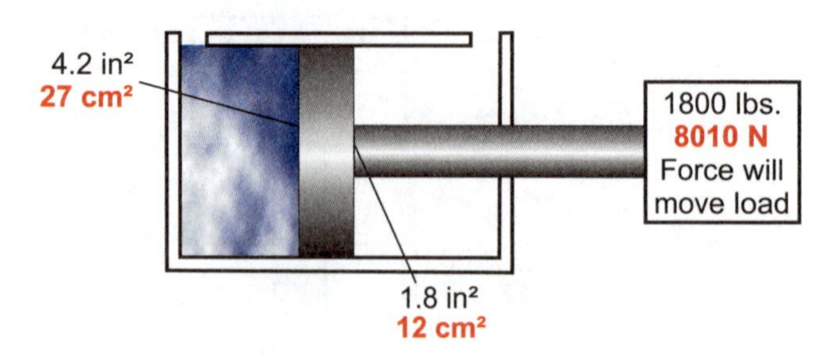

6. What pressure at the intensifier inlet will result in 13,000 psi at the outlet?

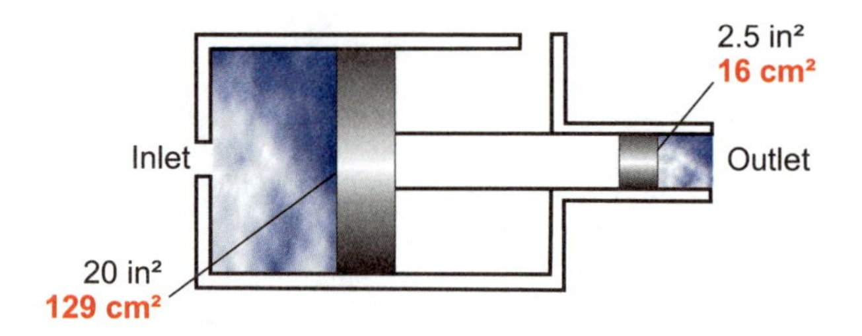

CHAPTER 3

Petroleum Base Hydraulic Fluid

A common fluid for a hydraulic system consists of paraffinic and naphthalenic petroleum oils which are blended for characteristics that make it suitable for use in a hydraulic system.

As was pointed out previously, hydraulic fluid is the substance used for transmitting energy from pump to actuator in a hydraulic system. The intent of this lesson is to concentrate on some characteristics of petroleum base fluid.

At the outset, the lesson is concerned with petroleum base fluid as a lubricant. Then it deals with the effects on a system of a petroleum base fluid's viscosity. Finally, we see some problems of a petroleum base fluid in service and some maintenance considerations with respect to hydraulic oil.

Besides acting as a medium for energy transmission, the second most important function of a petroleum base fluid is to act as a lubricant. The lesson begins by describing lubrication.

Lubrication

Lubrication is the process of reducing friction between relatively moving surfaces which are in contact.

Lubrication is a very important function of hydraulic fluid. Without lubrication, friction would cause system components to wear excessively and excessive heat to be generated.

Friction

Friction is a force which can stop or retard the motion of a moving object. Assume that one surface of a clean, dry steel block is at rest on a similar surface. Any attempt to slide the block across the contacting surface would be resisted by a frictional force. Friction occurs because of surface roughness and welding of minute metal surfaces.

If an apparently smooth surface of a typical component were magnified, it would appear to be quite irregular; even the best machining methods cannot eliminate these irregularities completely. As surfaces are rubbed together, material is plowed, ripped, and worn away at a considerable rate. The rougher the surface and the greater the sliding force, the more friction will be developed.

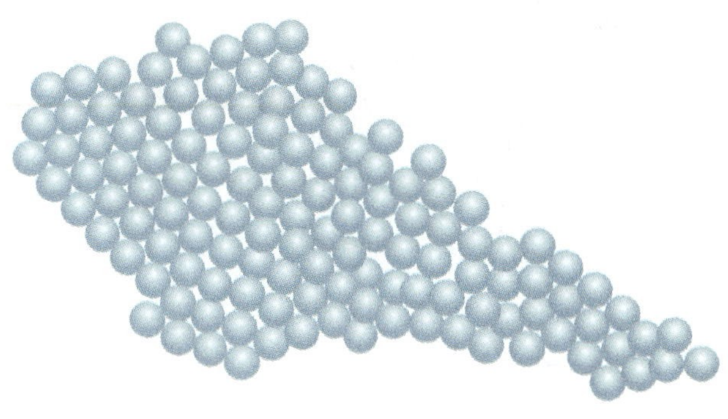

Figure 3-1

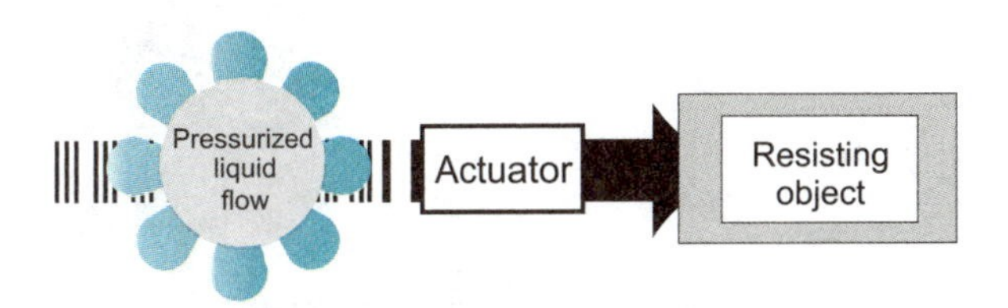

Figure 3-2

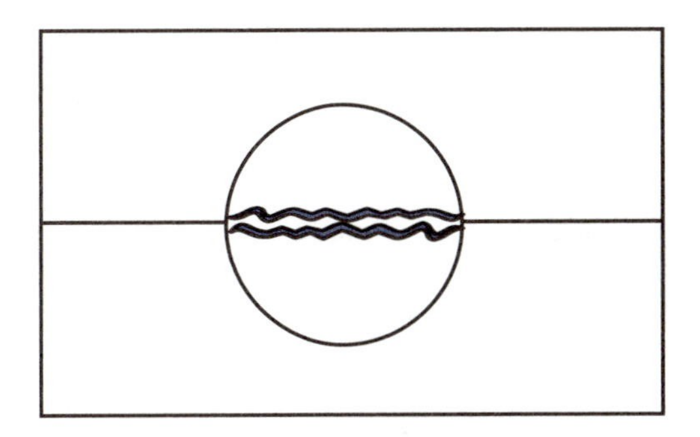

Figure 3-3

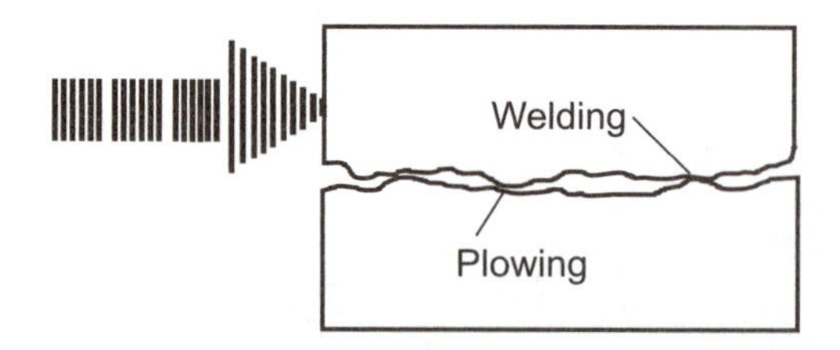

Figure 3-4

Friction can also be related to the infinitely small welds which commonly occur between contacting metal. As a force is applied to mating surfaces, high points of a metal are deformed until they acquire a large enough base to support a force. This action tends to bond the material at the contacting points. In moving the surfaces across one another, these tiny bonds must be ruptured, which action contributes to friction.

Previously, we saw that a liquid consisted of continuously-moving molecules which could take the shape of its container. We also learned that a liquid had a resistance to flow known as viscosity. In the following section, we shall see how a petroleum oil's capability of adhering to a surface and viscosity contribute to develop a lubricating film.

Figure 3-5

Fluid Film

Interaction between metal surfaces can be greatly reduced by introducing a lubricating film between them. Not having a lubricating film between moving parts is similar to rowing a boat on land.

Any liquid will form a lubricating film, but some do a better job than others. Water, for example, was the first hydraulic fluid, but it was a poor industrial lubricant because its fluid film is not durable. Petroleum oil is a good lubricant because it forms a durable fluid film.

Lubricity

Lubricity refers to a liquid's ability to form a durable fluid film between contacting surfaces. This ability is directly related to:

1. a fluid's natural film thickness
2. a fluid's tendency to adhere to a surface

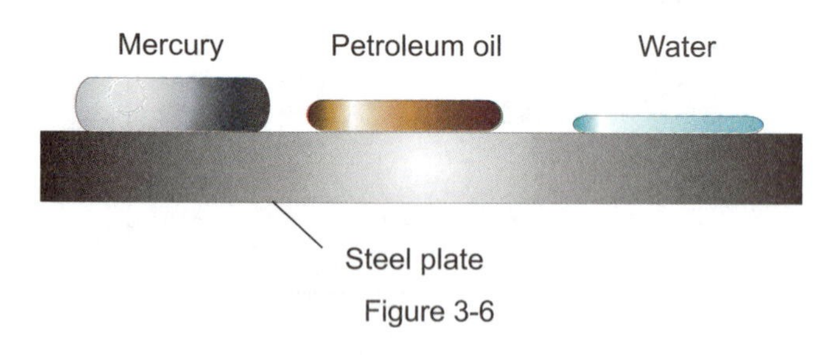

Figure 3-6

Petroleum oil has good lubricity. If at room temperature, a petroleum base hydraulic oil were poured on a steel plate, it would appear to wet, or adhere to the surface with a substantial fluid film. If water were likewise poured on the unprotected metal, it would appear to wet the surface, but its fluid film would be thin and therefore easily penetrated. For this reason, water has poor lubricity.

If the same procedure were followed with mercury, a thick fluid film would form, but it would show relatively little tendency to adhere to the steel. As a matter of fact, the mercury could be broken up into little balls or beads. Even though mercury does form a thick fluid film, it also has poor lubricity because it does not tend to stick to the steel (ferrous) surface.

A liquid with good lubricity adheres to a surface and also develops a substantial fluid film. Of the fluids used in a hydraulic system, petroleum oil has been found to exhibit the best lubricity.

Viscosity Affects a System

Up to this time, we have seen that a petroleum base hydraulic fluid has two important functions:

1. to act as a medium for energy transmission
2. to lubricate internal moving parts of a system

Both of these functions, and consequently hydraulic systems in general, are influenced by fluid viscosity which is probably one of the most significant characteristics of a petroleum base fluid. At this point we would like to review the characteristics of viscosity and then show how it can distinctively affect a system.

In the following section, we will redefine viscosity, see in what manner it is measured, and then determine how viscosity affects heat generation, lubricity, hydrodynamic lubrication, and clearance flow. The starting point for our review is at the molecular level.

Liquid Molecules

Just as all liquids, petroleum base hydraulic fluid is made up of molecules which are attracted to one another. This attraction is much greater than the molecules of a gas, but is less strong than the molecules of a solid, which are in a relatively fixed position.

Liquid molecules are free to slide past each other; and as a matter of fact, they are continuously moving.

Viscosity

Viscosity is the resistance of a liquid's molecules to flow or slide past each other; it is sort of an internal friction. An example of a high viscosity liquid is honey or molasses; water or cooking oil is an example of a low viscosity liquid.

Viscosity Affected by Temperature

As was indicated, a liquid is made up of molecules which are attracted to each other and continuously moving. Some experts feel that the more slowly molecules move, the greater are attractive forces resulting in an increased resistant to flow.

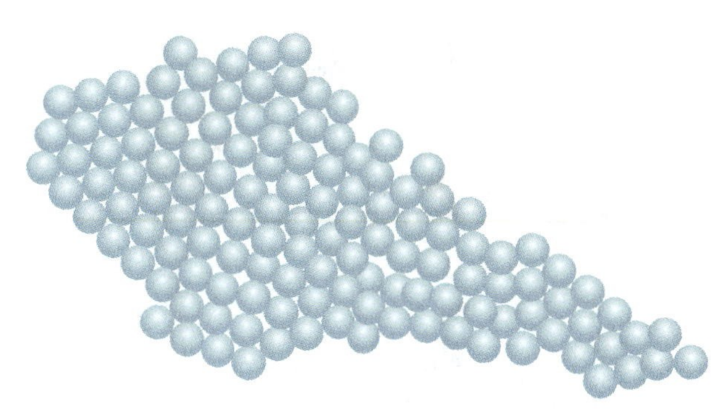

Figure 3-7

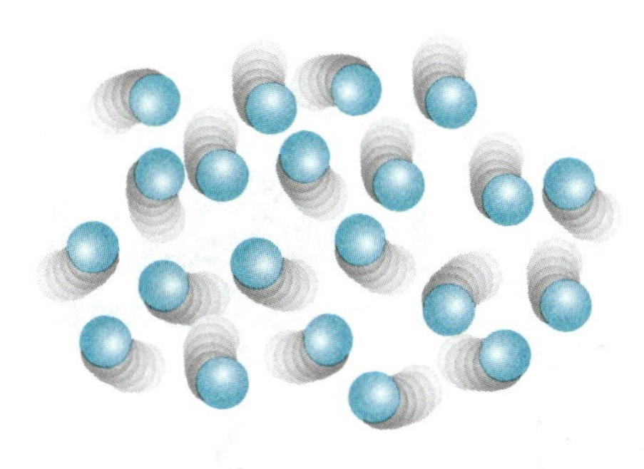

Figure 3-8

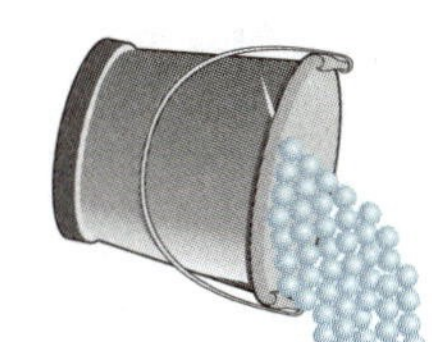

Figure 3-9

Figure 3-10

A bottle of molasses taken from a refrigerator consists of very slowly moving molecules which have large attractive forces; cold molasses has a high resistance to flow. Trying to pour this liquid through a funnel would be a time-consuming task.

Heating the molasses in a sauce pan adds energy to the molecules; molecular speed increases reducing attraction between molecules. With less of a resistance to flow, a reduced viscosity, heated molasses can more easily flow through the funnel.

Generally, as temperature increases, viscosity of a liquid decreases.

Saybolt Universal Second

One measure of liquid viscosity is the Saybolt Universal Second (abbreviated SUS or SSU). The SI unit of viscosity is the centistoke (CST).

The Saybolt Universal Second is named in honor of George M. Saybolt who in 1919 offered to the United States Bureau of Standards his Saybolt Viscosimeter - a device which measured viscosity.

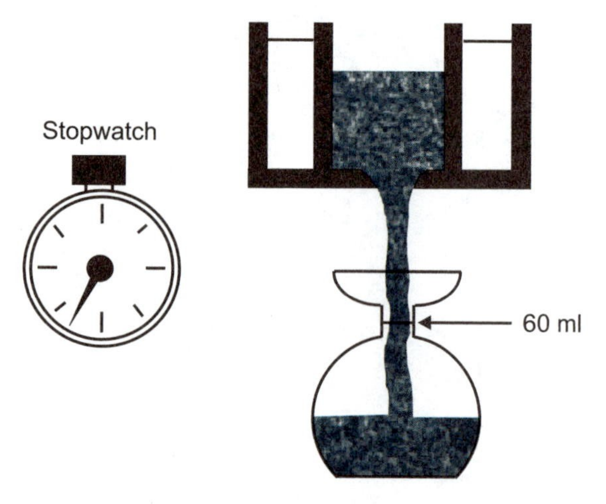

Figure 3-11

The use of the Saybolt Viscosimeter consisted of filling the apparatus with a liquid and heating the liquid to a specified temperature. With the liquid heated, a cork was pulled from the chamber bottom at the same time a stopwatch was started. The liquid then drained through an opening of a specific size until 60 milliliter (about 2 fluid ounces) was in the flask.

The stopwatch timed how long the liquid took to fill the flask. The result was a measure of viscosity in Saybolt Universal Seconds.

If an oil heated to a temperature of 100°F (37.7°C) took 143 seconds to fill the flask, its viscosity would be 143 SUS @ 100°F (37.7°C). If the same oil took 82 seconds to fill the flask when heated to 130°F (54.4°C), its viscosity would be 82 SUS (17.7 CST) @ 130°F (54.4°C). Viscosity is always associated with a temperature. Yet, it is common to hear someone say he is using "150 SUS (32 CST) @ 100°F (37.7°C)." The 100°F (37.7°C) temperature is assumed.

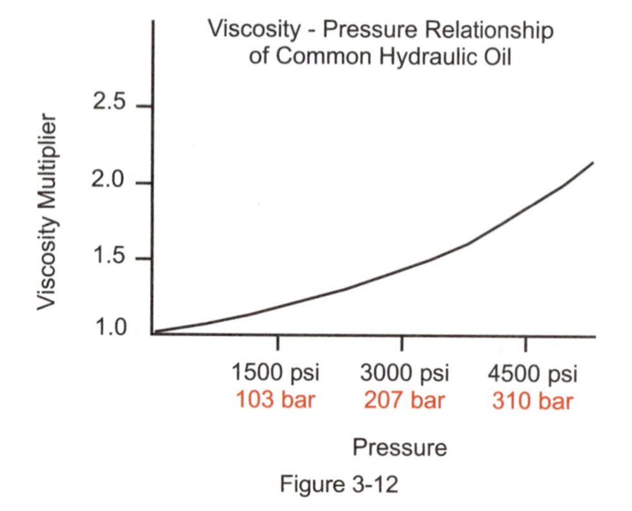

Figure 3-12

Viscosity Affected by Pressure

Viscosity is affected by system pressure. As pressure in a system increases, viscosity also increases. This is pointed out by the illustrated graph in Fig. 3-12.

The graph shows that for a common industrial hydraulic oil, viscosity increases 40% as pressure increases from zero to 3000 psi (207 bar).

Viscosity Affects Heat Generation

Viscosity of a petroleum base hydraulic fluid affects heat generation. We saw previously that this, along with friction and changing direction, was one of the liquid's resistance to pump flow.

A high viscosity liquid of 500 SUS (107.9 CST), having more internal resistance to flow, will cause more heat to be generated in a system than a low viscosity liquid of 150 SUS (32 CST).

In many hydraulic systems, viscosity of an oil is 150-250 SUS (32-53.9 CST) @ 100°F (37.7°C).

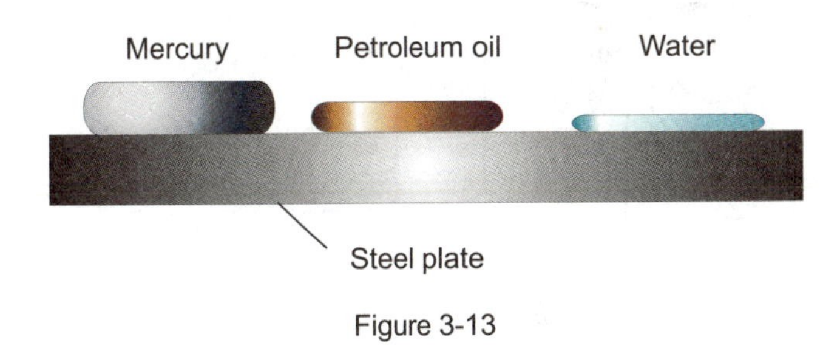

Figure 3-13

Viscosity Affects Lubricity

Since it is a resistance, viscosity may not at first appear to be a desirable characteristic. But, viscosity is a very important and desirable fluid characteristic since it affects lubricity.

It was illustrated that lubricity was dependent upon an oil adhering to a metal surface and developing a substantial fluid film; viscosity affects fluid film. The higher the viscosity, the thicker will be the fluid film. Of course, the fluid must be capable of readily flowing so determination of an appropriate viscosity for a system is a compromise between its ability to form a fluid film and its ability to flow.

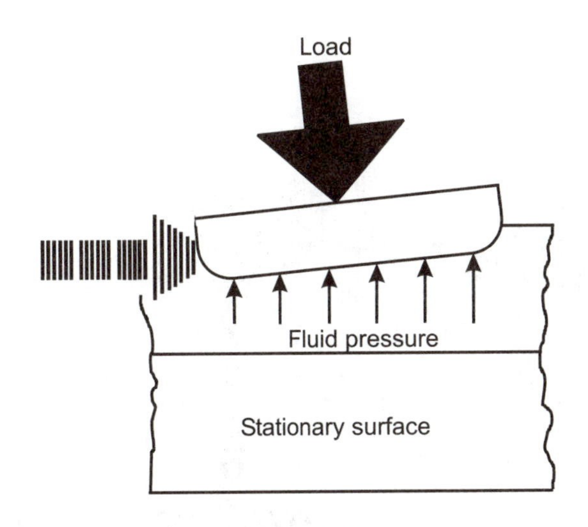

Figure 3-14

Viscosity Affects Hydrodynamic Lubrication

The ability to form a durable fluid film is an important characteristic of petroleum base fluid. We referred to this ability as lubricity.

One may feel that a fluid film would be difficult to maintain between moving parts since any rapid movement would tend to scrape a surface clean. But once parts begin to move, liquid viscosity does not usually allow this to happen.

A metal block immersed in oil and at rest on a stationary metal surface is separated by the oil's fluid film. As a force moves the block, the leading edge rises because the oil resists getting out of its way (viscosity). This action forms a fluid wedge under the block which floats it along like a boat planing on water. As long as pressure on the moving block does not become excessive, the fluid wedge would ward off normal attempts at penetration. This is known as hydrodynamic lubrication.

A low viscosity liquid like water would get out of the way too easily under low speed and high load conditions. The wedge would not fully form, resulting in a fluid film which could more easily be penetrated.

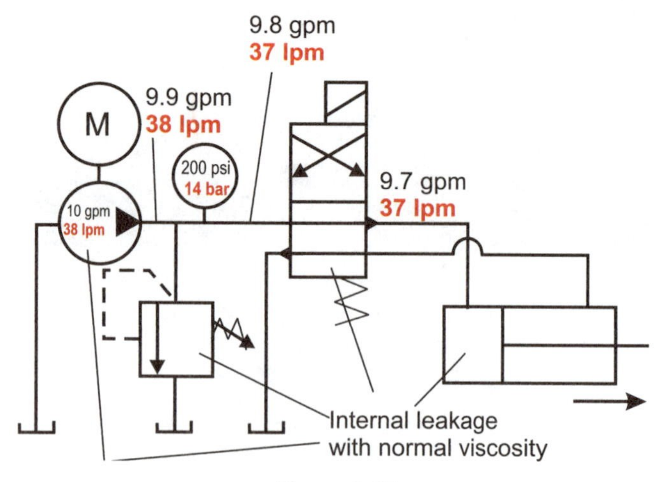

Figure 3-15

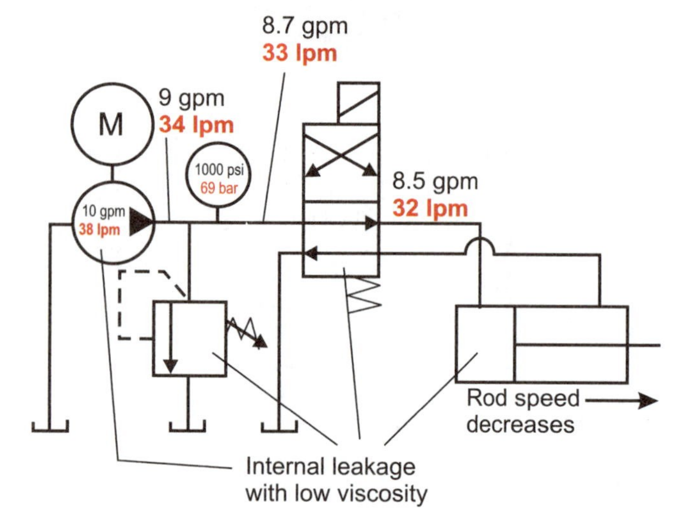

Figure 3-16

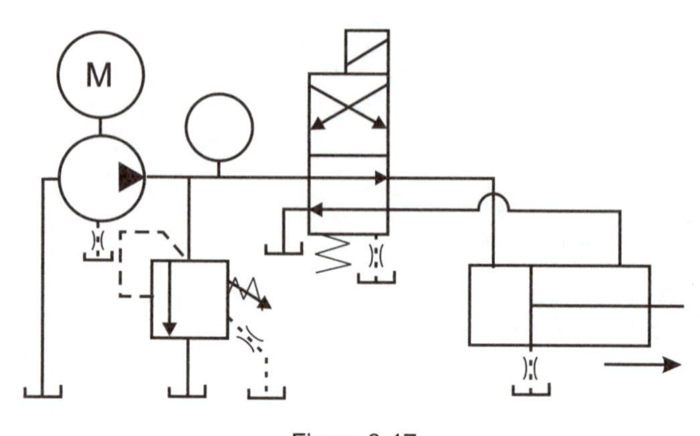

Figure 3-17

When system components are moving, they are lubricated by the hydrodynamic process. However, at startup or when excessive pressure pushes a moving part through a hydrodynamic wedge, a liquid's ability to form a durable fluid film (lubricity) becomes very important.

Viscosity Affects Clearance Flow

Another important effect of viscosity is its ability to help reduce leakage between clearances of close-fitting moving parts.

Many components of a hydraulic system do not have a zero-leakage seal between internal moving parts. Frequently, the seal is metal-to-metal through which a small portion of fluid is continuously flowing and lubricating. Examples of metal-to-metal seals could be found in piston pumps between pistons and piston bore; gear pumps between gears and housing; vane pumps between vane tip and cam ring; cylinders between piston rings and cylinder bore; and control valves between spool and valve body.

To achieve the best seal possible, clearances between moving parts are kept to a minimum. However, clearance size is not so small that a fluid cannot pass through and lubricate. Size of a clearance is frequently a compromise between sealing and lubrication.

Clearances between internal moving parts of a hydraulic component are in effect orifices; they continually meter lubricative-leakage flow. Just as any orifice, flow through a component clearance is affected by fluid viscosity.

With a viscous fluid, leakage and therefore lubricative flow through a clearance will be reduced. On the other hand, a fluid which is not viscous enough, or too thin, means that excessive fluid passes through component metal-to-metal clearances. This results in less flow passing into the system and an unnecessary buildup of heat.

Metal-to-metal clearances between component moving parts can be considered built-in fixed restrictions which are continually bleeding off flow. Too much leakage, then, can be harmful to the system. However, if too little flow passes through the clearance, the component may not be sufficiently lubricated. The system may become erratic and undependable as a result. A happy medium can be found with a fluid of the appropriate viscosity.

Viscosity Index

Since the viscosity of hydraulic oil does change with temperature and since viscosity is an important factor in a hydraulic system, systems which are not, or cannot be, maintained at a constant temperature need an oil whose viscosity remains relatively stable over a given temperature range.

An oil's viscosity index illustrates how viscosity is affected with changes in temperature. This relationship can be depicted by a straight line using ASTM (American Society for Testing Materials) standard viscosity-temperature charts for liquid petroleum products. When the viscosity of an oil at two temperatures is plotted on the paper, viscosity of the oil at any temperature can be determined by drawing a straight line connecting and running through the two points. (This can be done with any liquid petroleum product which does not have chemicals added that affect its natural viscosity-temperature relationship.)

If the viscosities of two oils were plotted on the graph paper, the oil with the more horizontal line would have the higher viscosity index. For example, oil A has a viscosity of 153 SUS (33 CST) @ 100°F (37.7°C) and 44 SUS ((9.5 CST) @ 210°F (98.9°C); while oil B has a viscosity of 165 SUS (35.6 CST) @ 100°F (37.7°C) and 42 SUS (9.1 CST) @ 210°F (98.9°C). Oil A has a more horizontal line and therefore has a higher viscosity index.

When the term viscosity index was first adapted, a certain oil showing a very rapid change in viscosity with respect to temperature was assigned a viscosity index of 0, while the best oil type showing a small change with temperature was arbitrarily given a viscosity index of 100. Therefore, oils at that time all had VIs between 0 and 100. With up-to-date refining practices, oils can now have VIs above 100.

In a modern industrial hydraulic system, an oil's viscosity index is generally required to be 90 or above. However, viscosity index means little for systems with a relatively constant temperature.

ASTM Standard Viscosity - Temperature Chart

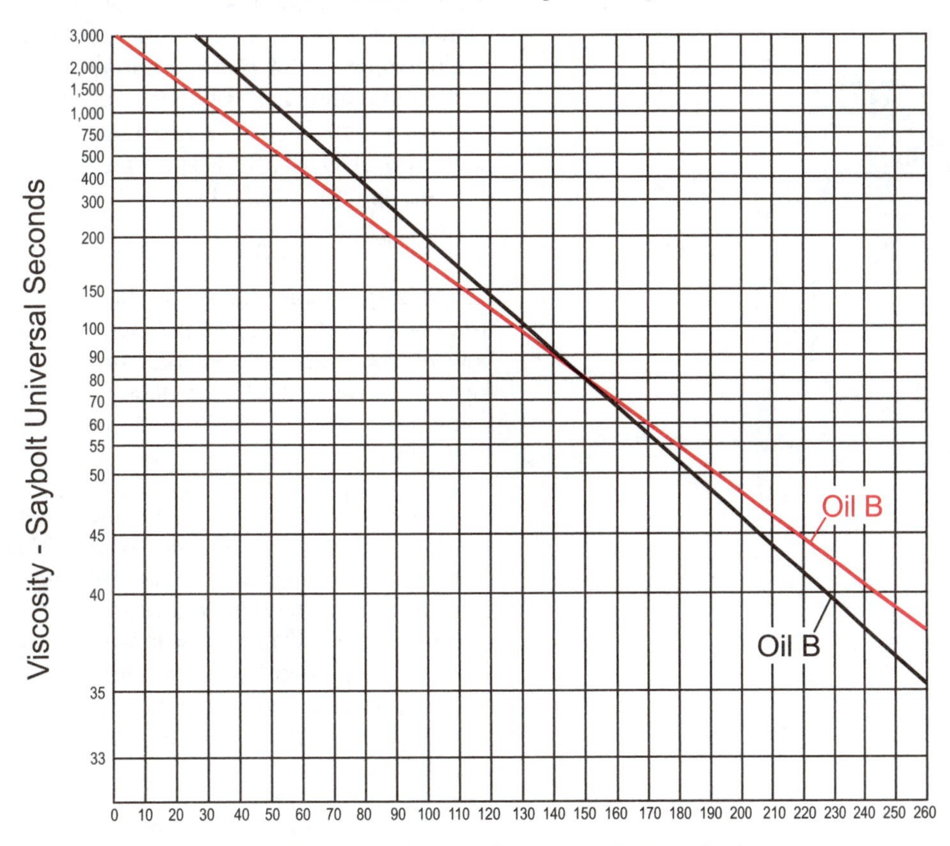

Figure 3-18

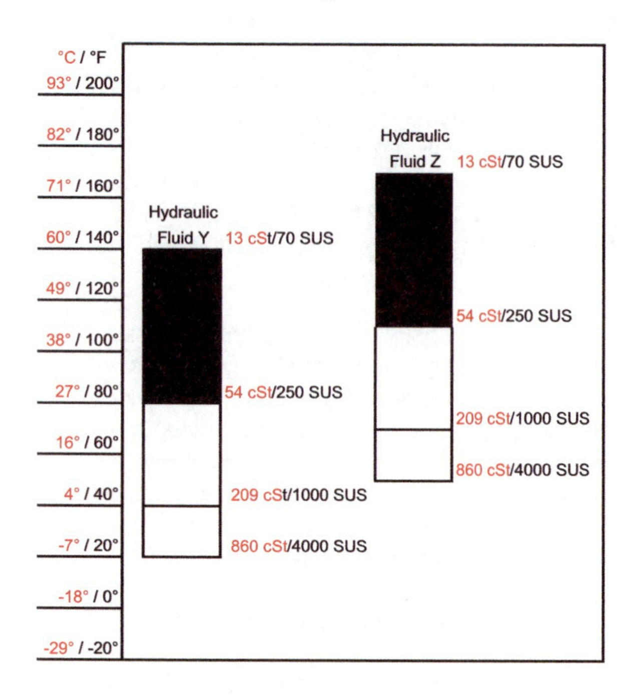

Figure 3-19

Figure 3-20

Oil Operating Range

Petroleum oil is an excellent lubricant for a hydraulic system, but not at all viscosities. If viscosity of an oil were too low, its fluid film would be like water and consequently too thin. If an oil's viscosity were too high, insufficient amounts of fluid would flow into bearings and component clearances. For this reason, manufacturers of rotating equipment (pumps, motors) which are especially dependent on proper bearing lubrication, specify the viscosity range at which their components are to be operated. When these components are sufficiently lubricated, it usually means the rest of the system is lubricated as well.

If components have a required viscosity range, then this information, along with the temperature range of the system, indicates the use of a specific oil. For example, a particular system at its operating temperature requires a minimum/ maximum viscosity of 70-250 SUS (15-54 CST). If the operating temperature range were 80-140°F (26.7-60°C), hydraulic fluid Y would be used. If the temperature range were 110-170°F (43.3-76.7°C), hydraulic fluid Z would be used.

Since temperatures can become quite low even in industrial environments, an oil can become extremely viscous. To ensure that their pumping mechanisms will fill, pump manufacturers also specify the maximum viscosity allowable at start-up. In general, these viscosities are 1000 SUS (216 CST) and 7500 SUS (1618 CST) for piston, vane and gear equipment, respectively.

Pour Point

ASTM graph paper does not point it out, but at extremely low temperatures petroleum oil does not flow. At low temperatures, wax structures begin to form in hydraulic fluids containing any paraffinic base crude. These wax formations hinder and may even stop flow.

Pour point of a hydraulic fluid is the lowest temperature at which it will pour under an ASTM laboratory test. In an actual system, if the maximum viscosity start-up specification is adhered to, the pour point of a fluid is generally not considered. But, when a system has the possibility of operating under extremely low temperature conditions, pour point of an oil should be at least 20°F below the lowest expected temperature.

Pour points for various oils are indicated in the manufacturer's data sheet of the specific oil.

Oil Problems and Additives

In the day-to-day operation of a system as petroleum base hydraulic fluid performs its function, certain problems can arise which may affect both fluid and system. Some of these problems are high pressure lubrication, oil oxidation, and oil contamination with water, air bubbles, and dirt. To correct some of these problems, hydraulic fluids are equipped with chemical additives.

The problems encountered with hydraulic fluids and the usual additives found to correct the problems, are dealt with below. It should be realized, however, that chemical additives cannot solve every fluid problem and that an oil cannot contain every available additive making it a super oil. Many fluid additives are not compatible with each other and therefore would give an unfavorable reaction if mixed.

High Pressure Lubrication

A good quality petroleum base hydraulic fluid is not a good enough lubricant for some systems. As pressures climb, the hydrodynamic fluid wedge between moving parts has more of a tendency to break down. This means lubrication is more dependent upon a fluid's inherent lubricity. To aid in lubricity or boundary lubrication at high pressures, hydraulic fluids are equipped with chemical additives.

Antiwear Additives

Antiwear (AW) or wear resistant (WR) additives can be divided into three types. One type, sometimes called an oiliness or lubricity agent, is a chemical made up of molecules that attach themselves vertically like blades of grass to metal surfaces. This creates a chemical film which acts as a solid when an attempt is made at penetration.

The additive molecules support the load, allowing a moving part to slip by. But this film is not very durable, tending to break down at high temperatures.

Another type of antiwear additive chemically combines with a metal surface to form a protective film. This film forms as low frictional heat is generated between contacting points of moving surfaces. They serve to smooth out or polish surfaces so that friction is reduced.

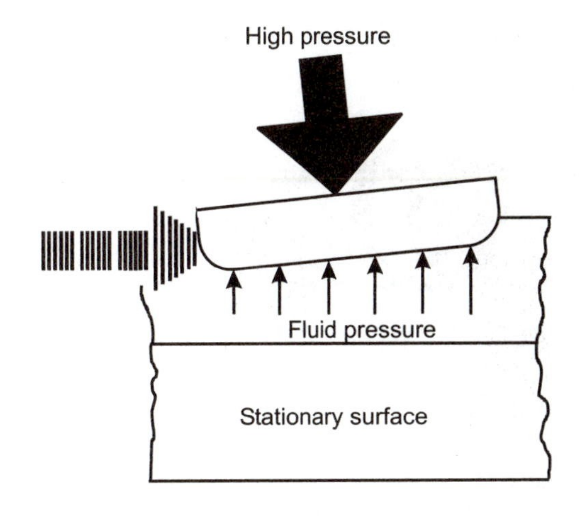

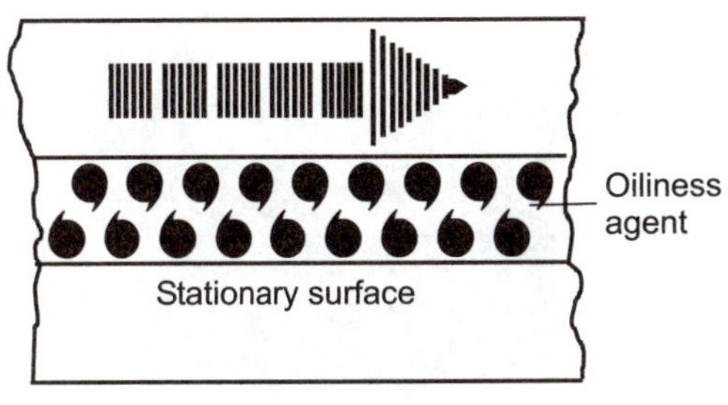

Figure 3-21

WR	Wear Resistant
AW	Anti-Wear additive
EP	Extreme Pressure additive

Figure 3-22

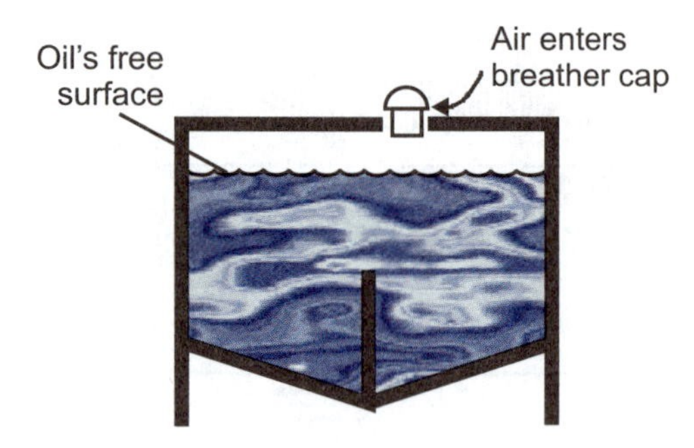

Figure 3-23

Another antiwear agent, known as an extreme pressure (EP) additive, forms a film on a metal surface as high frictional heat is generated. In a high pressure system, as mechanical interaction between surfaces becomes excessive, heat becomes excessive and the surfaces attempt to weld together. The extreme pressure additive comes out of solution at this point, keeping the surfaces apart.

All three types of antiwear additives are not found in the same fluid and are not used in the same applications. When oiliness agents are used, they are generally found in relatively low pressure systems (below 1000 psi/68.97 bar). When extreme pressure additives are found in a hydraulic system, the system will probably be operating above 3000 psi (207 bar), or the same fluid that is used to lubricate gears and machine ways is also used in the hydraulic system. A very common antiwear additive is the one which operates in the medium pressure range (1000-3000 psi/68.97-207 bar).

Check for High Pressure Lubrication

The check for a fluid's ability to give high pressure lubrication is the title of the oil or a manufacturer's catalog sheet. For example, with a Gulf Oil Co. fluid titled "Harmony 48AW", the AW stands for antiwear. Or, with a Sun Oil Co. fluid titled "Sunvis 816 WR", the WR indicates wear resistant.

Many refiners do not indicate the antiwear additive in an oil title. Consequently, the refiner's catalog or data sheet for a particular fluid must be referred to.

If excessive component wear has been a problem and the system's hydraulic fluid does not contain antiwear additives, switching to an oil with an antiwear additive will probably reduce the problem. This assumes, however, that the component wear was not the result of fluid contamination.

Oil Oxidation

Oxidation is a process by which material chemically combines with oxygen; this is a common occurrence.

If you have ever taken a bite out of an apple, you know that the pulp quickly turns brown as it is exposed to air. The same process happens when a car's fender is scraped down to bare metal; the exposed metal reacts with oxygen in the air and rusts. Many things on earth, including oil, oxidize in this manner.

Oxidation of hydraulic fluid can be pinned down to basically two system locations - reservoir and pump outlet. In both cases, oil reacts with oxygen but in different ways and the oxidation products are not the same.

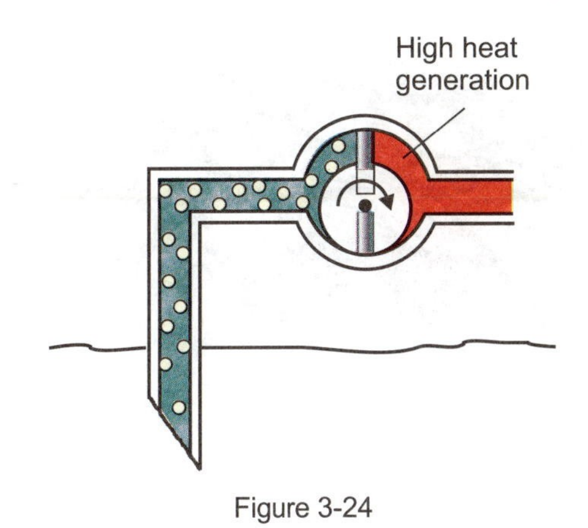

Figure 3-24

In a reservoir, the free surface of the oil reacts with oxygen in the air. The product of this reaction includes weak acids and soaps. Acids weaken and pit component surfaces; soaps coat surfaces and can plug pressure-sensing orifices and lubrication paths.

Heat is a major contributor to oil oxidation in a reservoir. As a rule, oil oxidizes twice as fast as normal for every 18-20°F (10-11°C) rise in temperature above an average reservoir temperature of 130°F (54.4°C). Reservoir oil also oxidizes more readily in the presence of iron and copper particles and water droplets.

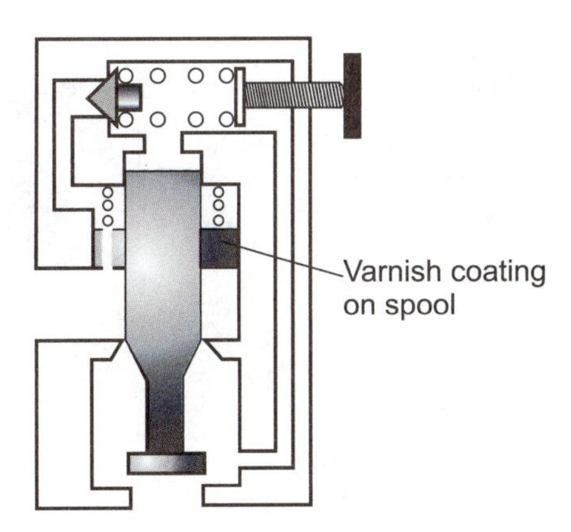

Figure 3-25

Besides the reservoir, another location where oil oxidation occurs is at pump outlet. If air bubbles are present in a pump suction line as a result of an air leak in the suction line or returning fluid velocity churning up the reservoir, they suddenly collapse upon being exposed to high pressure at pump outlet. This action generates a high temperature which according to some calculations can rise to 2100°F (1149°C) when the bubble is compressed from 0-3000 psi (0-207 bar). The high temperature fries the oil, forming resinous products, and causes the oil to acquire a characteristic burnt odor.

As high-temperature oxidation at pump outlet occurs, resinous materials are formed, but dissolve in the oil. When a hot surface (pump rotor, relief valve spool) is encountered, resins come out of solution forming a varnish or lacquer coating on the hot surface; this causes moving parts to stick.

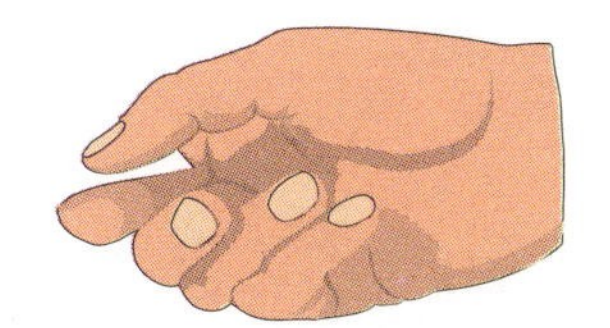

Figure 3-26

Resinous material can also form sludge which combines with dirt and floats around the system plugging small openings in valves and filters, and interferes with heat transfer to reservoir walls. Strong evidence exists that collapsing air bubbles at pump outlet is a major influence in rapid oil degradation.

Check for Oxidized Oil

A check for oxidized oil is performed by comparing a sample of the questionable fluid with a sample of new fluid out of a drum. With both fluids at the same temperature, fresh, new fluid will have a definite "body" and will tend to stick to your fingers as it is poured over your hands. And, if thumb and forefinger are rubbed together, the fluid will feel slippery.

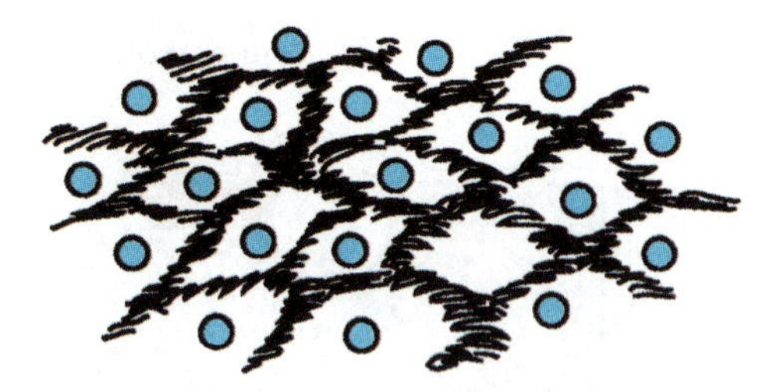

Figure 3-27

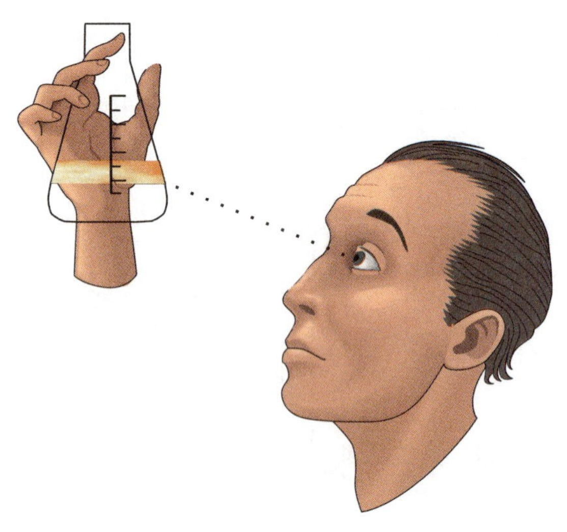

Figure 3-28

Oxidized fluid feels very much like water. As it is poured over your hand and fingers, oxidized oil runs off just as water. It exhibits little "body" and small tendency to adhere.

Oil which has been oxidized by the high temperature collapse of air bubbles will also have a characteristic pungent odor.

If any fluid sample exhibits the characteristics of oxidized oil, its condition is questionable. In this case, the fluid should be sent to a lab for further analysis. If this is impractical, the system should be drained and refilled with fresh fluid.

Water in Hydraulic Oil

All hydraulic oil contains water in varying amounts. In small quantities, however, water is broken into small droplets which are carried around by an oil.

We know from experience that water and oil do not mix (except for water soluble oils). Attempts to mix large amounts of water and oil will result in water settling out at the bottom of a tank. In small quantities, however, water is broken into small droplets which are carried around by an oil.

If an oil contains acidic and resinous products of oxidation, it has an increased tendency to take on water.

Check for Water in Hydraulic Oil

A check for water in hydraulic oil is performed by comparing a sample of the questionable fluid with a sample of new fluid.

Holding a beaker or glass of fresh oil up to a light, you will notice that it looks crystal; it sparkles a little. If a fluid sample contains .5% water, it will appear dull or smoky. If the sample contains 1% water, it will look milky.

An additional means of checking for water in oil is heating a fluid sample which appears milky or smoky. If the sample clears after a time, the oil probably contained water.

If an oil contains a small percentage of water (less than .5%), it is usually not discarded unless the system is critical. While in the fluid, water will hasten the oxidation process and reduce lubricity. After time, water will evaporate, but its products of oxidation will stay behind to cause further harm.

If an oil contains water in large quantities, much of it will eventually settle out. Centrifuging can be used to separate water and oil if time is important.

Rusting and Corrosion

In the context of a hydraulic system, corrosion refers to a deterioration of a component surface due to a chemical attack by acidic products of oil oxidation. Rusting refers to the process of a ferrous surface oxidizing due to the presence of water in oil.

The process of corrosion dissolves metal and washes it away, reducing the metal part size and weight. On the other hand, rusting adds materials to a ferrous surface, increasing its size and weight. Since the efficiency of precision components is affected when their parts are either too large or too small, rusting and corrosion cannot be tolerated in a hydraulic system.

Rust and Oxidation Inhibitors

R&O
Rust and Oxidation inhibitors

Rusting of ferrous component surfaces can be expected in a hydraulic system even if water is present in minute quantities; oil in its natural state does not provide adequate rust protection. Since it is impossible in actual practice to keep water out of a hydraulic system, hydraulic fluids are generally equipped with a rust inhibitor which coats metal surfaces with a chemical film.

Oxidation due to the interaction of air and fluid in a system's reservoir generates a chain of products which eventually attack metal surfaces and cause further fluid oxidation to occur. An oxidation inhibitor is a chemical which interferes with the oxidation chain.

The high-temperature oxidation which occurs as air bubbles collapse at pump outlet, cannot be reduced by a chemical. This form of fluid oxidation can be eliminated by removing air bubbles from the fluid stream to pump inlet.

Rust and oxidation inhibitors are the basic additives for most industrial systems. Hydraulic fluids equipped with these additives are sometimes referred to as R & O oils; the high grade is R & O turbine quality. Lower quality turbine oil is still suitable for many hydraulic applications and is designated R & O less-than-turbine quality."

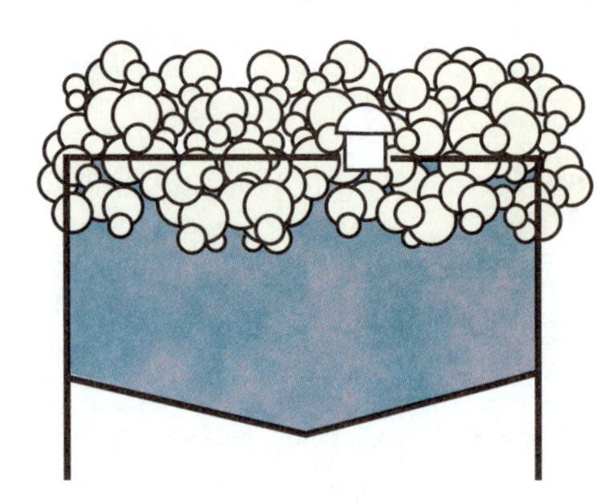

Figure 3-29

Foaming

As oil returns to a reservoir, it should release any entrained air bubbles which have been acquired in the system. In some systems where leaks are prevalent and/or returning oil is churned up as it enters a reservoir, foaming of the oil occurs. As a result, entrained air is pumped into the system, causing spongy, erratic operation, rapid oil oxidation and noise. In more severe cases, oil foam could bubble out of a reservoir creating a housekeeping problem.

Probably the best solution for alleviating foaming oil is to fix any system leaks and redesign the return part of the system with baffles or larger return lines which reduce fluid velocity. Sometimes, because of economics, convenience, or a lack of training, chemicals are used to solve the problem.

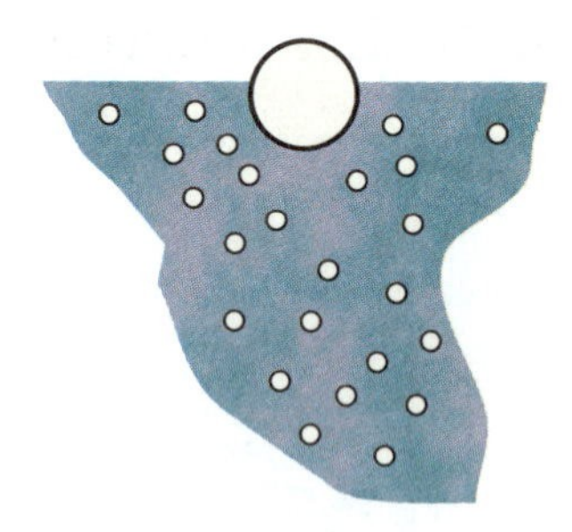

Figure 3-30

Anti-Foam Additives

In an attempt to discourage oil foaming hydraulic fluids can be equipped with anti-foam additives. In some cases, these additives work by combining small air bubbles into large bubbles which rise to a fluid surface and burst. In other cases, these additives function by interfering with air release which action reduces foaming, but increases the amount of air bubbles in the system. If an anti-foam chemical is desired in an oil, care should be taken that the agent selected does allow air to escape.

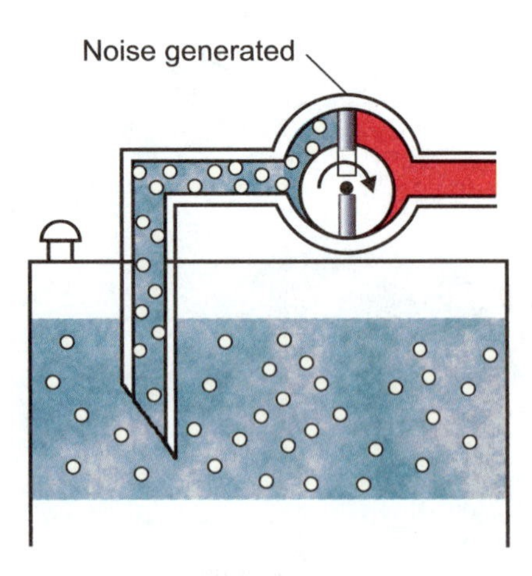

Figure 3-31

Check for Foaming

A check for foaming oil is performed by taking a fluid sample. By draining or drawing off fluid from a system's reservoir, you can tell by sight whether air bubbles are present in the fluid. The sample should be taken as close as possible to pump inlet line so that a representative sample of what is getting into the system can be taken.

Another indication that air bubbles are present in a system is noise. As air bubbles are swallowed by a pump, a high-pitched, erratic noise is emitted. In some cases, a pump will periodically emit a loud bang as if someone were exploding firecrackers inside the pump housing.

An additional indication of air bubbles is spongy system operation. This is evidenced by erratic actuator movements and erratic gage readings as a system is operating.

Dirt in Oil

The biggest problem with hydraulic oil in service is that it can easily become contaminated. The source of contamination can be water or air, but more frequently it is dirt.

Dirt in a hydraulic fluid can plug sensing orifices, cause moving parts to stick and wear excessively, and act as a catalyst to oxidize oil.

Dirt is an insoluble material in an oil which has several sources for contaminating oil. Dirt can be built into a system due to manufacturing, storing, and handling practices of system components and their assembly into a hydraulic system. Dirt can be generated within a system as a result of internal moving parts, flexing of component housings, and rust formation on reservoir walls. Dirt can also be added to a system as a result of servicing failed system components, not servicing reservoir breathers, and cylinder rods pulling in dirt as they retract. There is a continuous influx of dirt into hydraulic fluid.

At present, there is not a chemical additive which either keeps dirt out of, or removes dirt from, hydraulic fluid. Keeping dirt out of a system is the function of good system design and maintenance practices. Removing dirt from a fluid is the responsibility of filters and maintenance men.

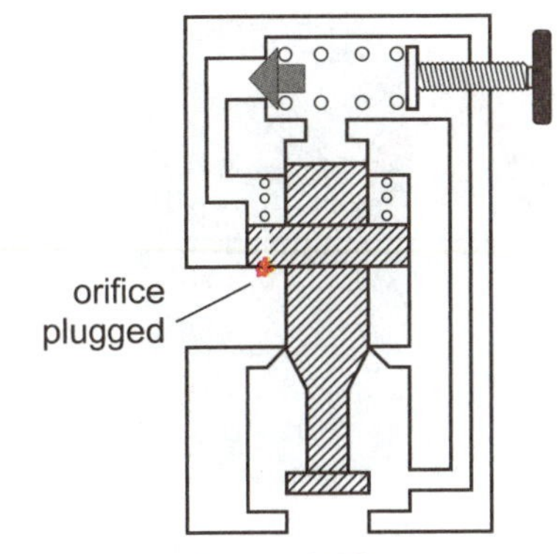

Figure 3-32

Figure 3-33

Check for Dirt in Oil

Trying to determine the dirt level of a fluid with the unaided eye is many times impossible. Holding a glass or beaker of hydraulic oil up to a light and inspecting for dirt is an inaccurate means of determining dirt contamination. Many harmful dirt particles for a hydraulic system are not normally visible. Determination of dirt contamination is best performed in a lab.

A check for dirt contamination in a hydraulic fluid is performed by checking indicators of a system's filters. Assuming that the filter element is appropriate for the system and that the indicator is functioning properly, the filter indicator will give an idea if the fluid is clean enough for the system. With an indication of "needs cleaning", the filter element should be serviced. If the indicator shows a bypassing condition, fluid is probably not clean enough and the filter should be serviced at once.

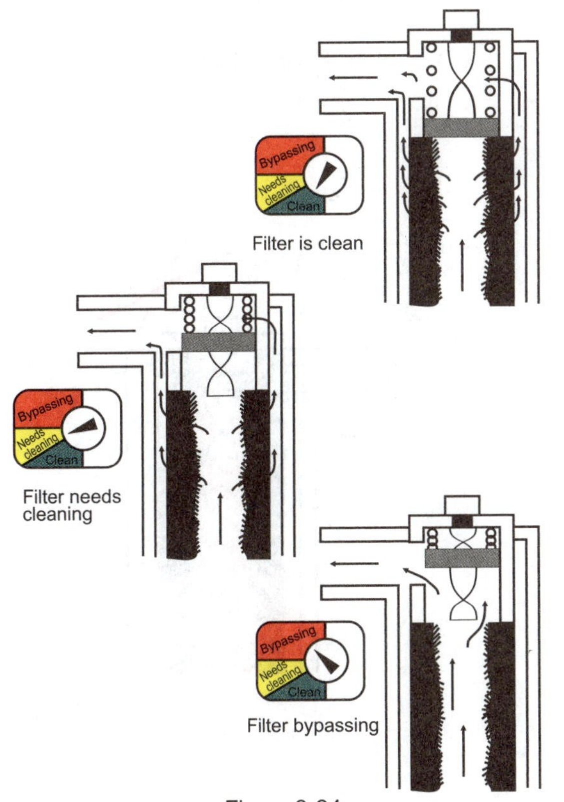

Figure 3-34

Figure 3-35

Figure 3-36

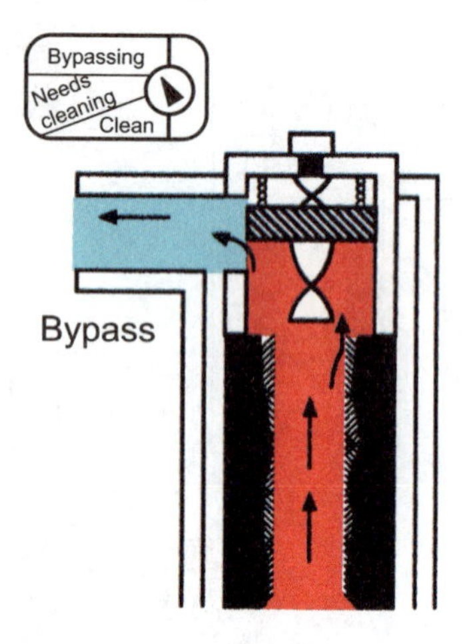

Figure 3-37

Hydraulic Oil Maintenance Considerations

As has been pointed out, hydraulic oil has several functions in a system and it contains additives which aid it in performing its function. Hydraulic oil is something special and it should be given special handling during storage, transfer to machine reservoir, and while operating in a system.

Keeping a fluid in top condition as it is stored, is a major consideration. Oil which becomes contaminated as it sits in a drum, is not only wasteful, but results in a false sense of security as oil supplies become depleted.

As a general rule, oil drums should be stored in a clean, dry place. If drums are stored outside, they should be stacked on their sides so that rain water does not collect on drum covers and leak past the seals into the oil.

Transferring oil from barrel to reservoir is another important consideration. Before the drum plugs are removed, the drum cover should be wiped clean. This procedure should also be followed for any apparatus or tools which will be used in the process such as hoses, pumps, funnels, reservoir filler hole, and the operator's hands.

Before the oil is actually dumped into the reservoir, check to see that the barrel contains the correct fluid by brand name and viscosity. All hydraulic fluids do not contain the same additives. Mixing additives is not recommended unless authorized by the oil manufacturer.

Once an oil is in a system, it should be monitored and maintained at regular intervals. Maintenance of the oil includes filling a reservoir when its minimum oil level has been reached (with fluid the same as or compatible with the fluid in the reservoir), fixing leaks, and servicing filters.

Servicing filter elements is very helpful in keeping a fluid in top condition. Dirt can be very harmful to a fluid because it acts as a catalyst for oil decomposition. This is especially true if the dirt particles are ferrous, lead, or copper. Filters usually remove a great percentage of dirt from a fluid stream. They do not remove dirt from the system, however; this is a maintenance function. Consequently, if filters are not maintained and cleaned when indicated, uncaught dirt not only passes to downstream components affecting their operation, but stored dirt on the filter element remains in the system contributing to oil decomposition.

Cleaning Wire Mesh Filter Elements

When servicing a filter with a wire mesh element, the element may be cleaned.

Wire mesh filter elements can be cleaned in several ways. With relatively coarse elements, no one way is by itself better than another. The degree to which an element becomes clean depends upon the care and effort used in the cleaning process, not to the specific cleaning method.

A common way of cleaning elements is washing in a clean solvent or a hot soap-water-ammonia solution and blowing off the element with clean air. A soft bristle brush (new paint brush) is helpful in scrubbing the element. At no time should a wire brush or any abrasive material be used.

To check the cleanliness of the element after cleaning, hold it up to a light. Any gray or dark areas indicate that the element must be re-cleaned.

Ultrasonic cleaning is a more expensive, but a more convenient way to clean elements. Dirty filter elements can be placed inside the ultrasonic device for a time and be removed clean and ready for re-use. Wire cloth elements with ratings of 40 micrometers or less need ultrasonic cleaning to effectively restore element life.

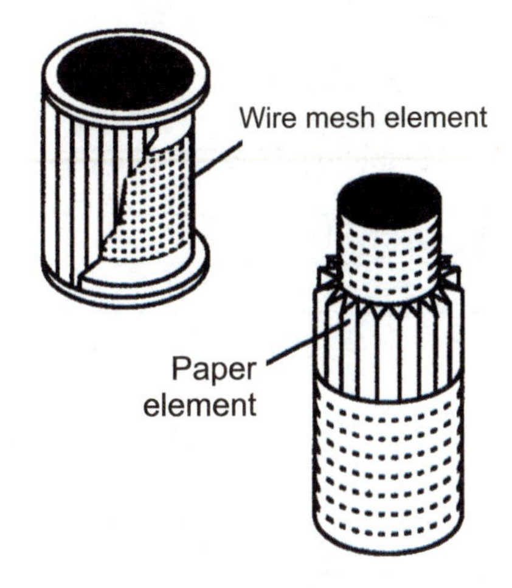

Figure 3-38

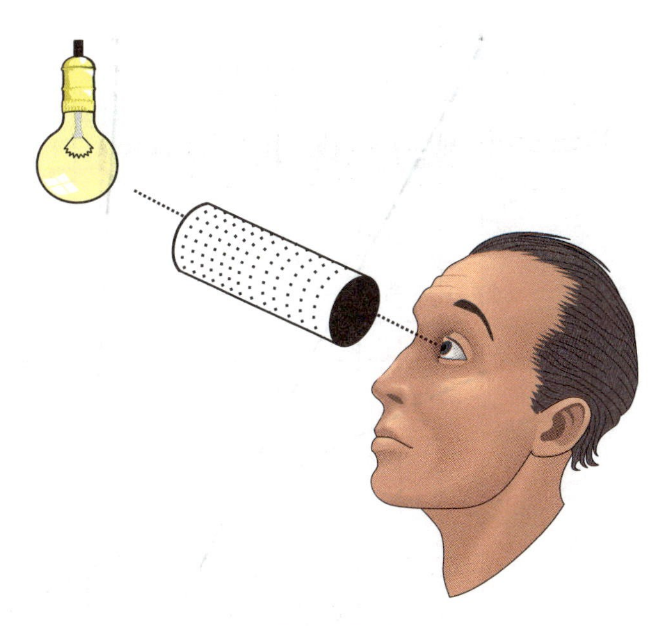

Figure 3-39

Chapter 3 Exercise
Problem 1

A system's hydraulic reservoir is filled with Mobil DTE 25 fluid. There is some concern by the maintenance supervisor that the fluid will be too viscous in the winter for his vane pumps. He expects the temperature to drop to 35°F (1.7°C) in the machine area during some winter months. Determine how viscous the fluid will become.

Brand Name	Fluid Type	Specific Gravity	Viscosity (SUS)	
Mobil DTE 25	PB	876	225 @ 100°F	49 @ 210°F

ASTM Standard Viscosity - Temperature Chart

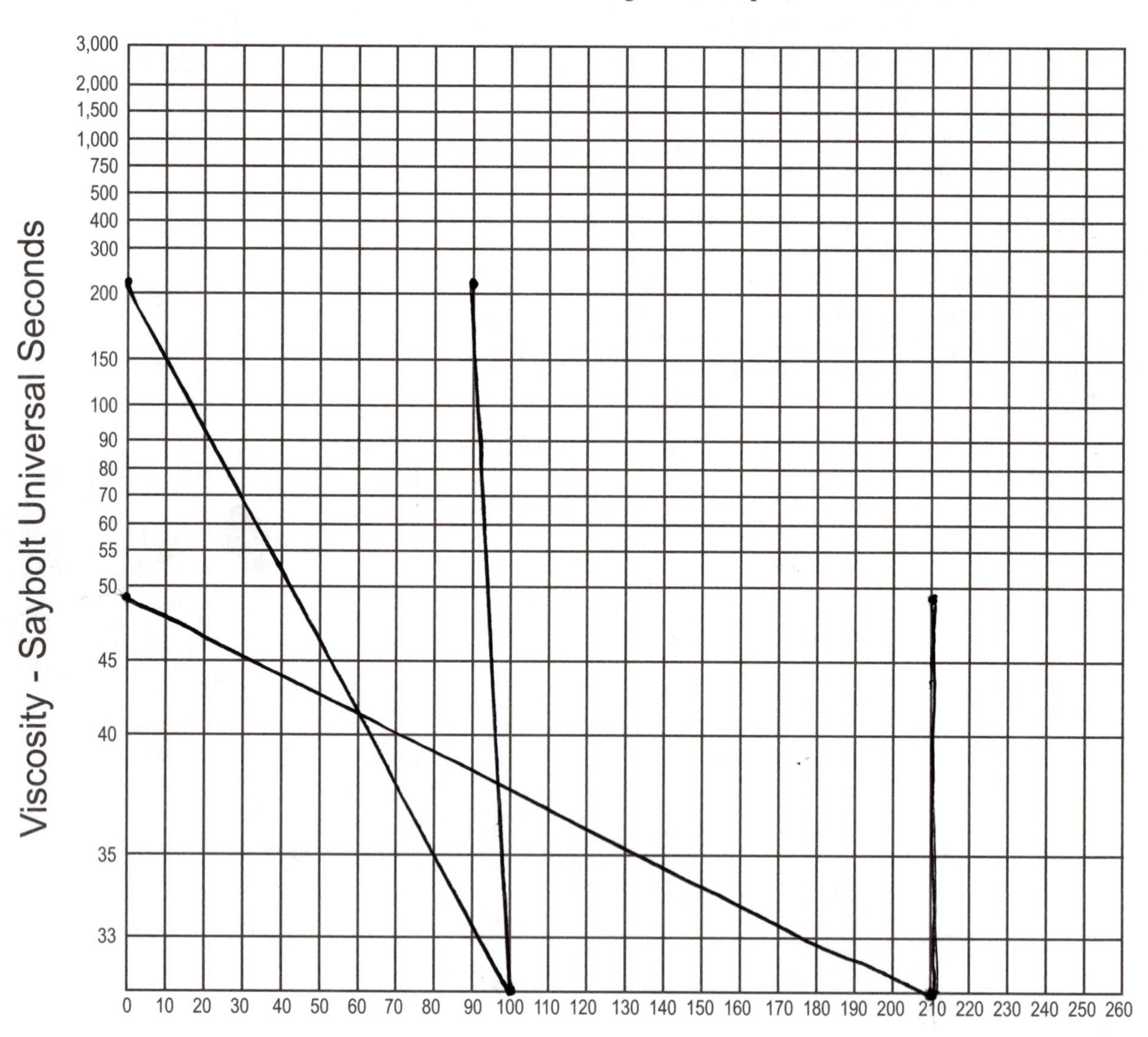

Chapter 3 Exercise
Problem 2

Observing that a system's fluid had a tendency toward foaming, it was decided to switch to an oil with an anti-foam additive. With the addition of the oil to the system's reservoir, foaming ceased but the oil appeared to oxidize quickly. Offer an explanation for the rapid oxidation of the oil. Because the oil reacts with oxygen in the air.

Chapter 3 Exercise
Problem 3

Coolant is falling onto a machine's hydraulic reservoir and into the fluid. When a maintenance man informs the machine operator of the condition, the operator points out that since water and oil don't mix, the water will settle to reservoir bottom and not harm a thing. Comment with respect to what harm the coolant could do in the oil.

Large quanities settle on the bottom and don't break down. Small quanities are broken down. If the oil contains regions of oxidation it has a increased tendency to take on water. Could cause oxidation.

Chapter 3 Exercise
Problem 4

Because of excessive leakage in a plant, a maintenance supervisor decides to switch to a straight mineral oil (an oil without additives) which is relatively inexpensive. The fluid will operate at 2200 psi (151.7 bar). Describe what probable results can be expected from this action.

Looking for EP Addatitive.

Chapter 3 Exercise
Problem 5

The spool of a directional valve of a particular system would periodically stick. Taking apart the valve, a maintenance man finds that the spool is covered with a brown coating which he can't wipe off with a rag. Explain what the coating is and where it came from.

CHAPTER 4

Fire Resistant Hydraulic Fluid

Since petroleum base hydraulic fluid is an excellent lubricant, systems which use it as their energy transmission medium can look forward to years of dependable life. But, in some systems or applications, petroleum oil has a major disadvantage; oil under pressure may spray (atomize) at a leak point - it has been the source of many industrial fires.

As we normally deal with petroleum oil in a system, it is not a high degree fire hazard. Petroleum oil is nonvolatile at room temperature and is capable of extinguishing a small flame like that of a match. However, a high pressure line with a pinhole leak spraying an oil mist into the air is a combustible mixture which can be easily ignited by an open flame. A leak of this nature can be considered a fuel nozzle.

Figure 4-1

In fire-hazardous industrial environments where undisturbed production and operator safety are of primary concern, and the ambient conditions are such that an accidental flame could strike, fire resistant fluids are employed. These fluids are used with the knowledge that operating expenses will increase because the fluid is more expensive than petroleum oil and component life will decrease.

The intent of this lesson is to identify common types of fire resistant fluids for a hydraulic system, to see some problems of a fire resistant fluid in service, and to indicate some maintenance considerations with respect to fire resistant hydraulic fluid.

Figure 4-2

Fire Resistance Determined

Fire resistant fluids are not fireproof. They are just as their name implies - resistant to fire. If fire resistant fluids are heated to a high enough temperature, they will burn.

Fire resistance of a particular fluid is determined by three test specifications: flash point, fire point, and auto ignition temperature. In describing the three tests below, petroleum base hydraulic fluid is used as the test fluid.

Flash Point

Flash point of a fluid is the temperature to which a fluid must be heated to give off sufficient vapor to ignite when a test flame is applied. As a petroleum oil or any liquid is heated, vapor is given off; liquid evaporates in other words. With a petroleum oil

Figure 4-3

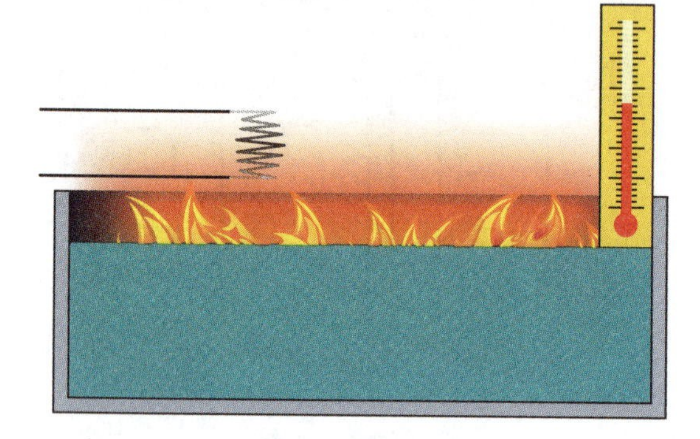

Figure 4-4

Figure 4-5

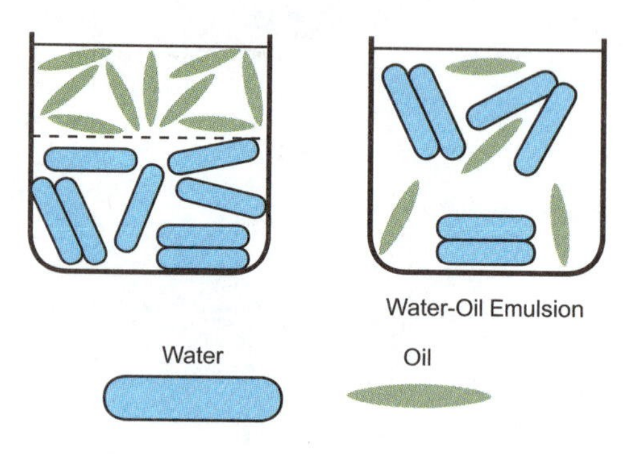

Figure 4-6

heated to 350-450°F (176.6-232.2°C), enough vapor is given off from the oil's surface to ignite when a flame is applied.

However, once the flame is removed, oil vapor ceases to burn.

Fire Point

Fire point of a fluid is the temperature to which it must be heated to burn continuously after a test flame has been removed. When a petroleum base hydraulic oil is heated above that temperature, enough vapor is given off from the oil's surface to ignite when a flame is applied and to remain lit after the flame is removed.

Auto Ignition Temperature

Auto ignition temperature of a fluid is the temperature at which it ignites without an external flame or spark.

Heating a petroleum base hydraulic oil between 500-700°F (260-371°C) will result in the fluid bursting into flame. This occurs without a flame present.

Types of Fire Resistant Fluid

Hydraulic fluids which are classified as fire resistant have higher flash, fire, and auto ignition temperatures than petroleum base oil. These fluids can be divided into two types - water base and synthetic.

Water Base Fluid

Water was the fluid used in the first hydraulic systems. Water had several disadvantages as far as lubrication was concerned, but it did not burn. When the need arose for a fire resistant hydraulic fluid, the initial action was to turn once again to water. However, since a certain amount of lubrication was demanded, oil was emulsified with the water.

Water-Oil Emulsion

A water base fire resistant fluid consisting of water and oil is not a mixture - oil and water do not mix. Since this is the case, oil is broken down into extremely small droplets, usually by a chemical emulsifier. Oil droplets are carried around by the water, increasing its lubricating qualities. If the fluid is exposed to fire, the water turns to steam extinguishing the flame.

A two-phase, water-oil fluid is known as an emulsion. At the time this fluid was popular, a normal ratio of water to oil in an emulsion of this type was 60% water to 40% oil. Water was the dominant fluid and carrier of oil droplets.

Soluble Oil Fluid (HFA)

Fire resistant hydraulic fluids which are predominantly water, are not normally found in present day hydraulic systems except where large amounts of fluid are lost due to leakage. In these systems, reduced component life is sacrificed for an economical fire resistant fluid. The fluid is relatively inexpensive because the percentage of water is at least 90%.

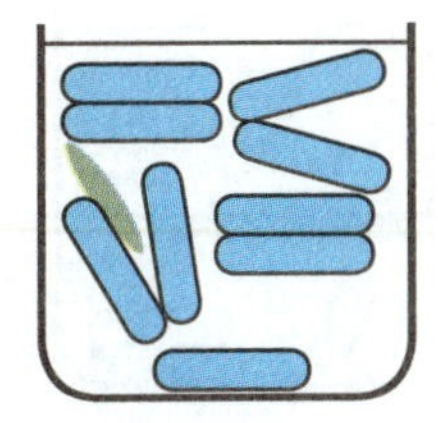

Soluble Oil Fluid

Figure 4-7

A water base hydraulic fluid made up of water emulsified with 1-10% oil, is an oil-in-water emulsion commonly referred to as soluble oil fluid. Anyone remarking that he is using 5% soluble oil in his system, is indicating that his fluid is made up of 95% water and 5% oil or chemical concentration.

Invert Emulsion (HFB)

Emulsion

Figure 4-8

A common water-oil emulsion of a modern hydraulic system is a creamy white liquid made up of 60% oil and 40% water. As compared to a previous emulsion (60% water - 40% oil), the ratio of this emulsion is turned around or inverted.

Since oil is the dominant liquid and carrier of water droplets, invert emulsions have increased lubricating characteristics with a slight decrease in fire resistance.

Viscosity of Water-Oil Emulsions

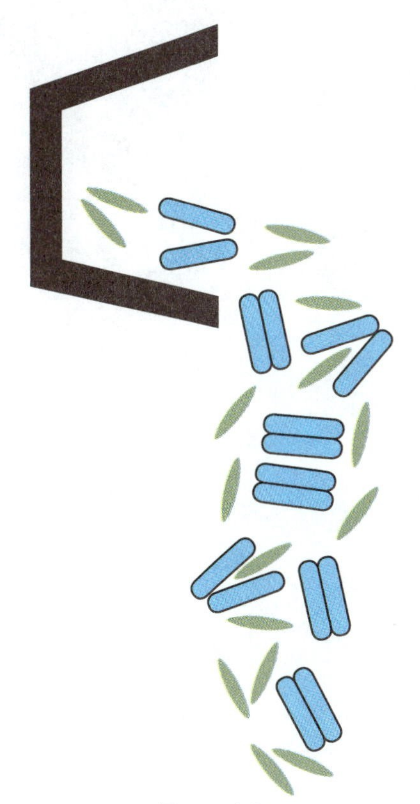

Figure 4-9

Viscosity of a water-oil emulsion is an important characteristic just as with petroleum hydraulic fluid. Since a soluble oil fluid contains a minimum of 90% water, its viscosity is basically that of water. Consequently, these fluids are rather poor lubricants.

An invert emulsion, on the other hand, normally consists of 60% oil. However, this does not mean that its fluid viscosity will be that of its base oil.

Viscosity of an invert emulsion operating in a typical hydraulic system will have a higher viscosity than a normal petroleum fluid for that system. For example, a system operating with an invert emulsion may have a viscosity of 375 SUS (80.9 CST) @ 100°F (37.7°C); whereas a petroleum oil would have a viscosity of 150 SUS (32 CST) @ 100°F (37.7°C).

Because of the shearing action between the two fluid phases as it moves through pump and system, invert emulsions exhibit a decrease in viscosity. To ensure that system components are properly lubricated, an invert emulsion with a higher than normal viscosity is used.

(ASTM graph paper does not properly depict the viscosity-temperature relationship of any invert emulsion or of fire resistant fluids in general.)

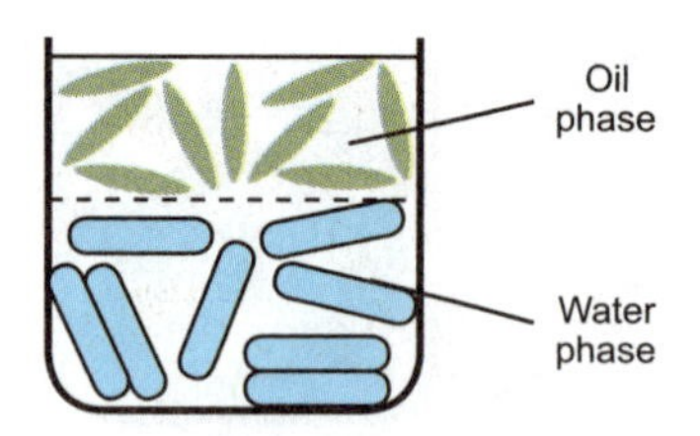

Figure 4-10

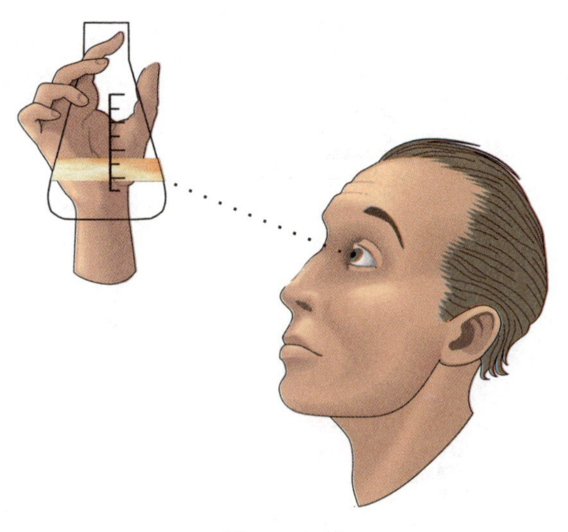
Figure 4-11

Problems with an Invert Emulsion

With a water base fire resistant fluid in a machine reservoir, certain problems can arise. Two problems specific to an invert emulsion are phase separation and bacteria formation.

Phase Separation

Invert emulsion fluids are not designed to be operated at low temperatures. At 32oF (0°C), ice slivers begin to form; at approximately -10°F (-23.3°C), the fluid freezes. Also, freezing and thawing of an invert emulsion cause the two phases to separate.

At the freezing point of water (32°F/0°C) some water droplets carried by the oil free themselves from the emulsion, forming ice crystals. As the system operates and temperature increases, ice crystals melt, but do not necessarily emulsify again. In this condition, the fluid has more tendency to rust system components and adversely affect lubrication.

Repeated freezing and thawing of an invert emulsion could cause water and oil phases to separate to a large degree. In this condition, it would be very difficult, if not impossible, to get the two liquids back together. As a result, fire resistance could be a serious problem.

Check for Phase Separation

A check for phase separation is performed by inspection. With the fluid in the reservoir mixed, it is difficult to determine whether oil and water phases have separated. Draining off a fluid sample into a jar and allowing the fluid to rest for a period of time, of time, you will note that any free water will settle to the bottom of the jar.

If you feel phase separation is severe, contact your fluid representative; he may recommend that the fluid be changed.

Bacteria Formation

In some situations, under the proper temperature conditions, an invert emulsion can support the growth of bacteria. Bacteria in large quantities can plug pressure sensing orifices of pressure control valves and pressure compensated flow controls. Bacteria can also plug filter elements. All these actions result in an undependable, nonproductive system.

Many invert emulsions are equipped with a bactericide additive to avoid this situation.

Check for Bacteria Formation

A check for bacteria formation is performed by sight and smell. If bacteria is present in an invert emulsion, inlet filters will appear to be coated by mucous or slime. And the bacteria will give off a very offensive odor.

If bacteria is present in an invert emulsion, it will most probably have to be changed.

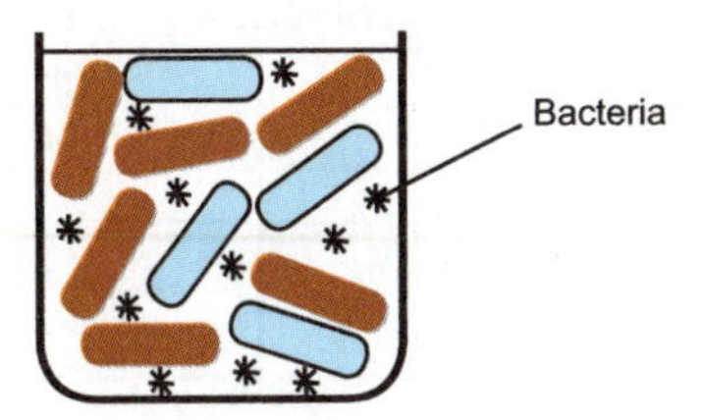

Figure 4-12

Water Glycol (HFC)

Water glycol is another type of water base fire resistant fluid; it consists of water and a glycol which has a chemical structure very similar to automotive antifreeze.

Water glycol is many times dyed red or pink, and normally consists of 60% glycol and 40% water, along with a chemical thickener to increase its viscosity. The glycol actually mixes with the water. The fluid is homogenous and not two-phase like an emulsion; that is, seen through a microscope, the fluid will not appear as separate droplets of water and glycol.

Water glycol fire resistant fluid works well at low temperatures.

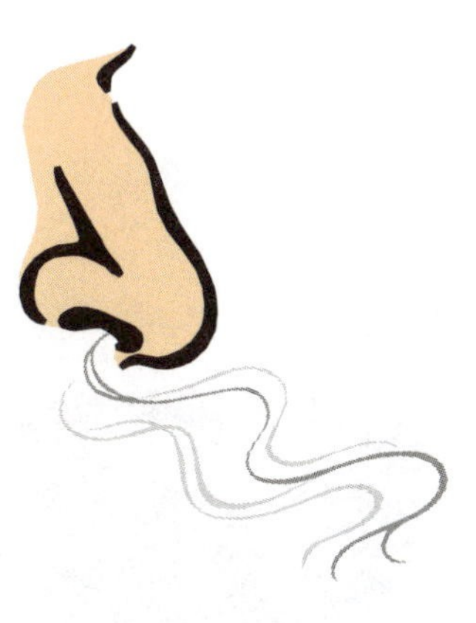

Figure 4-13

Comparing Invert Emulsion with Water Glycol

When comparing an invert emulsion with a water glycol fluid, we find that:

a. It is more difficult to keep an emulsion stable than to maintain a water glycol solution.
b. Stable invert emulsions have more lubricity.
c. Invert emulsions are less expensive.
d. Water glycol fluids are more fire resistant.
e. Water glycol fluids operate better at low temperatures.

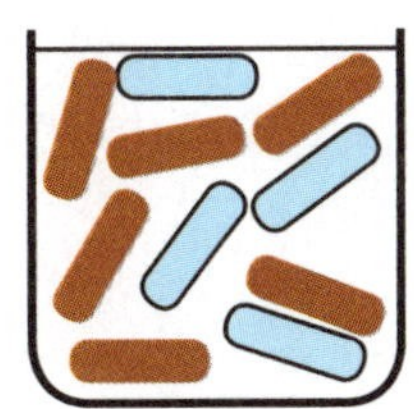

Figure 4-14

Problems with Water Base Fluids

With a water base fire resistant fluid in a hydraulic reservoir, certain problems can arise. Some of these problems are reduced component life and water evaporation.

Invert Emulsion

Water Glycol

Figure 4-15

Relative Lubricating Effectiveness	
Fluid	Lubricant Derating Factor
Petroleum oil	1.0
Invert emulsion	2.0
Water glycol	2.6

Figure 4-16

Figure 4-17

Lubricity of Water Base Fluids

Since water base hydraulic fluids contain a significant percentage of water for fire resistance, they have an inherent disadvantage. With respect to petroleum oil, these fluids have reduced lubricity.

Lubricity and oiliness agents are added to the fluids, but reduced component life is a realistic expectation when these fluids are used. Because of this handicap, water base fire resistant fluids are not normally used in systems which operate above 1800 psi (124 bar).

Of soluble oils, invert emulsions and water glycol fluids, stable invert emulsions have the best lubricity, followed by water glycol and soluble oil fluids, respectively.

Water Evaporation

Many fluid manufacturers recommend that water base fluids be operated at a maximum temperature of 140°F (60°C) with 120oF (49°C) being more desirable. Above 140°F (60°C), excessive water evaporation may occur.

As water evaporates from a water base fluid, some undesirable things can happen. Water vapor escaping from the fluid can condense on unprotected ferrous parts causing rust formation. After a time, rust scale flakes off the unprotected metal surface, becoming a source of dirt for the entire system.

Water base fluids are generally equipped with rust inhibitors, but any unprotected metal surface which is not bathed by the liquid is subject to the attack from escaping water vapor.

Water evaporation affects a fluid's fire resistance. Since the percentage of water determines the fire resistance of a water base fluid, water loss due to evaporation results in the liquid being less resistant to burn.

Loss of water from an invert emulsion or water glycol also affects fluid viscosity. Water loss in a water glycol increases fluid viscosity. In an invert emulsion, water loss causes viscosity to decrease or may even result in the fluid becoming unstable. Top ensure maximum fire resistance and the proper viscosity, water content of water base fire resistant fluid should be monitored by lab analysis at regular intervals.

Synthetic Fire Resistant Fluid (HFDR)

Synthetic fire resistant fluids are man-made liquids which are praised for their resistance to burning while performing close to petroleum oil with respect to lubrication. The most common type of synthetic fire resistant fluid is phosphate ester.

NOTE: Synthetic fire resistant fluids should not be confused with synthetic fluids such as silicones, silicate esters, dibasic acid esters, polyglycol ether compounds and polyol. These fluids have characteristics which are desirable for specific applications, but they are not normally considered fire resistant.

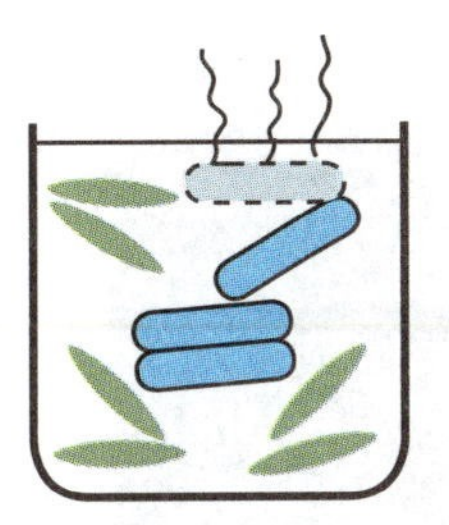
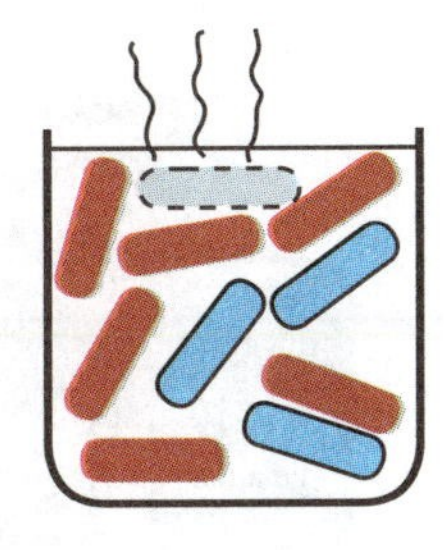

Water Evaporation

Figure 4-18

Synthetic

Figure 4-19

Phosphate ester fluids operate well at high pressure, provide excellent fire resistance, but they are very expensive. In high pressure systems where fire resistance is demanded, but the price of phosphate ester is prohibitive, a blend of phosphate ester and petroleum oil can be used. This fluid has the lubricity the systems demands, but gives a fire resistance less than a phosphate ester fluid.

Phosphate Ester - Petroleum Blend

Figure 4-20

Comparing Water Base with Synthetic Fire Resistant Fluid

When comparing a water base fluid with a synthetic fire resistant fluid, we find that:

a. Synthetic fluids exhibit more lubricity and can operate at higher pressures.
b. Synthetic fluids are more expensive.
c. Synthetic fluids are more fire resistant.
d. Flash point, fire point and auto ignition temperature for a phosphate ester fire resistant fluid are approximately 455°F (235°C), 665oF (352°C), and 1150°F (621°C) respectively.

Fire resistance for water base fluid is not indicated by flash and fire temperature points as long as water is present in the fluid. Auto ignition temperature for water glycol is approximately 1100°F (593°C); for an invert emulsion, it is approximately 825°F (440.6°C).

Problems with Fire Resistant Fluids

Using a fire resistant fluid in a hydraulic system can result in certain problems. Some of these problems are compatibility with seals and protective coatings, foaming and air retention, and dirt retention.

Synthetic

Blend

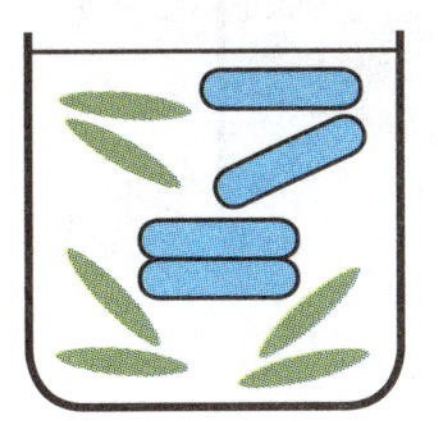

Invert Emulsion

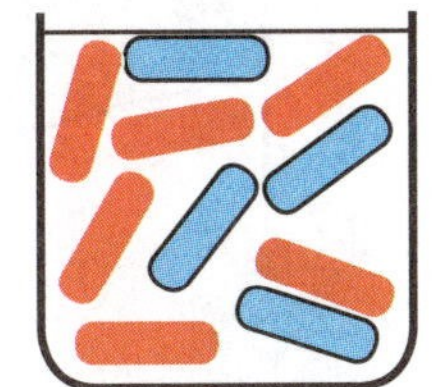

Water Glycol

Figure 4-21

Figure 4-22

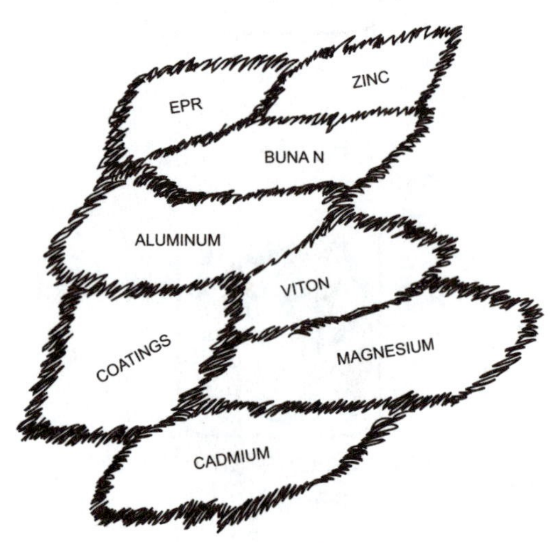

Figure 4-23

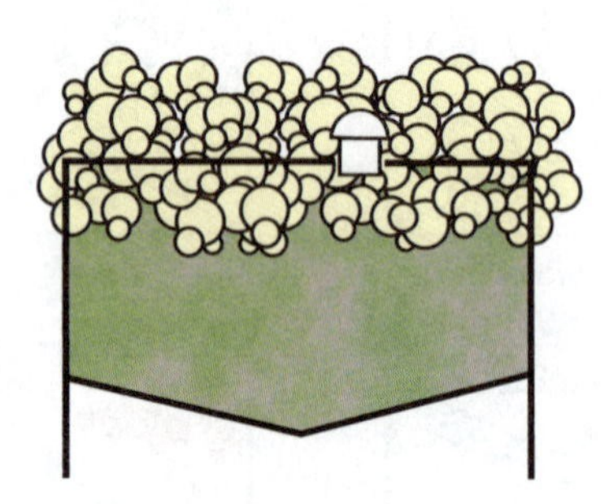
Figure 4-24

Compatibility with Fire Resistant Fluids

A common material for sealing petroleum oil is Buna N. This material is also compatible with an invert emulsion as well as a water glycol. When switching a system from petroleum oil to an invert emulsion or water glycol, system seals will not require changing if they are Buna N. These types of seals would require change if certain synthetic fluids, such as phosphate ester, were used.

When switching from petroleum oil to a water base fire resistant fluid, some problems could occur with protective coatings. If a reservoir interior is protected with petroleum compatible paints and varnishes, a water base fluid may dissolve the coating.

Water glycol fluids and some chemical concentrates are not compatible with some metals. They may attack zinc, cadmium, magnesium, and certain alloys of aluminum, generating gummy residues which plug orifices and filters and cause valve spools to stick. It is recommended that parts which are alloyed or plated with these metals not be used with water glycol. Examples of such parts might be galvanized pipe, and zinc or cadmium plated strainers, fittings and reservoir accessories.

The common Buna N seal material used for dynamic sealing of a petroleum base fluid is not acceptable for a phosphate ester or phosphate ester blend fluid. These fluids require Viton, EPR, or any other suitable material.

Synthetic fire resistant fluids tend to dissolve petroleum compatible paints and varnishes; however, these fluids do not attack common metals found in a hydraulic system.

Foaming and Air Retention with Fire Resistant Fluids

Water base and synthetic fire resistant fluids have more of a tendency to retain air and to foam compared to petroleum oil. After returning to the reservoir, fire resistant fluids require more time in a reservoir to give up any accumulated air bubbles.

Consequently, systems using fire resistant fluids should have larger reservoirs than comparable systems using petroleum oil.

Dirt Retention with Fire Resistant Fluids

As it returns to a reservoir, fire resistant fluids have more of a tendency to retain dirt in suspension compared to petroleum oil. A fluid is supposed to allow any reasonably sized dirt to settle to reservoir bottom, but a fire resistant fluid tends to hold the dirt.

When a fire resistant fluid is used in a system, good fluid filtration should be a prime consideration. And, the use of magnets should not be overlooked.

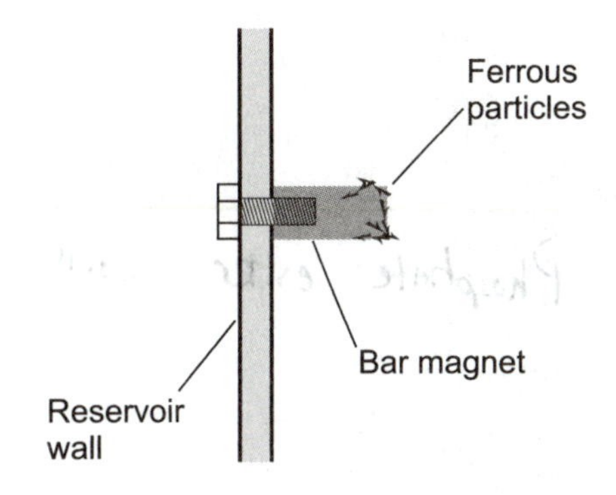

Figure 4-25

Maintenance Considerations

Maintenance considerations of fire resistant fluids with regard to storage are basically the same as for petroleum oil; that is, store barrels on their sides so that water does not collect on barrel tops and leak into the fluid.

Invert emulsion fluids have additional storage requirements. These fluids can have their stability affected by repeated freezing and thawing. Care should be taken then to ensure that the fluid does not freeze.

Transferring oil from barrel to reservoir is another important consideration. Before the drum plugs are removed, the drum cover should be cleaned. This procedure should also be followed for any apparatus or tools which will be used in the process such as hoses, pumps, funnels, reservoir filler holes, and the operator's hands. A check should be made to see that the barrel contains the correct fluid by brand name and viscosity. And, if a pump is used to transfer fire resistant fluid, care should be taken that the pump is not filled with a different fluid and that pump materials and connector assemblies are compatible with the fluid.

Figure 4-26

With the fire resistant fluid in the reservoir, it should be maintained and monitored at regular intervals. Maintenance of the fluid includes filling a reservoir when its minimum oil level has been reached, fixing leaks and servicing filters.

Water base hydraulic fluid should be regularly checked for its water content since its concentration must be kept in a narrow range as it affects viscosity and fire resistance.

Adding water to an invert emulsion is not normally recommended because of the critical mixing process which is demanded. Adding water to a water glycol solution is common but it is not a simple matter of running a hose to the reservoir from the nearest water tap. Makeup water should be free of mineral deposits which could contaminate a system. Distilled steam condensate or deionized water are suitable for use in a water glycol solution. The amount of water to be added is determined after analysis of the fluid sample by a lab.

Figure 4-27

Problem 1

Because of the danger of fire, a system operating at 2200 psi (151.7 bar) with petroleum base fluid is required to change to a fire resistant fluid.

Describe what type of fire resistant fluid might be recommended in this case and indicate its affect on existing system seals and protective coatings.

Phosphate ester will eat almost everything

Problem 2

A maintenance man, seeing that fluid level in a hydraulic reservoir is low, fills a pail with water from a shop faucet and dumps it into the reservoir. The reservoir is filled with a creamy white, water base fire resistant fluid. The water doesn't appear to mix with the existing fluid.

Point out two things the maintenance man has done wrong.

Problem 3

A certain individual cannot see why fire resistant fluid is needed in a system. "Oil doesn't burn," he says. "Anyone who has worked in a plant has seen someone extinguish a cigarette in a pool of oil."

Explain how hydraulic oil can be a fire hazard.

Oil under pressure can spray out and atomize becoming famable

Problem 4

A system which has recently been changed from petroleum oil to an invert emulsion, has a problem with excessive rusting on the inside reservoir surface.

Explain the cause of the rust formation.

Because it has more water content

Problem 5

On the night shift, a machine's hydraulic system was changed from petroleum oil to an invert emulsion. The day shift maintenance man knows that the viscosity of the petroleum oil was 150 SUS (32 CST) @ 100°F (37.7°C). He notices that the invert emulsion has a viscosity of 375 SUS (80.9 CST) @ 100°F (37.7°C). He feels someone has made a mistake.

Has a mistake been made? Explain.

CHAPTER 5

Operation at the Suction Side of a Pump

In a hydraulic system, much consideration is given to what is happening on the pressure side of a system. This is where all apparent action and work take place. The suction side of a pump is also a very important, but many times neglected, part of a system. We will concentrate on this area now.

Pump Location

Many times, pumps in an industrial hydraulic system are located on top of the reservoir which contains a system's hydraulic fluid. A suction or intake line connects the pump inlet with the liquid in the reservoir.

The liquid flowing from the reservoir to a pump can be considered a separate hydraulic system. But, in this system, the less-than-atmospheric pressure developed by the pump is the resistance to flow. Energy to move the liquid is applied by the atmosphere.

The atmosphere and fluid in the reservoir operate together as an accumulator.

Measuring Atmospheric Pressure

We generally think of air as being weightless. But, the ocean of air surrounding the earth does exert a pressure.

Torricelli, the inventor of the barometer, showed that atmospheric pressure could be measured by a column of mercury. Filling a tube with mercury and inverting it in a pan of mercury, he found that atmospheric pressure at sea level could support a column of mercury 29.92 inches (760 mm) high. Sea level atmospheric pressure, therefore, measures, or is equivalent to, 29.92 inches (760 mm) of mercury under standard conditions. Any elevation above sea level will of course measure less than this.

Hydraulic pressure is usually measured in units of psi or bar. But atmospheric pressure is typically measured in Hg or mm Hg. Vacuum pressures are also measured in in. Hg or mm Hg. Atmospheric pressure at sea level under standard conditions of 68°F (20°C) and 36% relative humidity is 29.92 in. Hg or 760 mm Hg. Those numbers are equivalent to 14.7 psia and 1.01 bar respectively. It is important to note that the bar is not defined as atmospheric pressure, but is defined as 100,000 N/m² or 100 kPa. Standard atmospheric pressure is 101,000 N/m².

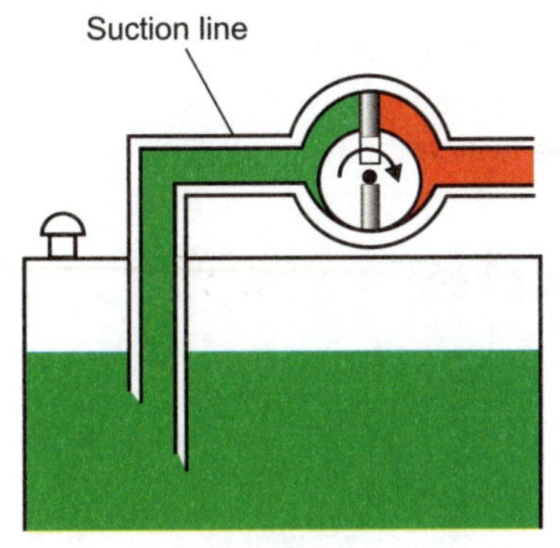

Figure 5-1

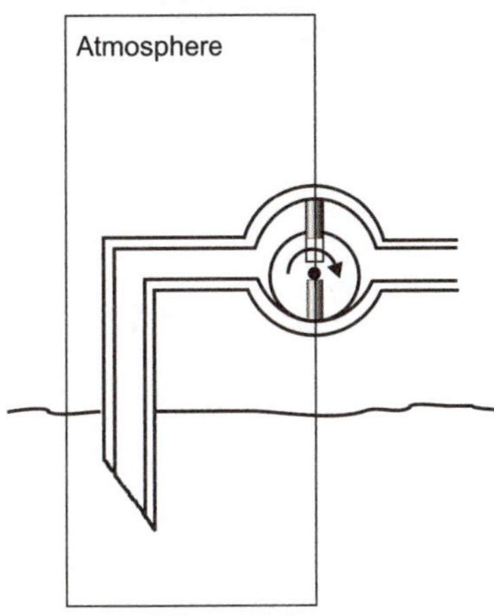

Figure 5-2

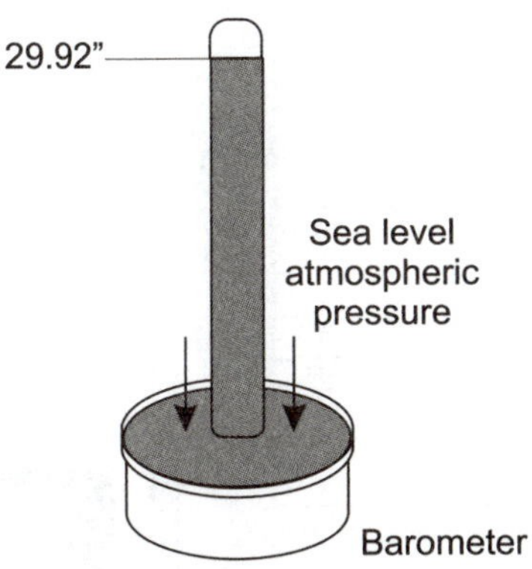

Figure 5-3

Altitude Above Sea Level		Barometer Reading		Atmospheric Pressure	
(ft.)	(m)	(in. Hg.)	(mm)	(psi)	(bar)
0	0	29.92	760	14.7	1
1000	304.8	28.8	732	14.2	.966
2000	609.6	27.7	704	13.6	.925
3000	914.4	26.7	678	13.1	.891
4000	1219.2	25.7	652.7	12.6	.857
5000	1524.0	24.7	627.3	12.1	.823
6000	1828.8	23.8	604.5	11.7	.790
7000	2133.6	22.9	581.6	11.2	.762
8000	2438.4	22.1	561.3	10.8	.735
9000	2743.2	21.2	538.5	10.4	.707
10000	3048.0	20.4	518.2	10.0	.680

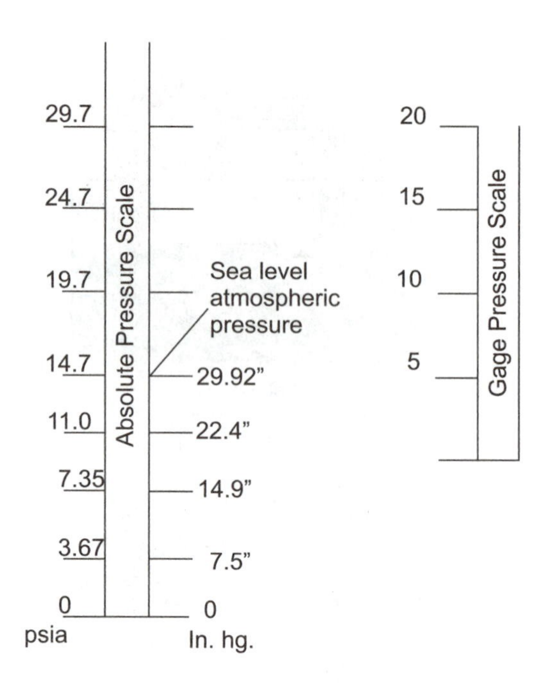

Figure 5-4

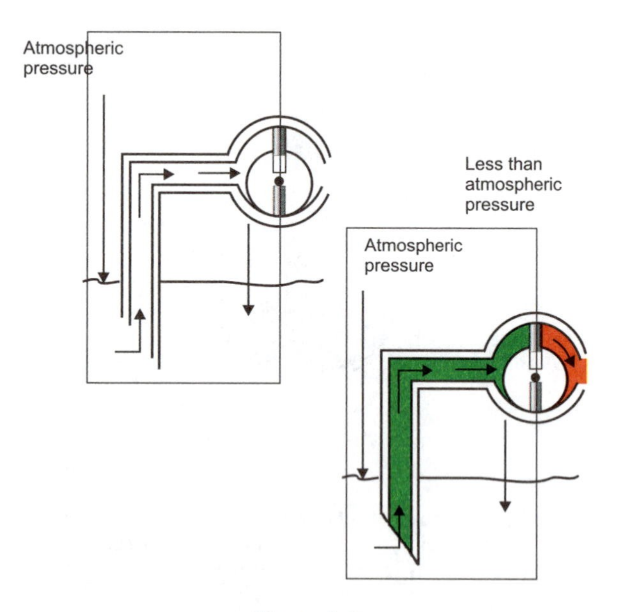

Figure 5-5

To convert between in. Hg and psia, notice that 1 psia is equal to 2.04 in. Hg. 1 bar is equal to 752 mm Hg. For approximation we can use 2 in. Hg per 1 psia and 750 mm Hg for 1 bar.

Absolute and Gage Pressure Scales

Either of two pressure scales are used to measure pressure in a hydraulic system--an absolute scale or a gage scale.

The absolute pressure scale begins at the point where there is a complete absence of pressure. The units of measurement can be either psi (bar) or inches of mercury (mm Hg).

The gage pressure scale begins at the point of atmospheric pressure. The units of approximate measurement are psi. To determine the absolute pressure from a gage, add standard atmospheric pressure to the gage reading. For example, if a pressure gage reads 100 psi (6.9 bar) and standard pressure is at 14.7 psi (1 bar), the equivalent absolute pressure is 114.7 psi (7.9 bar).

To differentiate between the two pressure scales, psig is used to denote gage pressure, and psia is used for absolute pressure.

Operation at Suction Side of Pump

When a pump is not operating, the suction side of a system is in equilibrium. A no-flow condition exists which is indicated by a zero pressure differential between pump and atmosphere. To receive a supply of liquid to its rotating group, the pump generates a less-than-atmospheric pressure. The system becomes unbalanced and a flow results.

Use of Atmospheric Pressure

The pressure applied to the liquid by the atmosphere is used in two phases:

1. supplying liquid to the pump inlet.
2. accelerating the liquid and filling the rapidly moving rotating group, 1200, 1800 rpm, are standard speeds.

The largest portion of atmospheric pressure is used in accelerating the liquid into the pump. However, the action of supplying liquid to the inlet port makes use of atmospheric pressure first. If too much is used in this phase, not enough pressure will be available to accelerate the liquid into the rotating group. The pump will starve, and something known as cavitation will occur.

Cavitation

Cavitation is the formation and collapse of gaseous cavities in a liquid. These cavities are harmful to pump life in two ways:

1. they interfere with lubrication
2. they destroy metal surfaces

On the suction side of a pump, bubbles form throughout the liquid. This results in a reduced degree of lubrication and an increased amount of wear.

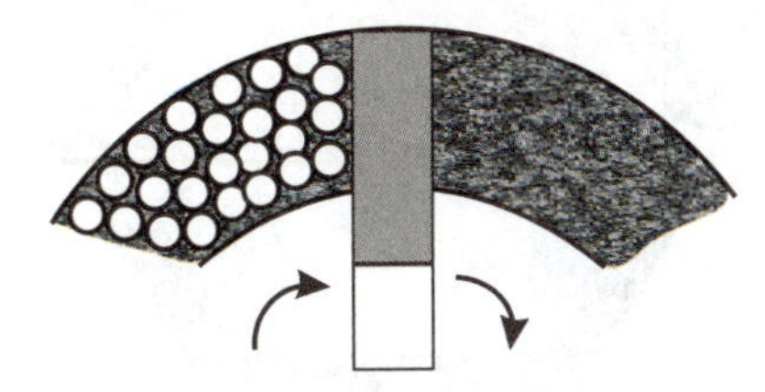

Figure 5-6

As these cavities are exposed to high pressure toward the outlet of the pump, the walls of the cavities collapse and generate tons of force per square inch. The release of energy generated by the collapsing cavities "eat away" metal surfaces and has the same effect as a sculptor's hammer and chisel on stone. If cavitation is allowed to continue, the pump's life will be reduced, and pieces of the pump will migrate to other areas of the system, harming other components.

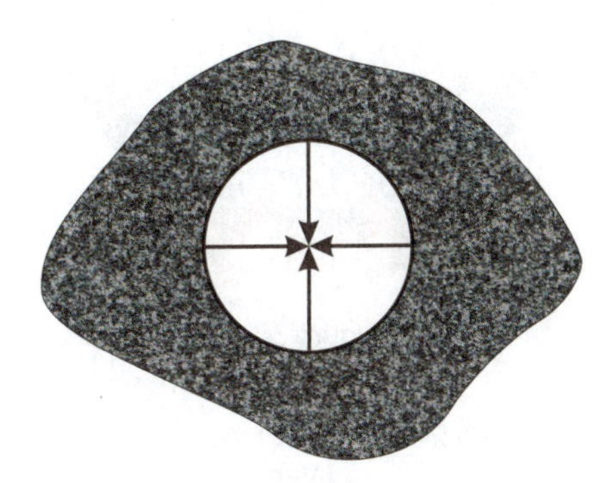

Figure 5-7

Indication of Cavitation

The most noticeable indication that cavitation is occurring is noise. The simultaneous collapse of cavities causes high-amplitude vibrations to be transmitted throughout the system, and a high shrieking sound to generate from the pump.

During cavitation, there is also a decrease in pump flow rate, because the pumping chambers do not completely fill with liquid, and system pressure becomes erratic.

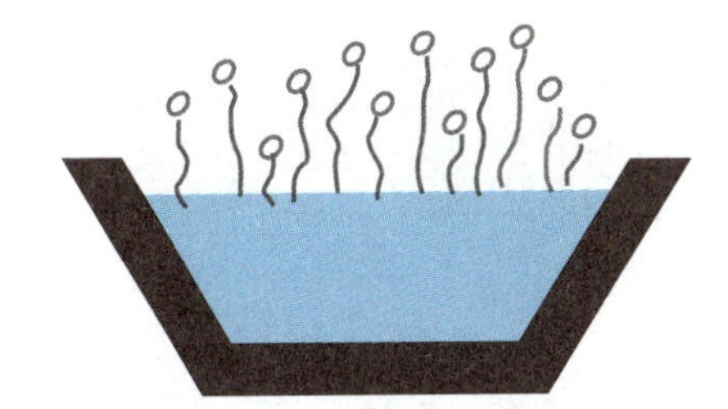

Figure 5-8

Cause of Cavity Formation

Cavities form within a fluid because the liquid is made to boil. The boiling in this instance is not caused by heating, but is brought about by reaching a low absolute pressure.

Liquid Vapor Pressure

All molecules in a liquid are moving continuously, but not all at the same speed. Molecules of liquid which move quickly try to escape from the liquid despite the strong attraction of neighboring molecules. These speedier molecules exert a force to enter the atmosphere. This force is the liquid's vapor pressure.

If a cover were placed on the liquid container, fast-moving molecules would enter the area above the liquid. When this area became saturated with vapor, molecules would collide with one another and be knocked back into the liquid. The action of molecules escaping from the liquid is evaporation. The action of molecules returning to the liquid is

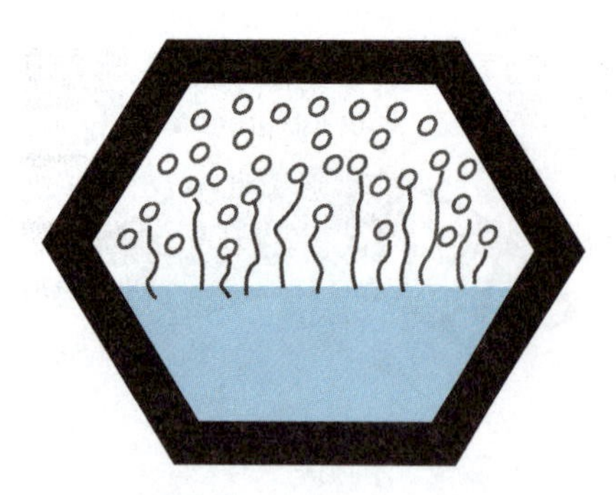

Figure 5-9

Vapor Pressure of Water

Temperature		Vapor Pressure Abs. Pressure	
°F	°C	inches	mm
100	37.8	2.0	50.80
110	43	2.6	66.04
120	49	3.5	88.90
130	54	4.5	114.30
140	60	5.9	149.86
150	66	7.7	195.58
212	100	29.92	760.00

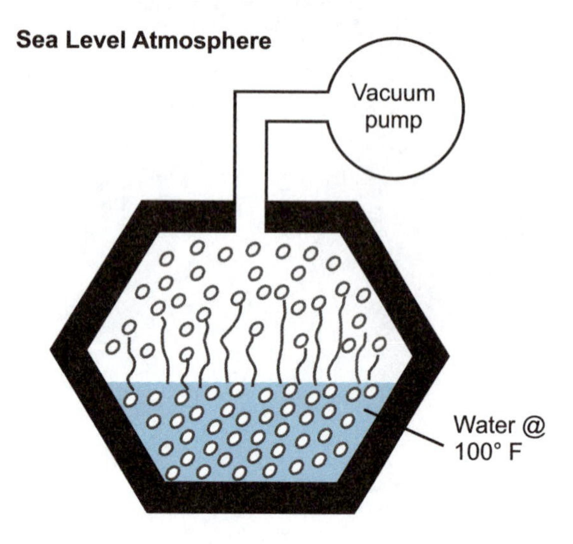

Figure 5-10

condensation. When the rate of evaporation equals the rate of condensation, the system will be in equilibrium. The pressure exerted by the vapor at this point will be the vapor pressure of the liquid. This pressure is often expressed in units of inches of mercury absolute pressure.

Vapor Pressure Affected by Temperature

The vapor pressure of a liquid is affected by temperature. With an increase in temperature, more energy is added to the liquid's molecules. The molecules move more quickly and the vapor pressure increases. When the vapor pressure equals the atmospheric pressure, liquid molecules freely enter the atmosphere. This is known as boiling.

The boiling point of water at sea level is 212°F (100°C). At this temperature the vapor pressure of water is equal to atmospheric pressure.

Boiling Affected by Pressure

A liquid can also be made to boil by decreasing the pressure acting on it. When this reduced pressure equals the vapor pressure of the liquid, molecules of liquid will freely enter the area above the liquid.

Water has a vapor pressure of 2 in. Hg (0.068 bar) at 100°F (37.2°C). If a container of 100°F (37.2°C) water were connected to a vacuum pump, the water would boil when the pressure in the container reached 2 in. Hg (0.068 bar) Abs. Press. This boiling action takes place in a cavitating pump.

Dissolved Air

Hydraulic fluid at sea level is made up of approximately 10% air. This air is dissolved in the liquid. It cannot be seen and does not apparently add to the liquid volume.

The capability of hydraulic fluid, or any liquid, to hold dissolved air decreases as the pressure acting on the liquid decreases. For example, if a beaker of hydraulic fluid, which has been exposed to the atmosphere, were placed in a vacuum chamber, the dissolved air would bubble out of solution.

Leading up to and during the action of cavitation, this dissolved air comes out of solution and contributes to the harm of the pump.

Entrained Air

Entrained air is air which is present in a liquid in an undissolved state. The air is in the form of bubbles.

If a pump happens to ingest fluid with entrained air, the air bubbles will have somewhat the same effect on a pump as cavitation. However, since this is not associated with a liquid's vapor pressure, we will refer to this action as pseudo-cavitation.

Many times, entrained air is present in a system because of a leak in a suction line or a bad pump shaft seal. Since the pressure at the suction side of a pump is usually less than atmospheric, any opening at this point will result in air being sucked into the fluid and to the pump. Also, any entrained air bubbles, which were not allowed to escape while the fluid was in the reservoir, will find their way to the pump.

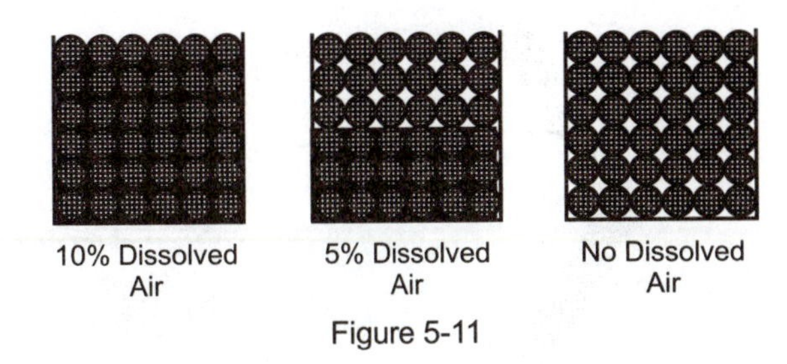

Figure 5-11

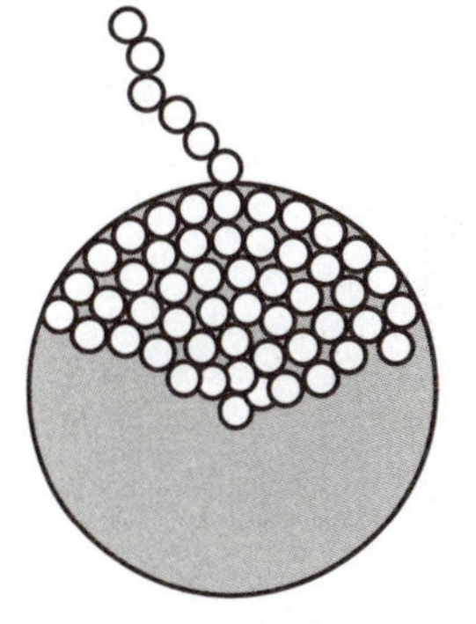

Figure 5-12

Suction Specification

Cavitation is very harmful to both pump and system. For this reason, manufacturers specify suction limitations for their pumps.

Manufacturers of positive displacement industrial hydraulic pumps usually specify the less-than-atmospheric pressure which must be present at the pump inlet to fill the pumping mechanism. However, this pressure specification is not given in terms of the absolute pressure scale, but in terms of the vacuum pressure scale.

Vacuum Pressure Scale

A vacuum is any pressure less than atmospheric. Vacuum pressure is a source of confusion since the scale begins at atmospheric, just as gage pressure, but works its way down in units of inches of mercury.

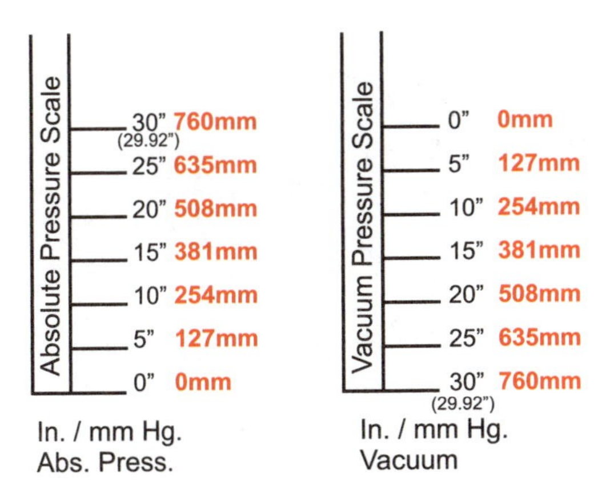

Figure 5-13

How Vacuum is Determined

In the illustration, a pan of mercury open to the atmosphere is connected by means of a tube to a flask which has the same pressure as the atmosphere. Since the pressure inside the flask is the same as the pressure acting on the pan of mercury, a column of mercury cannot be supported in the tube. Zero inches of mercury in the tube indicates a no-vacuum condition in the flask.

If the flask were evacuated so that the pressure inside were reduced by 10 inches (254mm) of mercury, the atmospheric pressure acting on the mercury in the pan would support a column of mercury 10 inches high (254 mm). The vacuum would measure 10 inches (254mm) of mercury.

If the flask were evacuated so that no pressure remained and a complete void existed, the atmospheric pressure acting on the mercury could support a column of mercury 29.92 inches (760 mm) high. The vacuum would measure 29.92 in. Hg. (760 mm).

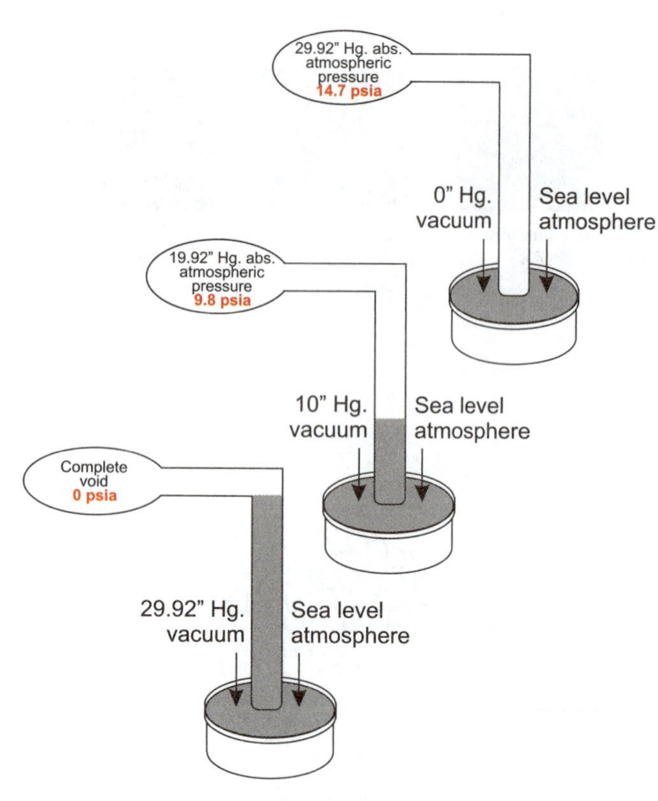

Figure 5-14

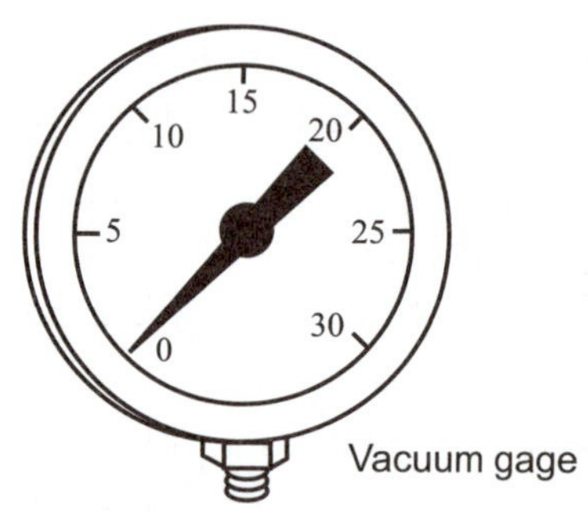

Figure 5-15

Zero inches (Zero mm) of mercury vacuum is atmospheric pressure or the absence of vacuum. 29.92 inches (760 mm) mercury vacuum indicates a perfect vacuum or zero absolute pressure.

Vacuum Gage

A vacuum gage is calibrated from 0-30 inches of mercury (0-760 mm Hg). Each division is one inch of mercury.

At sea level, to determine an absolute pressure from a vacuum gage reading, subtract the vacuum in inches (millimeters) of mercury from 30 (760 mm). For instance, a vacuum of 7 in. Hg. (177 mm) is actually an absolute pressure of 23 in. Hg. (583 mm).

Suction Specification Given in Terms of Vacuum

Leading pump manufacturers give their suction specification in terms of vacuum as it relates to sea level. When the pump is to be used at an elevation above sea level, the barometric pressure at that level must be taken into account.

If a pump manufacturer specifies no more than a vacuum of 7 inches of mercury (177 mm Hg.) be present at the pump's inlet port, this means the pump manufacturer desires an absolute or barometric pressure at the inlet of at least 23 inches (583 mm) of mercury in order to accelerate liquid into the pumping mechanism. If the absolute pressure at pump inlet were anything less than 23 in. Hg. (583 mm), the pump may be harmed. Of course, this depends on the design factor built into the allowable vacuum specification by the manufacturer.

A suction specification is also given for a pump while it is operating at a specified RPM and using petroleum base fluid. The specification is altered if the pump is operating at another speed or pumping a different liquid.

Effects of Different Fluids on Maximum Allowable Vacuum

A maximum allowable vacuum specification of a pump is affected by the fluid being pumped. A pump's suction specification is based on using petroleum base fluid which has a certain specific gravity and vapor pressure. If a fire resistant fluid is used, specific gravity and vapor pressure change affecting maximum allowable inlet vacuum.

High Specific Gravity Affects Maximum Allowable Vacuum

Specific gravity compares the weight of one liquid with another. More specifically, it is the ratio of weight between a volume of water and an equal volume of another liquid.

A cubic foot of water weighs 62.4 lbs. (28.3 kg) @ 60°F (15.6°C). A cubic foot of a common petroleum base hydraulic fluid weighs 56.4 lbs. (25.6 kg) @ 60°F (15.6°C). Dividing the weight of the hydraulic fluid by the water weight, we find that the weight of the oil is 90% of the water. The ratio of weights is 1 (water) to .90 (petroleum oil). The specific gravity (SG) of petroleum oil is then indicated as .90.

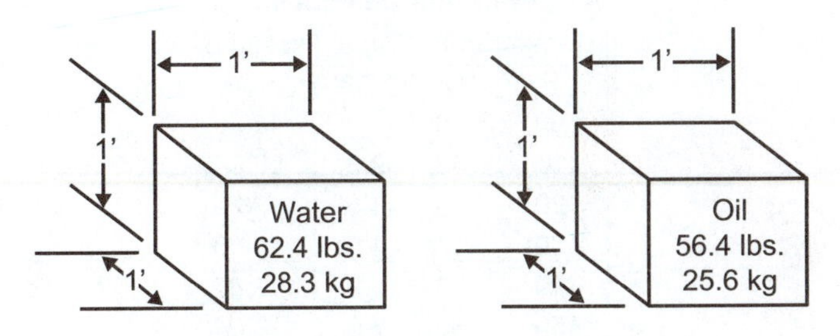

Figure 5-16

A pump's maximum allowable vacuum specification is based on using petroleum base fluid which has a specific gravity of approximately .87-.90. With a phosphate ester fire resistant fluid, specific gravity increases over 30% to approximately 1.15. Water base fluids have specific gravities ranging from .93 for an invert emulsion to 1.08 for water glycols. (See fluids chart in appendix.) With these heavier fluids, more pressure is required at pump inlet for fluid acceleration; consequently, maximum allowable vacuum may decrease appreciably.

Typical Vapor Pressures of Water Glycol Fluids

Temperature		Vapor Pressure	
°F	°C	inches	mm
100	37.8	1.3	33.02
110	43	1.9	48.26
120	49	2.5	63.5
130	54	3.3	83.82
140	60	4.1	104.14
150	66	5.5	139.7

High Vapor Pressure Affects Maximum Allowable Vacuum

Petroleum base and phosphate ester fire resistant fluids have very low vapor pressures at the operating temperature of a common hydraulic system. However, this is not the case with a water base fluid.

Since they contain a large percentage of water, invert emulsions and water glycol fluids can have vapor pressures of several inches of Hg where petroleum and synthetic fluids would have vapor pressures of a fraction of an inch of Hg. For this reason, they have more of a tendency to vaporize and cavitate.

To avoid cavitation with a water base fluid, a pump manufacturer ensures that sufficient pressure is present at pump inlet to accelerate the liquid into the pump without ever going below the fluid's vapor pressure. This is accomplished by reducing maximum allowable vacuum.

Typical Vapor Pressures of Water-on-Oil Fluids

Temperature		Vapor Pressure	
°F	°C	inches	mm
100	37.8	1.9	48.26
110	43	2.6	66.04
120	49	3.5	88.9
130	54	4.5	114.3
140	60	5.9	149.86
150	66	7.6	193.04

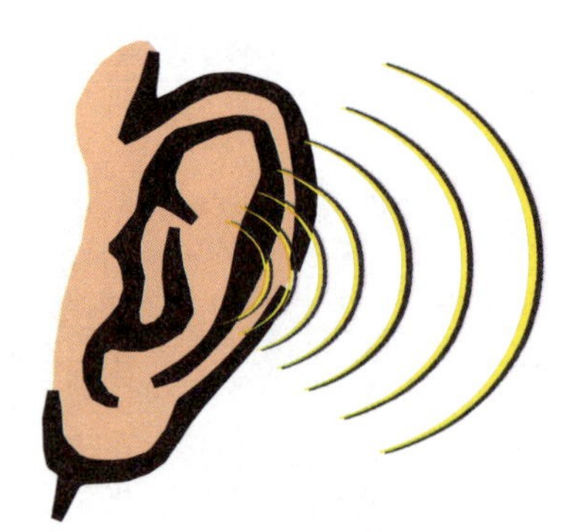

Figure 5-17

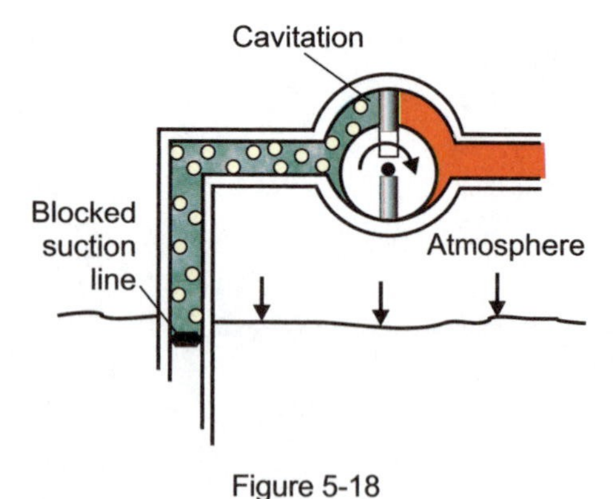

Figure 5-18

Check for Pump Cavitation

Maintenance men have the first opportunity to discover a cavitating pump or a pump which is sucking air. Being acquainted with their machinery, they receive the initial indication that something is wrong.

The most pronounced indication of a pump either cavitating or sucking air are high-pitched sounds, but they are of a slightly different character. A cavitating pump will have a steady, high-pitched sound which is probably due to collapsing bubbles of approximately the same size. On the other hand, a pump sucking air has somewhat of an erratic sound. If small amounts of air are finding their way to the pump, the noise may sound like a rattle or bad bearing. Large amounts of air will result in a very erratic banging and popping noise.

A more certain way to distinguish cavitation from air entrainment is to determine the absolute pressure at pump inlet by taking a vacuum gage reading and subtracting it from the barometric pressure. If this action indicates that insufficient pressure is available at pump inlet, cavitation may be occurring.

A new system with a cavitating pump is probably the result of poor suction line design or incorrect fluid viscosity. Changing to the appropriate viscosity or increasing the suction line size or reducing pressure differential will help the situation. A properly designed older system whose pump is cavitating may be the result of a plugged suction line due to a rag, newspaper, or animal. It may also be caused by a dirty filter without a bypass or an insufficient bypass.

Pump Priming

With respect to a pump, "prime" indicates that a pumping mechanism is filled with liquid. A pump which is not primed is filled with air or "air bound." Before pumping can occur, this air must be purged from suction line and pump cavities. If this is not done and the pump is allowed to operate in an unprimed condition for even a few minutes, it may be permanently damaged due to lack of lubrication.

At start up, a pump whose outlet is directly connected to tank through a directional valve, generally has an easy time of pushing trapped air to tank. A pump which must push its air over a relief valve has an impossible task; the reason being, an ordinary industrial hydraulic pump is a poor air compressor.

To release trapped air from an unprimed pump, a fitting at pump outlet can be loosened. The pump is then jogged until oil squirts out the fitting indicating the pump is primed. The fitting is then tightened. Trapped air may also be released by venting a relief valve.

Pump priming is frequently required on the new systems at start up and on systems with suction side leaks.

Terms and Idioms Associated with the Suction Side of a Pump

FLOODED SUCTION - situation in which pump inlet is below fluid level in reservoir.

HEAD PRESSURE - pressure exerted at the bottom of the fluid column. With a pump's inlet located below fluid level, head pressure is an additional source of energy to supply the pump. To calculate head pressure (in units of inches of mercury).

$$\text{head pressure (in. Hg.)} = \frac{\text{fluid column height (in.)} \times .036 \times \text{fluid specific gravity}}{.491}$$

$$\text{head pressure (mm Hg.)} = \text{fluid column height (mm)} \times \text{fluid specific gravity} \times .00288$$

LIFT - the height of a column or body of fluid below a given point expressed in linear units. To calculate lift pressure (in units of inches of mercury).

$$\text{lift pressure (in. Hg.)} = \frac{\text{fluid column height (in.)} \times .036 \times \text{fluid specific gravity}}{.491}$$

$$\text{lift pressure (mm Hg.)} = \text{fluid column height (mm)} \times \text{fluid specific gravity} \times .00288$$

SUCTION - act of the pump generating a pressure differential between itself and the atmosphere.

SUCTION PRESSURE - the absolute pressure of the fluid at the inlet pump.

Exercise
Operation at the Suction Side of a Pump
50 points

INSTRUCTIONS: Comment on the following statements.

1. STATEMENT -
   "We cavitated four pumps to death before we found the crack in the suction line." (Explain the difference between cavitation and the situation described by the statement).

2. STATEMENT -
   "Here in Denver, the atmospheric pressure is usually around 12.1 psia (0.8 bar). The pump manufacturer's specification is 5 in.Hg. (0.17 bar). Since one inch of mercury is approximately a half psi, we have plenty of pressure to work with." (Clarify this erroneous statement).

3. STATEMENT -
   "I don't understand what's wrong with my pump. It makes a high shrieking sound. And, I don't get the flow that is stated in the manufacturer's catalog. I thought the pump might not be developing enough vacuum, so I checked the vacuum with a gage. The gage reads 24 (0.81 bar) and the pump catalog says I only need 6" (0.204 bar). (What's happening to the pump? Clarify the last sentence in the statement).

4. STATEMENT -
   "Purchasing bought a different brand of pump and it cavitated right off the bat. We can't determine why, since the flow rate through the suction line is exactly the same." (What's the problem?)

5. STATEMENT -
   "Our machines were leaking like crazy so we switched to a heavier oil. Now the pump is noisy and doesn't last too long either." (What's happening to the pump and what caused the situation?)

CHAPTER 6

Hydraulic Actuators

Hydraulic actuators convert hydraulic working energy into mechanical working energy. They are the points where all visible activity takes place and one of the first things to be considered in the design of a machine.

Hydraulic actuators can be divided into basically two types: linear and rotary.

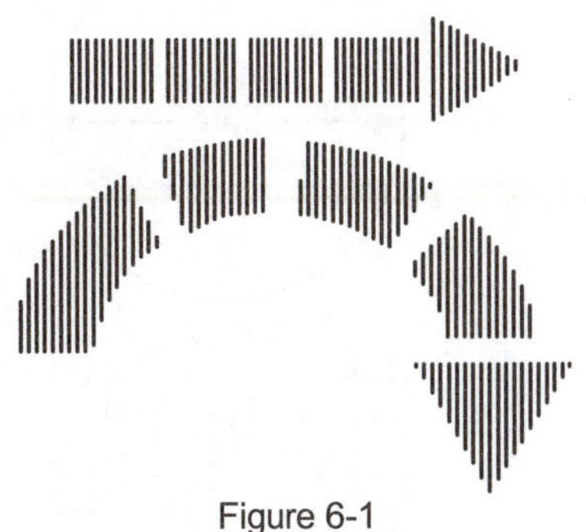

Figure 6-1

Cylinders

Hydraulic cylinders transform hydraulic working energy into a straight-line or linear mechanical energy which is applied to a movable resisting object to perform work.

What Cylinders Consist Of

Cylinders have been briefly touched on earlier. It was shown that a cylinder basically consists of a cylinder body, a closure at each end, and a movable piston attached to a piston rod. At one end, the cylinder body has an inlet port by which fluid enters the body, the other end is vented.

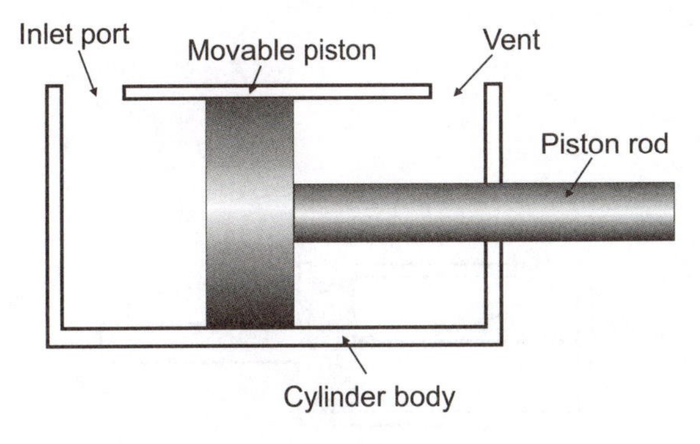

Figure 6-2

Cylinder Force

Through the stroke of a cylinder, hydraulic working energy is applied to the area of its movable piston. The pressure component of working energy applied to the piston will be no more than the resistance which a load offers.

Many times it is required to know what the pressure must be at a certain size cylinder to develop a particular output force. To determine this pressure, the following formula is used: (neglecting friction)

$$\text{pressure} = \frac{\text{force}}{\text{area}}$$

$$[\, P = F/A \,]$$

When the formula was used previously, the area and pressure, or the area and force were given. But, many times just the bore (diameter) of the cylinder is known and the area must be calculated. This calculation is as easy to make as the calculation for the area of a square.

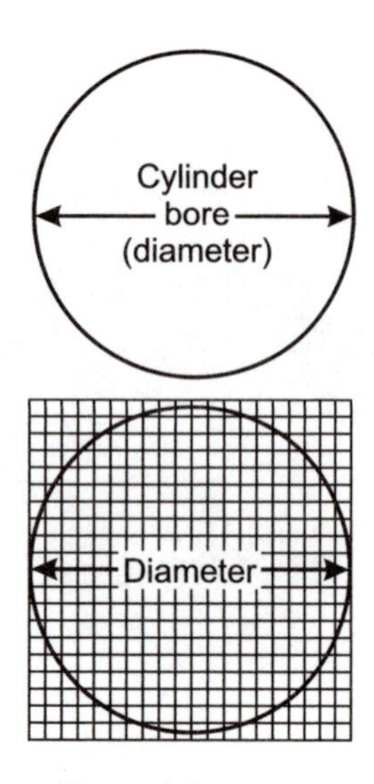

Figure 6-3

Area of a Circle

It is a fact that the area of a circle is exactly 78.54% of the area of a square whose sides are the length of its diameter (D).

To determine the area of a circle, multiply the circle diameter by itself and by .7854.

circle area = diameter² x .7854

Another commonly used formula is:

$$\text{circle area} = \frac{(\pi)\,(D^2)}{4}$$

Cylinder Stroke

The distance through which working energy is applied determines how much work is done. This distance is the cylinder stroke.

It has been illustrated that a cylinder can be used to multiply a force by the action of hydraulic pressure acting on a piston area.

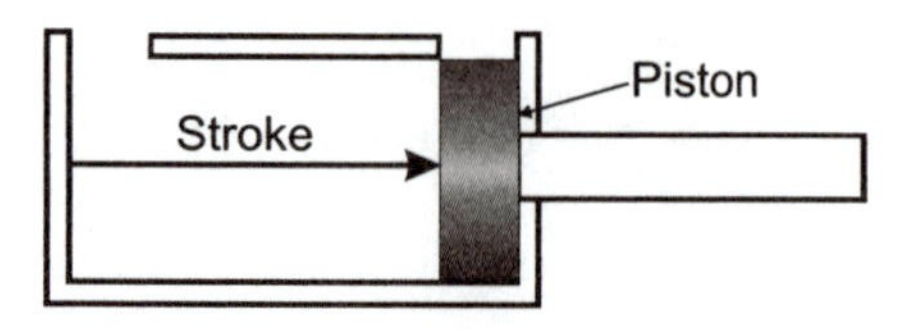

Figure 6-4

When multiplying a force hydraulically, it appeared that something was received for nothing. It appeared that a smaller force could generate a larger force under the right circumstances, and nothing was sacrificed. This is relatively true in a static system. But if the force were to be multiplied and moved at the same time, something would be sacrificed — distance.

Cylinder Volume

Each cylinder has a volume (displacement) which is calculated by multiplying its stroke in inches by the square inch area of the piston. This will give a volume of so many cubic inches (cubic centimeters).

cylinder volume	= piston area	x stroke
(in^3)	(in^2)	(in)
(cm^3)	(cm^2)	(cm)

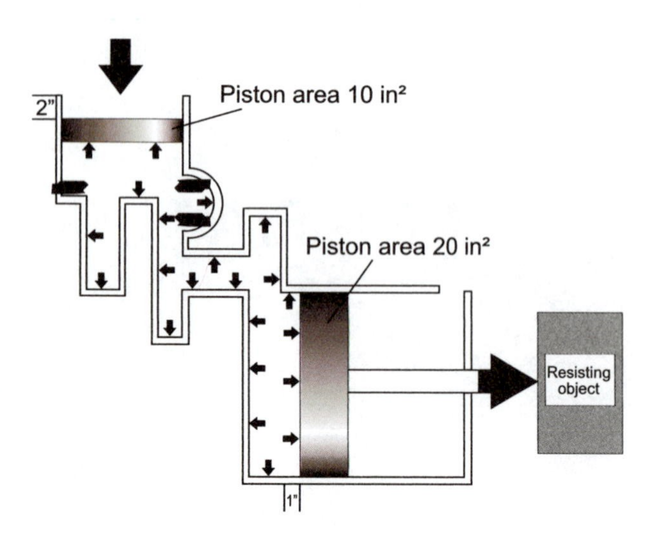

Figure 6-5

In the illustration, the top piston must move through a distance of 2" (5.08 cm) to make the cylinder piston move 1" (2.54 cm). In both instances, the work done is the same. The top piston displaces 20 in^3 (327.8 cm^3) of liquid, and the lower cylinder piston is displaced by 20 in^3 (327.8 cm^3) of liquid.

Piston Rod Speed

The rod speed of a cylinder is determined by how quickly the volume behind the piston can be filled with liquid. The expression which describes piston rod speed is:

$$\text{rod speed (in/min)} = \frac{\text{gpm} \times 231}{\text{piston area (in}^2\text{)}}$$

$$\text{rod speed (m/sec)} = \frac{\text{l/min} \times .1667}{\text{cm}^2}$$

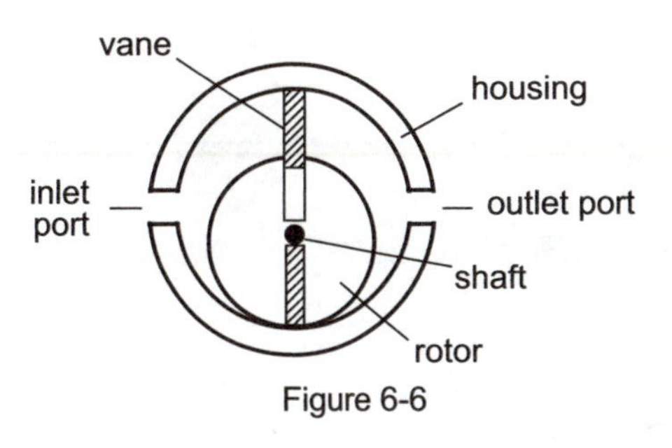

Figure 6-6

Hydraulic Motors

Hydraulic motors transform hydraulic working energy into rotary mechanical energy, which is applied to a resisting object by means of a shaft.

What Motors Consist Of

All motors basically consist of a housing with inlet and outlet ports and a rotating group attached to a shaft. The rotating group in the particular vane type motor illustrated consists of a rotor and vanes which are free to slide in and out.

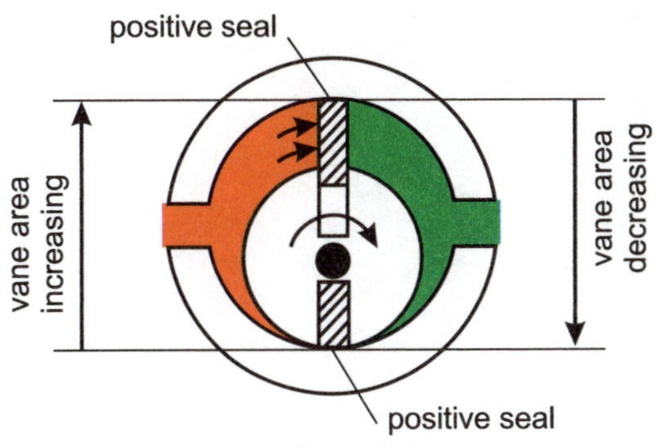

Figure 6-7

How Motors Operate

The rotating group of the motor is positioned off-center to the housing. The shaft in the rotor is connected to an object which offers a resistance. As fluid enters the inlet port, hydraulic working energy acts on any part of a vane exposed to the inlet port. Since the top vane has more area exposed to pressure, the force on the rotor is unbalanced and the rotor turns.

As the liquid reaches the outlet port where a decreasing volume is present, the liquid exits.

NOTE: Before a motor of this type will operate, the vanes must be previously extended and a positive seal must exist between vanes and housing. Centrifugal force cannot be depended on as in a pump. The manner in which these vanes are extended will be dealt with in a following section.

Torque

Torque is a rotary or turning effort. Torque indicates a force is present at a distance from a motor shaft. One unit for measuring torque is a lb. in. (Newton-meter).

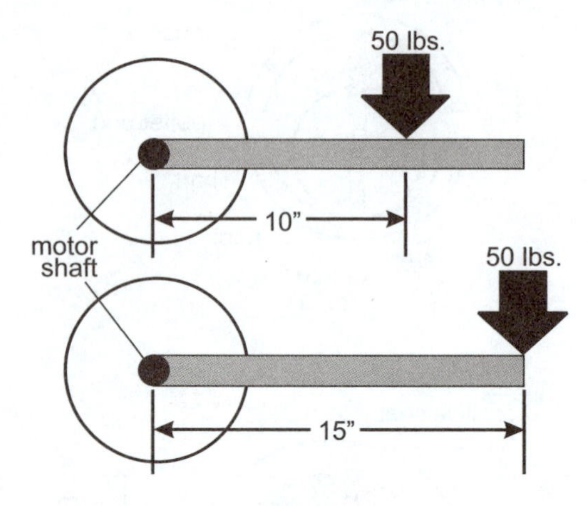

Figure 6-8

Description of Torque

Torque tells us where a force is in relation to the motor shaft. The expression which describes torque is:

torque =	force x	distance from shaft
(lb. in.)	(lb.)	(in.)
(N-m)	(N)	(m)

In the illustration, a force of 50 lbs. (222 N) is positioned on a bar which is attached to a motor shaft. The distance between the shaft and the force is 10 inches (0.254 m). This results in a torque or turning effort at the shaft of 500 lbs. in. (56.5 N-m).

If the 50 lbs. (222 N) were located 15 inches (0.38 m) along the bar, the turning effort generated at the shaft would be equal to a twisting effort of 750 lbs. in. (3330 N).

From these examples, we can see that the farther the force is from the shaft, the larger the torque at the shaft. It will also be noted that torque does not involve any movement.

A resisting object attached to a motor shaft generates a torque as described above. This, of course, is a resistance for the motor which must be overcome by hydraulic pressure acting on a motor's rotating group.

The expression used to describe the torque developed by a hydraulic motor is:

$$\underset{\text{(lb. in.)}}{\text{torque}} = \frac{\text{psi x motor displacement (in}^3\text{)}}{2\pi}$$

$$\underset{\text{(N-m)}}{\text{torque}} = \frac{\text{bar x motor displacement (cm}^3\text{)}}{20\pi}$$

Motor Shaft Speed

The speed at which the shaft of a hydraulic motor turns is determined by how quickly it is filled with fluid.

The expression which describes motor shaft speed is:

$$\underset{\text{(rpm)}}{\text{motor shaft speed}} = \frac{\text{gpm x 231}}{\text{motor displacement (in}^3\text{/rev.)}}$$

$$\underset{\text{(rpm)}}{\text{motor shaft speed}} = \frac{\text{l/min x 1000}}{\text{motor displacement (cm}^3\text{/rev.)}}$$

Power

We have seen earlier that the speed or rate of doing work is power. Power = ft.lbs./time or Joule/time = Watts.

The machine which performs its required work in three seconds produces more power than a machine which performs the same work in three minutes.

Mechanical Horsepower

We have also seen that horsepower, or watts, may be the unit of measure for power.

If a cylinder or hydraulic motor applies a mechanical force of 550 lbs. (2442 N) against a resisting object through a distance of one foot (0.30 m) in a time lapse of one second, one horsepower (746 watts) is used.

If the same work (550 ft. lbs./746 Joules) were completed in a half second, the work would be done at twice the rate, or two horsepower (1490 watts).

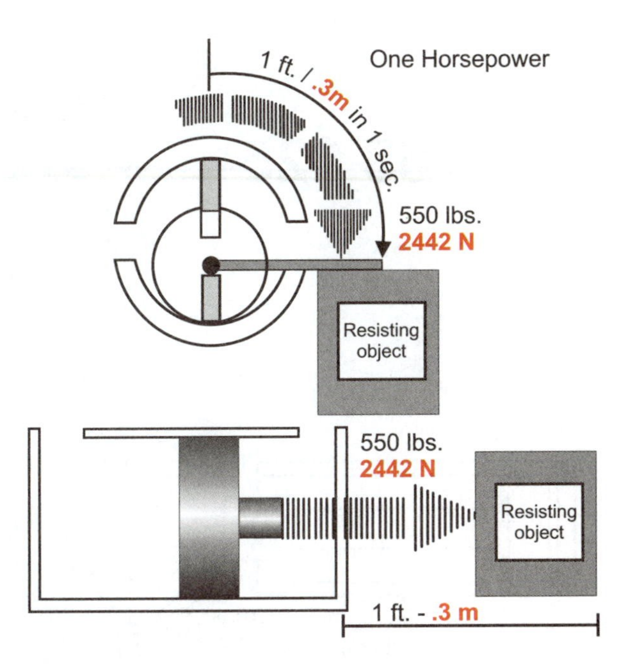

Figure 6-9

Hydraulic Horsepower

The mechanical horsepower transmitted by the cylinder or motor to the resisting object will also be the hydraulic horsepower requirement at the cylinder or motor.

A hydraulic system doing work at the rate of 550 ft. lbs. (746 Joules) per second is one hydraulic horsepower (746 Watts). However, instead of using the terms of "foot" (meter) and "lb" (Newton) found in the mechanical horsepower expression, these terms are converted to hydraulic terms of "psi" (bar) and "gpm" (l/min). Also, in the hydraulic horsepower formula, a conversion factor is used which shows the relationship between gpm, psi, feet and lbs. (l/min, bar, meter and Newtons).

gpm x psi

.000583

System and Cylinder Horsepower Calculation

To calculate horsepower developed by a hydraulic cylinder or the total hydraulic system, the following expression is used:

horsepower = gpm x psi x .00583
Watts = horsepower x 746
Watts = 5/3 l/min x bar

Motor Horsepower Calculation

To calculate horsepower developed by a hydraulic motor, the following expression is used:

$$\text{horsepower} = \frac{\text{rpm x torque (lb. in.)}}{63{,}025}$$

$$\text{kiloWatt} = \frac{\text{rpm x torque (N-m)}}{9543}$$

63,025

Rotary Actuators

Up to this point we have discussed rotary output of motors and straight line motion of cylinders. Now we are going to discuss actuators that have a limited degree of rotation. These devices, known as rotary actuators, are compact, simple, and efficient. They produce high torque and require small space and simple mounting.

Common applications for rotary actuators would be such things as machine tool indexing, bending operations, lifting or rotating heavy objects, rollover function, positioning, machining fixtures, ship rudder positioning, actuating valves, etc.

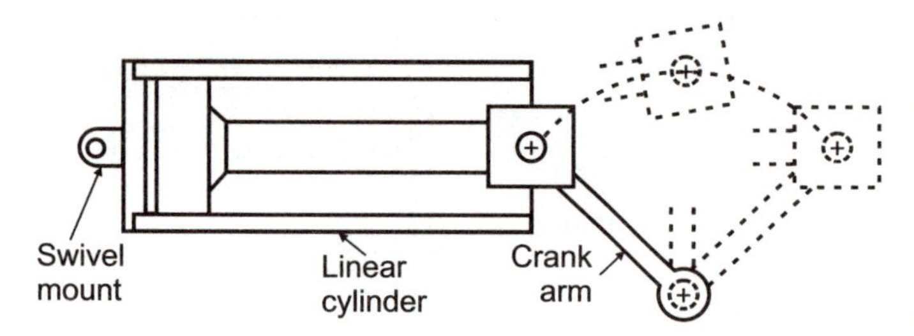

Figure 6-10

There are many types of rotary actuators. Perhaps the simplest of the rotary actuators is the linear cylinder. It is just a simple cylinder swivel-mounted at one end. The rod is attached to a crank arm that drives the shaft. The cylinder is controlled by a four-way directional valve that controls the direction of the cylinder. There are limits at the end of the drive stroke.

As with all mechanical devices, there are basic characteristics. One characteristic is that this cylinder based rotary actuator can be built up from commercially available components. This allows for design flexibility. It is inexpensive and repair parts are readily available.

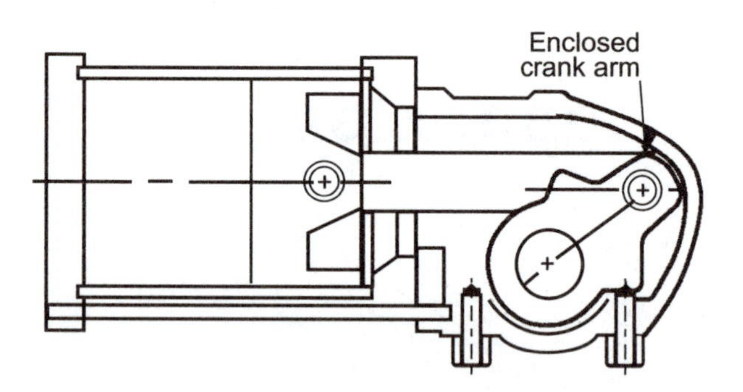

Figure 6-11

There are other characteristics that are not as favorable: the rod is unprotected and left exposed to the surrounding environment; the mechanism, the crank arm, is typically unshielded, and therefore is an unsafe condition. Also, tremendous side-loadings or sideward thrust is generally applied to the shaft causing premature failure, excess wear and galling.

With this particular type of rotary actuator the cylinder must be free to move. Therefore, flexible lines must be used as fluid conductors. With this type of rotary actuator the torque is not constant throughout the entire stroke of the cylinder.

The shrouded cylinder type of rotary actuator is very similar to the linear cylinder actuator. The shrouded cylinder has a shrouded enclosure around the cylinder rod and crank arm. Also, in this particular type there is generally additional bearing support to prevent the extreme side loading that may occur.

These particular types of rotary actuators can be equipped with solenoid operated valves, positioners or limit switches. Ordinarily their stroke is adjustable from a range of about 85 to 100 degrees.

Still another type of rotary actuator is called a "spring return" cylinder. This particular type uses a spring loaded cylinder to return the shaft to its normal position. The spring return actuators are available with torque outputs of up to 5000 pound inches (565 N-m).

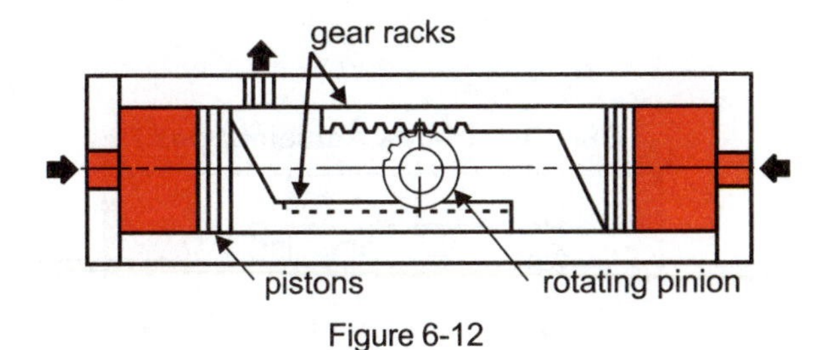

Figure 6-12

A very common type of rotary actuator is called the rack-and-pinion. This particular type of rotary actuator gives uniform torque in both directions and throughout the range of rotation. In this device fluid pressure will drive a piston which in turn is connected to a gear rack which rotates a pinion shaft. Standard rack-and-pinion units are available in rotations of 90°, 180°, 360° or more. Variations of the rack-and-pinion actuators can produce units with torque outputs of as much as fifty-two million pound-inches (5,876,000 N-m).

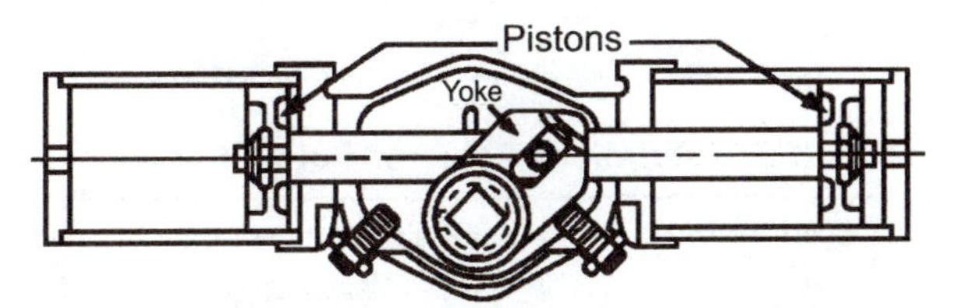

Figure 6-13

There is still another rotary actuator called a skotch yoke actuator. This type is very similar to two cylinders with a common rod. These are typically double acting or can be single acting; they could even be provided with a spring return. These actuators are rather large, with torque ranges as high as forty-five million pound-inches (5,085,000 N-m). These are limited to driving through very short arcs, generally 90° or less.

Vane type rotary actuators are also available. They can be single or multiple vanes. They can have rotations of 280° with single vane or 200° with dual vanes. The dual vanes would give twice the torque of the single vane. These have torque ratings of as high as five hundred thousand pound-inches (56,500 N-m).

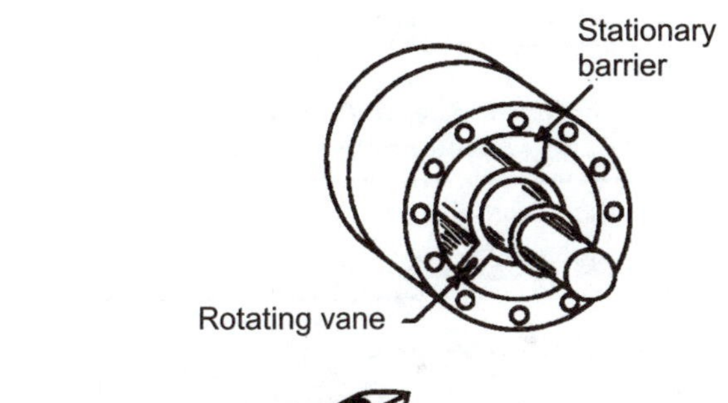

There is still another type rotary actuator that produces torque through the aid of a helical spline. Variation in length and pitch of the helix permit the rotation to be varied over a wide range. It consists of a helical splined shaft that rotates when a piston sleeve, which is restrained from rotating by sliders, moves within a cylinder. This particular type rotary actuator usually has 90°, 180°, 270° or 360° rotation and output torques of up to one million pound-inches (13,000 N-m).

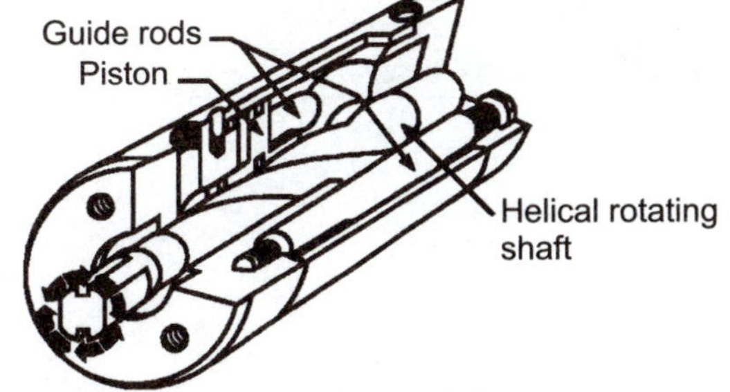

Figure 6-14

The sprocket type rotary actuator uses a drive piston, chain, and sprocket to rotate the shaft. In this type of actuator there is typically a large piston that acts as a driver or pulls the chain and a smaller piston that provides a seal to prevent fluid leakage past the return side of an endless chain. This type

GPM Determines Actuator Speed

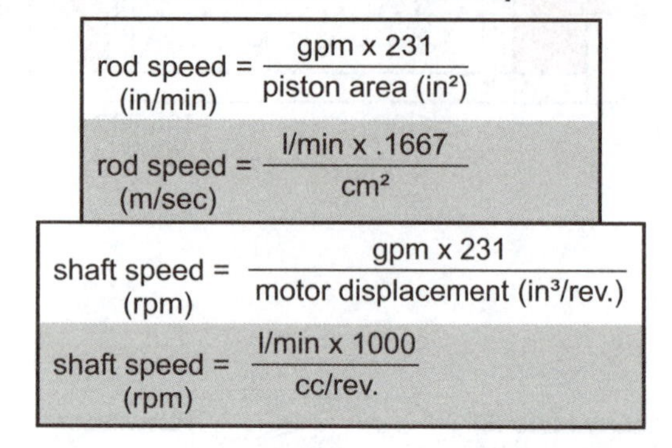

PSI Determines Actuator Output Force

cylinder force (lb.) = psi x piston area (in^2)

cylinder force (N) = 10 x bar x piston area (cm^2)

$$\text{motor torque (lb. in.)} = \frac{\text{psi x motor displacement (in}^3\text{)}}{2\pi}$$

$$\text{motor torque (N-m)} = \frac{\text{bar x motor displacement (cm}^3\text{)}}{20\pi}$$

Actuator Speed Multiplied by Actuator Output Force = Horsepower

cylinder horsepower = gpm x psi x .000583

Watts = 5/3 (l/min) (bar)

$$\text{motor horsepower} = \frac{\text{rpm x torque (lb. in.)}}{63{,}025}$$

$$\text{kWatt} = \frac{\text{torque (N-m) x rpm}}{9543}$$

of rotary actuator will have torque outputs to about 23 thousand pound-inches (2599 N-m). Shaft rotations with this type can be as high as five complete turns or 1800°.

Choosing the best rotary actuator for a particular application involves matching torques, speed and operating modes for that task. The actual choices of a rotary actuator will be covered in another text. In addition, deciding whether it is single or double acting, closed positioning, cushioned, and frequency or cycle rate will also be examined.

Generalizations About Hydraulic Actuators

Actuator speed is a function of flow gpm/l/min. The speed at which a cylinder's piston rod travels is determined by how quickly the flow gpm/l/min from the pump fills the volume behind the cylinder piston.

The speed at which the shaft of a hydraulic motor revolves is dependent on the rate at which the gpm/l/min from the pump fills the hydraulic motor.

Actuator output force is a function of pressure psi/bar. The output force at the shaft of a hydraulic motor is determined by the amount of hydraulic pressure acting on the exposed area of the motor's rotating group.

The horsepower developed by an actuator is a function of actuator speed multiplied by actuator output force.

For a cylinder, the output force is expressed by psi. The rod speed is denoted by gpm. The constant .000583 gives the relationship between psi, gpm, and horsepower.

For a hydraulic motor, the output force is expressed by torque. The motor's operating speed is denoted by rpm. The constant 63,025 give the relationship between rpm, torque, and horsepower.

Exercise
Hydraulic Actuators
25 points

Instructions: Solve the following problems:

1. A cylinder with a 3" (7.62 cm) bore and a stroke of 16" (40.64 cm) receives 18 gpm (68.22 l/min). What is the piston rod velocity?

2. Calculate how much output torque a motor will produce if it has a displacement of 13 in^3 (213 cm^3) and is subjected to 240 psi (16.55 bar)?

3. A cylinder with an 8" (20.32 cm) bore and a 36" (.9 m) stroke must extend in one minute. How much flow is required?

Exercise (cont'd.)
Hydraulic Actuators
25 points

4. A hydraulic motor develops 200 lb.ins. (2.25 N-m) of torque at 800 RPM. What is the horsepower?

2.54 HP

5. A cylinder with a 10" (25.4 cm) bore and a 24" (.609 m) stroke must move a 78,540 lb. (348717.6 N) load through its stroke in three seconds. How much hydraulic horsepower must be delivered to the cylinder?

10" Bore - 78.54 sq in

24" stroke 1000 psi

78,540 lbs Force

480 Rpm

24 · 60 ÷ 3 = 480

CHAPTER 7

Control of Hydraulic Energy

Working energy transmitted hydraulically must be directed and under complete control at all times. If not under control, no useful work will be done or a machine might be destroyed. One of the advantages of hydraulics is that energy can be controlled relatively easily by using valves.

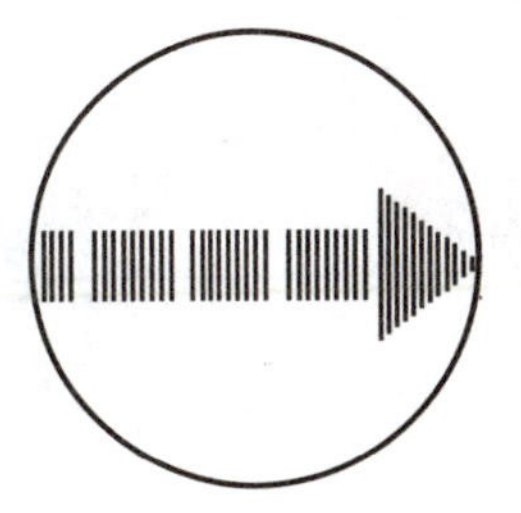

Figure 7-1

Valves

A valve is a mechanical device consisting of a body and an internal moving part which connects and disconnects passages within the body. The passages in hydraulic valves carry liquid. The action of the moving part controls maximum system pressure, direction of flow, and rate of flow.

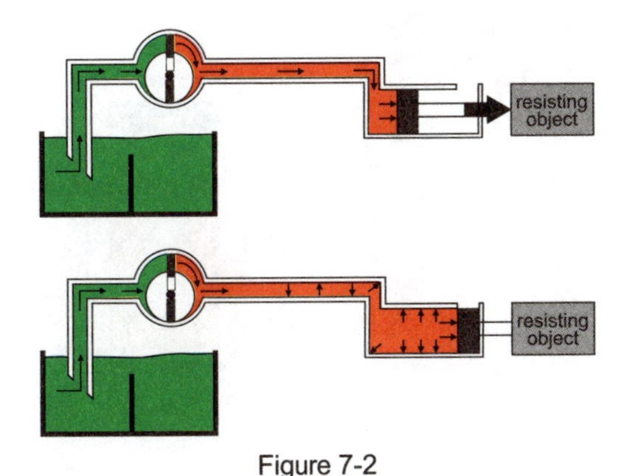

Figure 7-2

Control of Pressure

Hydraulic energy can be applied to a cylinder actuator, resulting in the performance of work. Once the cylinder is fully extended and the work is completed, a positive displacement pump will continue to absorb more energy from its prime mover and apply a higher pressure to the liquid. (Remember, the smallest resistance in the system signals the pump what pressure is to be applied.) With the cylinder fully extended, the smallest resistance becomes the physical strength of the system.

The pump will try to apply a pressure to overcome this resistance which would be damaging. One use of a pressure valve is to limit system pressure to a safe level.

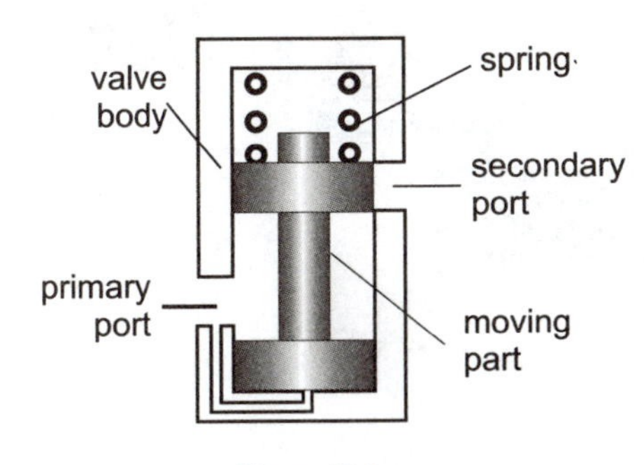

Figure 7-3

Pressure Control Valve

A pressure control valve has an internal moving part which is operated by pressure. When the pressure in a system reaches a certain level, the internal moving part connects or disconnects passages in a valve body, allowing the liquid to follow another path.

What a Pressure Control Valve Consists Of

A pressure control valve consists of a valve body with a primary and secondary passage and an internal moving part (spools). The external openings of the passages are known as primary and secondary ports, respectively.

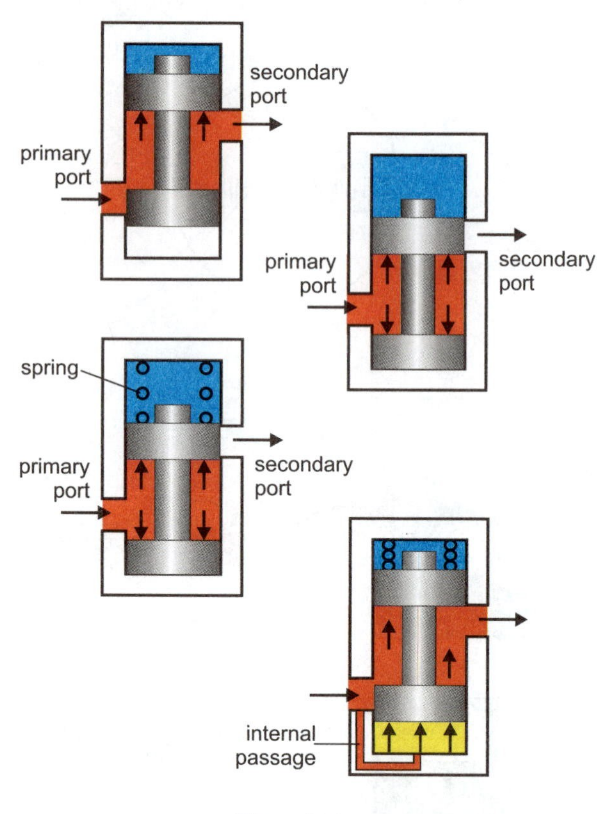

Figure 7-4

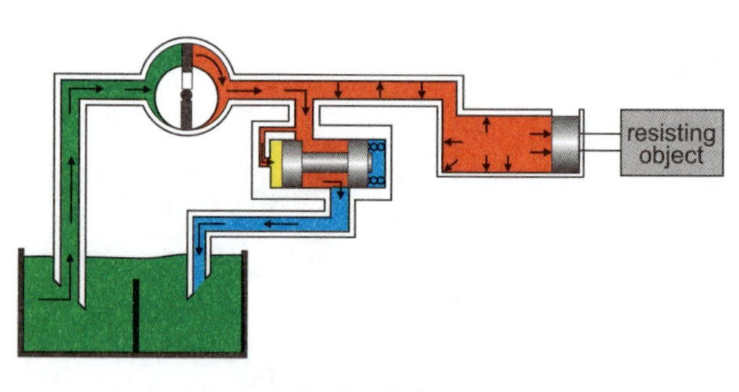

Figure 7-5

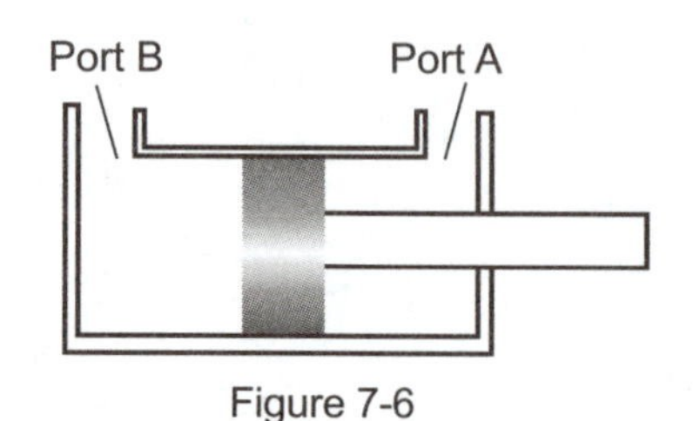

Figure 7-6

How a Pressure Control Valve Works

Many times, the internal moving part of a pressure control valve is a spool. In one extreme position, the spool connects the passages, allowing the fluid to flow through the valve. In the other extreme, the passages are disconnected and the flow path through the valve is blocked.

In pressure control valves, the spool is held biased in one extreme position by a spring. If the passages are disconnected and the flow path through the body is blocked in its normal condition, the valve is designated a normally non-passing pressure control.

Pressure is sensed at the bottom of the spool by an internal passage connected to the primary passage. When system pressure overcomes the force of the spring, the spool moves and the passages are connected. Fluid is free to flow through the valve.

(The fluid pressure used to operate the spool is known as pilot pressure. Pilot pressure is a common way of operating many types of hydraulic valves.)

If the primary port of this type pressure valve were connected to system pressure, and the secondary port were connected to the tank, the flow from the pump could be directed back to tank when pressure applied by the pump becomes excessive. A normally non-passing pressure valve used in this manner is called a relief valve.

Control of Actuator Direction

Once a cylinder is extended, it has to be retracted so that work can be done again.

To perform this task, a cylinder which is operated hydraulically in both directions is generally used, as well as something to change the direction of liquid flow.

Double-Acting Cylinder

A double-acting cylinder has a port at each end of the cylinder body by which fluid can enter and exit. This allows the piston rod to move in two directions (double-acting). To distinguish the ports on a double-acting cylinder, we will label one “A”, the other “B.”

Directional Control Valve

The moving part in a directional control valve connects and disconnects internal passages within the valve body, which action results in a control of fluid direction.

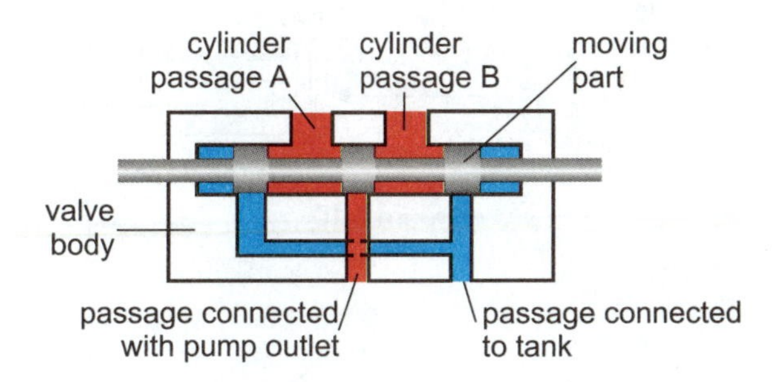

Figure 7-7

What a Directional Control Valve Consists Of

A typical directional control valve consists of a valve body with four internal passages and a sliding spool moving part which connects and disconnects the passages.

How a Directional Control Valve Works

With the spool in one extreme position, the pump passage is connected to cylinder passage B and tank passage is connected to cylinder passage A.

With the spool in the other extreme, the pump passage is connected to cylinder passage A and tank passage is connected to cylinder passage B.

With a directional control valve in a circuit, the cylinder's piston rod can be extended and work performed.

By shifting the spool to the other extreme, flow is directed to the other side of the cylinder. The piston rod retracts.

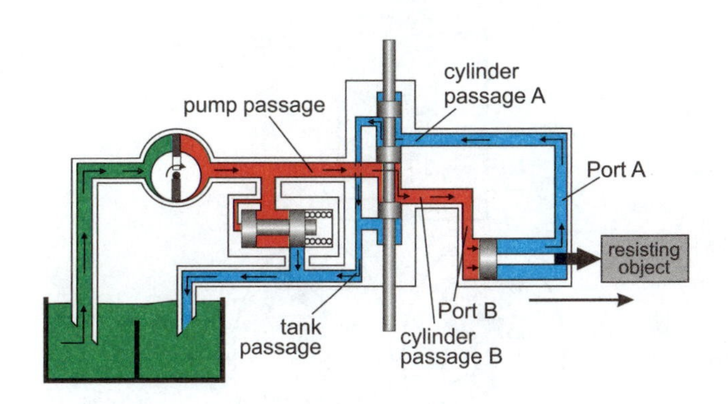

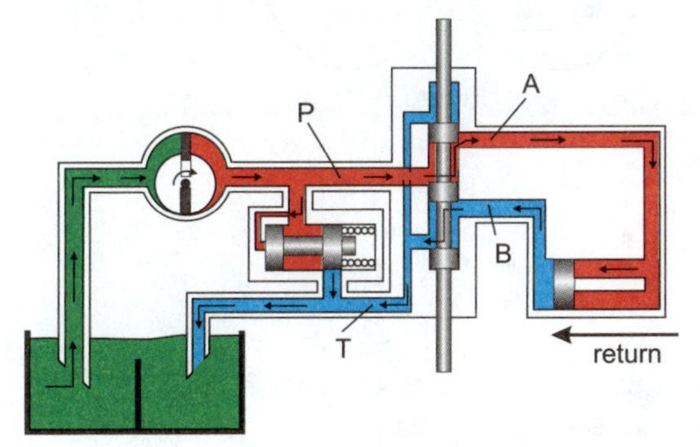

Figure 7-8

Control of Actuator Speed

In many applications, it is desirable and even necessary to control the speed at which an actuator does work.

It has been shown previously that the speed at which an actuator (cylinder, motor) does work is the direct result of how quickly it is filled. In other words, actuator speed is the result of the gpm flowing to the actuator.

Since the pump in a hydraulic system can be at constant displacement, it would make sense to select a pump with the desired flow rate. This is usually the case when only one actuator is used in a system.

Many times in a hydraulic system, there is more than one actuator. If the system is designed for the cylinders to act individually, then the pump's flow rate is selected for the required speed of the largest cylinder. This means the smaller actuators will move more quickly, which may be undesirable. To reduce the flow rate to these or any actuator, a flow control valve is used.

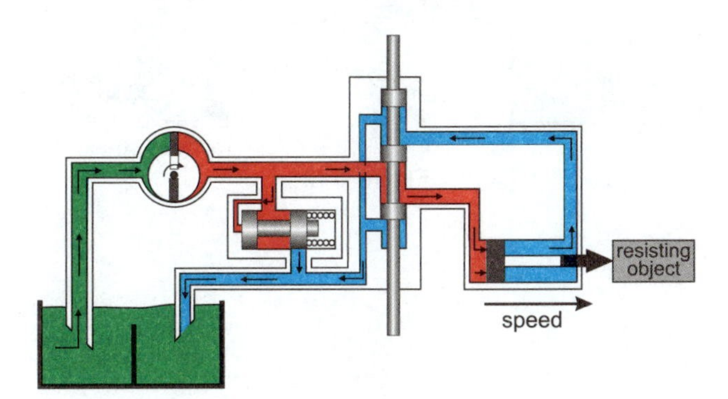

Figure 7-9

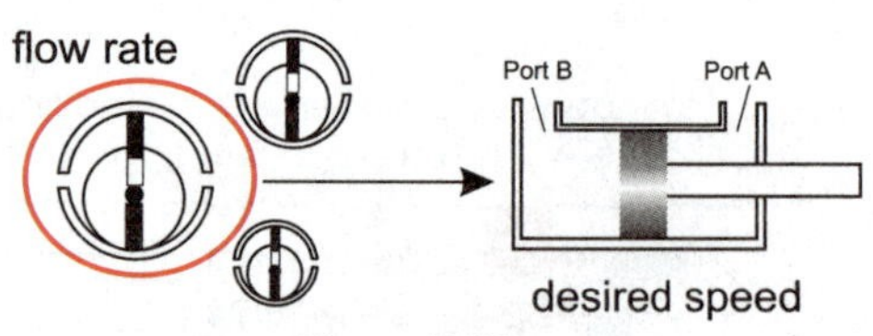

Figure 7-10

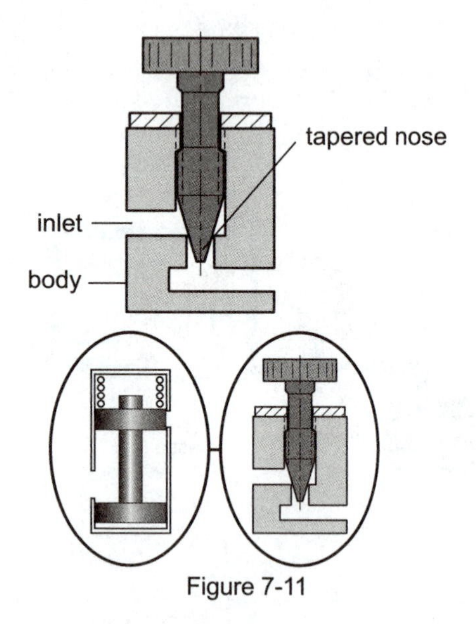

Figure 7-11

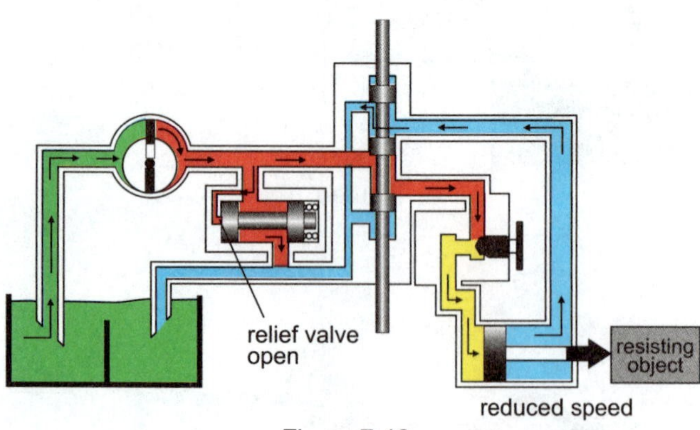

Figure 7-12

Flow Control Valve

A flow control valve, when properly used, always reduces the flow rate from a pump to an actuator.

What a Flow Control Valve Consists Of

A typical flow control valve consists of a valve body and movable part. The movable part in our example is a "tapered-nose" rod which is threaded into the valve body. The movable part in our flow control valve is better described as "adjustable" since no movement takes place while the valve is operating.

How a Flow Control Valve Works

In a hydraulic system, there is many times a direct working relationship between a flow control valve and a pressure valve functioning as a relief valve.

The flow control valve is a resistance which results in a higher pressure being applied by the pump. This pressure partially opens the relief valve. The result is some flow goes over the relief valve and less flow through the flow control and to the actuator.

A Simple Hydraulic System

The components which have been described make up a simple hydraulic system. The system can perform useful work because hydraulic working energy in the system can be controlled.

Hydraulic systems are found in many diverse fields, from aerospace, aircraft, and military operations, to industrial, mobile, and steel mill applications. All hydraulic systems operate on the same principles which have been discussed up to now. However, the difference between each "type" of hydraulics is in the components.

We shall, in the remaining text material, concentrate on some of the various types of components which are available in industrial hydraulics. We shall also design some elementary circuits to illustrate how these components may be used.

Hydraulic Symbols

The hydraulic components and elementary systems which have been shown to this point, have been illustrated in a pictorial manner. System diagrams have been cutaway to illustrate internal component operation. This technique is beneficial from an instructional point of view, but it is impractical from a workaday standpoint.

As in other technologies, hydraulic symbols are used to describe components and systems. The symbols for the components which have been discussed and the simple system which has been developed can be illustrated using either ANSI Y32.10 graphic symbols or ISO 1219 graphic symbols for fluid power.

NOTE: In addition to the components which have been discussed, the system also consists of an electric motor and a hydraulic filter. Hydraulic systems are generally driven by a prime mover like an electric motor. And, in order to achieve a degree of reliability, hydraulic systems should be protected from dirt with a hydraulic filter.

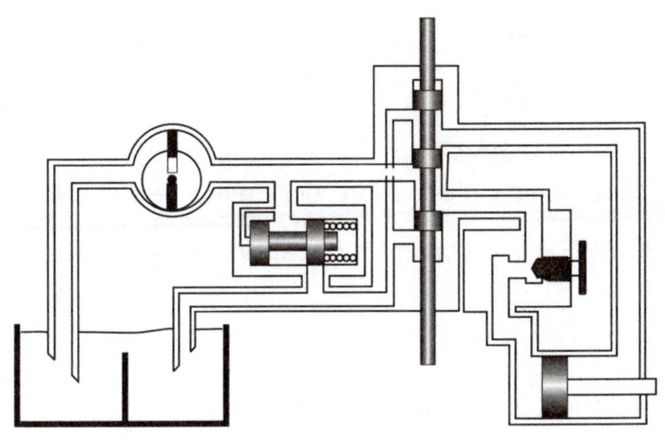

Figure 7-13

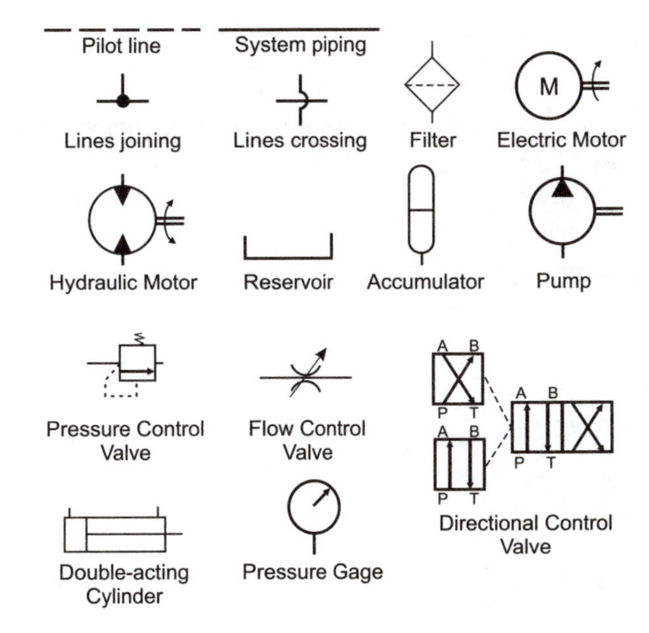

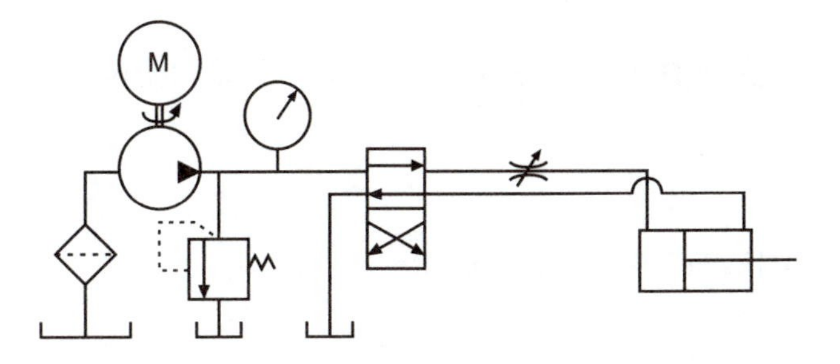

Figure 7-14

Exercise
Control of Hydraulic Energy
50 points

Instructions: In this assignment the answers are already given. You write the questions.

ANSWER 1. It operates by causing a resistance.

ANSWER 2. Under complete control.

ANSWER 3. Direction of flow, rate of flow, and maximum pressure.

ANSWER 4. It has a working relationship many times.

ANSWER 5. A body and an internal moving part.

ANSWER 6. Pump applies an additional pressure.

ANSWER 7. Has a normally closed condition.

ANSWER 8. A pump is selected with the desired flow rate.

ANSWER 9. Pilot pressure.

ANSWER 10. The physical strength of the system.

CHAPTER 8

Check Valves, Accumulators and Cylinders

What a Check Valve Consists Of

A check valve basically consists of a body with inlet and outlet ports and a movable member which is biased by spring force. The movable member can be a flapper, or plunger, but most often in hydraulic systems it is a ball or poppet.

How a Check Valve Works

Fluid flow passes through a check valve in one direction only.

When system pressure at the check valve inlet is high enough to overcome the spring force biasing the poppet, the poppet is pushed off its seat. Flow passes through the valve. This is known as the check valve's free flow direction. When fluid flow enters through the outlet, the poppet is pushed on its seat. Flow through the valve is blocked.

Check Valves in a Circuit

A check valve is a combination directional valve and pressure valve. It allows flow in only one direction and in that sense is a one-way directional valve.

A check valve is often used in hydraulic systems as a bypass valve. It allows flow to get around components, like flow control valves which would restrict flow in a reverse direction.

A check valve is also used to isolate sections of a system or a system component, such as an accumulator. The check valve keeps an accumulator from dumping its flow over a relief valve or through the pump.

NOTE: Whenever a check valve is used in an accumulator circuit, surges in pressure and flow must be accounted for if the valve is to be selected and used properly. Selecting a valve type and size based only upon pump flow and relief valve pressure could result in severe component damage and malfunction. Component selection for such circuits is beyond the scope of this text. The circuit shown is simplified and used for illustration only.

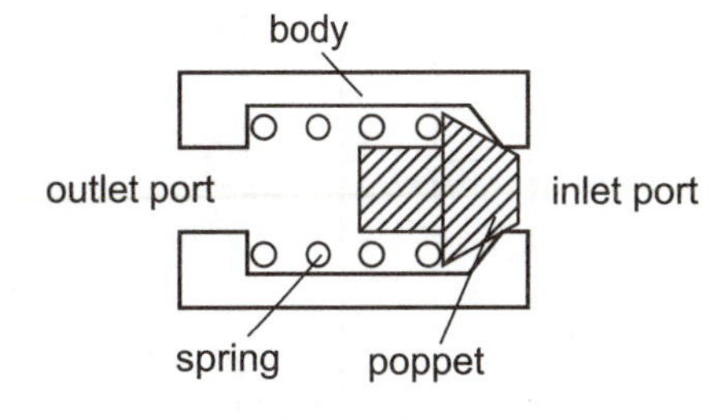

Figure 8-1

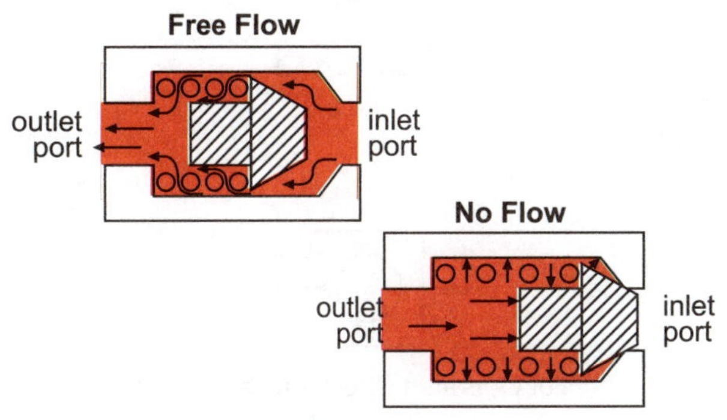

Figure 8-2

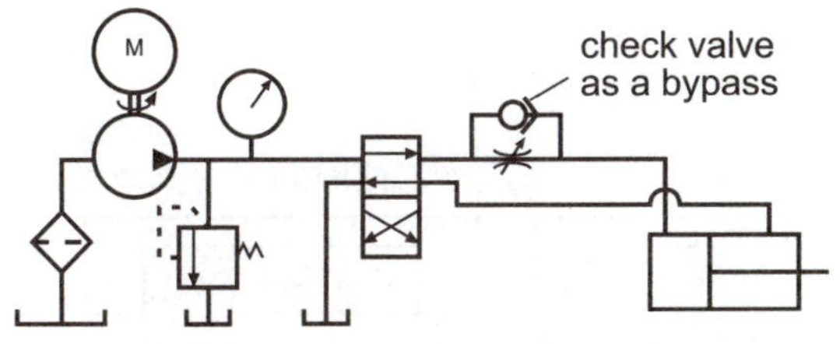

Figure 8-3

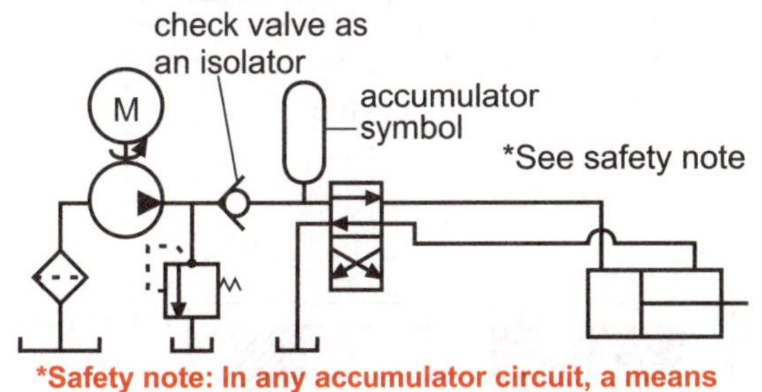

***Safety note: In any accumulator circuit, a means could be available of automatically unloading the accumulator when the machine is shut down. See page 8-6.**

Figure 8-4

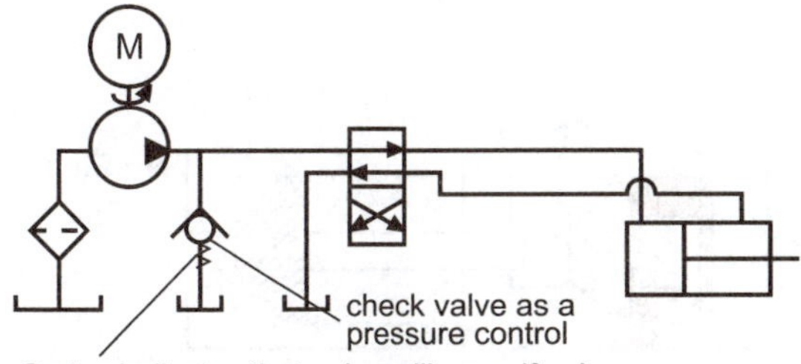

Spring indicates that valve will open if valve inlet pressure is greater thanoutlet pressure plug spring pressure.

Figure 8-5

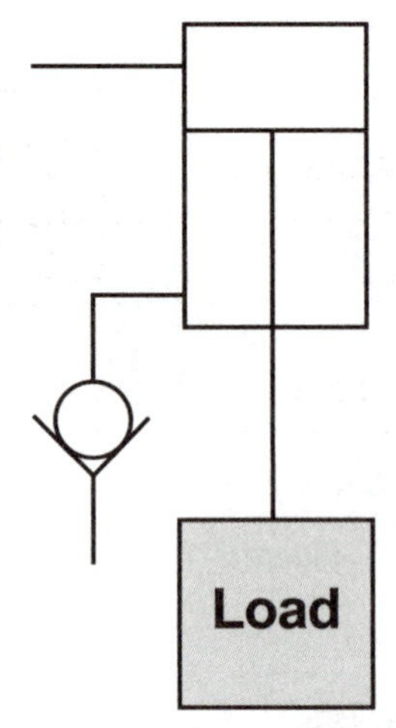

Figure 8-6

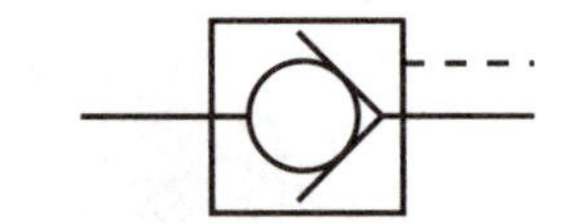
Pilot Operated Check Valve Symbol
Figure 8-7

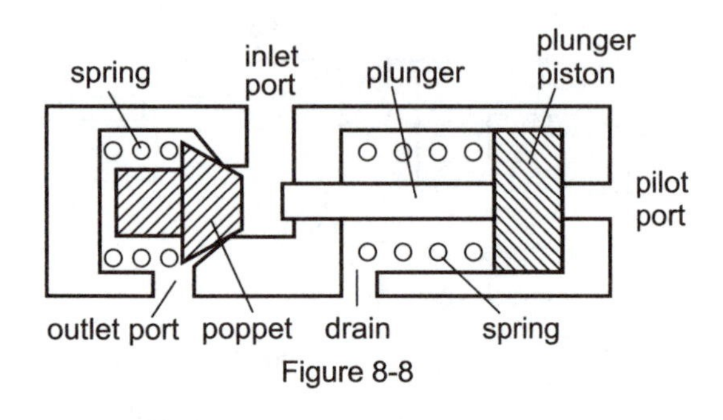

Figure 8-8

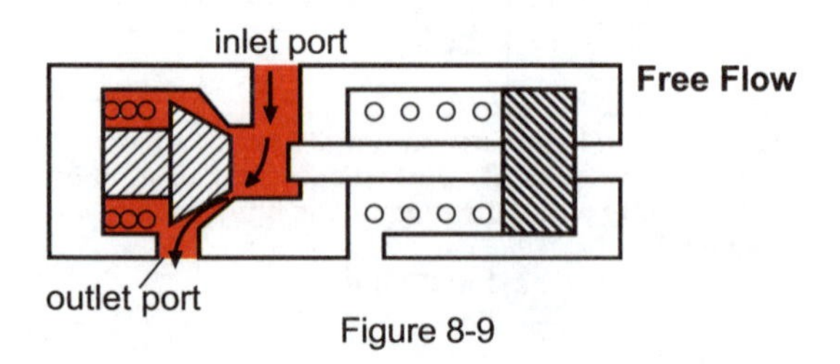

Figure 8-9

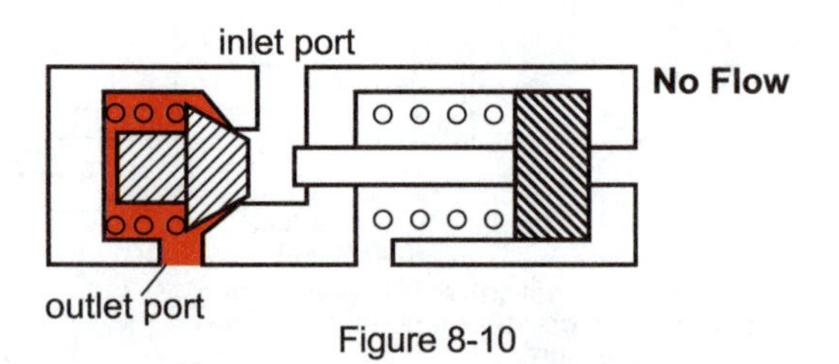

Figure 8-10

The movable part in a check valve is usually biased by very light spring force. When a heavier spring is used, a check valve can be used as a pressure control valve. (This is not commonly done).

Suspending a Load

Hydraulic components that have a spool construction generally have a small bypass flow. This does not necessarily indicate that hydraulic components are poor quality since much of this bypass flow is actually designed into the component for lubrication reasons. Leakage does become a problem, however, when a load attached to a cylinder is required to be suspended indefinitely without drifting down. In this application, a sealing type check valve is used.

A check valve is generally, a low leakage device. As a matter of fact, check valves can be designed to be practically zero leakage devices.

A check valve may keep a load suspended almost indefinitely. But remember, a check valve is a one-way valve. When the time comes for the load to descend, the valve's movable member must be forced off its seat. This requires a special valve known as a pilot operated check valve.

Pilot Operated Check Valve

A pilot operated check valve allows free flow in one direction. In the opposite direction flow may pass when pilot pressure unseats the valve's movable member.

What a Pilot Operated Check Valve Consists Of

A pilot operated check valve consists of a valve body with inlet and outlet ports and a poppet biased by a spring, just as an ordinary check valve. Directly opposite the check valve poppet is a plunger and plunger piston which is biased by a light spring. Pilot pressure is sensed at the plunger piston through the pilot port. The plunger spring chamber has a drain.

How a Pilot Operated Check Valve Works

A pilot operated check valve allows free flow from its inlet port to its outlet port just as an ordinary check valve.

Fluid flow attempting to pass through the valve from outlet to inlet port will force the poppet on its seat. Flow through the valve is blocked. When enough pilot pressure is sensed at the plunger piston, the plunger is moved and unseats the check valve poppet. Flow can pass through the valve from outlet to inlet as long as sufficient pilot pressure is acting on the plunger piston.

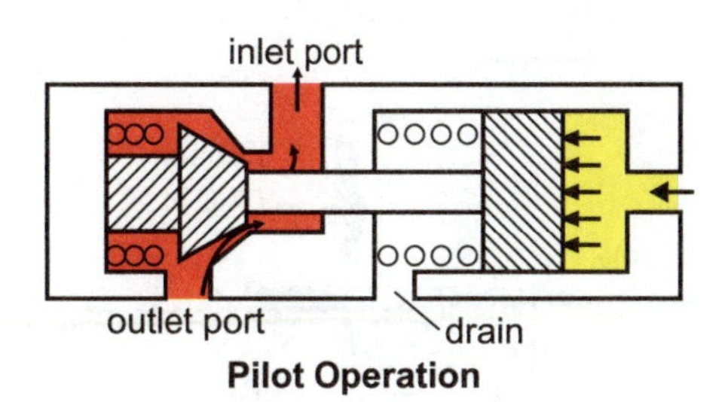

Figure 8-11

Pilot Operated Check Valves in a Circuit

With a pilot operated check valve blocking flow out of cylinder line B, the load will stay suspended as long as the cylinder seals remain effective. When it is time to lower the load, system pressure is applied to the cylinder piston through line A.

Pilot pressure to operate the check valve is taken from this cylinder line. The check valve will remain open as long as enough pressure is available in line A.

To raise the load, fluid can easily pass through the valve since this is the valve's free flow direction.

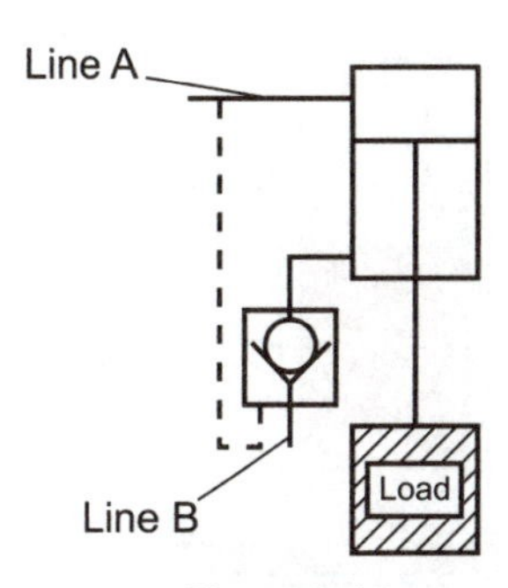

Figure 8-12

In some applications, it is required that a load attached to a cylinder's piston rod be locked in place. To fill this requirement, a pilot operated check valve can be positioned in each cylinder line. The pilot operated check valve will block flow out of the cylinder. The load will be held as long as the cylinder seals remain effective and there are no leaks (line, cylinder, check valve, etc.).

NOTE: For absolute hold — special locking cylinders are required. They contain mechanical locks. Mechanically locking the load is the safest means of holding a load.

Hydraulic Accumulators

An accumulator stores hydraulic pressure. This hydraulic pressure is potential energy since it can change to working energy.

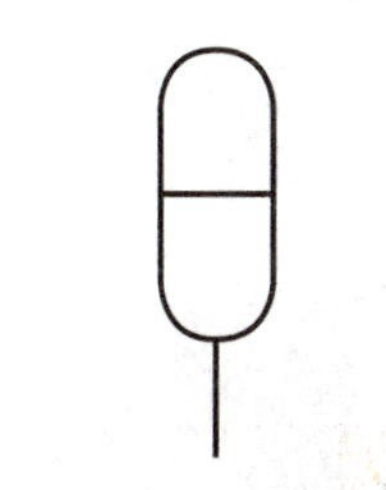

Accumulator Symbol

Figure 8-13

Accumulator Types

Hydraulic accumulators can be divided into weight-loaded, spring-loaded or hydro- pneumatic. Each classification identifies by what means an accumulator maintains a force on a liquid while it is stored.

Weight-Loaded Accumulator

A weight-loaded accumulator maintains a force on the liquid it stores by means of heavy weights acting on a piston or ram. The weights can be made of any heavy material such as iron, concrete or even water.

Weight-loaded accumulators are generally quite large in some cases holding hundreds of gallons. They can service several hydraulic systems at one time and are most often used in mill and central hydraulic systems.

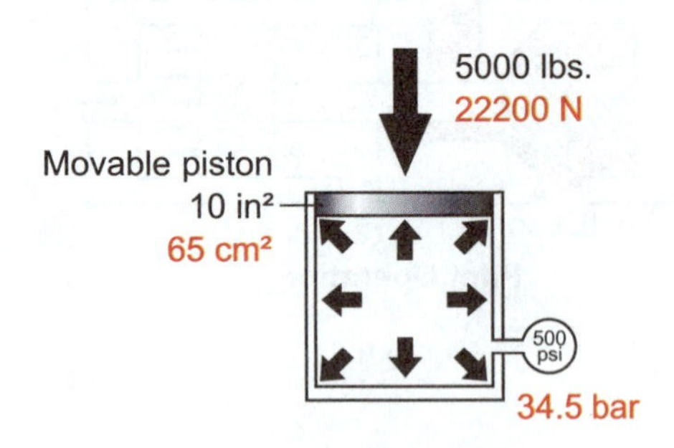

Figure 8-14

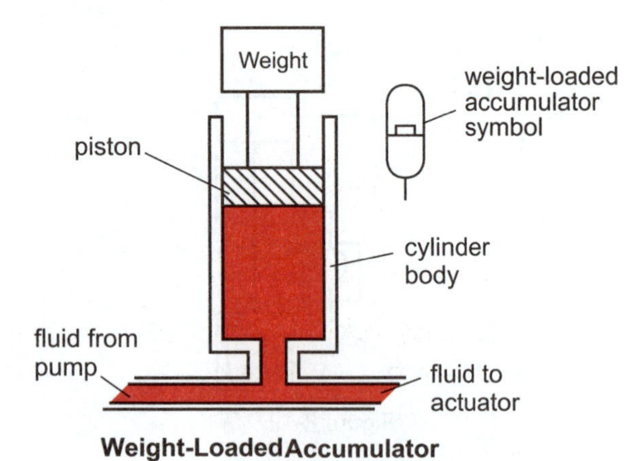

Weight-LoadedAccumulator

Figure 8-15

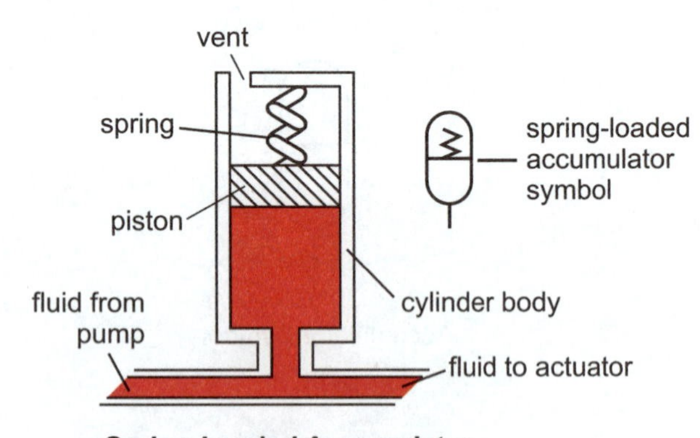

Spring-Loaded Accumulator

Figure 8-16

A desirable characteristic of a weight-loaded accumulator is that it stores fluid under a relatively constant pressure whether it is full or nearly empty; this will not be the case in other accumulator types. Because the weight applying the force to the liquid does not change, the same force is applied regardless of how much liquid is present in the accumulator.

An undesirable characteristic of a weight-loaded accumulator is shock generation. When a weight-loaded accumulator, discharging quickly, is suddenly stopped, the inertia of the weight could cause excessive pressure surges in a system. This can result in leaking fluid conductors and fittings, and early component failure due to metal fatigue.

Spring-Loaded Accumulator

A spring-loaded accumulator applies a force to its stored liquid by means of a spring acting on a piston.

Spring-loaded accumulators are generally much smaller than weight-loaded accumulators with sizes holding up to several gallons. Spring-loaded accumulators usually serve individual hydraulic systems and generally operate at low pressure.

As liquid is pumped into a spring-loaded accumulator, the stored fluid pressure is determined by the compression rate of the spring. An accumulator of this type will have more stored pressure with the piston moved up and the spring compressed 10 inches than if it were only compressed 4 inches.

To avoid accumulation of leakage fluid, the spring chamber of a spring-loaded accumulator is vented. Leakage fluid will eventually discharge from the vent hole.

Spring-loaded accumulators are not externally drained back to tank because they can cause oil foaming. With an external drain terminating either above or below fluid level, leakage accumulated above the piston will tend to foam during accumulator operation. As the accumulator discharges rapidly, fluid above the piston will be unable to keep up with piston movement. A less-than-atmospheric pressure will be generated in the spring chamber resulting in dissolved air coming out of the liquid. When the accumulator is recharged, the piston moves up pushing the aerated oil to tank. Since air bubbles in a reservoir are undesirable, spring-loaded accumulators are not generally externally drained.

With spring chamber vented, spring-loaded accumulators demand immediate attention once their piston seal wears. If maintenance is not performed on a spring-loaded accumulator with a poor seal, a housekeeping problem could arise.

Hydro-Pneumatic Accumulator

A hydro-pneumatic accumulator is the most commonly used accumulator in industrial hydraulic systems. This type accumulator applies a force to a liquid by using compressed gas.

NOTE: In all cases of hydro-pneumatic accumulators applied to industrial systems, dry nitrogen is used. COMPRESSED AIR SHOULD NEVER BE USED because of the danger of exploding an air-oil vapor.

Hydro-pneumatic accumulators are divided into piston, diaphragm and bladder types. The name of each type indicates the device separating gas from liquid.

Piston Type Accumulator

A piston type accumulator consists of a cylinder body and movable piston with resilient seals. Gas occupies the volume above the piston and is compressed as the cylinder body is charged with fluid. As fluid flows from the accumulator, gas pressure drops. When all liquid has been discharged, the piston has reached the end of its stroke and it covers the outlet keeping the gas within the accumulator.

Diaphragm Type Accumulator

A diaphragm type accumulator consists of two metal hemispheres which are bolted together, but whose interior volume is separated by a synthetic rubber, gas occupies the space. As fluid enters the other chamber, gas is compressed. Once all liquid has been discharged, the diaphragm covers the outlet retaining the gas within the accumulator. The diaphragm will not be pushed through the thickness of the diaphragm.

Bladder Type Accumulator

A bladder type accumulator consists of a synthetic rubber bladder inside a metal shell; the bladder contains the gas. As fluid enters the shell, gas in the bladder is compressed. Gas pressure decreases as fluid flows from the shell. When all liquid has been discharged, gas pressure attempts to push the bladder through the outlet. But, as the bladder contacts the poppet valve at the outlet, flow from the shell is automatically shut off.

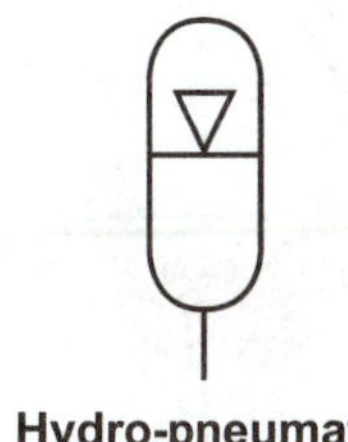

Hydro-pneumatic Accumulator Symbol

Figure 8-17

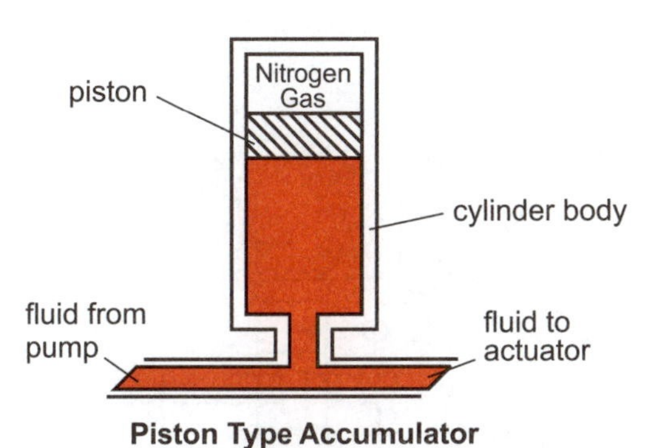

Piston Type Accumulator

Figure 8-18

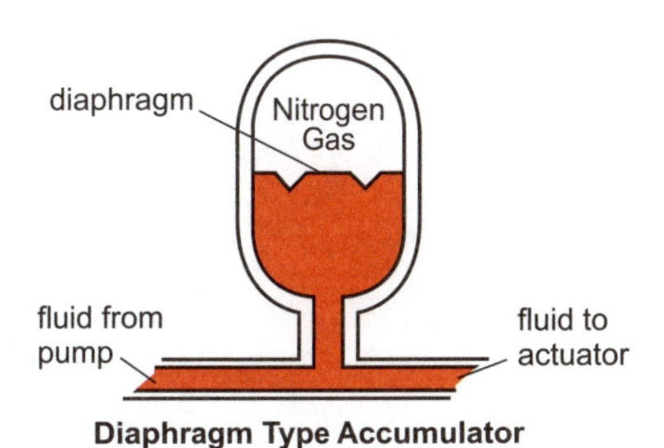

Diaphragm Type Accumulator

Figure 8-19

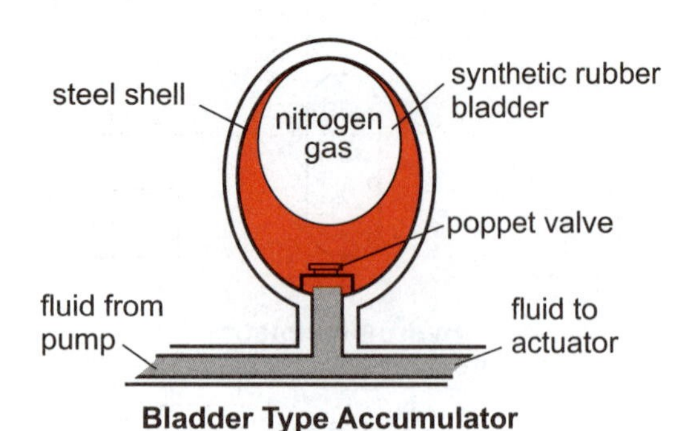

Bladder Type Accumulator

Figure 8-20

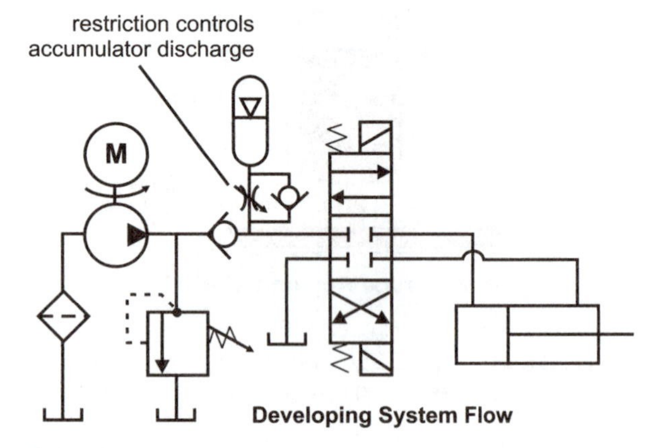

Developing System Flow

Safety Note: In any accumulator circuit a means should be available of automatically unloading the accumulator when the machine is shut down.

Figure 8-21

We have seen the various types of accumulators. In the next section, we find that all three types can be used to develop system flow and maintain pressure. And, hydro-pneumatic accumulators can be used to absorb shock.

Accumulators in a Circuit

Accumulators can perform a variety of functions in a hydraulic system. Some of these are maintaining system pressure, supplementing pump flow, and absorbing system shock.

Developing Flow

Developing liquid flow is one accumulator application. Since charged accumulators are a source of hydraulic potential energy, stored energy of an accumulator can be used to develop system flow when system demand is greater than pump delivery. For instance, if a machine is designed to cycle infrequently, a small displacement pump can be used to fill an accumulator over a period of time. When the moment arrives for the machine to operate, a directional valve is shifted downstream and the accumulator delivers the required pressurized flow to an actuator. Using an accumulator in combination with a small pump in this manner conserves peak horsepower. For instead of using a large pump/electric motor to generate a large horsepower all at once, the work can be evenly spread over a time period with a small pump/ electric motor.

Maintaining Pressure

Accumulators are used to maintain pressure. This can be required in one leg of a circuit while pump/ electric motor is delivering flow to another portion of the system.

In the circuit illustrated, two clamp cylinders are required to hold a part in place. As the directional valves are shifted, both cylinders extend and clamp at the pump's compensator setting. During this time, the accumulator is charged to the setting also.

System demands require that cylinder B maintain pressure while cylinder A retracts. As directional valve A is shifted, pressure at the pump as well as in line A drops quite low. Pressure at cylinder B is maintained because the accumulator has stored sufficient fluid under pressure to make up for any leakage in line B.

Accumulators not only maintain pressure by compensating for pressure loss due to leakage, but they also compensate for pressure increase due to thermal fluid expansion or external mechanical forces acting on a cylinder.

In the illustrated circuit, assume that the cylinder is operating near a furnace where ambient temperatures are quite high. This causes the fluid to expand. With an accumulator in the circuit, the excess volume is taken up keeping the pressure relatively constant. Without an accumulator, pressure in the line would rise uncontrollably and may cause a component housing, fitting or conductor to crack.

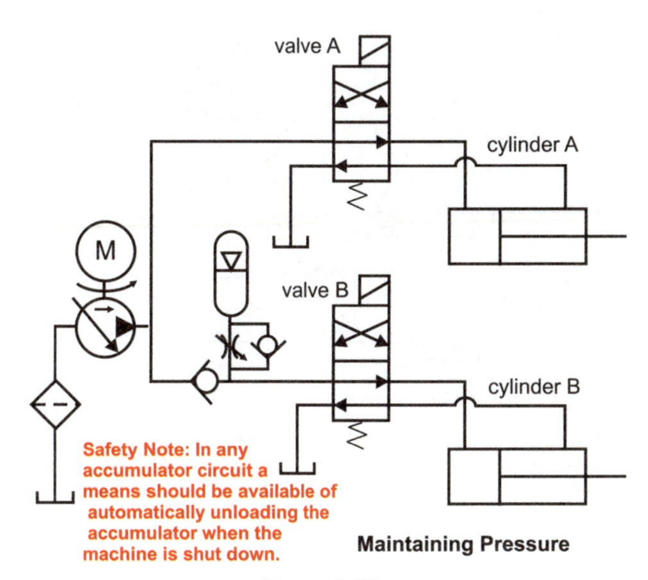

Figure 8-22

The same situation can also occur if an external mechanical force acts to retract the cylinder. Assume now that the cylinder is clamping a curing press. As curing occurs, heat within the press causes it to expand resulting in a force acting to retract the piston rod. The accumulator once again absorbs the additional volume, maintaining the pressure at a relatively constant level.

Absorbing Shock

Hydro-pneumatic accumulators are sometimes used to absorb system shock even though in this application they are difficult to properly design into a system.

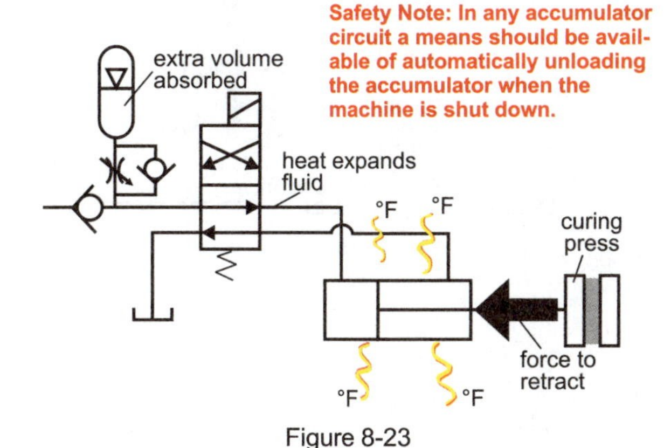

Figure 8-23

Shock in a hydraulic system may be developed from the inertia of a load attached to a cylinder or motor. Or, it may be caused by fluid inertia when system flow is suddenly blocked or changes direction as a directional valve is shifted quickly. An accumulator in the circuit will absorb some of the shock and not allow it to be transmitted fully throughout the system.

In the circuit illustrated, a pump/electric motor is delivering 100 gpm (379 l/min) to a cylinder at the required working pressure. If the closed center directional valve upstream from the cylinder is centered while work is occurring, the pressurized flow of 100 gpm (379 l/min) will be stopped all at once resulting in hydraulic shock reverberating in the system. An accumulator positioned ahead of the directional valve absorbs and reduces the shock effects.

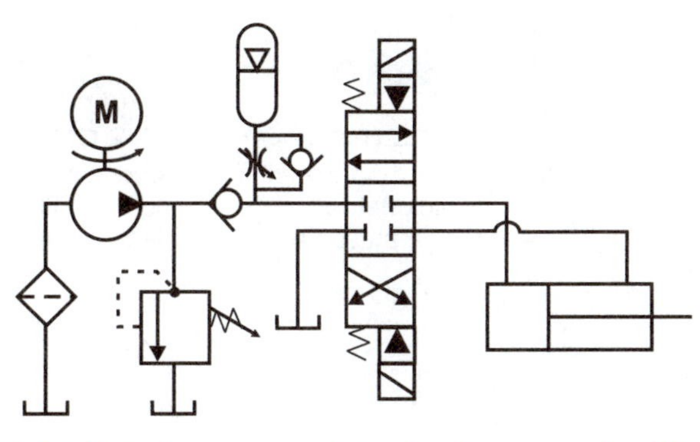

Safety Note: In any accumulator circuit a means should be available of automatically unloading the accumulator when the machine is shut down.

Figure 8-24

Shock may also occur in a hydraulic system due to external mechanical forces. In the circuit illustrated, the load attached to the cylinder has a tendency to bounce causing the rod to be pushed in and shock generated. An accumulator positioned in the cylinder line can help reduce the shock effects if properly precharged. If not, it could become overpressurized.

Since hydro-pneumatic accumulators are most common, we will for the remaining lesson deal with their operating characteristics. We will see what isothermal and adiabatic charging is and how precharge affects their operation.

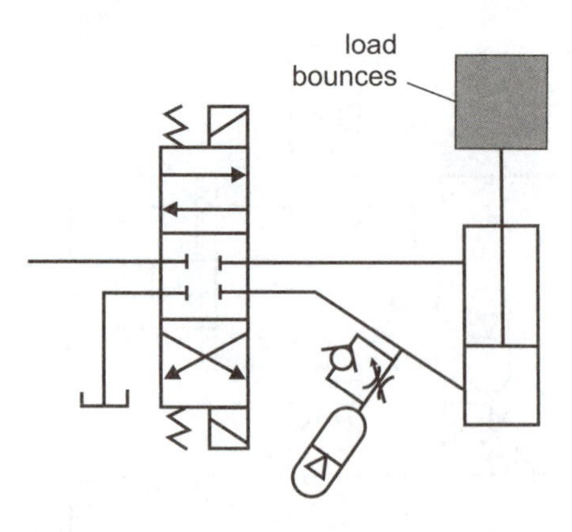

Safety Note: In any accumulator circuit a means should be available of automatically unloading the accumulator when the machine is shut down.

Figure 8-25

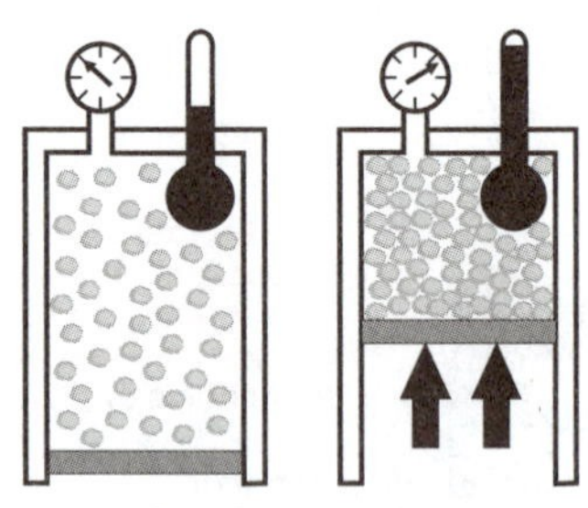

Gas Compression

Figure 8-26

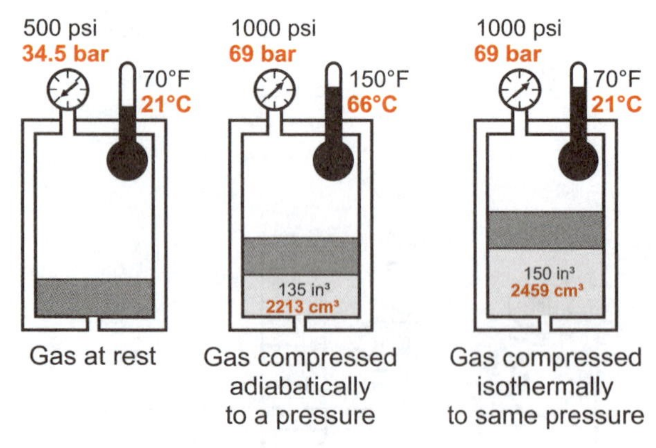

Figure 8-27

Isothermal and Adiabatic Charging

Since hydro-pneumatic accumulators use compressed gas to maintain pressure on a liquid, gas properties affect accumulator operation.

As a hydro-pneumatic accumulator is filled with liquid, gas is compressed, and that as a gas is compressed it heats up. With pressure remaining constant, a heated gas occupies more space than a gas at a lower temperature.

Isothermal describes the operation of an accumulator as the gas is maintained at a constant temperature. While an accumulator is being filled, isothermal operation indicates that the gas is being compressed slowly enough for the heat of compression to dissipate.

Adiabatic describes the operation of an accumulator as gas temperature changes. While an accumulator is being filled, adiabatic operation indicates that the gas is being compressed rapidly so that all heat of compression is retained.

A hydro-pneumatic accumulator which is being charged with liquid up to a certain pressure, will hold more liquid if it is charged isothermally rather than adiabatically.

Illustrated is a piston accumulator void of liquid. A pressure gage at accumulator top indicates a gas pressure of 500 psi (34.48 bar); a thermometer indicates a temperature of 70°F (21°C). Assume that the accumulator is going to be filled with liquid until a pressure of 1000 psi (68.97 bar) is reached. Fluid ceases to enter the accumulator at this point because this is the maximum pressure allowed to be developed by pump/electric motor.

As the accumulator is charged adiabatically, pressure and temperature begin to climb. When pressure reaches 1000 psi (68.97 bar), fluid ceases to enter accumulator inlet. At that point, the temperature is 150°F (65.6°C) and the accumulator holds 135 in^3 (2212.65 cm^3) of fluid.

Assume now, that the accumulator is charged isothermally. Pressure begins to climb, but temperature remains the same. Charging takes place so slowly that heat of compression dissipates. When a pressure of 1000 psi (68.97 bar) is reached, fluid ceases to enter accumulator inlet. At that point, gas temperature is still 70°F (21°C) and the accumulator holds 150 in^3 (2458.5 cm^3) of fluid.

A hydro-pneumatic accumulator operated isothermally (slowly) will be charged with more liquid than if it were operated adiabatically (quickly).

Isothermal and Adiabatic Discharging

As a hydro-pneumatic accumulator discharges liquid the gas expands and as a gas expands it cools. With pressure remaining constant, a cool gas occupies less space than a gas at an elevated temperature.

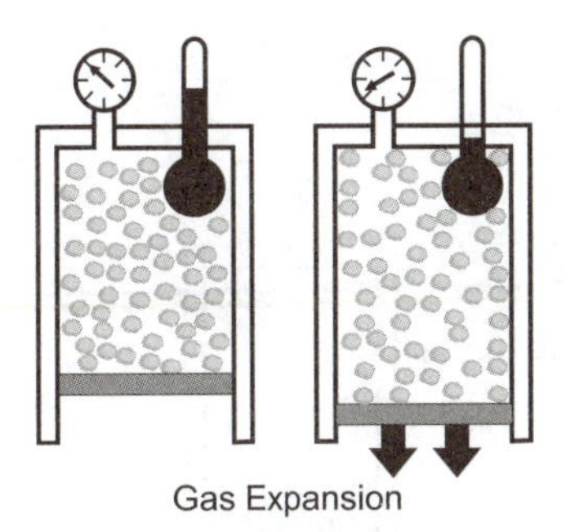

Figure 8-28

Isothermal and adiabatic describe the operation of an accumulator as it discharges fluid. An accumulator discharging fluid under isothermal conditions, indicates that discharge occurs slowly as gas expands; it is capable of acquiring heat from the ambient through accumulator walls or from the fluid. Adiabatic operation indicates that discharging occurs rapidly with no heat gain; as gas expands it cools.

A hydro-pneumatic accumulator which is discharging liquid until a lower pressure is reached, will discharge more liquid if it is discharged isothermally rather than adiabatically.

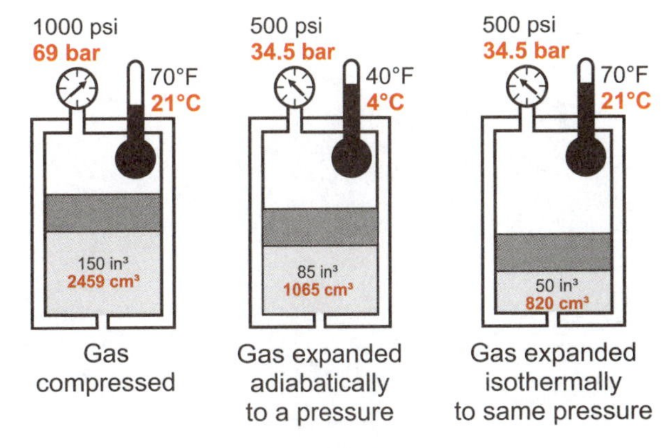

Figure 8-29

Illustrated is a piston accumulator which is charged with liquid to a pressure of 1000 psi (68.97 bar). A thermometer at accumulator top indicates a gas temperature of 70°F (21°C). Assume that the accumulator holds 150 in³ (2458.5 cm³) of fluid. Assume further that when a directional valve downstream is shifted, fluid will be discharged until accumulator pressure reaches 500 psi (34.48 bar). Fluid ceases to discharge from the accumulator at this time because 500 psi (34.48 bar) is the working pressure of the system. This means a pressure differential no longer exists at this point to develop a discharge flow.

As the accumulator discharges adiabatically, gas pressure and temperature begin to drop. When gas pressure reaches 500 psi (34.5 bar), fluid ceases to exit the accumulator. At that point, temperature of the gas is 40°F (4.4°C) and the accumulator holds 65 in³ (1065.35 cm³) of fluid. 85 in³ (1393 cm³) of fluid has therefore been discharged.

Assume, now, that the accumulator is discharged isothermally. Pressure begins to drop, but gas temperature remains the same. Since discharging takes place slowly, the gas is able to acquire heat from the walls of the accumulator. When 500 psi (34.5 bar) is reached, fluid ceases to exit the accumulator. At that point, temperature is still 70°F (21°C) and the accumulator holds 50 in³ (819.5 cm³) of fluid. 100 in³ (1639 cm³) of fluid has therefore been discharged.

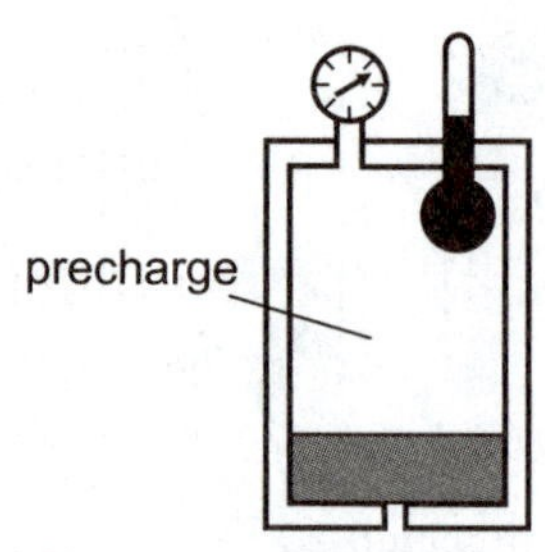

Safety Note: In an accumulator circuit, a means should be available of automatically unloading the accumulator when the machine is shut down.

Figure 8-30

We have seen that more fluid enters and exits a hydro-pneumatic accumulator as it is operated isothermally. But, this is usually an ideal situation. Ordinarily, accumulators are charged and discharged adiabatically. In the following section, we find that the biggest concern is not how much the accumulator holds, but how much fluid is discharged before a lower pressure is reached. This is largely affected by gas precharge.

Precharge

The gas pressure present in a hydro-pneumatic accumulator when it is drained of hydraulic fluid is the accumulator precharge. This pressure significantly affects a hydro-pneumatic accumulator's usable volume and operation as a shock absorber.

Precharge Affects Usable Volume

A hydro-pneumatic accumulator which is used to develop system flow or maintain pressure operates between a maximum and minimum pressure. An accumulator is filled with fluid until a maximum pressure is attained. At an appropriate time, it discharges fluid until a lower pressure is reached. The accumulator is then recharged until the max pressure is reached. The liquid volume discharged between the two pressures is the accumulator's usable volume.

Gas precharge affects usable volume of an accumulator. This may be best shown by an example.

Assume that a 231 in^3 (3786 cm^3) hydro-pneumatic accumulator is used to develop flow in a particular system. The accumulator is charged with liquid from a pump until a system pressure of 2000 psi (137.9 bar) is reached. To develop a flow, it is allowed to discharge to 1500 psi (103.4 bar). The gas precharge selected will determine the amount of fluid which the accumulator pushes into the system.

From the illustrated chart, it can be seen that a 231 in^3 (3786 cm^3) accumulator with a gas precharge of 100 psi (6.89 bar) holds 210 in^3 (3441.9 cm^3) of hydraulic fluid at 2000 psi (137.9 bar) if charged isothermally. (The upper value numbers in each row indicate isothermic operation.) When discharged to 1500 psi (103.4 bar), the accumulator holds 202 in^3 (3310.8 cm^3). Between the two points, 8 in^3 (131 cm^3) have been discharged. With a 100 psi (6.89 bar) precharge, this accumulator holds much liquid, but discharges little.

With gas precharge increased to 1000 psi (68.96 bar), accumulator holds 93 in³ (1524.3 cm³) of oil at 2000 psi (137.9 bar) and 59.5 in³ (975 cm³) at 1500 psi (103.4 bar). 33.5 in³ (549.1 cm³) have been discharged.

In this instance, the accumulator does not hold as much fluid as before; however, more fluid is discharged. Increasing the precharge to 1400 psi (96.6 bar), we find that the accumulator holds the least amount of fluid, but the most is discharged. At 2000 psi (138 bar), 53.5 in³ (876.9 cm³) of oil are held by the accumulator. At 1500 psi (103.5 bar), 11.6 in³ (190.1 cm³) are contained. Usable volume in this case is 41.9 in³ (686.7 cm³).

As a hydro-pneumatic accumulator develops pressurized flow or maintains pressure, usable volume is an important consideration. While developing a flow, a certain amount of fluid must discharge between two pressures in order for system demands to be met. While maintaining pressure, an accumulator must have available within a certain pressure range sufficient fluid to compensate for leakage. Consequently, correct precharge is a very important element in hydro-pneumatic accumulator operation.

Control of Usable Volume Discharge

The usable volume of an accumulator should be discharged at a controlled rate. If an accumulator is required to maintain system pressure, this controlled rate is automatically achieved by the leakage fluid it has to replace. However, an accumulator which is used to develop a pressurized flow can discharge its usable volume too rapidly as a downstream directional valve is shifted. For this reason, accumulators in this application are often equipped with a flow control and bypass check at their inlet-outlet port.

Adiabatic / Isothermal Accumulator Performance Chart (231 in³ Accumulator)

Liquid volume (in.³) stored in accumulator (each cell: Adiabatic / Isothermal)

Gas Precharge Pressure - psi (gage)	Operating Pressure - psi (gage): 100	200	300	400	500	600	700	800	900	1000	1100	1200	1300	1400	1500	1600	1700	1800	1900	2000	2100
100		86.6 / 112	113 / 154	144 / 174	158 / 187	168 / 196	175 / 202	182 / 207	186 / 211	190 / 214	192 / 216	196 / 218	198 / 220	200 / 222	202 / 223	204 / 224	206 / 225	207 / 226	209 / 227	210 / 227	211 / 228
200			57.4 / 76.6	39.7 / 116	112 / 141	126 / 157	138 / 168	147 / 178	155 / 184	161 / 190	166 / 195	170 / 198	174 / 202	178 / 204	181 / 207	184 / 209	186 / 211	188 / 213	190 / 214	192 / 215	194 / 216
300				43.4 / 58.5	71.4 / 94.0	91.1 / 118	105 / 134	118 / 148	127 / 158	136 / 166	143 / 173	148 / 176	153 / 184	157 / 188	162 / 191	165 / 194	169 / 197	172 / 199	174 / 202	177 / 203	179 / 205
400					34.2 / 46.7	58.8 / 78.5	77.3 / 101	92.0 / 118	103 / 132	114 / 143	121 / 151	128 / 159	135 / 165	141 / 171	145 / 175	149 / 179	153 / 183	157 / 186	160 / 189	163 / 191	165 / 194
500						28.5 / 39.3	50.2 / 67.5	67.0 / 88.6	80.5 / 105	91.8 / 119	102 / 130	110 / 139	117 / 146	123 / 153	128 / 159	134 / 164	138 / 169	142 / 173	146 / 176	149 / 179	152 / 182
600							24.6 / 33.8	43.6 / 59.0	58.8 / 78.8	72.1 / 95.0	83.2 / 108	92.4 / 119	101 / 128	108 / 136	114 / 143	120 / 149	126 / 154	130 / 159	132 / 164	136 / 168	140 / 171
700								21.7 / 29.9	38.6 / 52.5	53.0 / 71.1	65.1 / 86.3	75.5 / 99.4	84.6 / 110	92.6 / 119	99.5 / 127	106 / 134	112 / 141	117 / 146	121 / 151	125 / 155	129 / 160
800									19.1 / 26.2	35.0 / 47.7	48.0 / 64.5	59.3 / 79.4	69.4 / 91.9	78.1 / 102	85.8 / 111	92.5 / 119	99.8 / 127	105 / 133	110 / 139	114 / 144	119 / 148
900										17.4 / 24.1	31.6 / 43.2	43.6 / 59.4	54.7 / 73.3	63.9 / 84.9	72.5 / 95.5	80.0 / 104	86.8 / 112	92.8 / 120	98.5 / 126	104 / 132	108 / 137
1000											15.7 / 21.5	28.7 / 39.5	40.5 / 55.0	50.9 / 68.2	59.5 / 79.6	67.8 / 89.7	75.0 / 98.4	81.5 / 106	87.5 / 113	93.0 / 120	98.0 / 125
1100												14.2 / 19.8	26.8 / 36.6	37.4 / 58.3	47.2 / 63.9	55.9 / 74.7	63.4 / 89.4	70.4 / 93.1	76.9 / 101	82.6 / 108	88.0 / 114
1200													13.3 / 18.6	24.8 / 34.2	35.0 / 47.7	44.4 / 60.0	52.1 / 70.2	59.8 / 79.8	66.5 / 88.2	72.8 / 95.7	78.5 / 103
1300														12.3 / 17.1	23.1 / 31.8	32.5 / 44.6	41.0 / 55.9	49.6 / 66.3	56.4 / 75.5	63.1 / 83.9	69.1 / 91.1
1400															11.6 / 15.9	21.7 / 29.9	30.8 / 42.2	39.0 / 53.0	46.3 / 62.7	53.5 / 71.9	59.8 / 80.0
1500																10.6 / 15.0	20.2 / 28.0	28.9 / 39.8	36.9 / 50.1	44.4 / 59.8	51.9 / 68.5

Figure 8-31

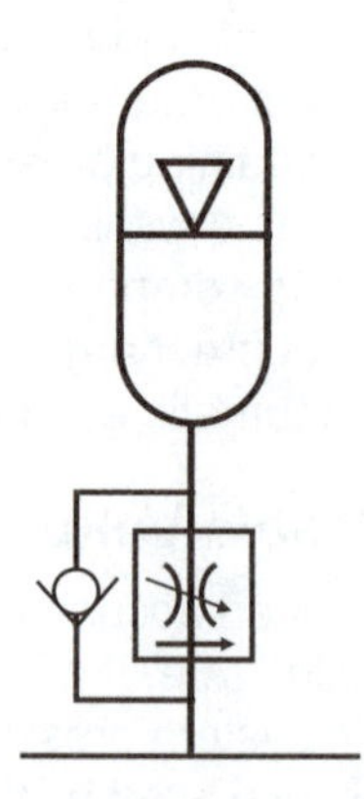

Safety Note: In any accumulator circuit, a means should be available of automatically unloading the accumulator when the machine is shut down.

Figure 8-32

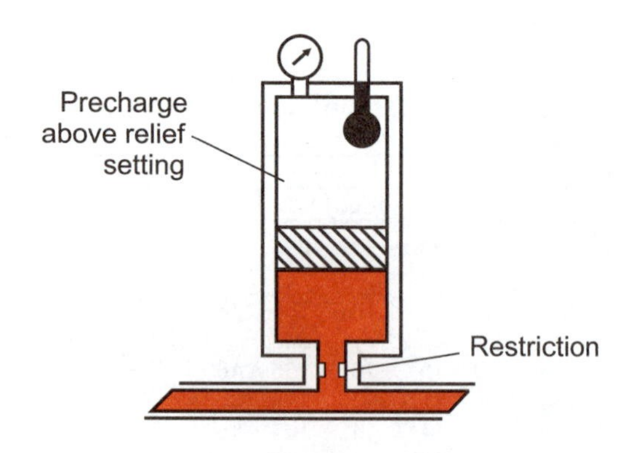

Figure 8-33

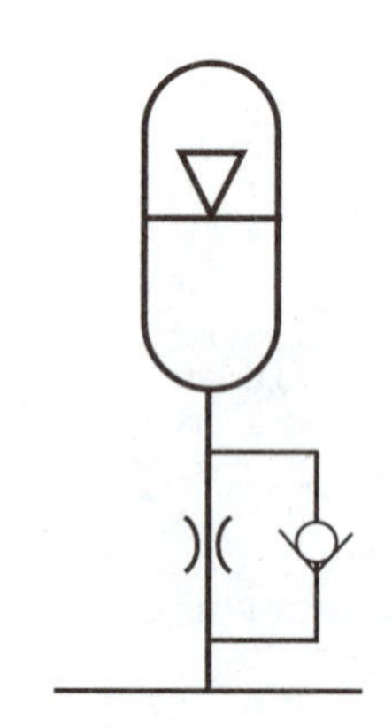

Figure 8-34

Precharge Affects Shock Absorber Operation

Precharge of a hydro-pneumatic accumulator affects its operation as a shock absorber.

Shock generation in a hydraulic system is the result of fast pressure rises due to an external mechanical force acting on a cylinder or hydraulic motor, or the result of liquid crashing into a component as a valve is suddenly closed. An accumulator acts to reduce a shock effect by limiting pressure rise.

In a hydraulic system, as shock pressures develop, they attempt to displace or push the fluid to another part of the system. But, since liquid is relatively incompressible, it won't move or compress.

Without an accumulator in the line, shock pressures can climb to a high value because they have a relatively solid base on which to build. With an accumulator in the line, the base for shock pressure development becomes soft.

Above a certain system pressure as a shock pressure begins to build, an accumulator absorbs the volume of liquid the shock attempts to compress or displace. The line in which the accumulator is located becomes compressible above a certain point.

Gas precharge for a hydro-pneumatic accumulator used as a shock absorber is generally set slightly above the maximum working pressure of the line in which it is located. If the maximum pressure happens to be determined by the relief valve setting, gas precharge might be 100 psi (6.896 bar) above this.

For example, if a relief valve were set for a maximum pressure of 2000 psi (137.9 bar), accumulator precharge would be 2100 psi (144.8 bar). This means the accumulator would not function in the circuit until this pressure was reached. The accumulator later would be void of liquid until a pressure of 2100 psi (144.8 bar) was generated.

Assume that in a particular system, a shock pressure is defined as any pressure of 2100 psi (144.8 bar) and a 58 in^3 (950.6 cm^3) accumulator was required to absorb 4 in^3 (65.6 cm^3) of oil in order to dissipate the pressure. From this, it can be seen that a 58 in^3 (950.6 cm^3) accumulator operating adiabatically with a precharge of 2100 psi (144.8 bar) will allow pressure to climb to 2300 psi (158.6 bar) as it absorbs 4 in^3 (65.6 cm^3) of oil.

Adiabatic / Isothermal Accumulator Performance Chart (58 in³ Accumulator)

Liquid volume (in.³) stored in accumulator (each cell: adiabatic / isothermal)

Gas Precharge Pressure - psi (gage)	Operating Pressure - psi (gage)																			
	1100	1200	1300	1400	1500	1600	1700	1800	1900	2000	2100	2200	2300	2400	2500	2600	2700	2800	2900	3000
1000	4.4 / 6.1	8.1 / 11.2	11.4 / 15.5	14.3 / 19.2	16.8 / 22.4	19.1 / 25.3	21.2 / 27.8	23.0 / 29.9	24.7 / 32.0	26.2 / 33.7	27.6 / 35.4	28.9 / 36.8	30.2 / 38.2	31.2 / 39.4	32.3 / 40.5	33.2 / 41.6	34.2 / 42.6	34.9 / 43.4	35.8 / 44.3	36.5 / 45.1
1100		4.0 / 5.6	7.6 / 10.3	10.5 / 14.4	13.3 / 18.0	15.8 / 21.1	17.9 / 23.8	19.9 / 26.2	21.7 / 28.4	23.3 / 30.4	24.9 / 32.2	26.2 / 33.7	27.5 / 35.2	28.6 / 36.6	29.8 / 37.8	30.8 / 39.0	31.8 / 40.1	32.8 / 41.1	33.7 / 41.9	34.4 / 42.8
1200			3.7 / 5.2	7.0 / 9.7	9.9 / 13.5	12.5 / 16.9	4.7 / 19.8	16.9 / 22.5	18.8 / 24.9	20.5 / 27.0	22.2 / 29.0	23.6 / 30.7	25.0 / 32.4	26.2 / 33.7	27.4 / 35.1	28.5 / 36.4	29.7 / 37.7	30.7 / 38.8	31.5 / 39.7	32.3 / 40.6
1300				3.5 / 4.8	6.5 / 9.0	9.2 / 12.6	11.6 / 15.8	13.9 / 18.7	15.9 / 21.3	17.8 / 23.7	19.5 / 25.7	21.0 / 27.6	22.5 / 29.3	23.8 / 30.9	25.1 / 32.5	26.2 / 33.8	27.3 / 35.0	28.3 / 36.2	29.2 / 37.2	30.2 / 38.3
1400					3.3 / 4.5	6.1 / 8.4	8.7 / 11.9	11.0 / 14.9	13.1 / 17.7	15.1 / 20.3	16.9 / 22.6	18.6 / 24.6	20.1 / 26.4	21.4 / 28.1	22.8 / 29.7	24.0 / 31.2	25.1 / 32.5	26.2 / 33.8	27.3 / 34.9	28.2 / 36.0
1500						3.0 / 4.2	5.7 / 7.9	8.2 / 11.2	10.4 / 14.2	12.5 / 16.9	14.4 / 19.3	16.1 / 21.5	17.7 / 23.5	19.2 / 25.3	20.5 / 27.0	21.9 / 28.6	23.0 / 30.0	24.2 / 31.4	25.2 / 32.6	26.2 / 33.8
1600							2.9 / 4.0	5.4 / 7.5	7.8 / 10.7	9.9 / 13.5	11.8 / 16.0	13.6 / 18.4	15.2 / 20.5	16.9 / 22.5	18.3 / 24.3	19.6 / 26.0	21.0 / 27.6	22.2 / 28.9	23.3 / 30.3	24.3 / 31.7
1700								2.7 / 3.7	5.1 / 7.1	7.3 / 10.1	9.3 / 13.0	11.3 / 15.4	13.0 / 17.8	14.7 / 19.7	16.2 / 21.6	17.5 / 23.3	18.9 / 25.0	20.2 / 26.6	21.4 / 28.0	22.4 / 29.3
1800									2.5 / 3.5	4.9 / 6.8	7.0 / 9.7	9.0 / 12.2	10.8 / 14.8	12.5 / 16.9	14.0 / 18.9	15.5 / 20.8	16.9 / 22.6	18.2 / 24.2	19.3 / 25.6	20.5 / 27.1
1900										2.4 / 3.3	4.7 / 6.5	6.7 / 9.2	8.6 / 11.8	10.3 / 14.1	11.9 / 16.2	13.5 / 18.2	14.8 / 19.9	16.2 / 21.7	17.5 / 23.3	18.6 / 24.8
2200											2.3 / 3.2	4.5 / 6.2	6.4 / 8.8	8.3 / 11.4	9.9 / 13.5	11.6 / 15.6	13.0 / 17.5	14.7 / 19.2	15.6 / 20.9	16.9 / 22.5
2100												2.2 / 3.1	4.2 / 5.9	6.1 / 8.4	7.8 / 10.7	9.5 / 13.0	11.0 / 15.0	12.5 / 16.9	13.8 / 18.6	15.1 / 20.2
2200													2.1 / 3.0	4.1 / 5.6	5.9 / 8.2	7.6 / 10.4	9.2 / 12.5	10.7 / 14.5	12.0 / 16.3	13.4 / 18.1
2300														2.0 / 2.9	3.9 / 5.4	5.6 / 7.8	7.3 / 10.0	8.7 / 12.0	10.3 / 14.0	11.6 / 15.8
2400															2.0 / 2.7	3.7 / 5.2	5.4 / 7.5	7.0 / 9.7	8.5 / 11.7	9.9 / 13.5
2500																1.8 / 2.6	3.6 / 5.0	5.2 / 7.3	6.8 / 9.4	8.3 / 11.4
2600																	1.8 / 2.5	3.5 / 4.8	5.0 / 6.9	6.5 / 9.6
2700																		1.7 / 2.4	3.4 / 4.7	4.8 / 6.7

Figure 8-35

If the precharge were set too low, for instance 1500 psi (103.4 bar), the accumulator would hold 14.4 in³ (236 cm³) at 2100 psi (144.8 bar). In order to absorb 4 in³ (65.6 cm³) pressure would climb to over 2300 psi (158.6 bar). If the precharge were at a high value of 2500 psi (172.4 bar), pressure would climb to almost 2800 psi (193 bar) before 4 in³ (65.6 cm³) could be absorbed. Precharge of a hydro-pneumatic accumulator used as a shock absorber is quite important.

As an accumulator operates in a system as a shock absorber, it is generally required to get rid of the fluid it has accumulated in a controlled fashion. Commonly, accumulators in these applications are once again equipped with a restriction and bypass check valve. With this arrangement, an accumulator can accept its required fluid, yet any fluid accumulation can bleed off through the restriction.

Losing Gas Precharge

Just because a hydro-pneumatic accumulator is charged once to the proper gas precharge, it does not mean it will remain charged to that pressure indefinitely. As accumulators operate, gas pressure can seep out through the gas valve. This can be due to a faulty or deteriorated seal in the valve, or an improperly seated poppet in the valve core.

Hydro-pneumatic accumulators also lose gas precharge when discharging fluid. With bladder and diaphragm type accumulators, this usually occurs in a catastrophic manner as a result of a rupture in the synthetic rubber separator. When a piston type accumulator discharges, gas pressure can escape across the piston due to worn seals. Piston type accumulators give an indication of wear as gas precharge gradually dissipates.

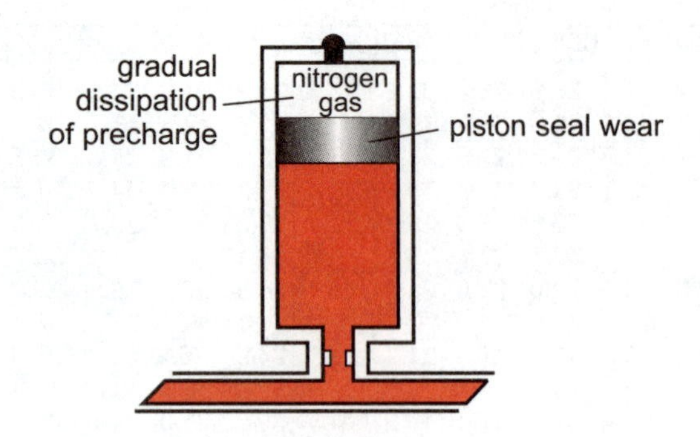

Figure 8-36

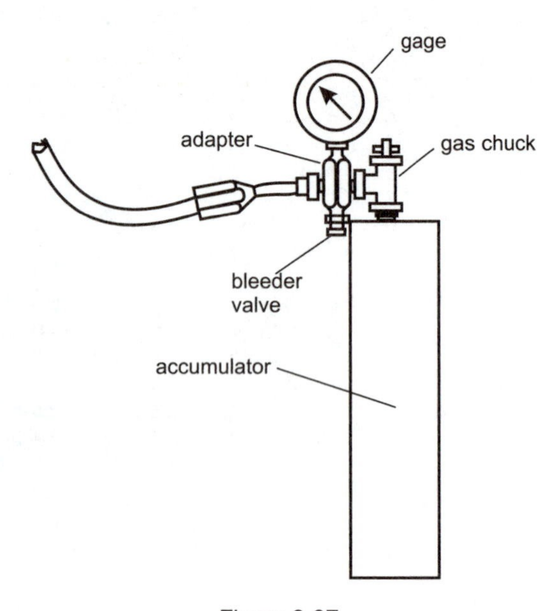

Figure 8-37

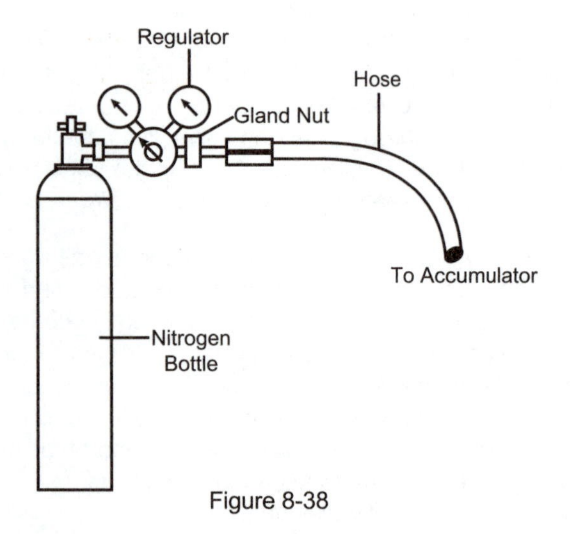

Figure 8-38

Checking Gas Precharge

Since proper gas precharge is an important consideration in hydro-pneumatic accumulator performance, precharge should be checked periodically. A necessary piece of equipment for checking gas pressure is a precharging and gaging assembly. This assembly is primarily made up of a gas chuck, bleeder valve and pressure gage.

To check precharge, discharge the accumulator of fluid and remove the protective cap which frequently is found on the valve in accumulator top.

Screw the gas chuck handle of the assembly all the way out; check to see that the bleeder valve is closed. Attach the assembly to the accumulator gas valve at the gas chuck. Using a wrench, tighten gas chuck swivel nut securely onto gas valve. Turn gas chuck stem in; this depresses the core in the accumulator gas valve registering a gage pressure. This is the accumulator precharge.

If the accumulator is properly charged, back the gas chuck handle out and open the bleeder valve venting the assembly. Loosen the gas chuck swivel nut and remove the assembly. Replace the accumulator gas valve protective cover.

If it is found that the accumulator is overcharged, excess pressure can be bled off through the bleeder valve.

If it necessary to increase gas pressure, back the gas chuck handle out. Open the bleeder valve venting the assembly; then reclose the bleeder valve. At this point, the charging assembly will have to be connected to a nitrogen bottle.

With the nitrogen bottle gas valve off, connect a hose from the bottle gas valve to the charging assembly gas valve. Turn the gas chuck completely in depressing the accumulator core. Crack open the gas valve of the nitrogen bottle to slowly fill the accumulator. Shut off the valve when the gage indicates the desired pressure.

Once the gage indicates the appropriate pressure, shut off the bottle gas valve, turn out the gas chuck handle, open the gas bleeder valve. The charging hose and gaging assembly can then be removed.

In the following section, we will be concerned with unloading pump/electric motor once an accumulator is charged. We will see how this might be done with an unloading valve. But, we will find that venting a relief valve or using a differential unloading relief valve is a better means.

NOTE: Precautions must be taken to protect against accumulator overpressure during precharge and operation.

Pump Unloading in Accumulator Circuits

In a typical circuit, when an accumulator is charged and work is not required from any part of the system, pump/electric motor flow is unloaded to tank with the least possible pressure.

In the circuit illustrated, an unloading valve is used to dump flow back to tank once an accumulator is charged to the unloading valve setting. Frequently, unloading in this manner is only good for a few seconds at the most. With any leakage in the system downstream from the check valve, pressure in the accumulator will drop as fluid is discharged to compensate. This results in the valve gradually closing. The opening to tank through the valve progressively becomes more and more restrictive until accumulator pressure drops below the valve cracking pressure. Up to this point, pump/electric motor has to develop more and more power as the valve closes.

Once the valve closes, pump/electric motor must therefore generate power to recharge the accumulator to the unloading valve setting. This means the pump/electric motor works when it is supposed to be unloaded.

To keep a pump/electric motor fully unloaded until it is required to recharge an accumulator, an electric pressure switch can be used.

In the circuit illustrated, a pressure switch senses accumulator pressure sending and cutting-out electrical signals at various pressure levels. The electrical signals are transmitted to a normally non-passing, solenoid operated 2-way valve which vents a pilot operated relief valve. When the accumulator is charged to the pressure switch setting an electrical signal to the 2-way valve solenoid venting the relief valve and unloading pump/electric motor is sent. The setting of the pressure switch determines the pressure range within which a pump/electric motor works. Using a pressure switch to vent a relief valve, results in a pump/electric motor being fully unloaded when system conditions dictate.

The wiring and additional piping required in venting a relief valve can be eliminated by using a differential unloading relief valve. In the following section, this valve is described. We find that its operation is dependent on a differential piston.

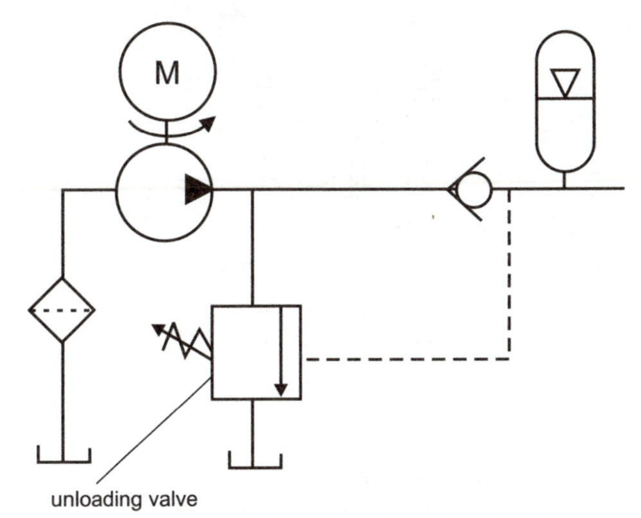

Safety Note: In any accumulator circuit, a means should be available of automatically unloading the accumulator when the machine is shut down.

Figure 8-39

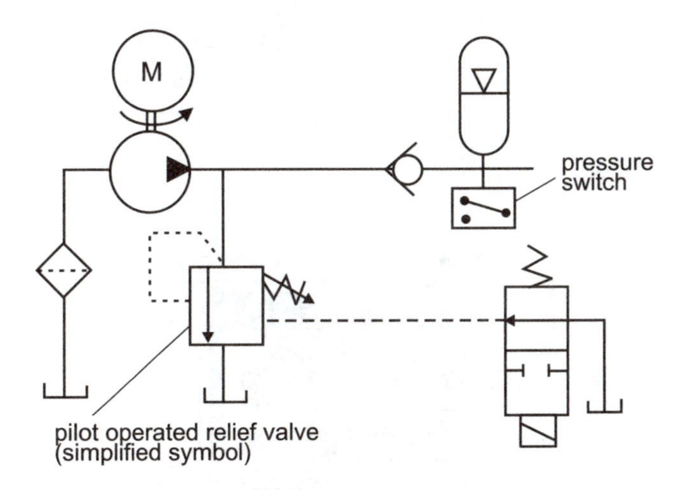

Safety Note: In any accumulator circuit, a means should be should be available of automatically unloading the accumulator when the machine is shut down.

Figure 8-40

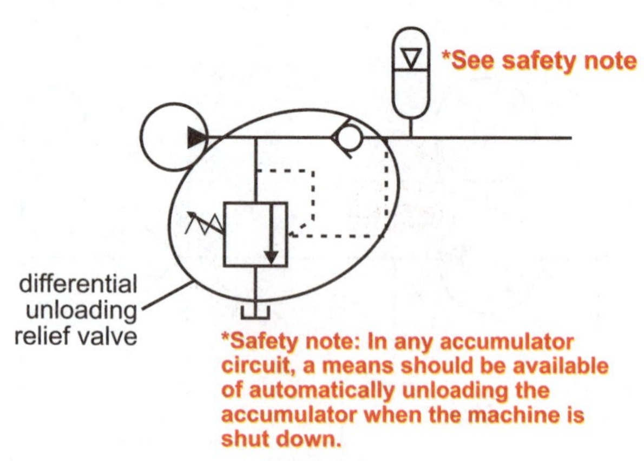

Figure 8-41

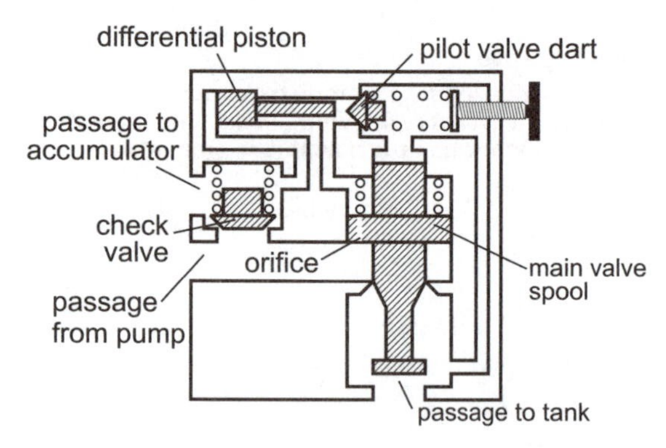

Figure 8-42

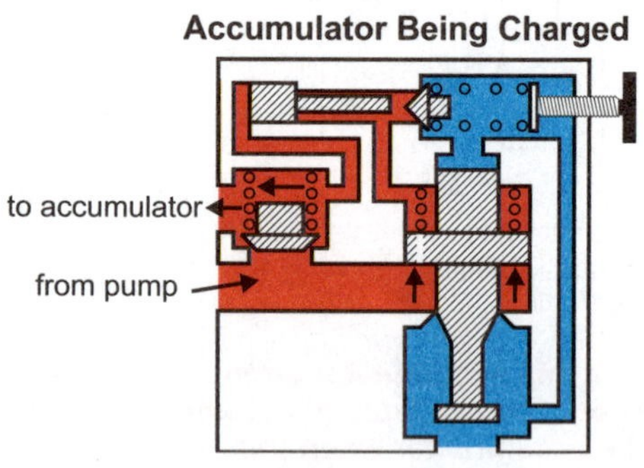

Figure 8-43

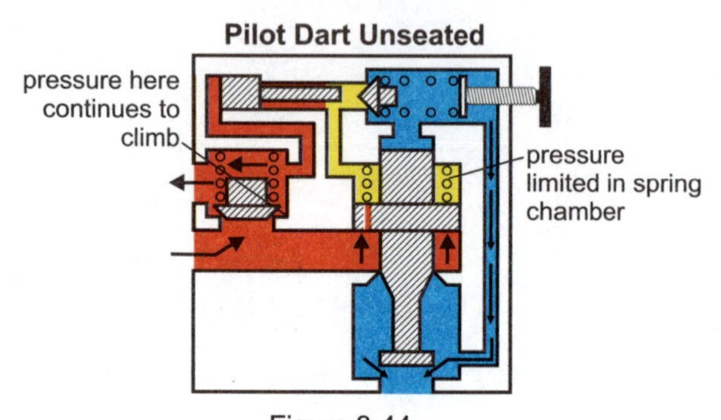

Figure 8-44

Differential Unloading Relief Valve

Instead of using a pressure switch and solenoid valve to vent a relief valve while an accumulator is charged, one hydraulic component can be used a differential unloading relief valve. A differential unloading relief valve is specifically designed for use with accumulators. As its name implies, the valve unloads a pump/electric motor over a differential pressure range.

What a Differential Unloading Relief Valve Consists Of

A differential unloading relief valve consists of a pilot operated relief valve, check valve, and differential piston in one valve body. The valve body includes pump, tank and accumulator passages.

How a Differential Unloading Relief Valve Works

In a differential unloading relief valve, check valve and pilot operated relief valve operate in their usual manner. The accumulator charges through the check valve.

The differential piston is free to move within a bore opposite the pilot valve dart. Areas exposed to pressure at each end of the piston are equal. During the time the accumulator is being charged, pressure at each end of the piston is relatively equal. (Consider the pressure differential across the check valve to be negligible.) As a result, the piston does not move.

When a large enough pressure is present at the pilot valve dart, the dart is pushed off its seat. We have seen earlier that this action limits the pressure in the spring chamber of the main valve. With pressure limited in the spring chamber and also at one end of the differential piston, the piston is moved toward the pilot dart forcing the dart completely off its seat. This in effect releases the main spool spring chamber of pilot pressure, venting the relief valve and unloading the pump/electric motor. At the same time the check valve closes so that accumulator flow cannot discharge through the relief valve.

At this point, the accumulator's maximum pressure has been achieved and pump/electric motor is unloaded. The differential piston is the key to unloading the pump until the accumulator discharges to a lower pressure.

The differential piston has a 15% greater area exposed to pressure than the area of the pilot dart. Since force = pressure x area, the piston holds the pilot dart off its seat with a 15% greater force than the force which unseated the dart. This means that in order to reseat the pilot dart, the spring must acquire a 15% greater force from somewhere; or, it must wait until the pressure in the system falls off 15%. Of course, the dart reseats when system pressure falls off 15%.

In this way, a differential unloading relief valve allows a pump/electric motor to unload when an accumulator is charged and to remain unloaded until recharging is required.

A limitation of a differential unloading relief valve is that the valve's secondary pressure is fixed because the difference in areas between piston and pilot dart is fixed. This is frequently 15% and in some cases 30% of the pilot valve setting. For example, a differential unloading relief valve with a 15% differential will unload between 1000 psi (69 bar) and 850 psi (59 bar) with its pilot valve adjusted for 1000 psi (69 bar). Or, with its pilot valve adjusted for 2000 psi (138 bar), it will unload between 2000 psi (138 bar) and 1700 psi (113 bar).

Hydraulic Cylinders

In all applications, hydraulic working energy must be converted to mechanical energy before any useful work can be done. Hydraulic cylinders convert hydraulic working energy into straight-line mechanical energy.

We have seen previously basic cylinder operation. Now we shall deal with cylinder construction.

What a Cylinder Consists Of

A hydraulic cylinder consists of a cylinder body, a movable piston and a piston rod attached to the piston. End caps are attached to the cylinder body barrel by threads, keeper rings, tie rods or a weld. (Industrial cylinders use tie rods.)

As the cylinder rod moves in and out, it is guided and supported by a removable bushing called a rod gland or rod bearing.

The end through which the rod protrudes is called the "head" or "rod head." The opposite side without the rod is termed the "cap". Inlet and outlet ports are located at the head and cap ends.

Seals

For proper operation, a positive seal must exist across a cylinder's piston as well as at the rod gland. Cylinder pistons are typically sealed by using lipseals, cast iron piston rings, or a single bidirectional sealing element.

Relief Valve Vented by Differential Piston

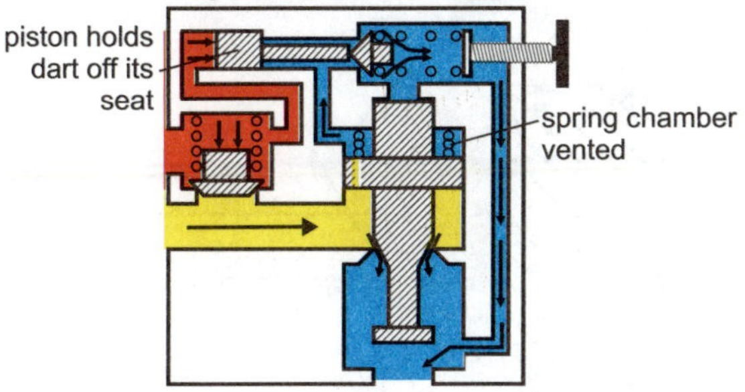

piston holds dart off its seat

spring chamber vented

Figure 8-45

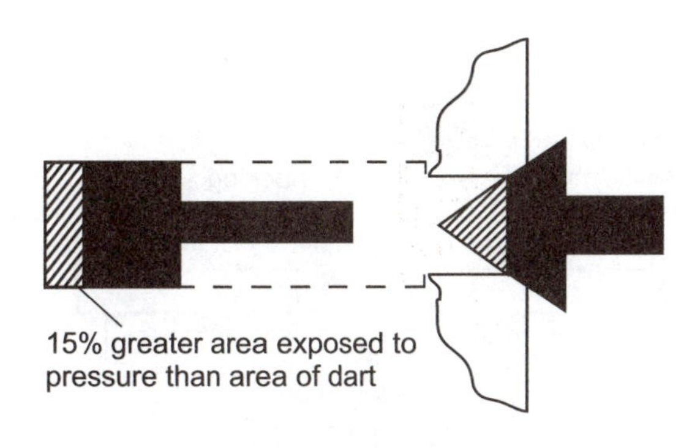

Figure 8-46

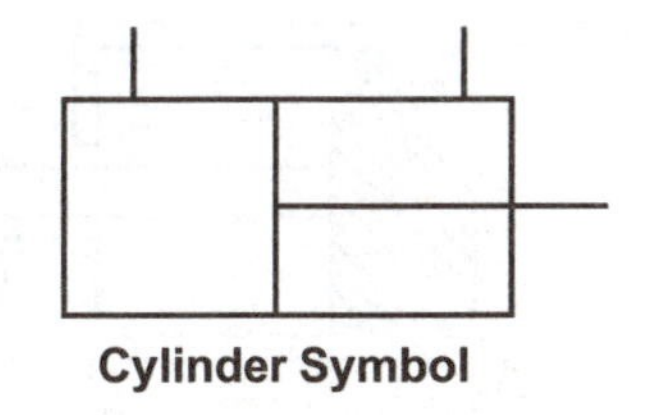

Figure 8-47

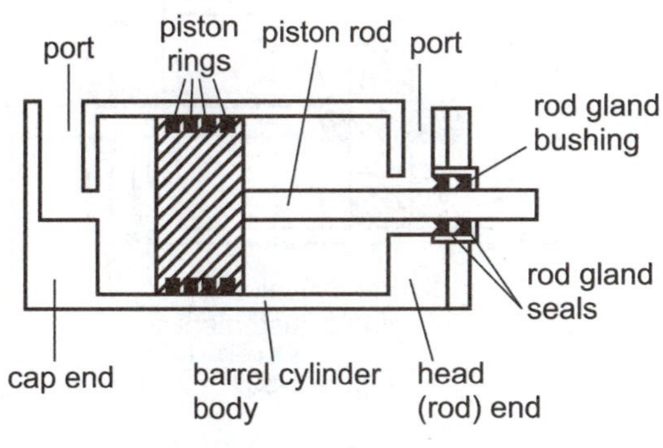

Figure 8-48

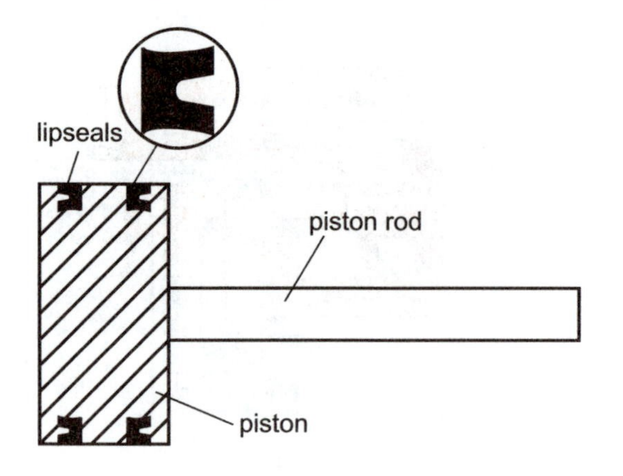

Figure 8-49

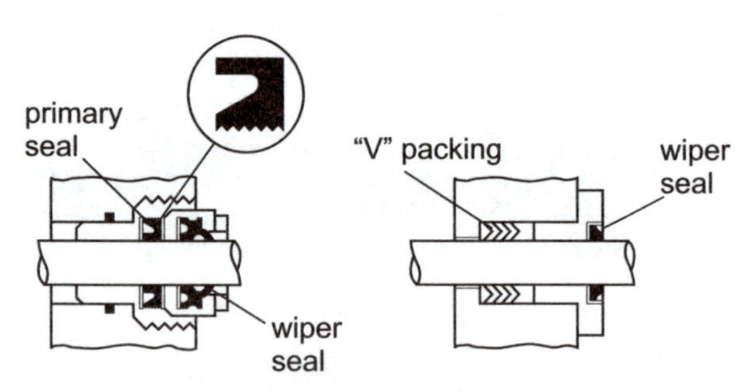

Figure 8-50 & 8-51

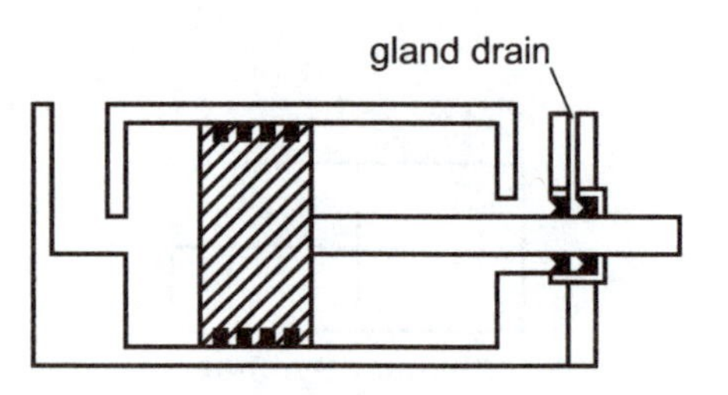

Figure 8-52

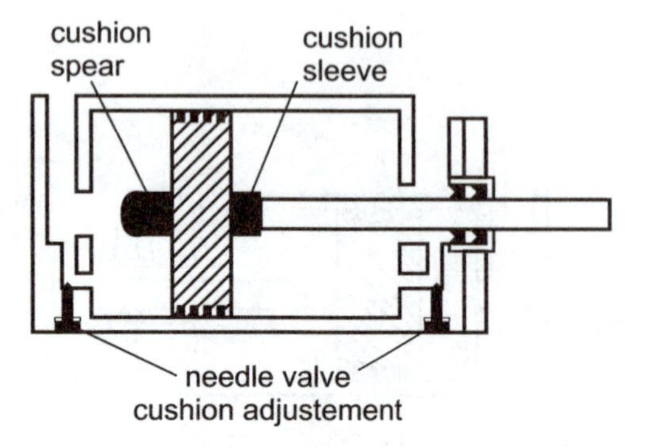

Figure 8-53

Piston rings are durable, but may exhibit some leakages under normal conditions. Lipseals and bidirectional seals offer a more positive seal, but may be less durable in some applications.

Rod gland seals come in several varieties. Some cylinders are equipped with a "V"-shaped or cup-shaped primary seal made of leather, polyurethane, nitrile or viton; and a wiper seal which prevents foreign materials from being drawn into the cylinder.

The seal material or compound should be verified to be compatible with the fluid media and operating conditions. One effective type of rod gland seal consists of a primary seal with serrated edges along its inside surface. The edges contact the rod continuously and scrape it clean of fluid. A second wiper seal, catches any fluid which may get by the primary seal and also wipes the rod of foreign material when the rod retracts.

Gland Drain

During the operation of the latter rod gland seal described above, any fluid which collects in the chamber between the primary and wiper seals is drawn back into the cylinder during the retraction stroke. But, on extremely long stroke cylinders (10 ft./3.05 m or more), there is a possibility that too much fluid may collect in the chamber causing a leaking rod gland. In these applications, or in any application where excessive amounts of fluid may collect in the gland, the rod gland should be externally drained.

Hydraulic Shock

When hydraulic working energy moving a cylinder's piston runs into a dead end (as at the end of a cylinder's stroke), the liquid inertia is changed into a concussion known as "hydraulic shock." If a substantial amount of working energy is stopped, the shock may damage the cylinder.

Cushions

To protect against excessive shock a cylinder can be equipped with cushions. Cushions slow down a cylinder's piston movement just before reaching the end of its stroke. Cushions can be applied at either or both ends of a cylinder.

What a Cushion Consists Of

A cushion consists of a needle valve flow control, a cushion spear attached to the cap end side of the piston and a cushion sleeve over the rod. These devices act like plugs at their own respective ends.

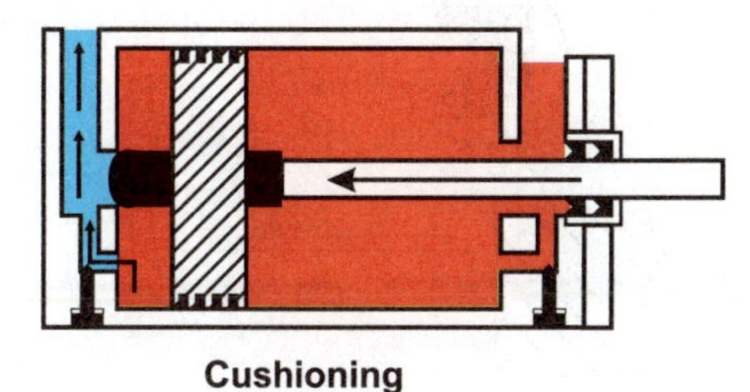

Cushioning

Figure 8-54

How a Cushion Works

As a cylinder piston approaches the end of its travel, the plug blocks the normal exit for a liquid and forces it to pass through the needle valve. At this point, some flow goes over the relief valve at the relief valve setting. The remaining liquid ahead of the cylinder piston is bled off through the needle valve and slows down the piston. The adjustment of the needle valve will determine the rate of deceleration.

In the reverse direction, flow bypasses the needle valve by means of a check valve within the cylinder (Not shown).

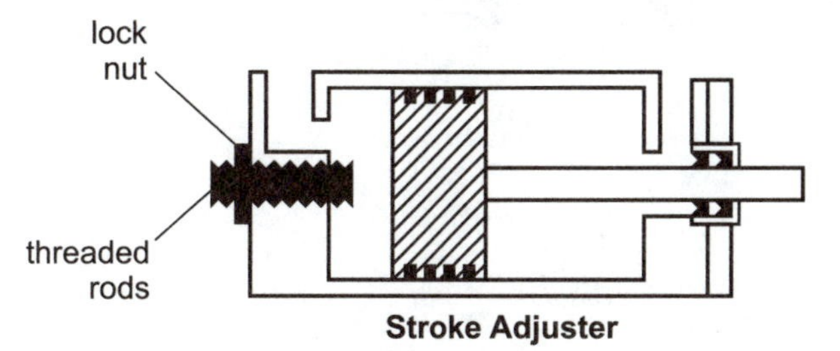

Stroke Adjuster

Figure 8-55

Stroke Adjusters

Sometimes the stroke length of a cylinder must be externally controlled. Periodic adjustment is accomplished with a threaded rod which can be screwed in or out of the cylinder cap. Each style of stroke adjuster must be examined for its effect on stopping, impact, shock, etc.

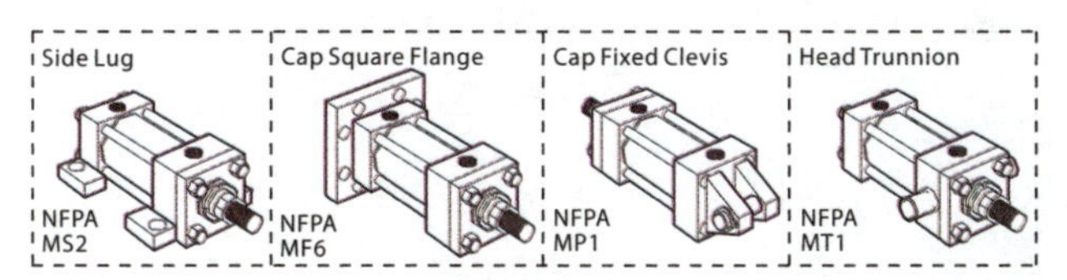

Figure 8-56

Cylinder Mounting Styles

Cylinders can be mounted in a variety of ways, among which are flange, trunnion, side lug and side tapped, clevis, tie rod, and bolt mounting. Centerline mount would be a very good choice to minimize leakage due to cylinder movement.

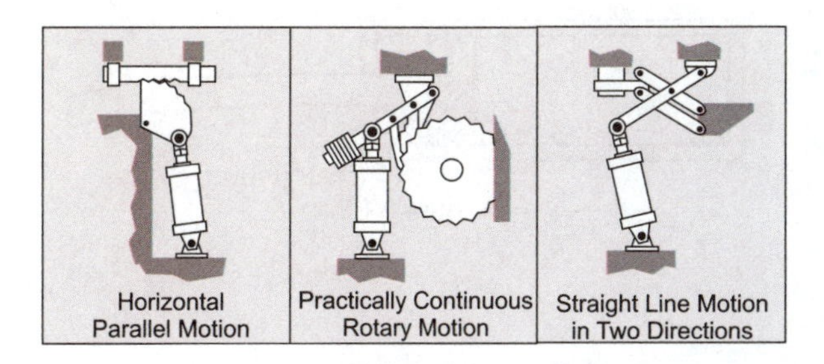

Figure 8-57

Mechanical Motions

Cylinders convert hydraulic working energy into straight line, or linear, mechanical motion. But, depending on the way in which they are attached to mechanical linkages, cylinders provide many different mechanical motions.

Types of Cylinder Loads

Cylinders can be used in an unlimited number of applications to move various types of loads. But, in general, a load which is pushed by a cylinder rod is termed a thrust load. A load which is pulled by a cylinder rod is called a tension load.

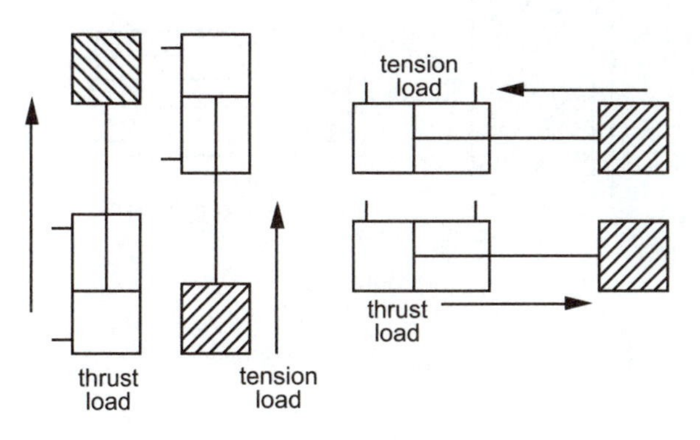

Figure 8-58

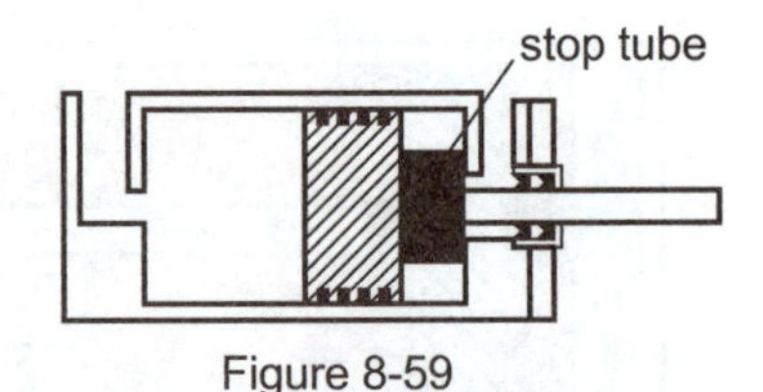

Figure 8-59

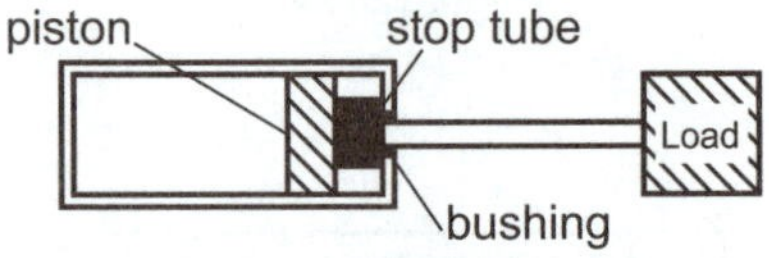

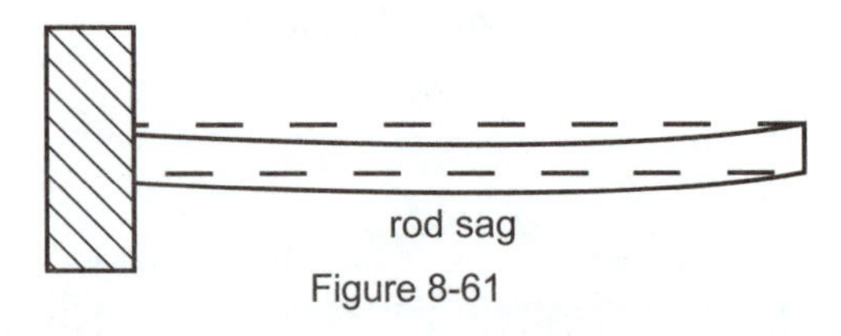

Figure 8-60

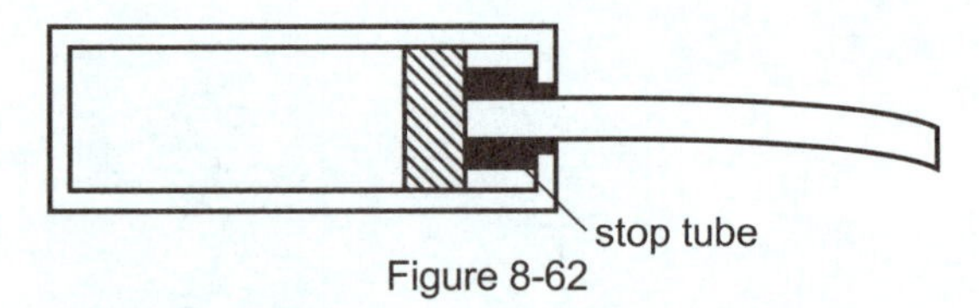

Figure 8-61

Figure 8-62

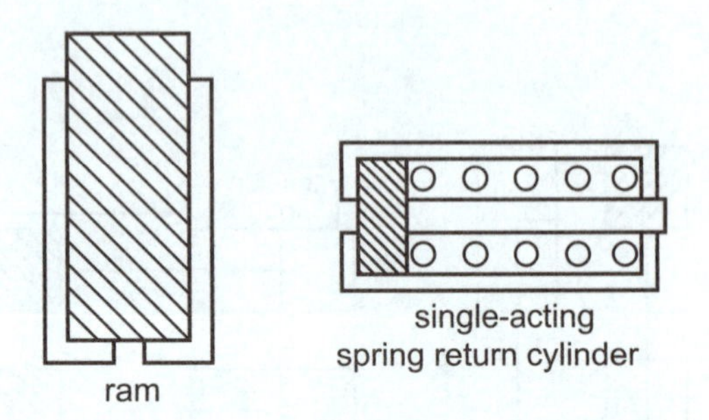

Figure 8-63

Stop Tube

A stop tube is a solid, metal collar which fits over the piston rod. A stop tube keeps the piston and rod gland bushing separated when a long-stroke cylinder is fully extended.

Since it is a bearing, a rod gland bushing is designed to take some loading when supporting the rod as it moves in and out of the cylinder.

Along with being a bearing, a rod gland bushing is also a fulcrum for the piston rod. If the load attached to the piston rod of a long-stroke cylinder is not rigidly guided, then at full extension, the rod will tend to teeter-totter or jackknife at the bushing causing excessive loading. A stop tube in effect protects the rod gland bushing by reducing the bearing load at full extension between both piston and bushing.

Believe it or not, but the heavy, steel rods of long-stroke cylinders sag just because of their weight. A 5/8" (1.59 cm) diameter piston rod weighs 1 lb. per foot (.4536 kg per .3040 m) and will sag over 1 in. (2.54 cm) at the center of a 10 ft. (3.05 m) span.

On long-stroke, horizontally mounted cylinders, undesirable bearing loads are generated at the rod gland bushing because of the rod weight when the rod is fully extended. On these cylinders, a stop tube is used to separate bushing and piston when the rod is extended. This reduces the load on the rod gland bushing.

Most cylinders do not need stop tubes. To determine when a stop tube is required, or what the length of a stop tube should be, consult the cylinder manufacturer's catalog.

Hydraulic Cylinder Types

Hydraulic cylinders can be of various types. Some common cylinders are described below and in later lesson material are seen as they are applied in a circuit.

Single rod cylinder: cylinder with a piston rod extending from one end.

Double rod cylinder: cylinder with a single piston and a piston rod extending from both ends.

Double-acting cylinder: cylinder in which fluid pressure is applied alternately to both sides of a cylinder piston to effect extension and retraction of a piston rod.

Telescoping cylinder: cylinder with nested multiple tubular rod segments which provide a long working stroke in a short retracted envelope.

Tandem cylinder: cylinder consisting of two or more cylinder bodies mounted in line with their piston rods connected to form a common piston rod; rod seals are installed between cylinder bodies to permit double acting operation of each.

Duplex cylinder: cylinder consisting of at least two cylinder bodies to permit double acting operation of each.

With single rod, double rod, double acting, tandem and duplex cylinders defined, let's now look at their operation in a system. We begin with a double acting, single rod cylinder.

Double-Acting Single Rod Cylinder Operation

A very common cylinder in industrial hydraulic applications is a double acting, single rod cylinder. Just as other hydraulic cylinders, this type cylinder is concerned with accepting gpm and psi and converting it into mechanical force and piston rod motion.

In the following material, it will be illustrated how rod speed of this type cylinder is determined by gpm, how mechanical force is affected by psi and how both of these elements are affected by major and minor piston areas.

Piston and Effective Piston Area

Piston and effective areas generally refer to a double acting, single rod cylinder. Piston major area indicates the piston area exposed to pressure at the cylinder cap side. Effective minor area (also called the annulus area) refers to the piston area exposed to pressure at the cylinder rod side. Since the rod covers a portion of the piston at this point, effective area is always less than piston area.

Illustrating Rod Speed While Extending

Rod speed of a cylinder is determined by how quickly the volume behind a piston can be filled with

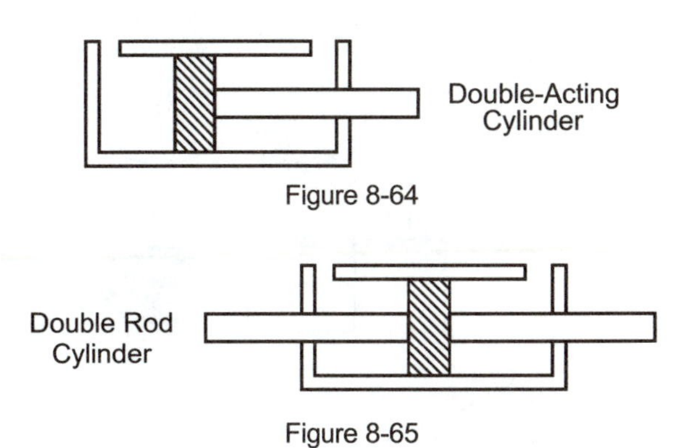

Figure 8-64

Figure 8-65

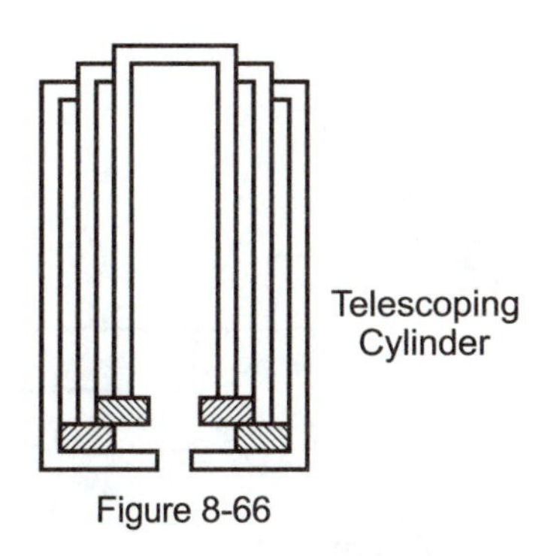

Figure 8-66

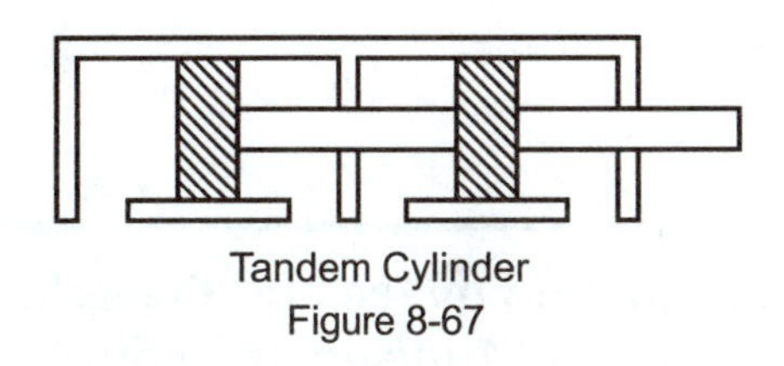

Tandem Cylinder
Figure 8-67

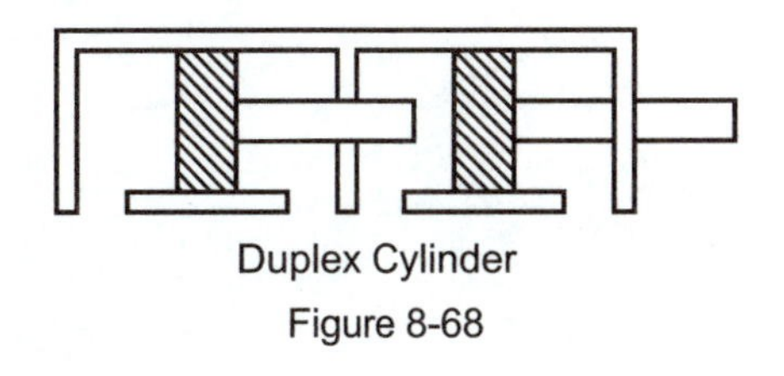

Duplex Cylinder
Figure 8-68

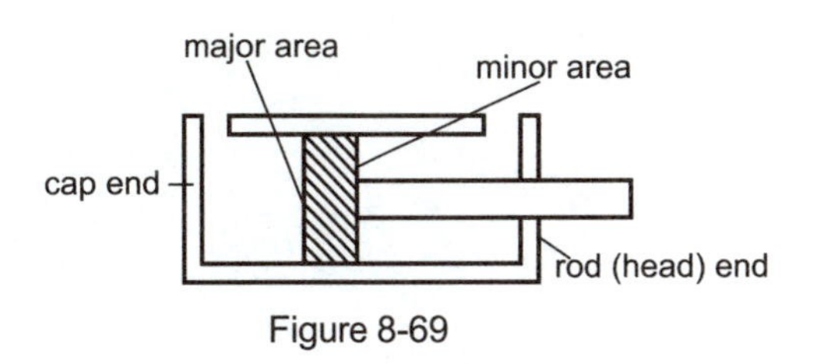

Figure 8-69

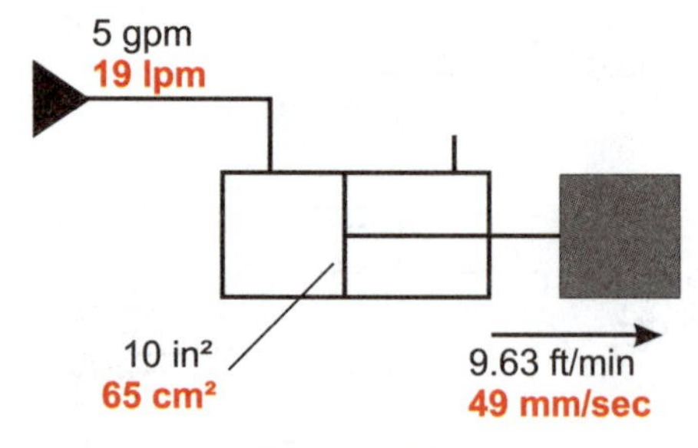

Figure 8-70

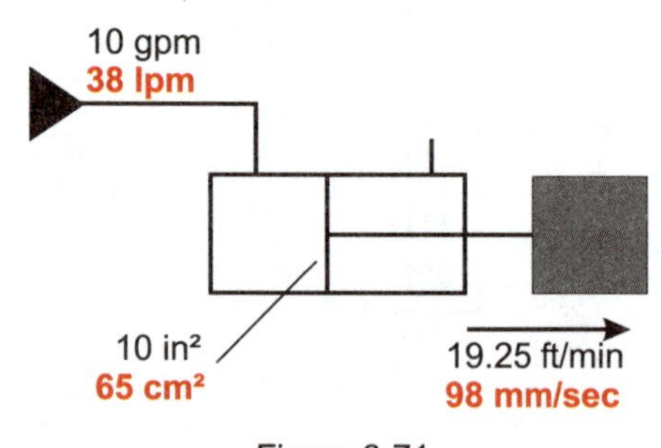

Figure 8-71

*** If calculating m/sec and the solution is less than .1 m/sec, the solution should be expressed in mm/sec.**

liquid. Rod speed is many times measured in units of feet per minute or (ft/min) or meters per minute (m/min). This is described by the expression:

$$\text{Rod speed (ft/min)} = \frac{\text{gpm} \times 19.25}{\text{piston area (in}^2\text{)}}$$

$$\text{* (m/sec)} = \frac{\text{l/min} \times .167}{\text{piston area (cm}^2\text{)}}$$

Assume that a cylinder with a 10 in² (64.5 cm²) piston area receives a flow rate of 5 gpm (18.95 l/min). Using the above expression, we find that the rod speed is 9.63 ft/min (49 mm/sec).

	(ft/min)	* (m/sec)
Rod speed	$= \frac{\text{gpm} \times 19.25}{\text{piston area (in}^2\text{)}}$	$= \frac{\text{l/min} \times .167}{\text{piston area (cm}^2\text{)}}$
	$= \frac{5 \times 19.25}{10}$	$= \frac{18.95 \times .167}{64.5}$
	$= \frac{96.25}{10}$	$= \frac{3.165}{64.5}$
	$= \frac{96.25}{10}$	$= \frac{3.165}{64.5}$
	= 9.63 ft/min	= 49mmm/sec

This means that 5 gpm (18.95 l/min) fills the volume behind the piston quickly enough to move piston and piston rod a distance of 9.63 ft. (49 mm) in one minute. If the same cylinder received twice the flow rate (10 gpm/37.9 l/min), rod speed would be doubled (19.25 ft/min or 97.8 mm/sec).

The more flow a cylinder receives, the more quickly it will fill with liquid and the faster it will extend.

Illustrating Discharge Flow While Extending

Flow entering the cap end of a double-acting, single rod cylinder determines the rate at which a cylinder piston rod will extend. While this is occurring, of course, flow also discharges from the rod side. Discharge flow is an important concern in many systems and can be calculated by using the same basic formula.

With the same cylinder as in the previous examples, assume that the rod has a cross sectional area of 2 in² (12.9 cm²) and that piston and piston rod extension speed is 9.63 ft/min (49 mm/sec) as calculated above. Using the same expression for rod speed, we find that rod speed and the minor area of the cylinder are known, but gpm (l/min) is not.

Rearranging the formula, we then solve for gpm (l/min) which is the discharge flow rate at the rod side.

(gpm)	(l/min)
$= \frac{9.63 \text{ ft/min} \times 8 \text{ in}^2}{19.25}$	$= \frac{49 \text{ mm/sec} \times 52\text{cm}^2}{.167}$
$= \frac{77.04}{19.25}$	$= \frac{2528.4}{.167}$
= 4.0 gal/min	= 15 l/min

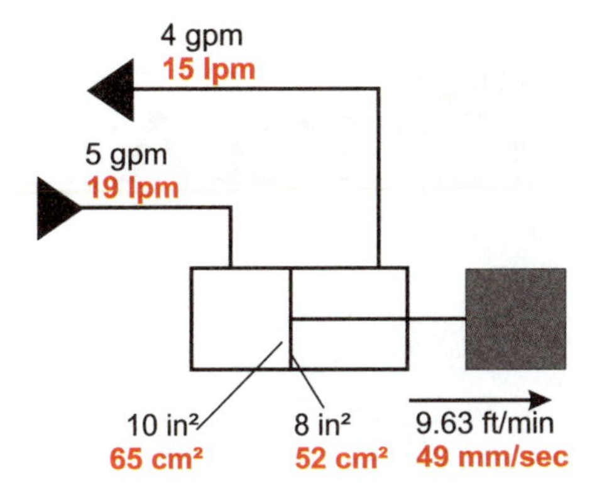

Figure 8-72

This indicates as 5 gpm (19 l/min) enters the cap end of the cylinder, 4 gpm (15 l/min) discharges back to tank from the cylinder rod side.

While extending, discharge from a single rod cylinder is always less than the flow rate entering the cap end.

Illustrating Rod Speed While Retracting

During retraction, when full pump flow is directed to the rod side of a single rod cylinder, a piston rod will retract faster than it extended. This can be calculated by using the expression for rod speed. Except in this instance, instead of using the major piston area, the minor area is used.

Assume that 10 gpm (38 l/min) enters the rod side of our cylinder during retraction:

(ft/min)	(m/sec)
Retracting rod speed $= \frac{\text{gpm} \times 19.25}{\text{minor piston area (in}^2)}$	$= \frac{\text{l/min} \times .167}{(\text{cm}^2)}$
$= \frac{10 \text{ gpm} \times 19.25}{8 \text{ in}^2}$	$= \frac{38 \times .167}{52 \text{ cm}^2}$
$= \frac{192.5}{8}$	$= \frac{6.35}{52}$
= 24.06 ft/min	= .12 m/sec

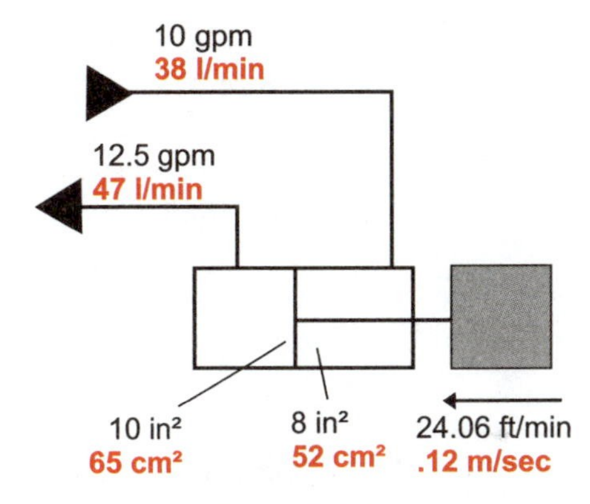

Figure 8-73

F = P x A

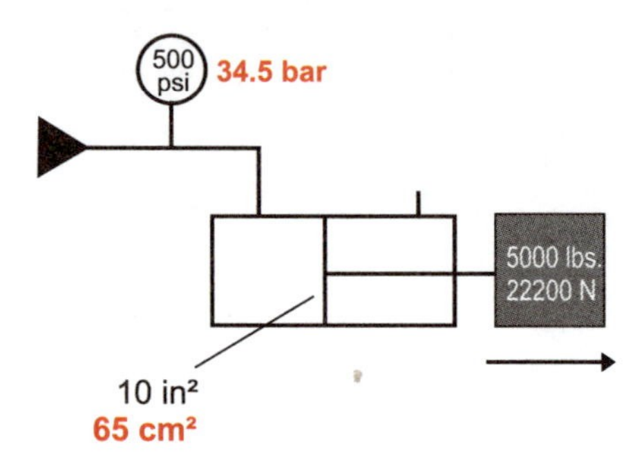

Figure 8-74

With flow to a cylinder remaining constant, double acting, single rod cylinders always retract faster than they extend.

Illustrating Discharge Flow While Retracting

During retraction, when full pump flow is directed to the rod side of a single rod cylinder, discharge flow will be greater than incoming flow. This can be calculated by using the same expression for gpm (l/min) used previously. In this instance, however, the major piston area is used.

Assume once again, that 10 gpm (38 l/min) enters the rod side of our cylinder during retraction.

$$\text{gpm} = \frac{\text{rod speed (ft/min) x major piston area (in}^2\text{)}}{19.25}$$

$$= \frac{24.06 \text{ ft/min x } 10 \text{ in}^2}{19.25}$$

$$= \frac{24.06}{19.25}$$

$$= 12.5 \text{ gpm discharge}$$

$$\text{l/min} = \frac{\text{rod speed (m/sec) x major piston area (cm}^2\text{)}}{.167}$$

$$= \frac{.12 \text{ m/sec x } 65 \text{ cm}^2}{.167}$$

$$= \frac{7.74}{.167}$$

$$= 46 \text{ l/min discharge}$$

Anytime a double acting, single rod cylinder is retracting, more flow is discharging from the cap end than entering the rod end. Therefore, pump flow is not necessarily the maximum flow rate in a system.

When the return side of a system is designed, consideration is given to the discharge flow from retracting single rod cylinders. This is one reason why piping, valving and filters at the return side are sized larger than their counterparts in the main system.

If an occasion arises where return side parts must be disassembled, ensure that a smaller sized component is not substituted for an original part upon reassembly.

Cylinder Force While Extending

While extending, the mechanical force developed by a cylinder is the result of hydraulic pressure acting on the cap end of the cylinder piston. This is expressed by the formula:

Force (lbs.) = Pressure (psi) x Area (in²)

(Newtons) = Pressure (bar) x Area (m²)

For example, if a load offers a resistance to move of 5000 lbs. and the area of the cylinder piston is 10 in², then a hydraulic pressure of 500 psi is required to equal the load as calculated by the formula:

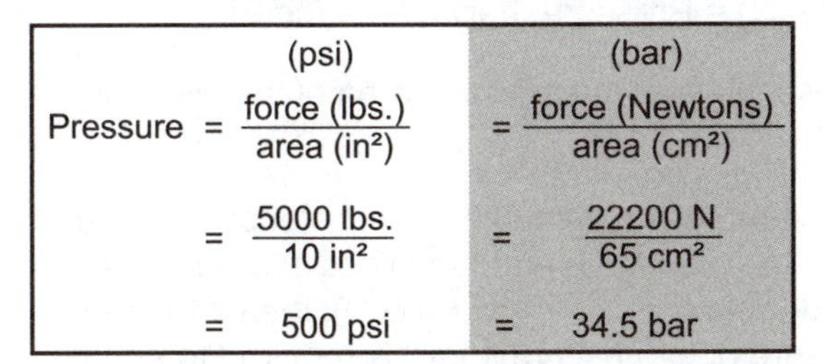

	(psi)	(bar)
Pressure	$= \frac{\text{force (lbs.)}}{\text{area (in}^2\text{)}}$	$= \frac{\text{force (Newtons)}}{\text{area (cm}^2\text{)}}$
	$= \frac{5000 \text{ lbs.}}{10 \text{ in}^2}$	$= \frac{22200 \text{ N}}{65 \text{ cm}^2}$
	= 500 psi	= 34.5 bar

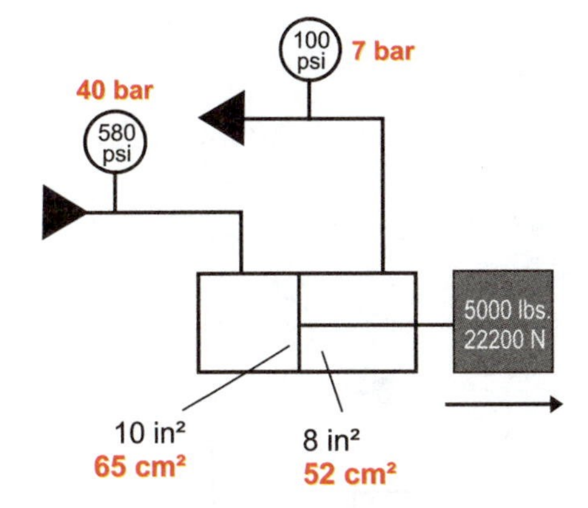

Figure 8-75

However, the assumption was made that pressure was zero on the other side of the piston. Even though the piston minor area is drained to tank while extending, tank line pressure, or back pressure, can be as high as 100 psi (7 bar) in some systems.

In our example, assume a back pressure of 100 psi (7 bar) during extension. This generates a force on the effective area of the piston to retract piston and piston rod. This force, together with the resistance offered by the load, must be overcome before the cylinder will extend at full speed. If the effective area of the piston were 8 in² (52 cm²) and the load offered 5000 lbs. (22200 N), then the force which must be overcome is 5800 lbs. (25752 N), which is made up of 5000 lbs. (22200 N) of the load and 800 lbs. (3552 N) (100 psi x 8 in²) generated by the back pressure. Pressure required at the cylinder piston is, therefore, 580 psi (40 bar).

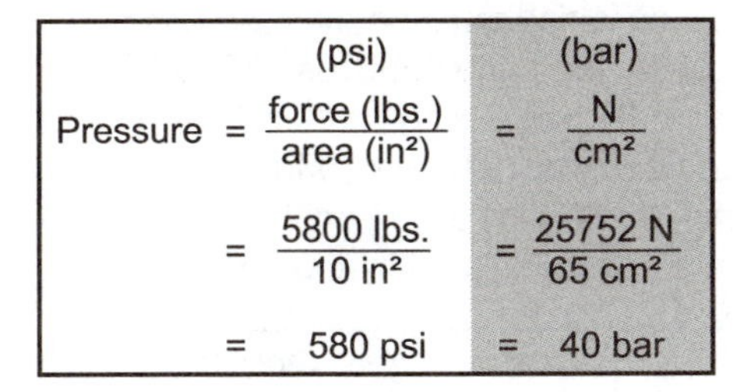

	(psi)	(bar)
Pressure	$= \frac{\text{force (lbs.)}}{\text{area (in}^2\text{)}}$	$= \frac{\text{N}}{\text{cm}^2}$
	$= \frac{5800 \text{ lbs.}}{10 \text{ in}^2}$	$= \frac{25752 \text{ N}}{65 \text{ cm}^2}$
	= 580 psi	= 40 bar

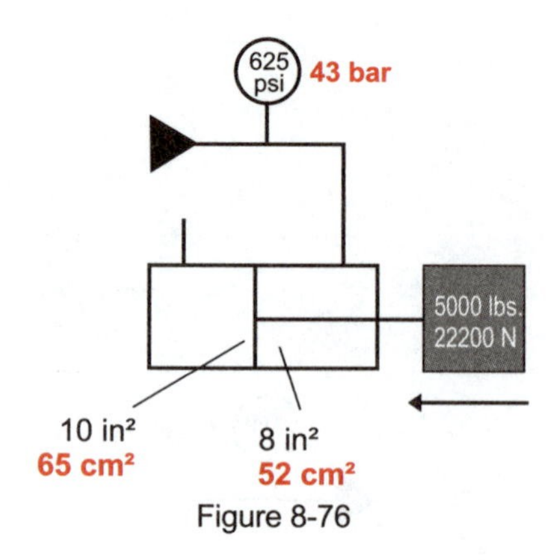

Figure 8-76

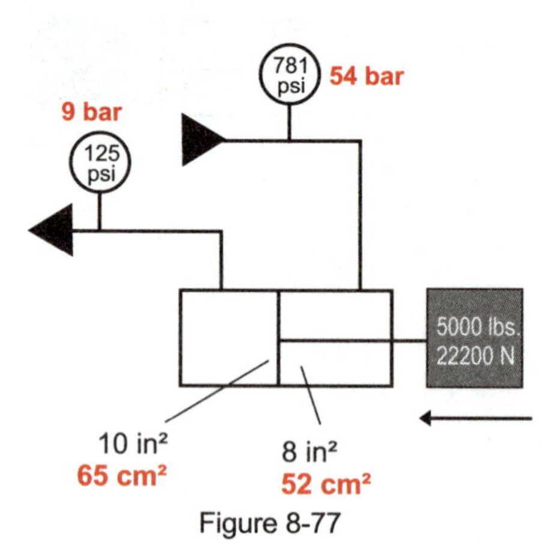

Figure 8-77

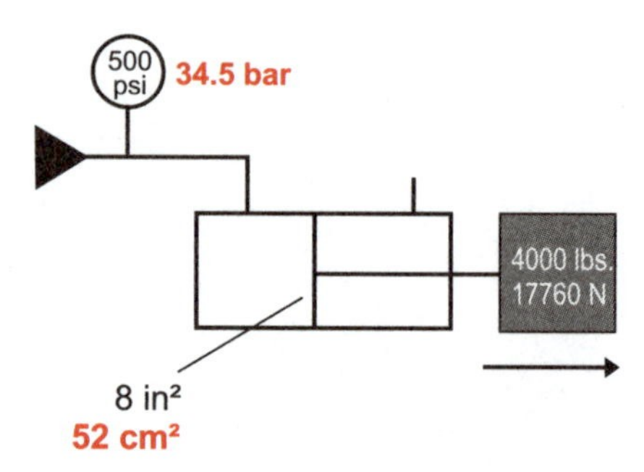

Figure 8-78

Anytime a load is to be extended by a cylinder through a distance at a certain speed, pressure required at a cylinder piston is used to equal load resistances as well as liquid resistances flowing back to tank.

Cylinder Force While Retracting

While retracting, the mechanical force developed by a cylinder is the result of hydraulic pressure acting on the effective area. This is also expressed by the formula:

Force (lbs.) = Pressure (psi) x Area (in^2)

Force (N) = Pressure (bar) x Area (cm2)

In this instance, the effective area of the piston is used.

For example, if a load offers a resistance to move of 5000 lbs. (22200 N) and the effective area of the cylinder piston is 8 in^2 (51.60 cm^2), then a hydraulic pressure of 625 psi (43.1 bar) is required to equal the load as calculated by the formula:

	(psi)	(bar)
Pressure	$= \frac{\text{force (lbs.)}}{\text{area (in}^2\text{)}}$	$= \frac{N}{cm^2}$
	$= \frac{5000 \text{ lbs.}}{8 \text{ in}^2}$	$= \frac{22000 \text{ N}}{51.6 \text{ cm}^2}$
	= 625 psi	= 43.1 bar

This also assumes that backpressure is not present at the cap end. But, backpressure is even more pronounced while retracting than extending.

Assume once again that the load offers a resistance to move of 5000 lbs. (22200 N) and that backpressure is 125 psi (8.62 bar). This is a higher backpressure than seen during extension because discharge flow during retraction is usually more than during extension.

The total resistance which must be overcome is 5000 lbs. (22200 N) of the load and 1250 lbs. (5550 N) as a result of 125 psi (8.62 bar) acting on 10 in^2 (64.5 cm^2). This is a total of 6250 lbs. (27750 N) which is overcome by 781 psi (53.86 bar) acting on the 8 in^2 (51.6 cm^2) effective area as calculated by the formula:

	(psi)	(bar)
Pressure	$= \frac{\text{force (lbs.)}}{\text{area (in}^2\text{)}}$	$= \frac{N}{cm^2}$
	$= \frac{6250 \text{ lbs.}}{8 \text{ in}^2}$	$= \frac{27750 \text{ N}}{51.6 \text{ cm}^2}$
	= 781 psi	= 53.86 bar

While retracting a single rod cylinder, back pressure at the cap end will usually be higher than while extending.

We have been shown how cylinder rod speed is affected by gpm (l/min) and piston area, and how cylinder force is affected by psi (bar) and piston area. In the next section, we see what is done when one or more of these elements is restricted.

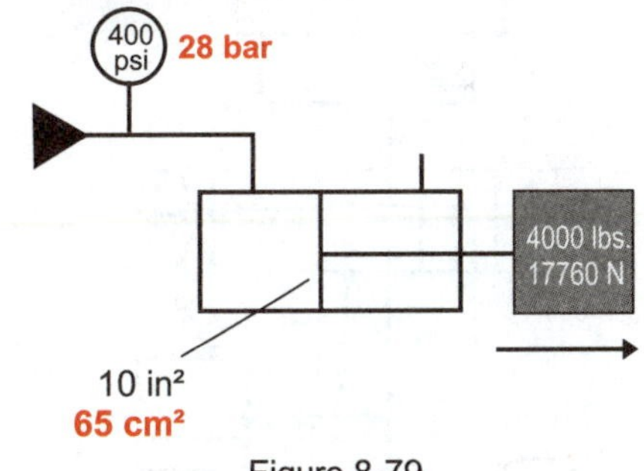

Figure 8-79

Affecting Cylinder Force

As illustrated, the force generated by a cylinder is a function of fluid pressure acting on the cylinder piston area. If a particular cylinder is required to generate more output force than it is currently developing, frequently fluid pressure is increased to the appropriate level.

In some cases, pressure generated by pump/electric motor is limited to a low value. In order to increase output force in this instance, cylinder size can be increased. This allows pump/electric motor to develop less pressure to achieve the same mechanical output force.

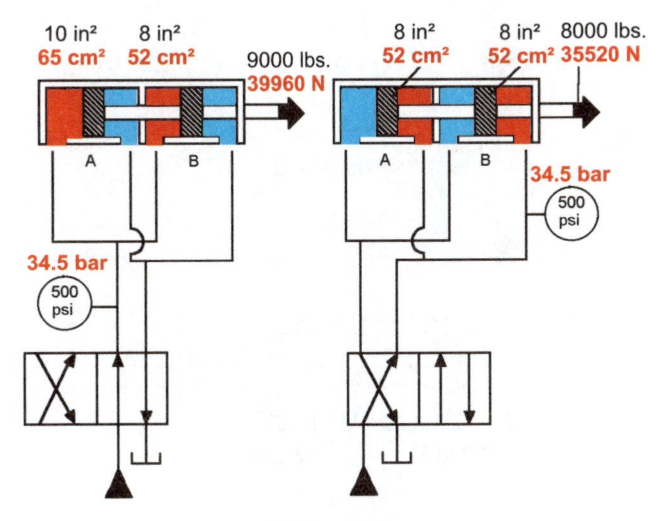

Figure 8-80

In some situations, both system pressure and cylinder bore size are not allowed to be increased. Machine dimensions are so fixed that a certain bore size cylinder can only be used. In these applications, a tandem cylinder can be employed.

Tandem Cylinder Circuit

A tandem cylinder consists of two or more cylinder bodies mounted in line. Piston rods are connected to form a common piston rod, and rod seals are installed between cylinders to permit double acting operation of each. A tandem cylinder gives increased output force when cylinder bore size is limited, but not its overall length.

Assume that a cylinder with a piston area of 10 in² (64.5 cm²) is the largest size which can be physically mounted on a machine. Yet, the maximum working pressure available to equal load resistances is only 500 psi (34.48 bar). This must move a load which offers a resistance of 9000 lbs. (39960 N).

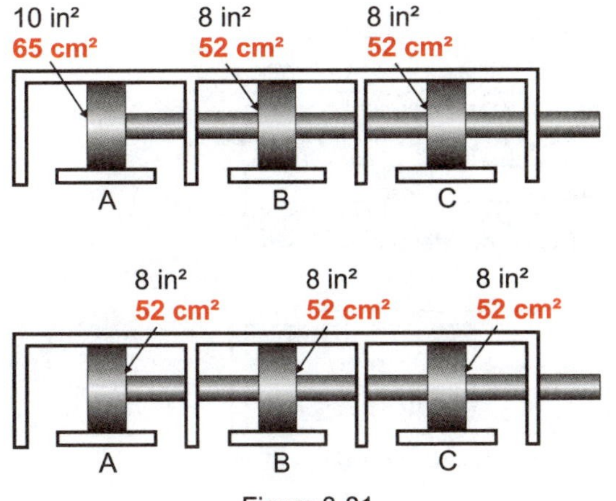

Figure 8-81

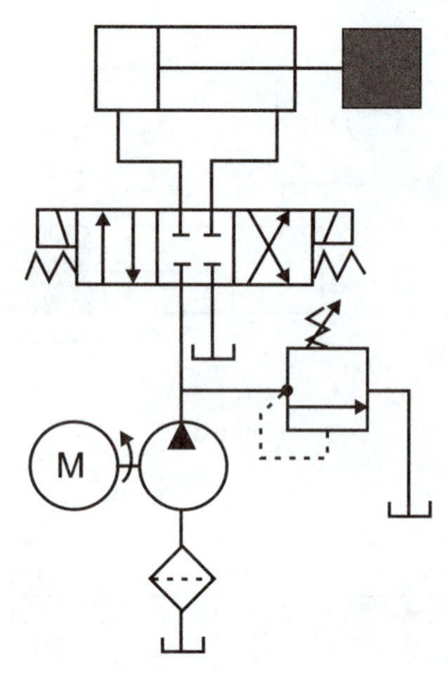

Figure 8-82

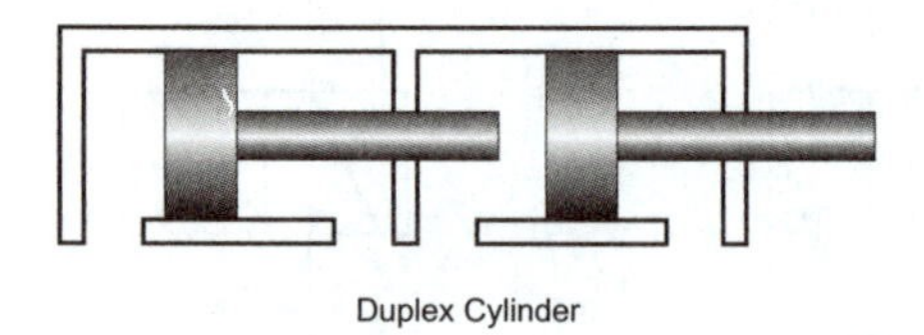

Duplex Cylinder

Figure 8-83

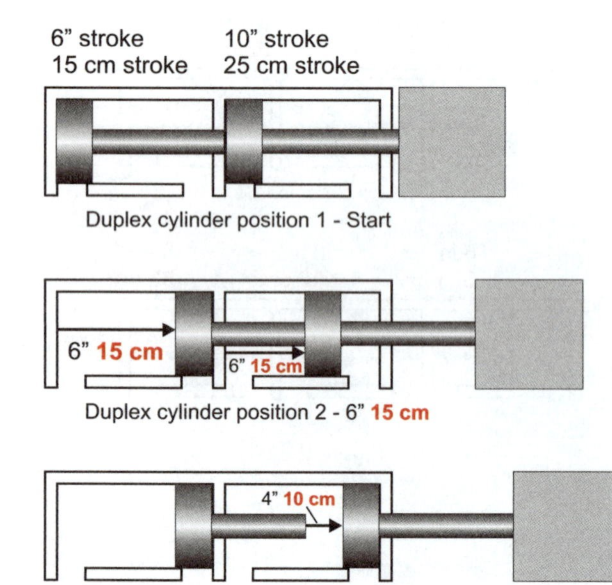

Figure 8-84

In this application, a tandem cylinder is used which is made up of two cylinders with a major area of 10 in² (64.5 cm²) and a minor area of 8 in² (51.6 cm²). However, since one cylinder has a rod connected to both sides, 10 in² (64.5 cm²) of piston A plus 8 in² (51.6 cm²) of piston B are exposed to fluid pressure during extension. During retraction, 16 in² (103.2 cm²), the total minor areas of the pistons, are exposed to fluid pressure.

During extension with 500 psi (34.48 bar) acting on 10 in² (64.45 cm²) of Piston A and 8 in² of Piston B, 9000 lbs. (39960 N) of force is developed. In the retraction stroke, 500 psi (34.48 bar) acts on the total 16 in² (103.2 cm²) minor area resulting in a force of 8000 lbs. (35520 N). With 500 psi (34.48 bar), this cylinder can develop a maximum force of 9000 lbs. (39960 N) extending and 8000 lbs. (35520 N) retracting.

If another cylinder section were added to the group, the tandem set would have 26 in² (167.7 cm²) of area exposed during extension and 24 in² (154.8 cm²) during retraction. With a maximum working pressure capability of 500 psi (34.48 bar) in both directions, 13000 lbs. (57720 N) could be developed during extension and 12000 lbs. (53280 N) during retraction.

By using cylinder pistons in tandem, increased output force can be achieved when cylinder bore size and maximum working pressure are limited.

Achieving Mechanical Positions

Besides transforming hydraulic power into linear mechanical power, cylinders are also used to achieve a mechanical position.

After a cylinder has moved a load through its stroke, a definite mechanical position has been achieved. As long as cylinder mounting and piston rod are not altered, a load can be depended on to reach this position time after time.

In some cases, it is desirable to have a load stop in an intermediate position. This can be accomplished by centering a closed center directional valve as the load reaches a definite point along the cylinder stroke.

In some applications, a simple method of providing an accurate repeatable position is to use a duplex cylinder.

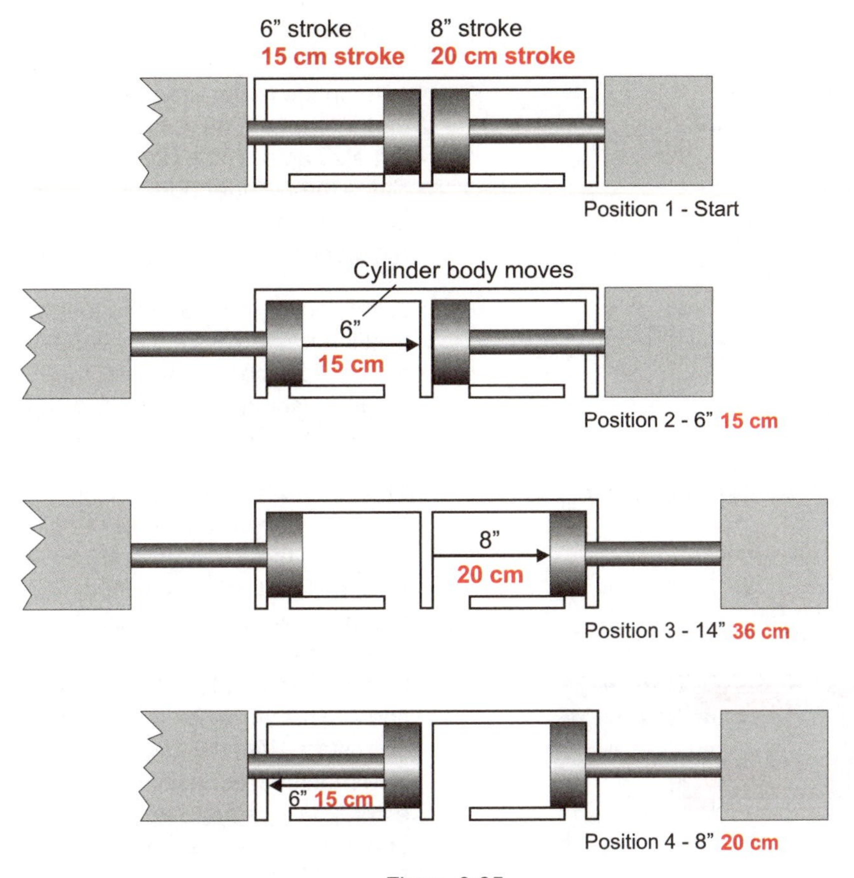

Figure 8-85

Duplex Cylinder Circuit

Whereas a tandem cylinder consists of two or more cylinder bodies of equal stroke, a duplex cylinder is made up of at least two cylinder bodies with different strokes. Cylinder bodies are joined together, but piston rods are not connected. Rod seals are installed between cylinder bodies to permit double acting operation of each.

A duplex cylinder gives three possible mechanical positions for a load. One position is with both piston rods retracted. Another position is with the shorter rod extended its length. Extending the short piston rod in turn pushes the rod of the long stroke body and the load an equal distance. A third position is reached when the long piston rod is stroked the remainder of its length.

In the illustration, a duplex cylinder consists of 6 in. (15.2 cm) and 10 in. (25.4 cm) stroke cylinder bodies. Position 1 for the load is with both rods fully retracted. As a directional valve connected to the 6 in. (15.2 cm) stroke cylinder body is shifted, the piston rod extends its length moving the 10 in. (25.4 cm) stroke piston rod and the load 6 in. (15.2 cm) to position 2.

After the work operation has been performed at position 2, a directional valve connected to the 10 in. (25.4 cm) stroke cylinder body is shifted. This action extends the rod its remaining 4 in. (10.16 cm) achieving position 3 at 10 in. (25.4 cm).

When the directional valves are shifted to retract, both rods return to the start position.

The three mechanical positions achieved with this particular duplex cylinder are start, 6 in. (15.2 cm) and 10 in. (25.4 cm).

A duplex cylinder can also be arranged so that cylinder bodies are connected at their cap ends. In this case, one piston rod is attached to a machine member which remains stationary. With this set up, four positions can be achieved — three extending and one retracting.

A duplex cylinder with cylinder bodies connected at their cap ends is illustrated. One piston rod is mechanically attached to a machine member which does not move. One cylinder body has a stroke of 6 in. (15.2 cm); the other has an 8 in. (20.32 cm) stroke.

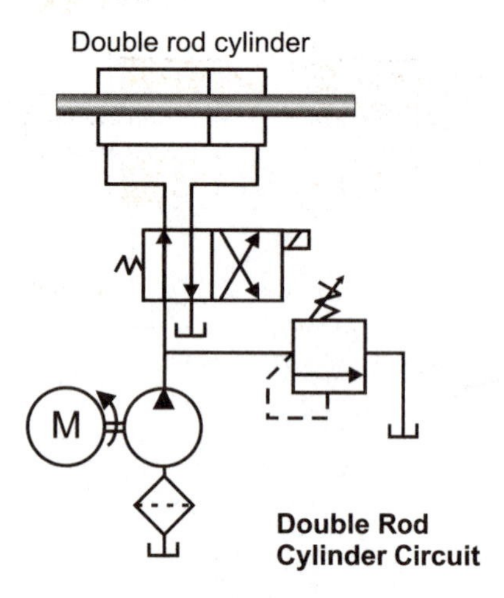

Figure 8-86

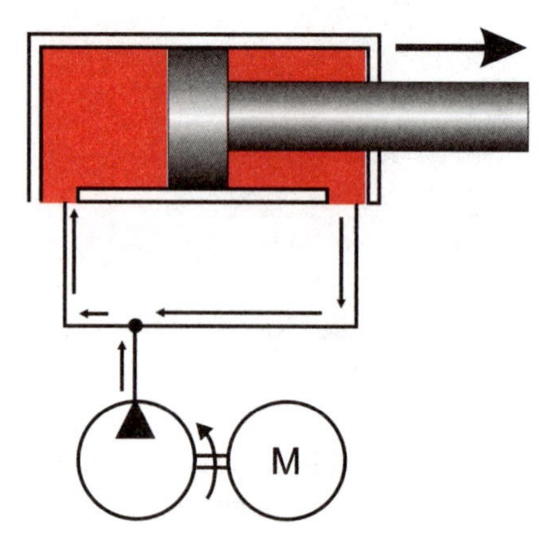

Figure 8-87

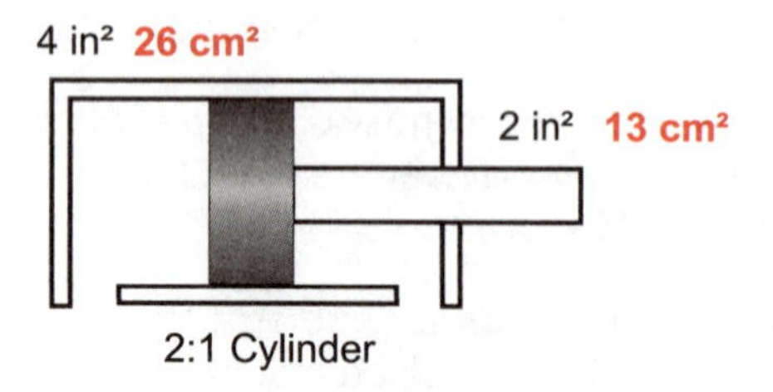

Figure 8-88

With both piston rods fully retracted, position 1 is achieved. As a directional valve connected to the 6 in. (15.2 cm) stroke body is shifted, fluid pressure acts on the major area of its piston; but, the piston cannot move. The same pressure acting on the piston also acts on the cap end of the cylinder body. Since the cylinder body is not attached to anything, both cylinder bodies, along with the load, move out a distance of 6 in. (15.2 cm). This is position 2.

At this point, the rod of the long stroke cylinder body is still retracted. As a directional valve connected to the cylinder body is shifted, the rod extends its length pushing the load out an additional 8 in. (20.32 cm). Position 3 is therefore 14 in. (35.56 cm).

This duplex cylinder has achieved three positions during its extension stroke. Position 4 is realized by retracting the 6 in. (15.2 cm) stroke cylinder first. As its directional valve is shifted, fluid pressure acts on the minor area of its piston. Since it is stationary, pressure acts on the rod end of the body pulling both bodies and load back 6 in. (15.2 cm) to position 4. This would be a distance of 8 in. (20.32 cm) from start. With this duplex arrangement, three positive mechanical positions are achieved during extension — start, 6 in. (15.2 cm), 14 in. (35.56 cm); and, one position is achieved during retraction — 8 in. (20.32 cm).

Applications of this type require the use of hoses as fluid conductors since cylinder bodies move.

Affecting Cylinder Speed

The speed at which a cylinder extends and retracts is frequently of great concern. As was illustrated, this is a function of how quickly the volume behind the cylinder piston is filled with liquid.

In the following section some typical attempts used to affect cylinder speed will be shown.

Double Rod Cylinder Circuit

As shown previously, a double acting, single rod cylinder retracts faster than it extends. In some applications, it is required that a cylinder extend and retract at the same speed. One means of accomplishing this is with a double rod cylinder.

Since a double rod cylinder usually has the same diameter rod on both sides of the piston, piston areas exposed to system flow are equal. With the rate of flow to each side remaining constant, rod speed is the same whether extending or retracting.

Regeneration with a 2:1 Cylinder

In some systems, a cylinder's speed is increased by taking the discharge flow from the rod end of a cylinder and adding it to the flow to the cylinder's cap end. If a 2:1 cylinder is used in the system, the cylinder's speed will be the same whether extending or retracting.

A 2:1 cylinder has a rod with a cross- sectional area equal to one half of the piston area. In other words, the rod side of the piston has one half the area exposed to pressure as the cap end side. (In actual practice the rod area is not precisely half the piston area. However, we will consider it so in order to facilitate our calculations.)

With the cylinder in a circuit, flow and pressure are directed to both sides of the piston at the same time. It may appear that the cylinder would be hydraulically locked. But, the difference in piston areas exposed to pressure results in a larger force being generated on the major piston area to extend the rod. With 20 psi (1.4 bar) on both sides of the piston, 80 lbs. (355.2 N) of force would be generated to extend the cylinder rod; and 40 lbs. (177.6 N) of force would be generated to retract the rod. This is a 40 lb. (177.6 N) differential in favor of extending the rod. The rod does extend. As the piston moves out, fluid which is displaced from the rod end switches position to the other side of the piston. This means pump flow is not required to fill the total volume behind the piston. Pump flow only has to fill behind an area equal to the cross sectional area of the rod.

With a 2:1 cylinder, this means the cylinder will extend twice as fast as normal with pump remaining the same. For example, if our 2:1 cylinder received 5 gpm (18.95 l/min) at its cap end, the following expression could be used to calculate rod speed:

	(ft/min)	(m/sec)
Rod speed =	$\frac{\text{gpm x 19.25)}}{\text{rod area (in}^2\text{)}}$	= $\frac{\text{l/min x .167}}{\text{cm}^2}$
=	$\frac{5 \times 19.25}{2}$	= $\frac{18.95 \times .167}{12.9}$
=	$\frac{96.25}{2}$	= $\frac{3.165}{12.9}$
	= 48.13 ft/min	= .24 m/sec

Regeneration can only occur during rod extension.

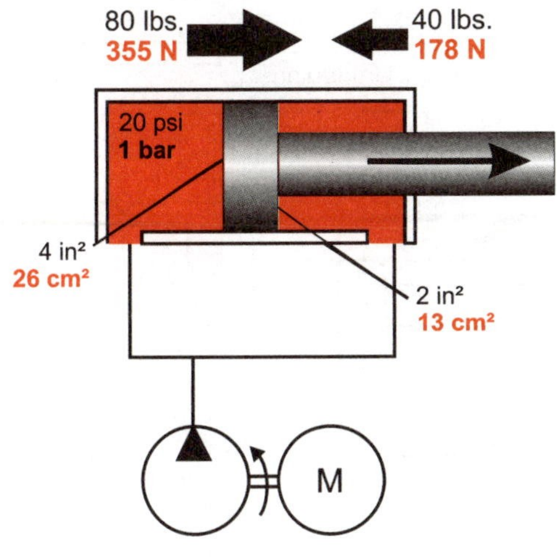

Figure 8-89

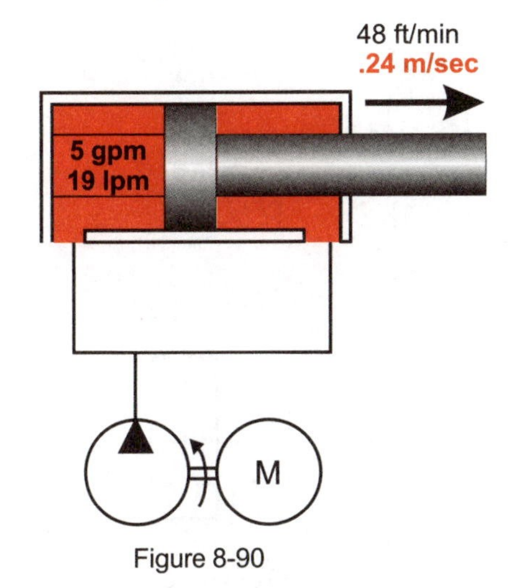

Figure 8-90

F = P x A

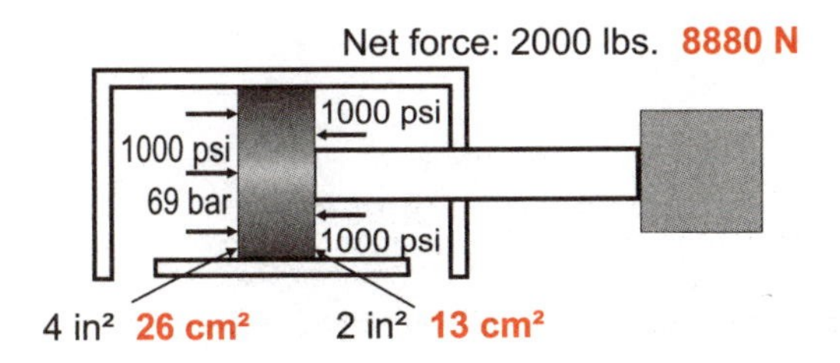

Figure 8-91

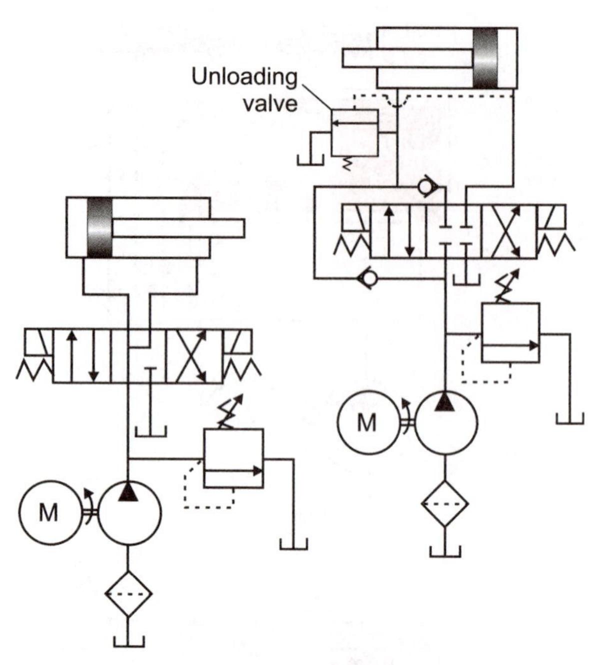

Figure 8-92

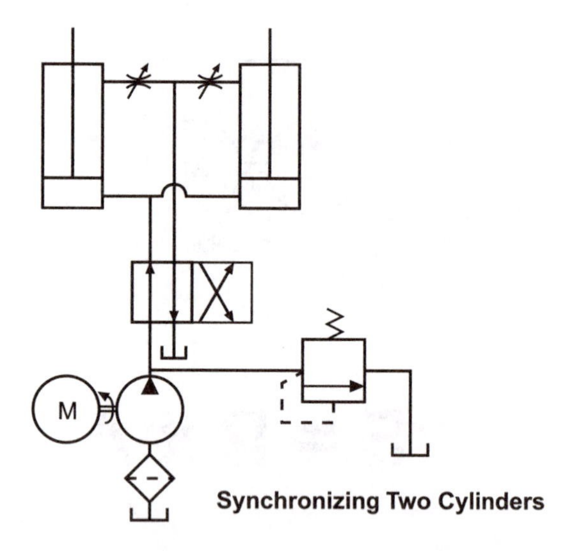

Figure 8-93

A characteristic of a 2:1 cylinder in a regenerative circuit is that rod speeds extending and retracting are basically the same.

To retract the cylinder rod, the directional valve is shifted. The cap end of the cylinder is drained to tank. All pump flow and pressure is directed to the rod end side. Since the pump is filling the same volume as at the cap end side (half cap end volume), the rod retracts at the same speed.

Cylinder Force During Regeneration

A disadvantage of regeneration is that output force is reduced. Since fluid pressure is the same on both sides, the effective area on which the force is generated is the cross-sectional rod area.

In our example, as the cylinder contacted the work load, assume the pressure climbed to a relief valve setting of 1000 psi (68.97 bar). This results in 1000 psi (68.97 bar) on 4 in² (25.81 cm²) equalling 4000 lbs (17760 N). on the major piston area. At the rod side, 1000 psi (68.97 bar) acting on the 2 in² (12.9 cm²) minor area develops 2000 lbs. (8880 N) This is a net force of 2000 lbs. (8880 N) to extend the rod.

Net force in regenerative circuits is usually determined by using the cross-sectional area of the rod in calculations. Since areas on either side of the piston are balanced except for the rod area, net force is a result of pressure acting on a piston area equal to the rod cross-sectional area. In our example, rod area equals 2 in² (12.9 cm2). 1000 psi (68.97 bar) acting on 2 in² (12.9 cm2) results in a net force of 2000 lbs. (8880 N) to extend the rod as calculated previously.

In calculations for speed and force during regeneration, the cross-sectional area of the rod is used, not the major piston area.

Connecting a cylinder in a regenerative circuit results in a faster rod speed while extending, but output force is reduced. Cylinder force is sacrificed for rod speed. And, with a 2:1 cylinder in a regenerative circuit, rod speeds extending and retracting are basically the same.

Sample Regenerative Circuits

Since a force disadvantage exists while a cylinder regenerates, regeneration is frequently employed only to extend the rod to the work load. When work is to be done, the rod side of the cylinder is drained so that full force can be realized.

Illustrated are two common examples of regenerative circuits in which the rod side of the cylinder can be drained when necessary. In one circuit, this is accomplished through the center position of a directional valve. The other circuit employs an unloading valve.

Now that we have seen how cylinders act in a circuit, in the next section we see how they are affected by wear.

Synchronizing Two Cylinders

One of the most difficult, if not impossible, things to accomplish in hydraulic systems is synchronizing the movement of two cylinders, even when using the most sophisticated types of flow control valves. Typical values of synchronization range from 1/8" to 1/16" (3 to 1.5 mm) depending upon cylinder stroke.

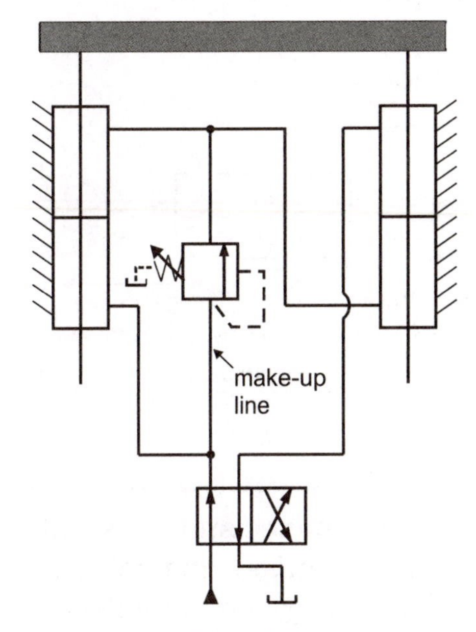

Figure 8-94

Even after the cylinders have been synchronized to within usable limits, in a relatively short period of time they will be out of synchronization because of the different wear characteristics of the cylinders and slightly different reactions of the flow control valves to the same set of conditions.

In order to achieve a more positive control of speed, some systems are designed so that the discharge flow from one cylinder is used as the input flow to another cylinder. Systems of this nature may still not be in perfect synchronization because of leakage. A characteristic of these circuits is that they are equipped with makeup and replenishing lines for the piping between cylinders.

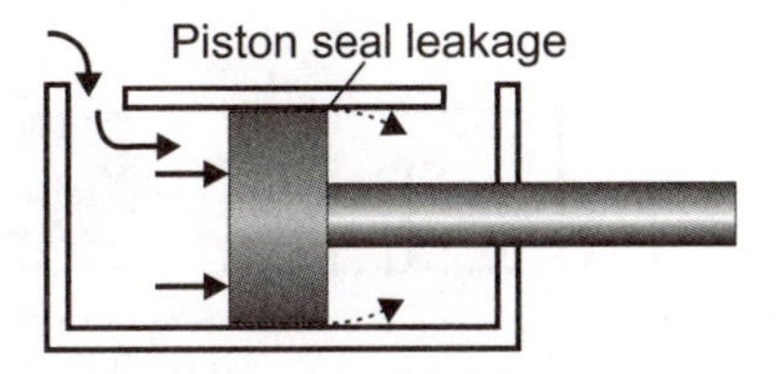

Figure 8-95

NOTE: When two or more cylinders must stroke together, it is recommended that their piston rods be mechanically connected to one another. The connection must be rigid and could be a strong beam or the load itself.

Piston Seal Leakage

As a cylinder operates in a system, cylinder seals wear resulting in leakage at the piston rod and across the cylinder piston. Leakage at the rod seal results in a housekeeping problem and can readily be detected. Seal leakage across the cylinder piston is not as easily determined.

In the next section, we find how piston leakage causes rod speed to decrease and may even cause intensification in some cases. We then learn what checks can be made to determine if a piston seal is leaking excessively.

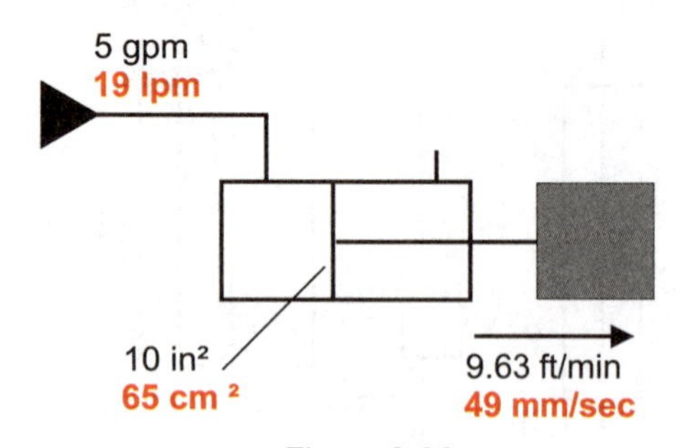

Figure 8-96

*** If calculating m/sec and the solution is less than .1 m/sec, the solution should be expressed in mm/sec.**

Piston Leakage Affects Rod Speed

Piston seals are commonly cast iron rings or a resilient, synthetic compound. Leakage past piston rings or worn lip seals can reduce cylinder speed. However, a noticeable reduction in cylinder rod speed would require .5 gpm (1.89 l/min) or more leakage. At this point, the cylinder probably has extensive internal damage. In a clamping application, leakage past one piston is not too much of a problem. However, when there are many such cylinders on a machine, the clamping pressure may not be there because all the pump flow is leaking across the many cylinder pistons.

Rod speed of a cylinder is determined by how quickly pump flow can fill the volume behind a cylinder piston. This is the case whether the rod is extending or retracting. A cylinder piston with an excessively worn seal allows fluid to bypass. This fluid does not fill a volume behind a cylinder piston and therefore does not contribute to rod speed.

In the illustration, a cylinder with a major area of 10 in² (64.5 cm²) receives 5 gpm (18.95 l/min) at its cap end. Rod speed extending is a function of 5 gpm (18.95 l/min) filling the volume behind 10 in² (64.5 cm²) and is calculated by the following:

	(ft/min)	(m/sec)
Rod speed	$= \frac{\text{gpm x 19.25)}}{\text{rod area (in}^2\text{)}}$	$= \frac{\text{l/min x .167}}{\text{cm}^2}$
	$= \frac{5 \text{ x } 19.25}{2}$	$= \frac{18.95 \text{ x .167}}{12.9}$
	$= \frac{96.25}{2}$	$= \frac{3.165}{12.9}$
	= 48.13 ft/min	= .24 m/sec

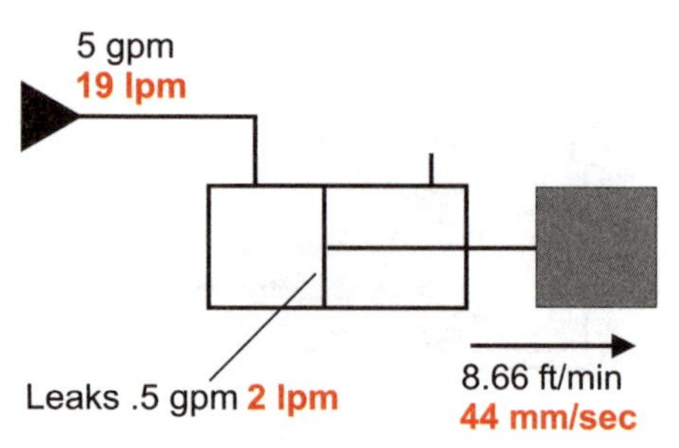

Figure 8-97

	(ft/min)	* (m/sec)
Rod speed	$= \frac{\text{gpm x 19.25)}}{\text{piston area (in}^2\text{)}}$	$= \frac{\text{l/min x .167}}{\text{cm}^2}$
	$= \frac{4.5 \text{ x } 19.25}{10}$	$= \frac{17.05 \text{ x .167}}{64.5 \text{ cm}^2}$
	$= \frac{86.63}{10}$	$= \frac{2.847}{64.5 \text{ cm}^2}$
	= 8.66 ft/min	= 44mm/sec

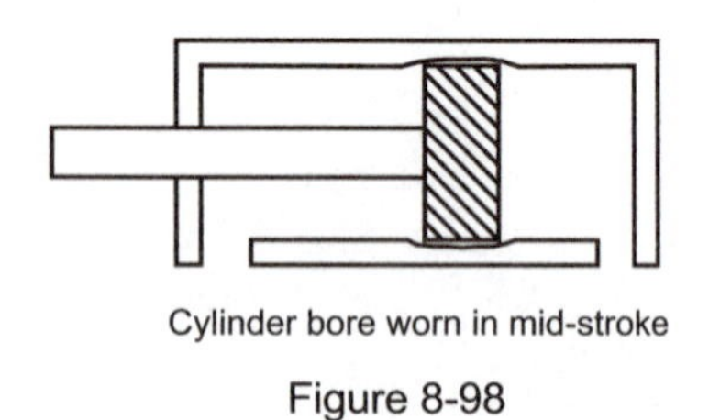

Figure 8-98

If a piston seal were worn excessively, bypass fluid would subtract from the 5 gpm (18.95 l/min) causing rod speed to decrease.

Assume that worn piston seals in the cylinder allow .5 gpm (1.89 l/min) to bypass during rod extension. Consequently, 4.5 gpm (17.05 l/min) fills the volume behind the piston even though 5 gpm (18.95 l/min) enters the cap end. 4.5 gpm (17.05 l/min) is used in the calculations resulting in a slower rod speed:

	(ft/min)	(m/sec)
Rod speed =	$\frac{\text{gpm} \times 19.25)}{\text{rod area (in}^2)}$ =	$\frac{\text{l/min} \times .167}{\text{cm}^2}$
=	$\frac{5 \times 19.25}{2}$ =	$\frac{18.95 \times .167}{12.9}$
=	$\frac{96.25}{2}$ =	$\frac{3.165}{12.9}$
	= 48.13 ft/min	= .24 m/sec

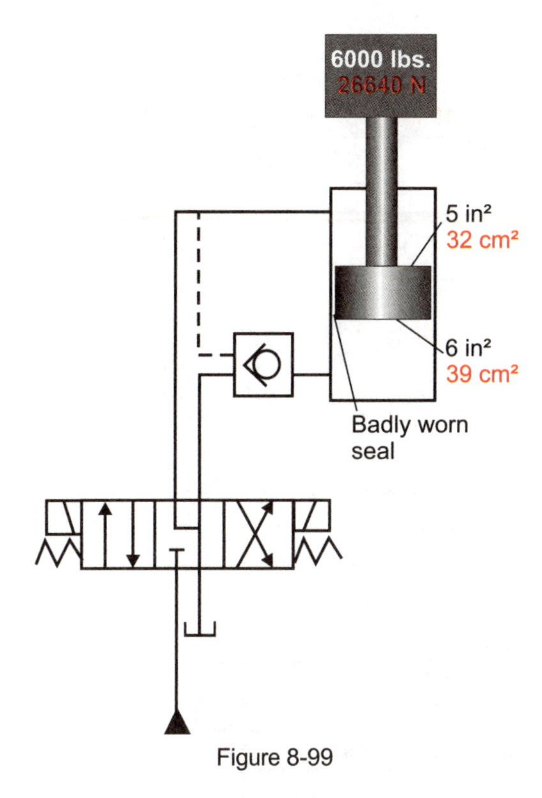

Figure 8-99

Piston seal wear and the leakage which results causes cylinder rod speed to decrease even though full pump flow enters the cylinder. This means work will be done over a longer period of time. System operating temperature will increase because of the wasted hydraulic power flowing across the piston.

Cylinders can leak excessively in only portions of their stroke. If a cylinder is primarily cycled in mid-stroke and excessive contamination is present in a fluid, piston seal and cylinder bore can become scored. This causes excessive leakage and reduced rod speed in mid-stroke only. During its cycle, once piston passes the worn portion of the bore, bypass fluid is reduced and rod speed increases.

Intensification from Piston Leakage

In some cases, piston seal leakage can cause pressure intensification.

In the circuit illustrated, a cylinder in conjunction with a directional valve and pilot operated check valve is required to raise and hold a load in mid-stroke. The load is 6000 lbs. (26640 N) The cylinder piston has a major area of 6 in² (38.7 cm²) and a 5 in² (32.25 cm²) minor area.

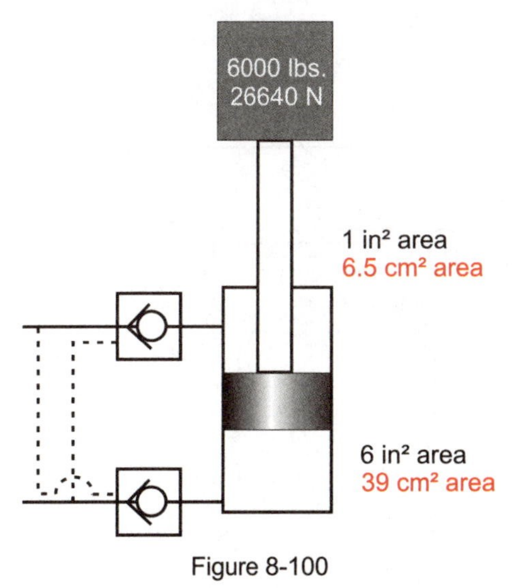

Figure 8-100

The cylinder in the circuit has an excessive leakage problem both at rod gland and piston seal. Since excessive leakage at a piston rod is usually quite obvious and a housekeeping problem, assume the rod gland seals are replaced. This still leaves the piston seal leakage.

Assume when the directional valve is shifted, flow at a pressure of 1000 psi (68.96 bar) enters the cap end of the cylinder raising the load. As the float center directional valve is centered, the load gradually falls because of leakage across the piston.

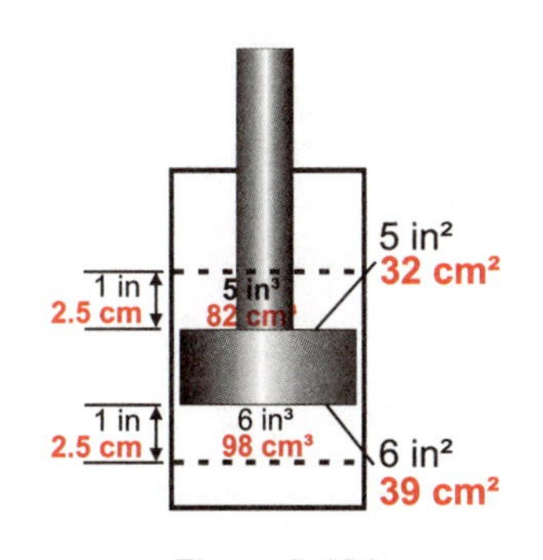

Figure 8-101

To remedy the problem, a pilot operated check valve is added to the rod-side cylinder line. Now when the directional valve is centered, fluid is not allowed to escape from the cylinder. Since a seal

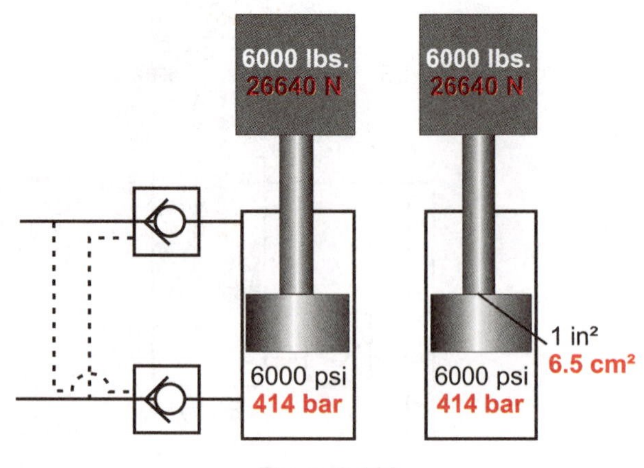

Figure 8-102

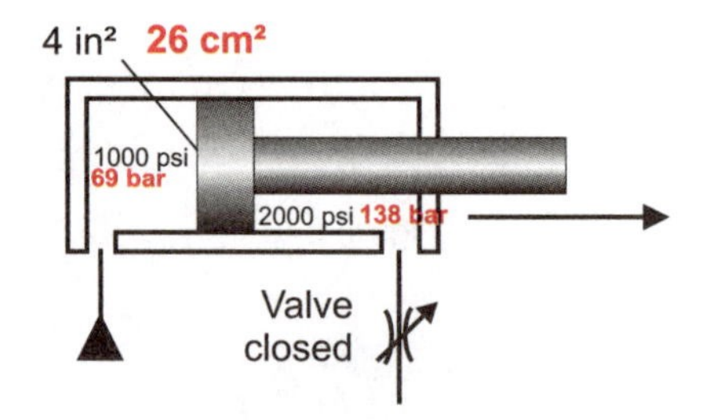

Figure 8-103

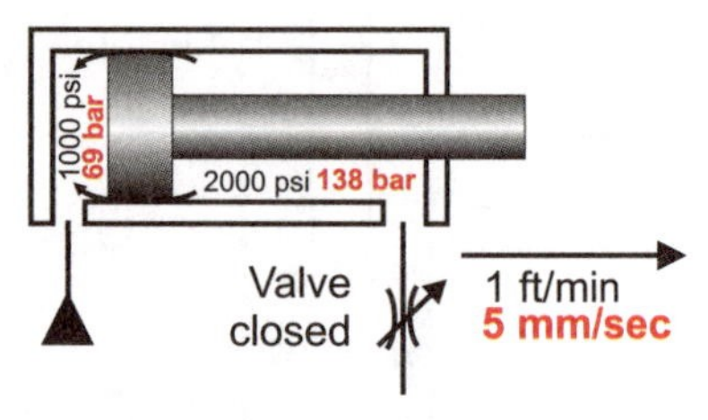

Figure 8-104

basically does not exist across the piston, liquid volumes on either side of the piston can communicate. Without a piston seal, piston and piston rod can be considered immersed in the liquid.

It may be felt that since a piston has no seals, the load, piston rod and piston will drift down even though fluid is trapped in the cylinder. This is not the case and may be explained by an example.

A piston and piston rod are immersed halfway in a cylinder filled with liquid. If the situation is analyzed, it can be seen that drift is impossible as long as liquid does not leak out of the cylinder. The piston has a piston area of 6 in² (38.7 cm²) and a 5 in² (32.25 cm²) minor area. If the piston moved down 1 in. (2.54 cm), 6 in³ (98.29 cm³) of oil would be displaced [6 in² (38.7 cm²)] piston area x 1 in./2.54 cm stroke). At the other side of the piston, only a 5 in³ (81.91 cm³) space would be evacuated. Since 6 in³ (98.29 cm³) of oil does not fit into a 5 in³ (81.91 cm³) space, the piston cannot drift down unless some fluid leaks out of the cylinder.

Returning to our problem, we find that with piston seals worn excessively, pressures on either side of the piston are equal. Pressure acting on the piston area is offset by pressure acting on the effective area except for an area equal to the rod cross-sectional area. This is the area which supports the load. Since rod area is 1 in² (6.45 cm²) and the load is 6000 lbs. (26640 N), pressure generated in the cylinder is 6000 psi (413.79 bar) (6000 lbs./1 in²).

Another way of thinking of it would be to consider a rod immersed part way into a cylinder filled with liquid. If the rod had a cross-sectional area of 1 in² (6.45 cm²) and supported 6000 lbs. (26640 N), 6000 psi (413.79 bar) would be generated in the cylinder. If a smaller-than-bore piston were attached to the rod, pressure would still be the same since any area of the piston not in contact with the rod would be cancelled out by equal pressures on both sides.

The generation of high pressure in this manner can be a source of ruptured seals and external leakage.

Checking for Piston Seal Leakage

Checking for piston seal leakage can be accomplished by seeing the effect of bypass flow on rod speed.

To check for piston leakage, a needle or shut off valve is piped into the rod side cylinder line. With the valve closed and the piston bottomed against the cap end, the cap end of the cylinder is subjected to full system pressure. The valve is then cracked open allowing the piston to move a short distance along its stroke. The valve is then closed. At this point, full system pressure will act on the major area of the piston resulting in an intensified pressure at the rod side.

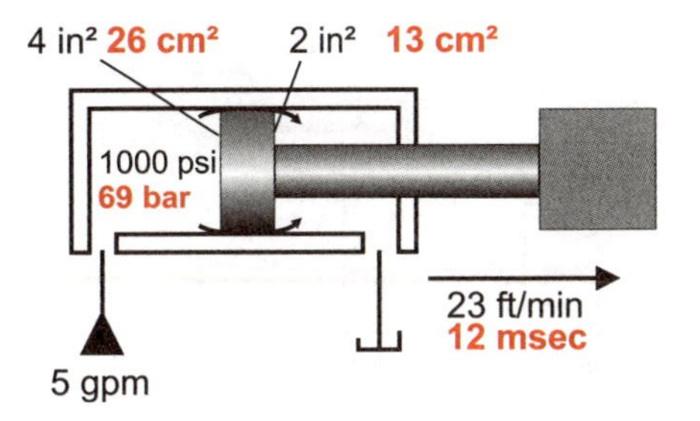

Figure 8-105

In the illustration, a 2:1 cylinder has a piston area of 4 in² (25.8 cm²) and an effective area of 2 in² (12.9 cm²). With the relief valve set at 1000 psi (68.96 bar) and the rod end cylinder port blocked, 4000 lbs. (17760 N) (4 in² x 1000 psi) of force is generated on the piston area to extend the rod. This 4000 lbs. (17760 N) acts on the 2 in² effective area of the piston resulting in 2000 psi (137.9 bar) (4000 lbs./2 in²) back pressure at the rod side.

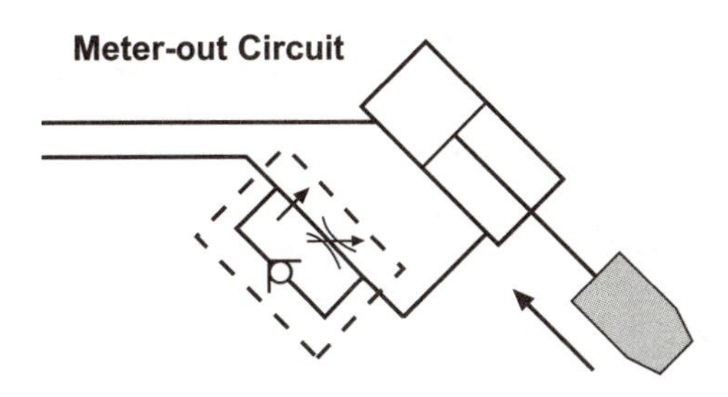

Figure 8-106

With 1000 psi (68.96 bar) at the cap end and 2000 psi (137.9 bar) at the rod end, any fluid leakage will transfer from the rod side to the cap end side causing the piston rod to drift out. This check is performed at regular intervals along the cylinder stroke.

As a piston seal check is performed, the rate at which the rod drifts determines the reduction in rod speed as the cylinder operates in a system. In our illustrated 2:1 cylinder, assume that under test conditions the rod drifted out at a rate of 1 ft/min (5.08 mm/sec) as a 1000 psi (68.96 bar) differential existed across its piston. With the cylinder operating in a system at a differential of 1000 psi (68.96 bar), a reduction in speed of 1 ft/min (5.08 mm/sec) can be expected. With the cylinder receiving 5 gpm (18.95 lpm), the piston rod would extend at a rate of 23.06 ft/min (.12 m/sec). Whereas, if piston seals were in perfect condition, the rod would extend at 24.06 ft/min (.122 m/sec). If the reduced rod speed caused by piston seal leakage cannot be tolerated, the cylinder should be repaired or replaced.

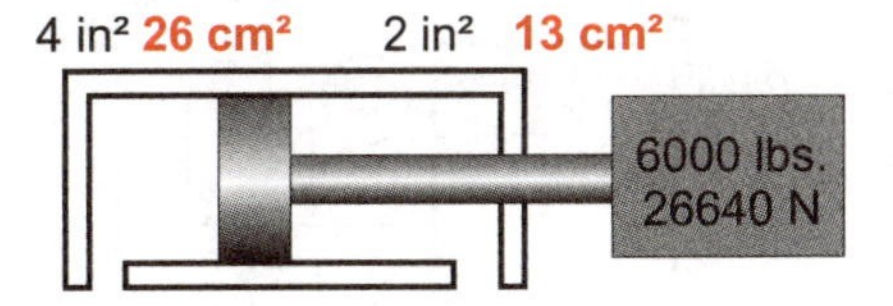

Figure 8-107

Keep in mind that cast iron piston rings can leak 1-3 in³ (16.39-49.17 cm³) of oil per minute at a pressure of 1000 psi (68.96 bar). They are designed to leak somewhat for purpose of lubrication. This flow should not be confused with leakage flow due to wear. Leakage flow (bypass) should be considered when studying or troubleshooting a hydraulic system.

In the next section, we find that an intensified pressure is present at the rod side of an unloaded, extending, single rod cylinder while flow is being metered out. This pressure can cause harm to a cylinder.

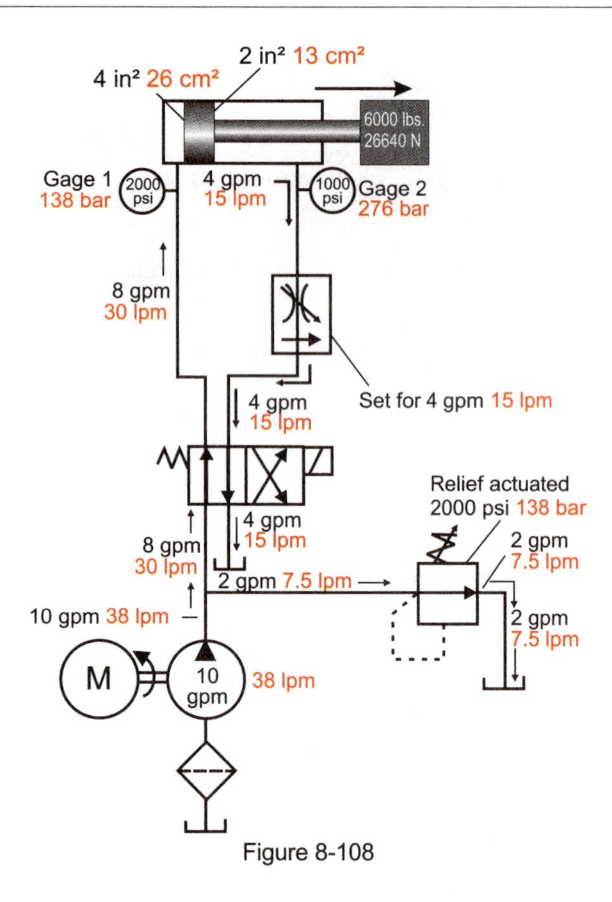

Figure 8-108

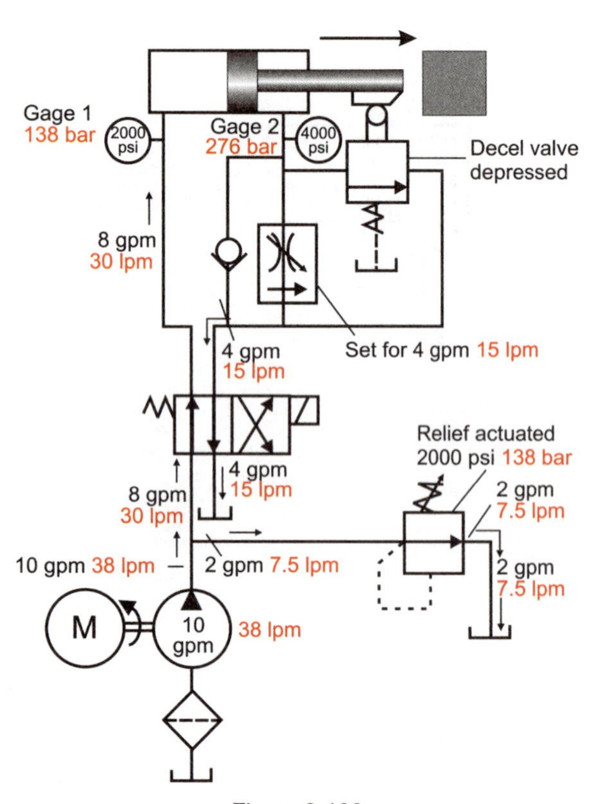

Figure 8-109

Intensification at Cylinder Rod Side

A flow control valve could be positioned at the rod side of a cylinder. This valve would then restrict flow from the cylinder and as a result the cylinder would not be allowed to runaway from pump flow. This is known as a meter-out circuit.

The flow control valve kept the cylinder from running away by causing a back pressure to be generated on the minor area of piston. The resultant force kept the piston and piston rod under control.

In any meter-out circuit in which a single rod cylinder is pushing out a load, whenever the pressure acting on the piston major area and its resultant force is more than is required to equal the load. The excess force develops a backpressure on the piston effective area.

In the illustration, a 2:1 cylinder is required to move a 6000 lb. (26640 N) load. The area of the piston is 4 in² (25.8 cm²) and the effective area is 2 in². From the formula: pressure (psi) - force (lbs)/area (in²), it can be calculated that 1500 psi (6000 lbs/4 in²) must act on the piston area of the piston in order to equal the load.

In the circuit illustration, assume that the pump flow is 10 gpm (37.9 l/min), relief valve setting is 2000 psi (137.9 bar), and the flow control valve is set to meter 4 gpm (15.16 l/min) out of the cylinder as the rod extends.

With a 2:1 cylinder, if 10 gpm (37.9 bar) enters the cap end, 5 gpm (18.95 l/min) discharges from the rod end. With the flow control set for 4 gpm (15.16 l/min), only 8 gpm (30.32 l/min) is allowed into the cylinder.

When pump/electric motor is turned on, pressure begins to build in the system. (This, of course, happens almost instantly.) When pressure reaches 1500 psi (103.4 bar), the load has been equalled. But, pump/ electric motor cannot put its flow of 10 gpm (37.9 l/min) into the system at 1500 psi (103.4 bar). More pressure is developed — 1600 (110.3 bar), 1700 (117.2 bar), 1800 (124.1 bar), 1900 psi (131 bar). When pressure reaches 2000 psi (137.9 bar), the relief valve cracks open enough to accept 2 gpm (7.58 l/min). Ten gpm (37.9 l/min) discharges from the pump at 2000 psi (137.9 bar). Eight gpm (30.32 l/min) at 2000 psi (137.9 bar) heads toward the cylinder; 2 gpm (7.58 l/min) at 2000 psi (137.9 bar) dumps over the relief valve.

At this point in time, gage 1 at the cylinder cap end indicates 2000 psi (137.9 bar). 2000 psi (137.9 bar) on 4 in^2 (25.8 cm^2) results in 8000 lbs. (35520 N) extending the piston rod. 6000 lbs. (26640 N) is used to equal the load. The 2000 lbs. (8880 N) in excess is offset by a 1000 psi (68.96 bar) back pressure on the 2 in^2 (12.9 cm^2) minor area. The back pressure is generated as a result of the flow control valve restriction. Gage 2 indicates 1000 psi (68.96 bar).

If the same situation occurred with no load, the restriction of the flow control valve would cause an extremely high pressure to be generated.

In the circuit illustrated, our 2:1 cylinder extends rapidly to the work load. At a point near the work, a deceleration valve is closed off forcing fluid to pass through a flow control valve as it exits cylinder rod side. The rod continues to extend for a short distance and then contacts the work load.

Between the time the deceleration valve closes and the load is contacted, 8000 lbs. (35520 N) of force is generated on the piston major area, but a load is not present to make use of it. This results in 8000 lbs. (35520 N) being absorbed by a backpressure acting on 2 in^2/12.9 cm^2 (8000 lbs/2 in^2 = 4000 psi). Gage 2 indicates 4000 psi (275.9 bar) and continues to indicate 4000 psi (275.9 bar) until the load is contacted. Pressure intensification at the rod side of a cylinder can cause rod seals to leak or rupture.

Intensification can be expected to occur in the above manner any time an extending, single rod cylinder is being metered out without a load. Since cylinder cushions are also metered-out restrictions, pressure intensification will occur any time an extending, single rod cylinder goes into cushion. This does not affect the rod seal, but can cause leakage fluid to discharge from the needle valve cushion adjustment.

From previous illustrations, it has been shown that discharge flow from pump/electric motor is not necessarily the maximum flow rate in a system. The above example points out that a relief valve setting is not necessarily the maximum pressure in a system.

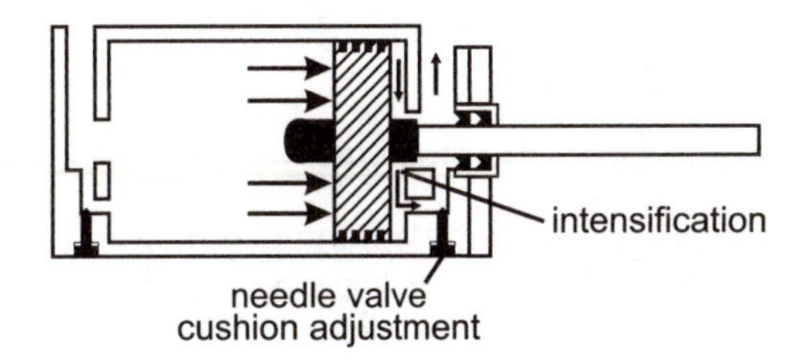

Figure 8-110

Terms and Idioms Associated with Check Valves

BACK PRESSURE CHECK - A check valve used to cause the generation of a system pressure level required for the operation of other valves.

LOAD LOCK VALVE - Two pilot operated check valves in one valve body

P O CHECK - pilot operated check valve

Exercise
Check Valves, Accumulators and Cylinders
50 points

Instructions: Solve the problems.

1. **PROBLEM:** Connect the hand pump, cylinder and reservoir so that with each stroke of the hand lever (forward and backward), the load is raised the same amount. The load must not be allowed to drop when the hand pump is not pumping. To solve the problem, three check valves are needed. These are the only components which may be added to the circuit.

   Do not be concerned with lowering the load or refilling the reservoir.

   In the hand pump, the rod side of the piston has half the area as the cap end side.

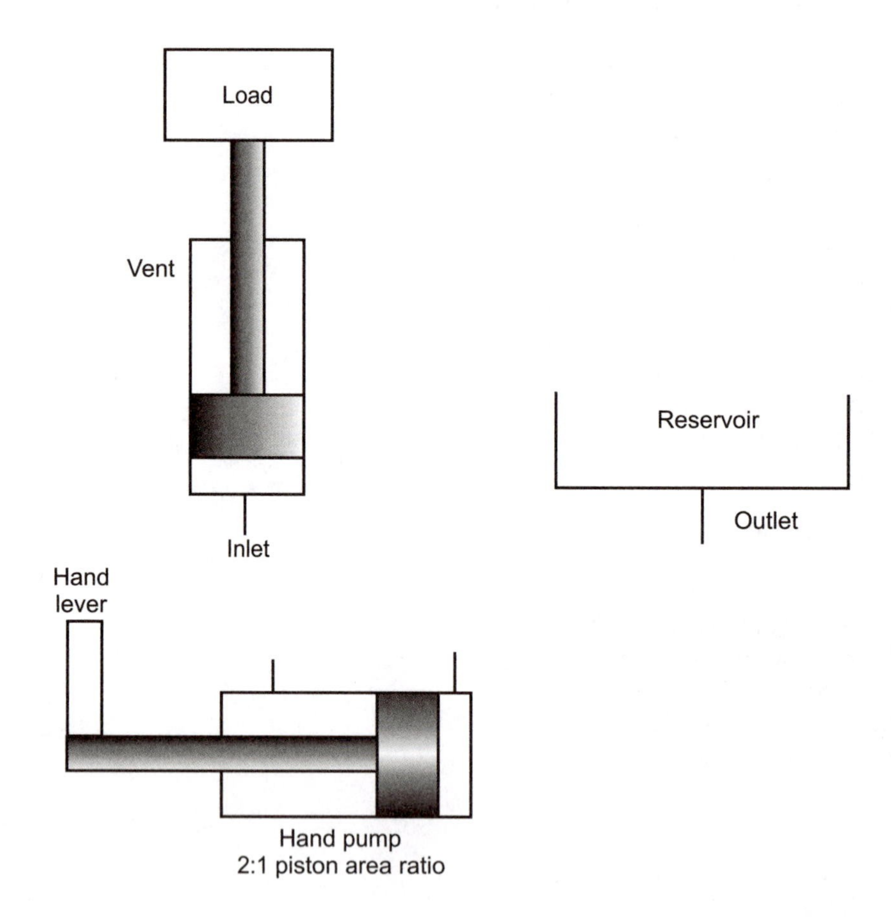

Check Valves, Accumulators and Cylinders (cont'd.)

2. **SITUATION:** The 6000 lb. (26640 N) load must be suspended, but excessive leakage across the piston causes the load to drift down. A pilot operated check valve was placed in each cylinder line to control the drift.

   PROBLEM: The cylinder's rod seal was blown out. Why?

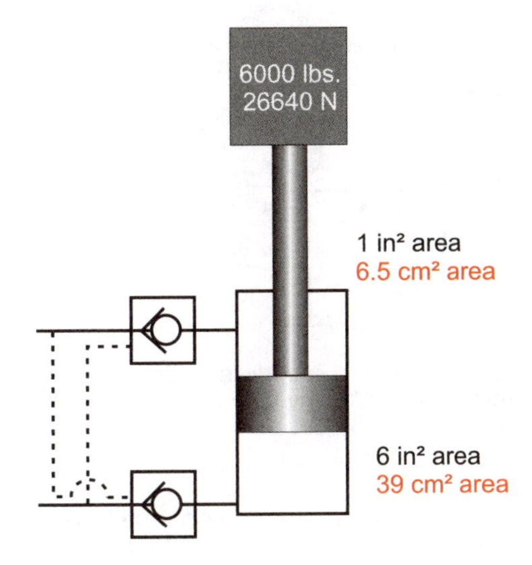

3. **PROBLEM:** Why can the load still drift down?

   (Assume pilot operated check valves seal perfectly.)

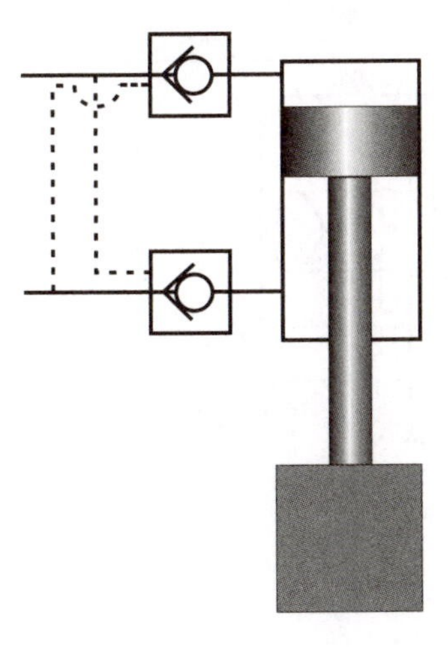

Check Valves, Accumulators and Cylinders (cont'd.)

4. **SITUATION:** For each machine cycle, a system requires a large volume flow for a short period of time. To conserve horsepower, pump flow should be allowed to return to tank at the least possible resistance when the accumulator is filled. This can be accomplished by connecting the pressure valve's pilot line to a certain section of the system.

   PROBLEM: Complete the design of the circuit by connecting the pressure valve's pilot line to the appropriate part of the system.

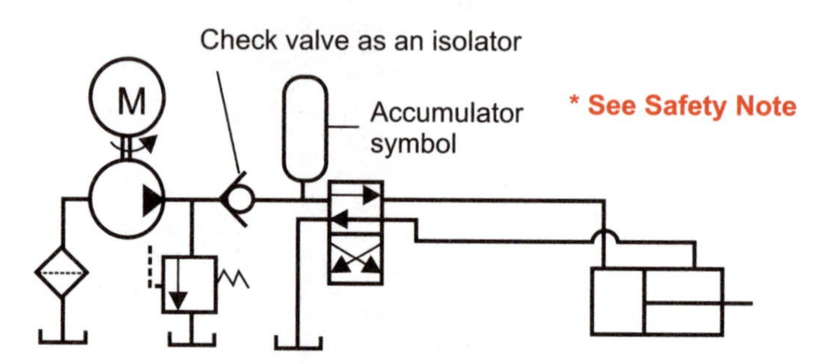

***Safety Note:** In any accumulator circuit, a means should be available of automatically unloading the accumulator when the machine is shut down.

5. **PROBLEM:** In the regenerative circuit below, the components are connected so that the cylinder piston rod travels at the same speed extending and retracting. System pressure = 1000 psi (68.98 bar), gpm = 10 (37.9 l/min).

   If the cylinder rod has a 2" (5.08 cm) diameter, what is the rod speed? ______________

   What is the maximum force which can be developed by this cylinder? ______________

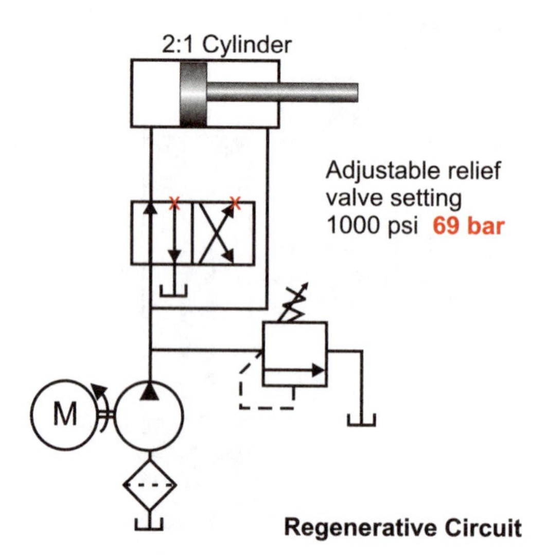

Regenerative Circuit

Chapter 9

Flow Control Valves

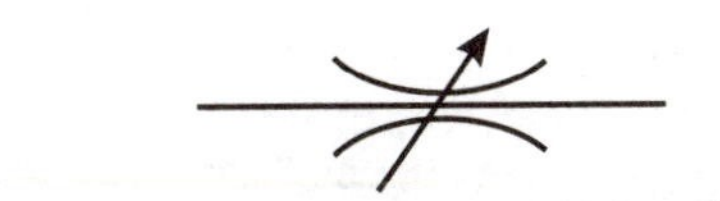

Adjustable Flow Control Valve Symbol

Figure 9-1

The function of a flow control valve is to reduce a pump's flow rate in its leg of a circuit. It performs its function by being a higher than normal restriction for the system. To overcome the restriction, a positive displacement pump applies a greater pressure to the liquid which causes some of the flow to take another path. This path is usually over a relief valve, but may be to another leg of a system.

In the circuit illustrated, a 5 gpm (18.95 lpm) pump applies to the liquid whatever pressure is necessary to get its flow out into the system. If there is no load at the cylinder, this pressure energy will be changed into heat because of a liquid's viscosity, friction and changing direction.

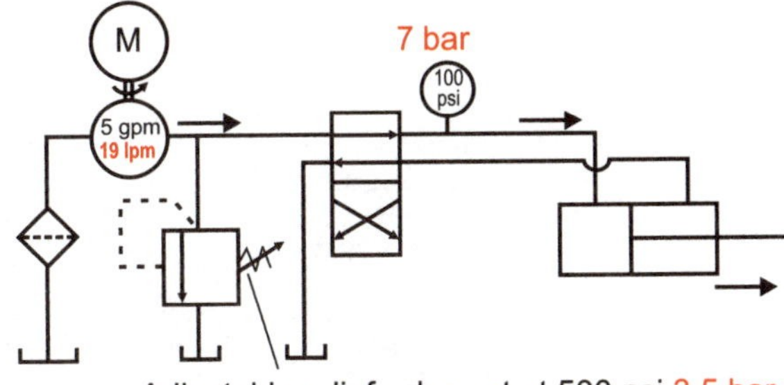

Adjustable relief valve set at 500 psi 3.5 bar

Figure 9-2

If a flow control valve which restricted pump flow were placed in the circuit, the pump would still attempt to push its total volume through the valve. When the pressure ahead of the valve reached relief valve setting, some flow would be diverted over the relief valve. The rate of flow through the flow control valve at that time will be something less than 5 gpm (18.95 lpm), yet the pressure ahead of the flow control will be up to the relief valve setting. (In the example circuits used in this section, we will assume that the pressure differential through a directional valve and any associated piping is zero.) As far as the flow control valve is concerned, this increased pressure is a source of potential energy which it changes into kinetic energy (rate of flow). The degree to which this happens is dependent on the flow control's orifice.

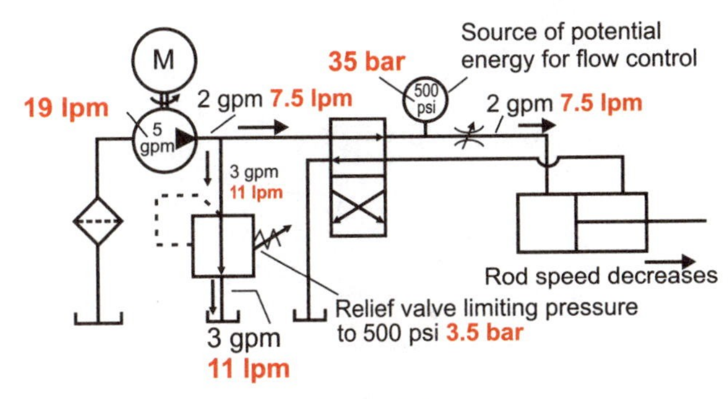

Figure 9-3

Orifice

An orifice is a relatively small opening in a fluid's flow path. Flow through an orifice is affected by several factors, three of these factors are:

1. size of the orifice
2. pressure differential across the orifice
3. temperature of the fluid

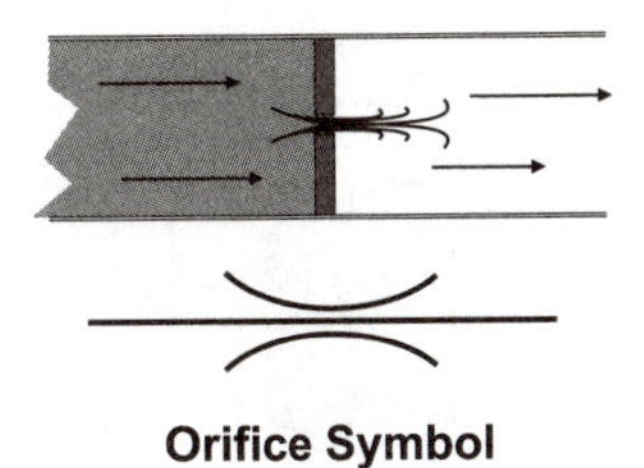

Orifice Symbol

Figure 9-4

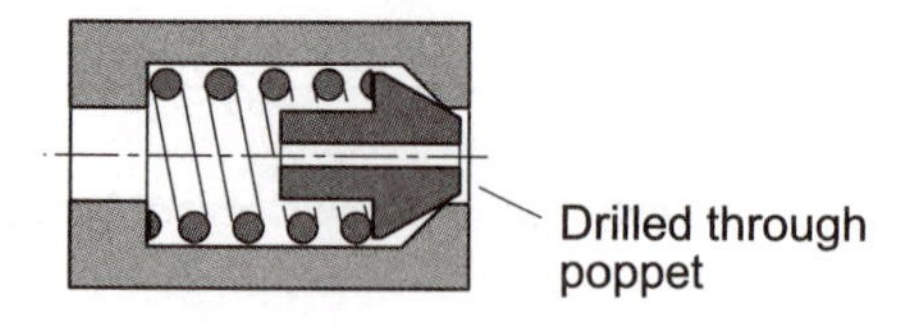

Figure 9-5

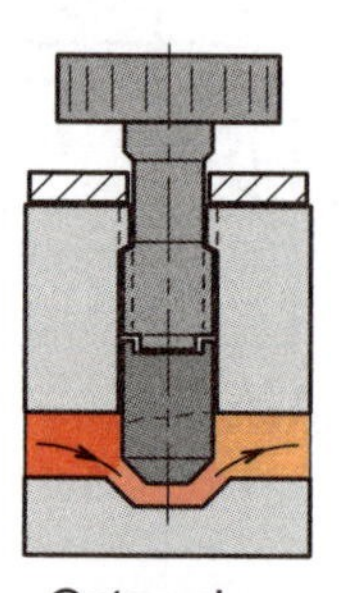

Gate valve
Figure 9-6

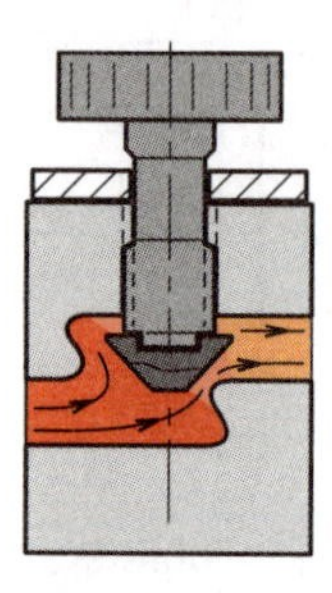

Globe valve
Figure 9-7

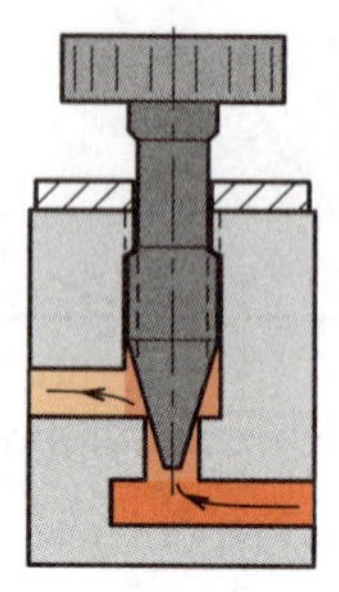

Needle valve
Figure 9-8

Orifice Size Affects Flow

The size of an orifice controls the flow rate through the orifice. A common, everyday example of this is a garden hose which has sprung a leak. If the hole in the hose is small, the leak will be in the form of a drip or spray. But, if the hole is relatively large, the leak will be an stream. In either case, the hole in the hose is an orifice which meters a flow of water to the surrounding outside area. The amount of flow which is metered depends upon the size of the opening.

Fixed Orifice

A fixed orifice is a reduced opening of an unadjustable size. Common examples of fixed orifices used in hydraulics are a pipe plug or check valve with a hole drilled through its center, or a commercial, factory pre-set flow control valve.

Variable Orifice

Many times, a variable orifice is more desirable than a fixed orifice because of its degree of flexibility. Gate valves, globe valves and needle valves are examples of variable orifices.

Gate Valve

A gate valve has a flow path straight through its center. The size of the orifice is changed by turning the handle which positions a gate or wedge across the fluid path.

Although gate valves are not designed to restrict flow, they are found in some systems where coarse metering is required.

Globe Valve

A globe valve does not have a straight-through flow path. Instead, the fluid must bend 90° and pass through an opening which is the seat of a plug or globe. The size of the opening is changed by positioning the globe.

Needle Valve

The fluid going through a needle valve must turn 90° and pass through an opening which is the seat for a rod with a cone-shaped tip. The size of the opening is changed by the positioning of the cone in relation to its seat. The orifice size can be changed very gradually because of fine threads on the valve stem and the shape of the cone.

A needle valve is the most frequently used variable orifice in an industrial hydraulic system.

Needle Valves in a Circuit

The circuit illustrated consists of a 5 gpm (18.95 l/min) positive displacement pump, relief valve, directional valve, a adjustable orifice and a cylinder which has a piston area of 3 in² (19.35 cm²).

With the relief valve set a 500 psi (34.48 bar), the pump attempts to push its 5 gpm (18.95 l/min) flow through the orifice. Because of the size of the orifice opening, only 2 gpm (7.58 l/min) passes through the orifice before the pressure reaches the relief valve setting of 500 psi (34.48 bar). (This, of course, happens instantly.)

Two gpm passes through the orifice and out to the actuator. Three gpm (11.37 l/min) goes over the relief valve. The piston rod moves at the rate of 13 ft/min. (3.96 m/min).

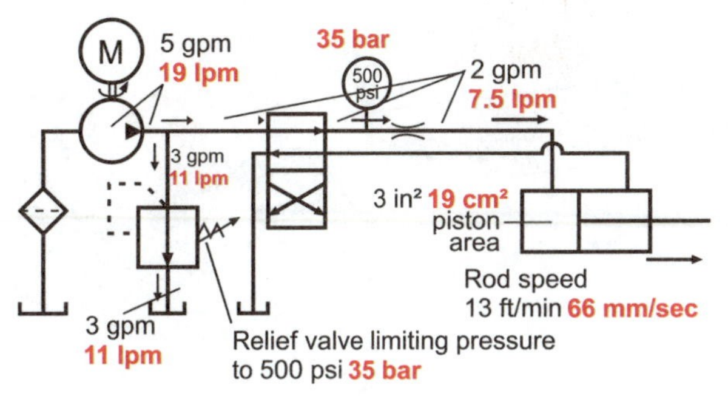

Figure 9-9

$$\text{Rod speed (ft/min)} = \frac{\text{gpm} \times 231\ (\text{in}^3/\text{gal})}{\text{piston area (in}^2)\ (12\ \text{in/ft})}$$

$$\text{Rod speed (m/min)} = \frac{\text{l/min} \times 10}{\text{piston area (cm}^2)}$$

Needle Valve Orifice Increased

Turning the knob out and opening the needle valve orifice allows more flow to pass through the valve and out to the cylinder before the relief valve setting is reached. Rod speed increases.

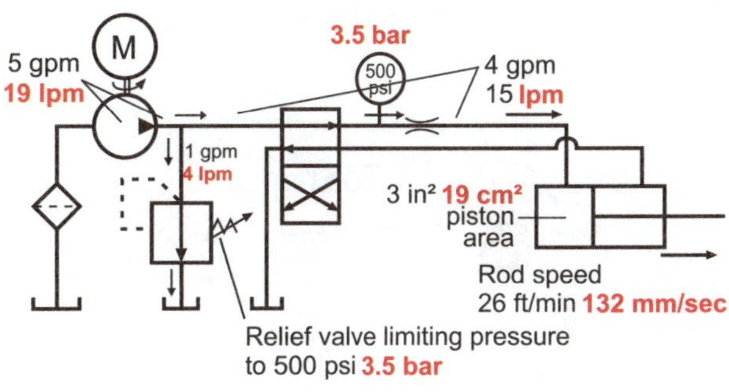

Figure 9-10

Needle Valve Orifice Decreases

Turning the knob in and decreasing the needle valve orifice allows less flow to pass through the needle valve before the 500 psi (34.48 bar) relief setting is reached. Rod speed decreases since the cylinder receives less flow.

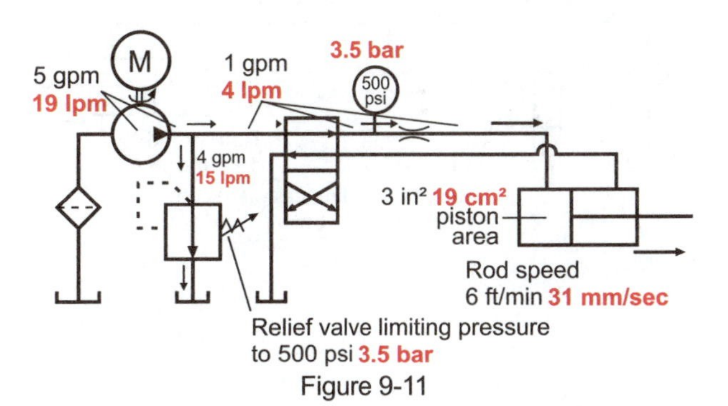

Figure 9-11

Pressure Differential Affects Flow

Flow through an orifice is affected by pressure differential. Since pressure in a hydraulic system is potential energy, the greater the difference in pressure across an orifice, the more flow will be developed.

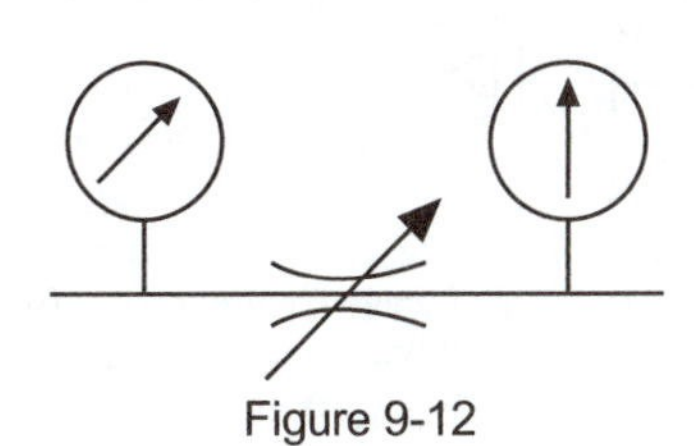

Figure 9-12

Examples from Everyday Life

After a day at the beach or camping in the woods, a plug is removed from an air mattress so air can escape.

If air is allowed to escape by itself, the mattress takes a while to collapse because the pressure differential is small.

Figure 9-13

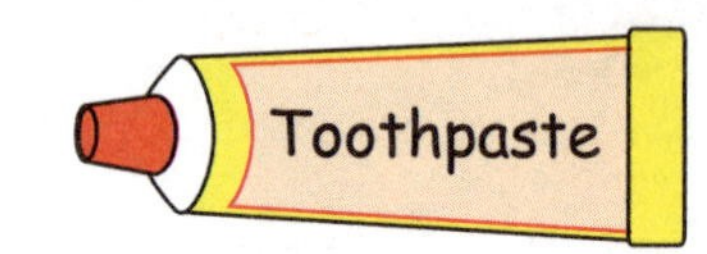

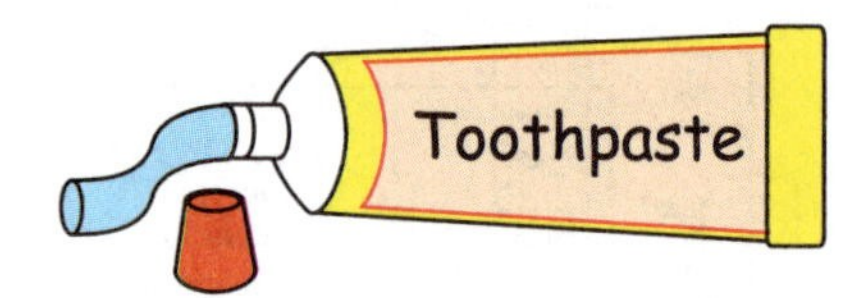

Figure 9-14

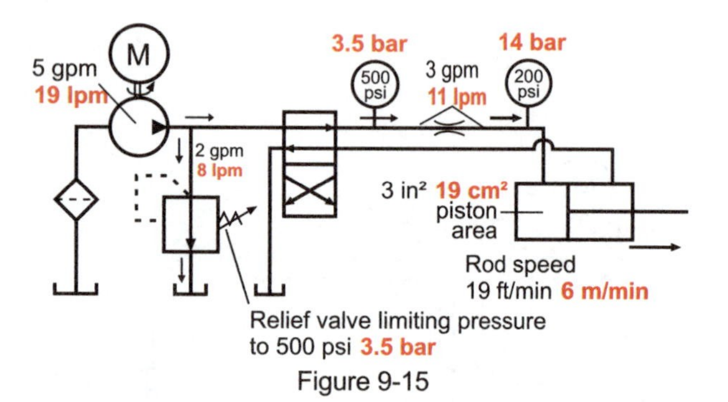

Figure 9-15

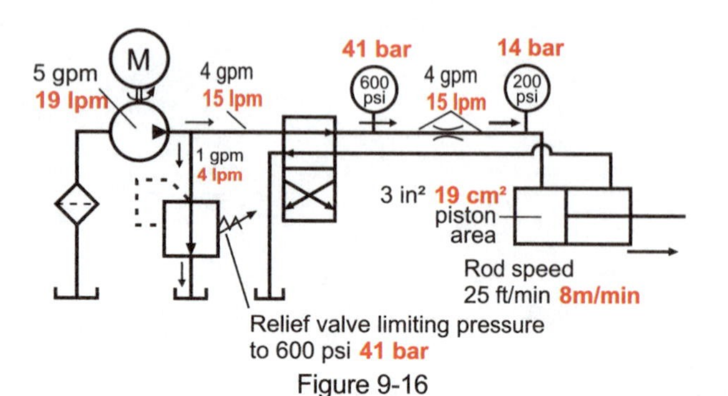

Figure 9-16

If the mattress were squeezed, air comes rushing out. Squeezing results in the development of a higher pressure inside the mattress. Pressure differential from within the mattress to the atmosphere has increased. The harder the mattress is squeezed, the more pressure is developed, and the larger the rate of air flow.

Gently squeezing a full tube of toothpaste results in a small amount of toothpaste on the toothbrush.

If a full tube of toothpaste were squeezed with more force, a large amount of toothpaste would be on the floor. Pressure differential from within the tube to the atmosphere is greater when the tube is stepped on than when it is squeezed.

Flow Through Needle Valves in a Circuit Affected By Pressure Differential

In the circuit illustrated, the needle valve is adjusted to restrict the 5 gpm (19 l/min) pump flow. Relief valve is set at 500 psi (34.5 bar). Pressure ahead of the needle valve is the 500 psi (34.5 bar) relief valve setting. The pressure required to overcome the resistance of the work load is 200 psi (14 bar). 200 psi (14 bar) of the 500 psi (34.5 bar) ahead of the needle valve is used to overcome the resistance of the load. The remaining 300 psi (21 bar) is used to develop a fluid flow through the needle valve. In this particular instance, the 300 psi (21 bar) pressure differential across the needle valve results in a flow of 3 gpm (11.3 l/min) and a rod speed of 19.25 ft/min (5.87 m/min). Two gpm (7.58 l/min) returns to tank through the relief valve.

Relief Valve Setting Increased

With the work-load pressure and the setting of the needle valve remaining the same, relief valve setting is increased to 600 psi (41.38 bar).

Now 600 psi (41.38 bar) is the pressure ahead of the needle valve. 200 psi (14 bar) of the 600 psi (41.38 bar) is used to overcome the resistance of the load. The resulting 400 psi (28 bar) pressure differential develops a flow of 4 gpm (15 l/min) through the needle valve. One gpm (3.79 l/min) passes over the relief valve. Rod speed increases to 26 ft/min (7.92 m/min).

Work-Load Pressure Increased

With the relief valve setting again at 500 psi (34.5 bar) and with the same needle valve adjustment, work-load pressure has increased to 400 psi (28 bar) because of a larger load. 500 psi (34.5 bar) is the pressure ahead of the needle valve. 400 psi (28 bar) of the 500 psi (34.5 bar) is used to overcome the resistance of the load. The remaining 100 psi (6.9 bar) develops a 1 gpm (3.79 l/min) flow through the needle valve. Four gpm (15 l/min) is dumped over the relief valve. Rod speed decreases to 6 ft/min (30 mm/sec).

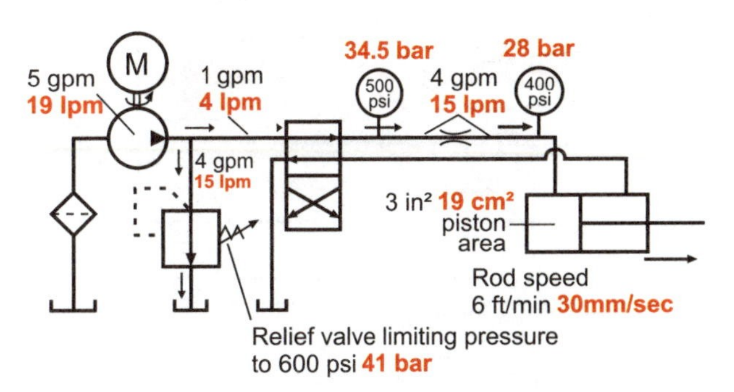

Figure 9-17

Pressure Compensated Flow Control Valves

As can be seen from the previous examples, any change in pressure ahead of or after a metering orifice affects the flow through the orifice resulting in a change of actuator speed. These pressure changes must be neutralized, or compensated for, before an orifice can precisely meter fluid.

Needle valves are designated non-compensated flow control valves. They are good metering devices as long as pressure differential across the valve remains constant and the needle stays centered. If more precise metering is required, a pressure compensated flow control valve is used; that is, a flow control which makes allowances for pressure changes ahead or after the orifice.

Pressure compensated flow control valves are classified as either restrictor or bypass types.

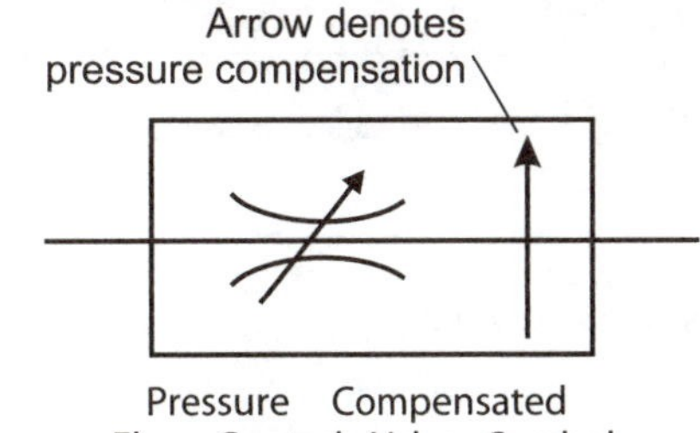

Pressure Compensated Flow Control Valve Symbol

What a Restrictor Type Pressure Compensated Flow Control Valve Consists Of

A restrictor type pressure compensated flow control valve consists of a valve body with inlet and outlet ports, a needle valve, a compensator spool and a spring which biases the spool.

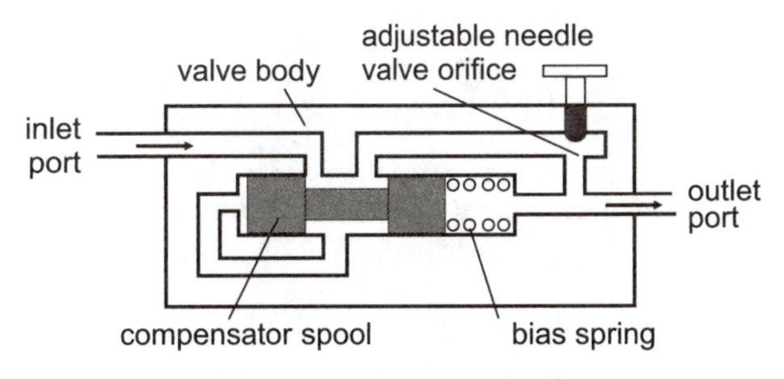

Pressure Compensated Flow Control Valve - Restrictor Type

Figure 9-18

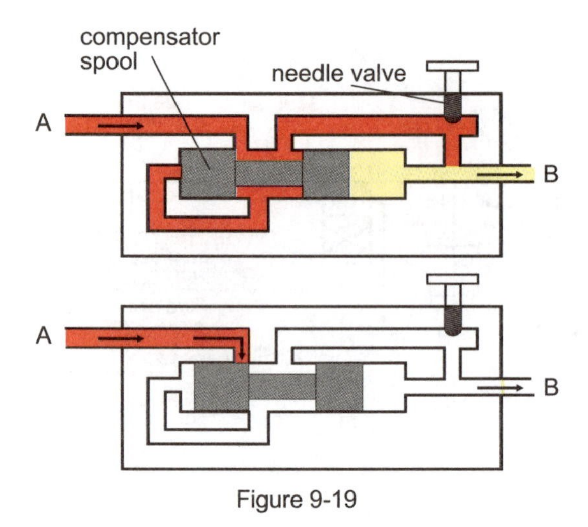

Figure 9-19

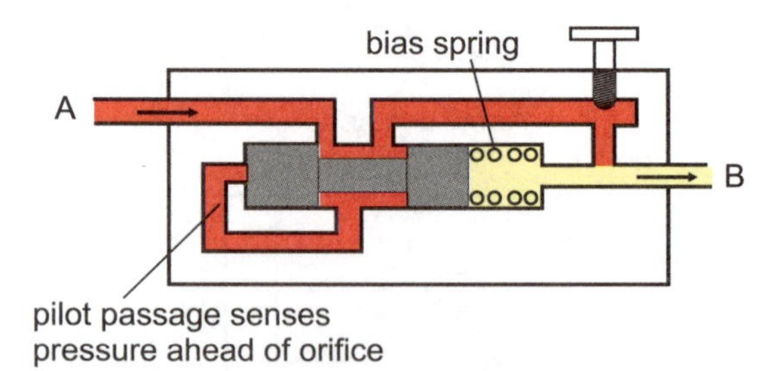

Figure 9-20

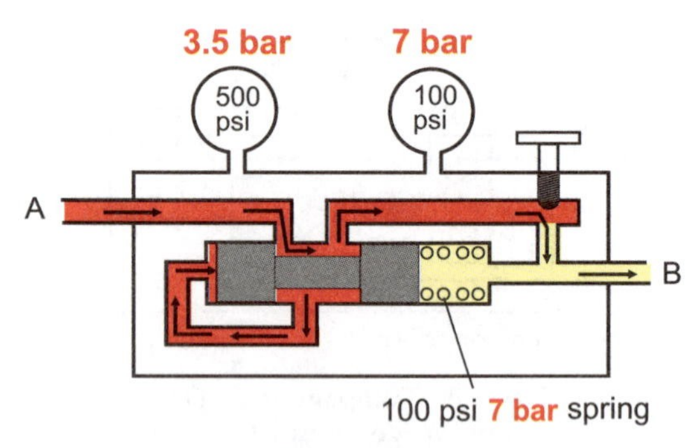

Figure 9-21

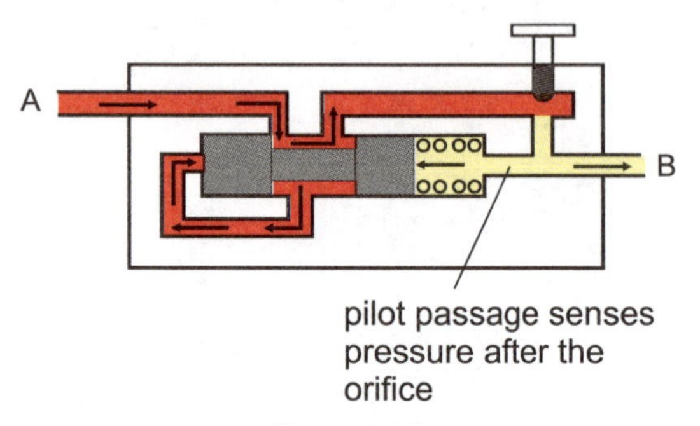

Figure 9-22

How a Restrictor Type Pressure Compensated Flow Control Valve Works

To determine how a restrictor type valve works, we will examine its operation step by step.

With the compensator spool fully shifted toward side A, any pressurized fluid flow entering the inlet port will arrive at the needle valve.

With the spool shifted slightly toward side B, pressurized fluid flow is blocked through the valve.

To keep the flow path through the valve open, a spring biases the compensator spool toward side A.

Pressure ahead of the needle valve is sensed at the A side of the spool by means of an internal pilot passage. When fluid pressure at this point tries to become more than the pressure of the spring, the spool will move toward side B.

With the needle valve orifice adjusted for something less than pump flow, pressure ahead of the needle valve wants to climb to the relief valve setting. When the pressure attempts to rise above the value of the compensator spring, the spool moves and restricts flow to the needle valve. As the fluid passes over this restriction, all of the pressure energy in excess of the value of the spring is turned into heat. For example, if the spring had a value of 100 psi (6.89 bar) and the relief valve were set at 500 psi (34.48 bar), fluid pressure at the valve's inlet will be 500 psi (34.48 bar). But, the compensator spool reduces the pressure before it gets to the needle valve by transforming 400 psi (27.59 bar) into heat energy as the fluid passes through the restriction. This means that regardless of what the pressure is at the flow control inlet, the pressure ahead of the needle valve to develop flow will always be 100 psi (6.89 bar).

As we have seen from previous circuits using needle valves, controlling the pressure ahead of the orifice is only half the battle. A fluctuation in pressure after the orifice must also be compensated for. In other words, a constant pressure differential is required. To accomplish this, a pilot passage which senses pressure downstream from the needle valve is directed to the bias spring chamber. Now two pressures bias the spool toward side A — spring pressure and fluid pressure after the needle valve.

If the spring has a value of 100 psi (6.89 bar), fluid pressure ahead of the orifice would be limited to 100 psi (6.89 bar) above the pressure after the orifice. As long as the relief valve is set high enough, the pressure differential across the needle valve orifice will always be the value of the spring which in this case is 100 psi (6.89 bar). In this way, the same amount of pressure is available to develop a flow through the orifice regardless of pressure fluctuations.

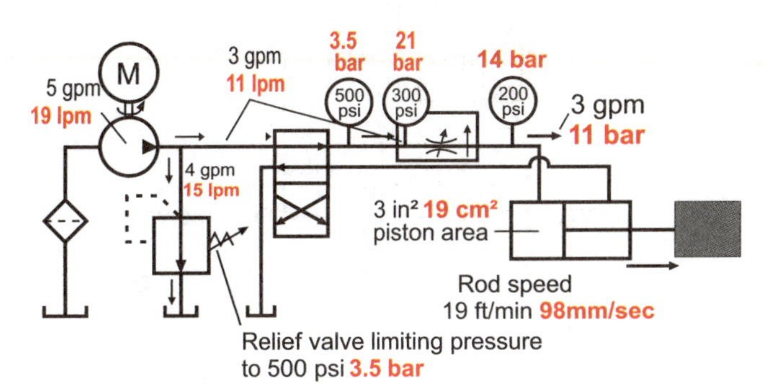

Figure 9-23

Restrictor Type Pressure Compensated Flow Control Valves in a Circuit

In the circuit illustrated, the restrictor type pressure compensated flow control valve is set for 3 gpm (11.37 l/min). Relief valve setting is 500 psi (34.48 bar). Work-load pressure is 200 psi (13.79 bar). The spring biasing the compensator spool has a value of 100 psi (6.89 bar).

During system operation, the workload pressure of 200 psi (13.79 bar), plus the 100 psi (6.89 bar) spring, bias the compensator spool.

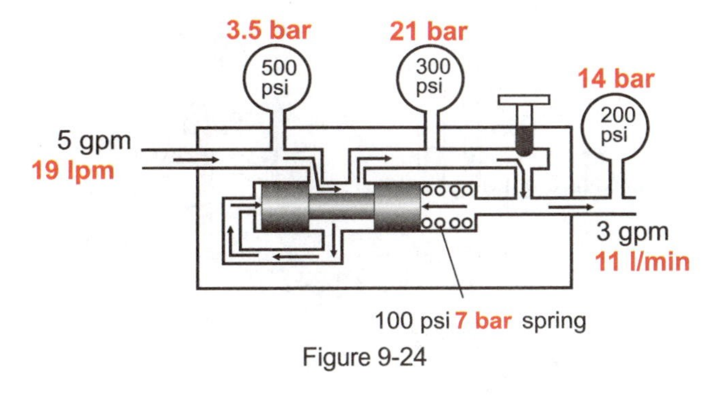

Figure 9-24

The pump attempts to push its total flow of 5 gpm (18.95 l/min) through the needle valve orifice. When pressure ahead of the needle valve reaches 300 psi (20.69 bar), the compensator spool moves and causes a restriction for the incoming fluid. The pressure at the flow control inlet rises to the relief valve setting of 500 psi (34.48 bar). As the fluid passes over the restriction made by the compensator spool, 200 psi (13.79 bar) of the 500 psi (34.48 bar) is transformed into heat. The pressure ahead of the needle valve is limited to 300 psi (20.69 bar). Of this 300 psi (20.69 bar), 200 psi (13.79 bar) is used to overcome the resistance of the load; 100 psi (6.89 bar) is used to develop a flow rate through the needle valve orifice. The flow rate in this case is 3 gpm (11.37 l/min). The remaining 2 gpm (7.58 l/min) is dumped over the relief valve. See Appendix B-18.

Workload Pressure and Relief Valve Setting Increased

If the workload pressure were increased to 400 psi (27.58 bar), or if the relief valve setting were re-set to 600 psi (41.38 bar), 100 psi (6.89 bar) would still be available to develop a flow rate through the needle valve. As long as the relief valve setting is 100 psi (6.89 bar) higher than the workload pressure, or in other words high enough to operate the compensator spool, a constant rate of flow of 3 gpm (11.37 l/min) will be delivered to the cylinder.

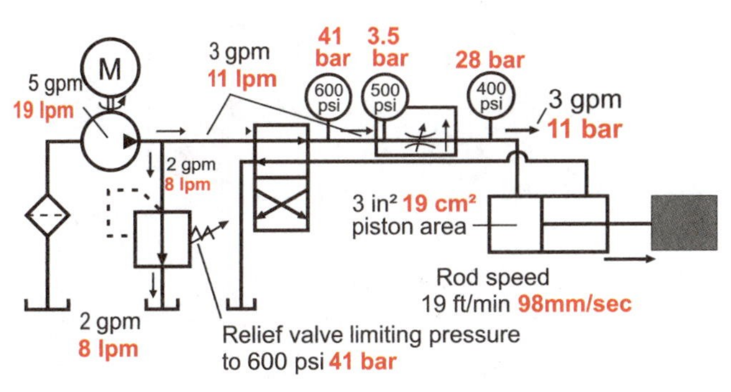

Figure 9-25

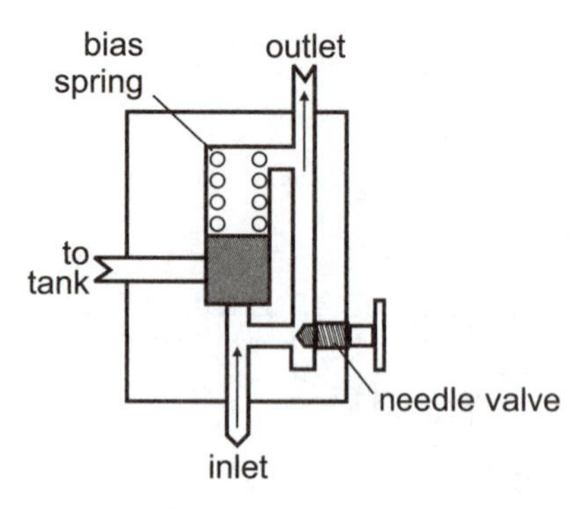

Figure 9-26

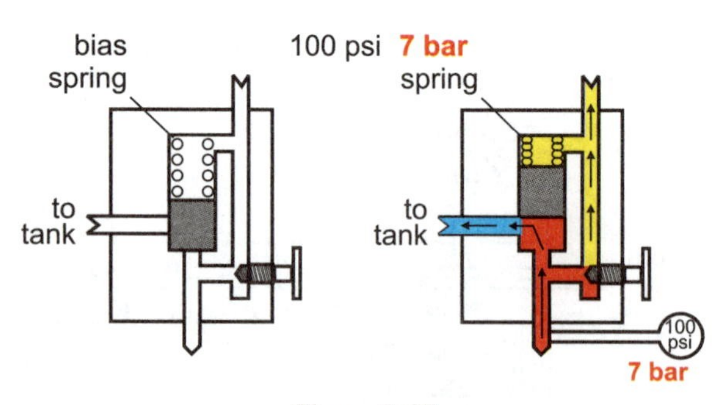

Figure 9-27

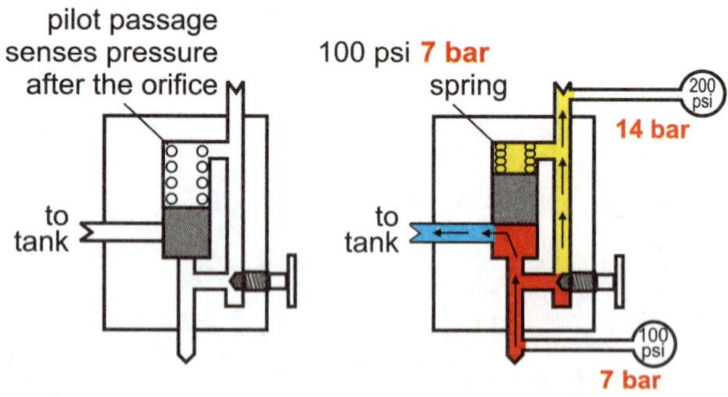

Figure 9-28

What a Bypass Type Pressure Compensated Flow Control Valve Consists Of

A bypass type pressure compensated flow control valve consists of a valve body with inlet, outlet and tank ports; a needle valve; a compensator spool and a spring which biases the spool.

How a Bypass Type Pressure Compensated Flow Control Valve Works

To determine how a bypass type valve works, we will examine its operation step by step.

The compensator spool in this valve develops a constant pressure differential across a needle valve orifice by opening and closing a passage to tank.

With the compensator spool fully seated in the down position, the passage to tank is blocked. With the compensator spool in the up position, the passage to tank is open.

In this condition, any flow coming into the valve will return to tank. In its normal condition, the compensator spool is biased in the closed position by a spring. If the spring as assembled has a value of 100 psi (6.89 bar), the pressure ahead of the needle valve will be limited to 100 psi (6.89 bar).

During system operation, the pressure ahead of the needle valve attempts to rise to the relief valve setting. When the pressure reaches 100 psi (6.89 bar), the spool uncovers the passage to tank thus limiting the pressure ahead of the needle valve to 100 psi (6.89 bar). A constant pressure ahead of the needle valve orifice does not necessarily guarantee a constant flow rate. If the pressure after the orifice changes, the pressure differential across the orifice changes and consequently so does the flow.

To compensate for this situation, pressure after the needle valve orifice is added to the top of the piston by means of a pilot passage. Two pressures now bias the spool spring pressure and fluid pressure after the needle valve.

If the spring had a value of 100 psi (6.89 bar), fluid pressure ahead of the needle valve orifice would be limited to 100 psi (6.89 bar) above the pressure after the orifice. As long as the relief valve setting is high enough, pressure differential across the needle valve will always be the value of the spring which in this example is 100 psi (6.89 bar). In this way, the same amount of pressure is available to develop a flow through the orifice regardless of changes in pressure.

Bypass Type Pressure Compensated Flow Control Valves in a Circuit

In the circuit illustrated, the bypass type pressure compensated flow control valve is set for 3 gpm (11.37 l/min). Relief valve setting is 500 psi (34.48 bar). Workload pressure is 200 psi (13.79 bar). The spring biasing the compensator spool has a value of 100 psi (6.89 bar).

During system operation, the workload pressure of 200 psi (13.79 bar), plus the 100 psi (6.89 bar) spring, bias the compensator spool. The pump attempts to push its total flow of 5 gpm (18.95 l/min) through the needle valve orifice. When pressure ahead of the needle valve reaches 300 psi (20.68 bar), the compensator spool uncovers the passage to tank. Pressure ahead of the needle valve is therefore limited to 300 psi (20.68 bar). Of this 300 psi (20.68 bar), 200 psi (13.79 bar) is used to overcome the resistance of the load; 100 psi (6.89 bar) is used to develop a flow rate through the needle valve. The flow rate in this case is 3 gpm (11.37 l/min).

The remaining 2 gpm (7.58 l/min) is bypassed to tank through the passage in the valve.

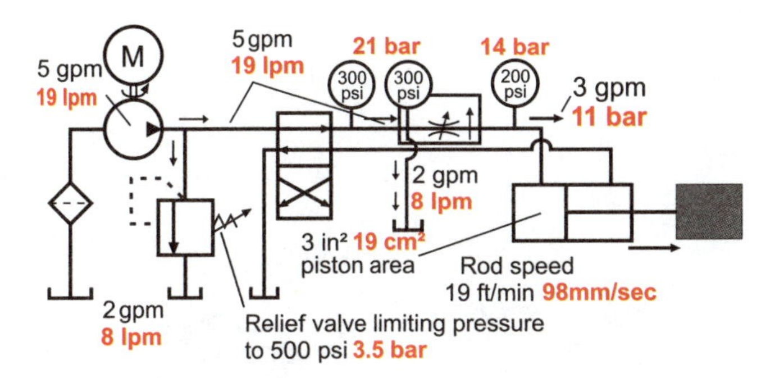

Figure 9-29

Workload Pressure and Relief Valve Setting Increased

If the workload pressure were increased to 400 psi (27.59 bar), pressure ahead of the needle valve orifice would then be limited to 500 psi (34.48 bar). 100 psi (6.89 bar) would still be available ahead of the orifice to develop the 3 gpm (11.37 l/min) flow rate.

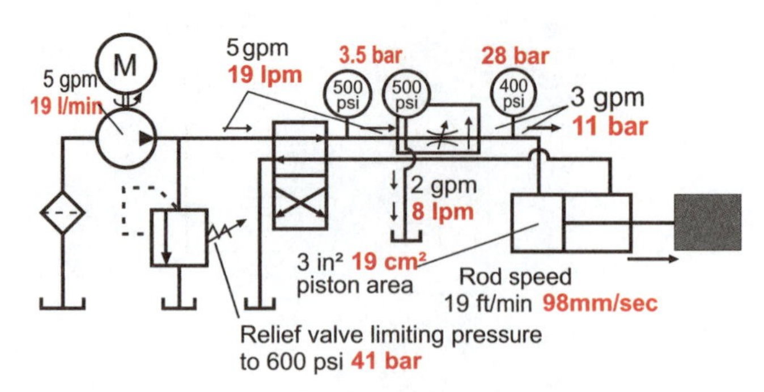

Figure 9-30

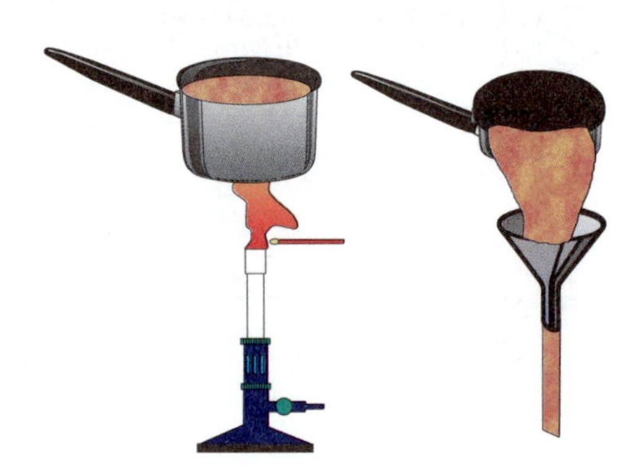

Figure 9-31

Note that when a bypass type pressure compensated flow control valve is used in a circuit, excess flow is not diverted over the relief valve. The operating pressure for the flow control is 100 psi (6.89 bar) above the workload pressure.

Since excess flow does not return to tank over the relief valve, a higher setting has no effect on the operation of the flow control. However, if the relief valve were not set at least 100 psi (6.89 bar) above the workload pressure, the compensator spool in the flow control would not operate. Flow through the needle valve would not be pressure compensated.

Even though a restrictor type valve generates more heat, it is the more popular type of pressure compensated flow control in an industrial hydraulic system. This popularity is probably the result of its accuracy and its flexibility in controlling flow in any part of a system. (A bypass flow control can only control flow to an actuator.)

Temperature Affects Flow

So far, it has been shown that flow through an orifice is affected by the size of the orifice and the pressure differential across the orifice. Flow through an orifice is also affected by temperature which changes a liquid's viscosity.

For example, pouring a viscous liquid like cold molasses from a sauce pan, through a funnel is a time-consuming job. Heating the sauce pan results in the molasses flowing readily through the funnel. Rate of flow through the funnel increases because heating reduces a liquid's viscosity.

Just like any mechanical, electrical or pneumatic system, hydraulic systems are not 100% efficient. While in operation this inefficiency shows up in the form of heat which reduces a liquid's viscosity. Like heated molasses, the warmed fluid flows more readily through an orifice. If the pressure differential across a metering orifice and the size of an orifice are kept constant, the flow rate through the orifice and to the actuator will increase with a rise in temperature. If precise actuator speed is necessary, a change in fluid temperature must be compensated for.

Temperature Compensation with a Metal Rod

One method of temperature compensation is the use of an aluminum rod. The rod is attached to the movable section of a variable orifice and controls the size of the orifice with a change in temperature.

How Temperature Compensation with a Metal Rod Works

The flow rate through an orifice tends to become greater as temperature increases. The heat expands the rod which pushes the movable section of the orifice toward its seat, decreasing the opening. The flow rate for the heated fluid through the smaller orifice is the same as the flow rate through the normal orifice before heating. Consequently, flow rate is not affected by an increase in temperature.

If temperature is decreased, flow rate tends to become less. The decreased temperature causes the rod to contract which pulls the movable section away from its seat, increasing the opening.

The flow rate for the cooled fluid with the larger orifice is the same as the flow rate through the normal orifice before it was cooled. Therefore, flow is not affected by a decrease in temperature.

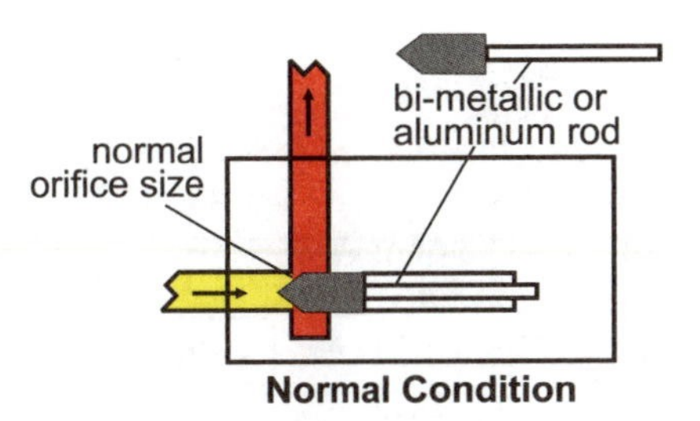

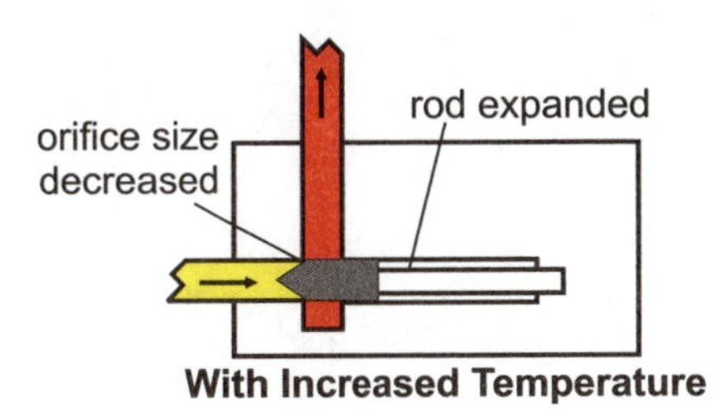

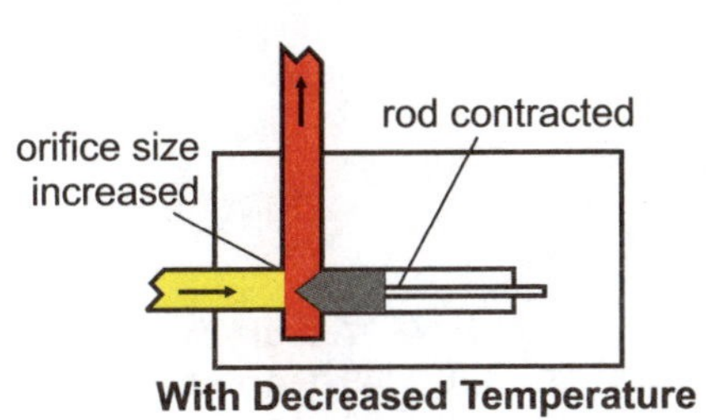

Figure 9-32

Temperature Compensation with a Sharp Edge Orifice

Laboratory experiments have shown that when liquid passes through a properly shaped orifice with a sharp edge, rate of flow is not affected by temperature. The manner in which liquid is sheared, while moving across a sharp edge, is of such a character that it actually cancels out or neutralizes the effect of a fluid's viscosity. The reason this occurs is not clearly understood, but its effect results in very accurate control.

Temperature-Pressure Compensated Flow Control Valve

Temperature compensation using a sharp edge orifice is a non-moving type compensation which disregards the effects of temperature over a given range. It is very difficult to design and manufacture an orifice of this type because the characteristics of the orifice must fall within certain mathematical boundaries and the orifice must be precision machined and held to very close tolerances. Some manufacturers still use the aluminum or bi-metallic rod method of temperature compensation because of this difficulty.

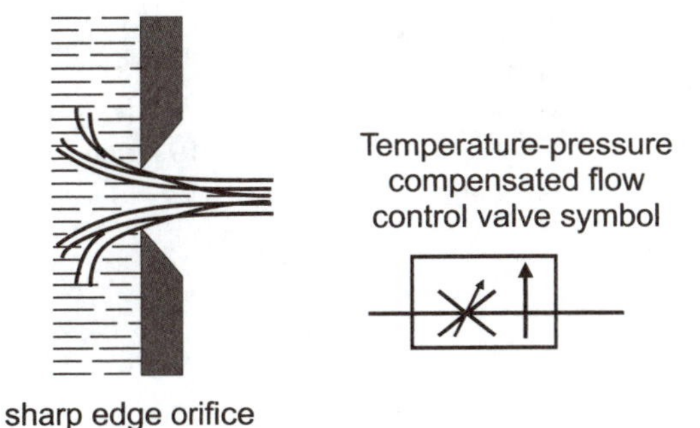

Figure 9-33

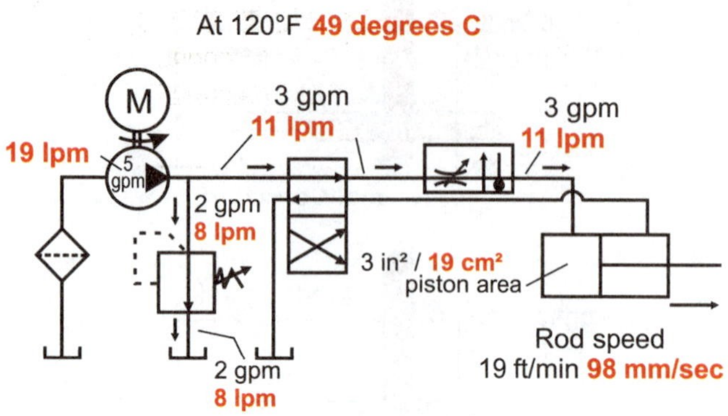

At 120°F 49 degrees C
M
19 lpm
5 gpm
3 gpm
11 lpm
3 gpm
11 lpm
2 gpm
8 lpm
3 in² / 19 cm²
piston area
Rod speed
19 ft/min 98 mm/sec
2 gpm
8 lpm

Figure 9-34

At 80°F 27 degrees C
120°F 49 degrees C
140°F 60 degrees C
M
19 lpm
5 gpm
3 gpm
11 lpm
3 gpm
11 lpm
2 gpm
8 lpm
3 in² / 19 cm²
piston area
Rod speed
19 ft/min 98 mm/sec
2 gpm
8 lpm

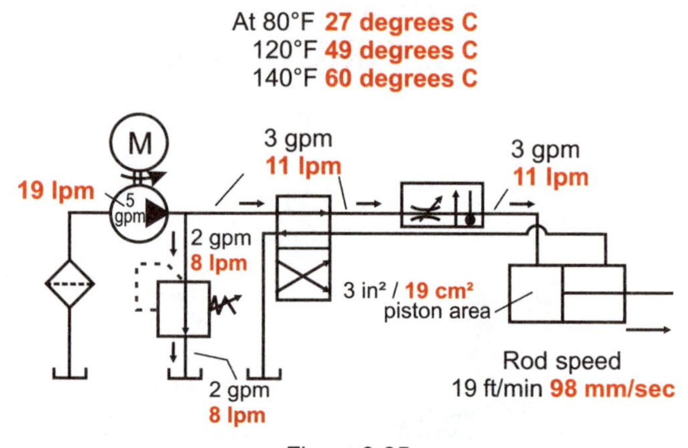

Figure 9-35

Temperature-Pressure Compensated Flow Control Valve in a Circuit

In the circuit illustrated, a pressure compensated flow control valve will control the operating speed of the cylinder effectively as long as the temperature remains at a constant 120°F (48.9°C).

The operating temperature of industrial hydraulic systems may range from 80°F (26.7°C) in the morning to 140°F (60°C) in the afternoon. As a result the operating speed of the actuator changes over the course of the day.

If the speed of an actuator must be precise throughout the workday, a temperature-pressure compensated flow control could be used.

Lunge Control

The proper operation of a pressure compensated or temperature-pressure compensated flow control valve depends upon the compensator spool partially restricting flow to the orifice and transforming excess pressure energy into heat. This was pointed out previously in the description of a restrictor type flow control.

When flow is not being metered through the valve, the compensator spool is fully seated toward side A and the path to the orifice is completely opened. When it comes time for the valve to operate, the orifice is directly subjected to the pump's pressurized flow for an instant before the compensator can react. This causes a burst of fluid to be pushed through the orifice and results in a jump or lunge at the actuator. In some precision applications, this jump or lunge may damage the machine, the tools, or the product being machined. To avoid this situation, a pressure compensated or temperature-pressure compensated flow control valve can be equipped with a lunge control.

A lunge control is a screw adjustment which pre-positions the compensator spool at the A side of the valve. While the valve is operating, the screw is turned in until it contacts the compensator spool, and then is backed off slightly. When the valve is not functioning, the spool does not reseat, but remains near its compensating position. Now when flow through the valve resumes, the spool compensates immediately and does not allow the actuator to jump. Lunge controls should be used only in circuits in which the load pressure remains relatively constant.

Flow Control Valves in a Circuit

Up to this time, when the operation of a particular flow control was described in a circuit, the flow control was positioned in the circuit directly before the actuator whose speed it was controlling. In this arrangement, all the flow is measured as it enters the actuator. This is termed a meter-in circuit.

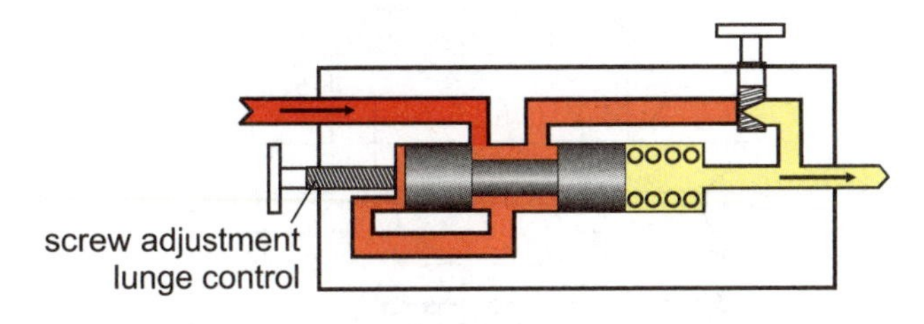

Figure 9-36

Meter-In Circuit

A meter-in circuit is used to control the speed of an actuator which works against a positive load. In other words, while the orifice is metering fluid to an actuator, the workload pressure is continuously a positive value. An example of a constant load would be any load which is vertically lifted.

A meter-in circuit is the only type of circuit in which a bypass type PC flow control can be used.

For reasons which will be explained in a later section, a meter-in circuit is not used to accurately control the speed of a hydraulic motor. See Appendix B-18.

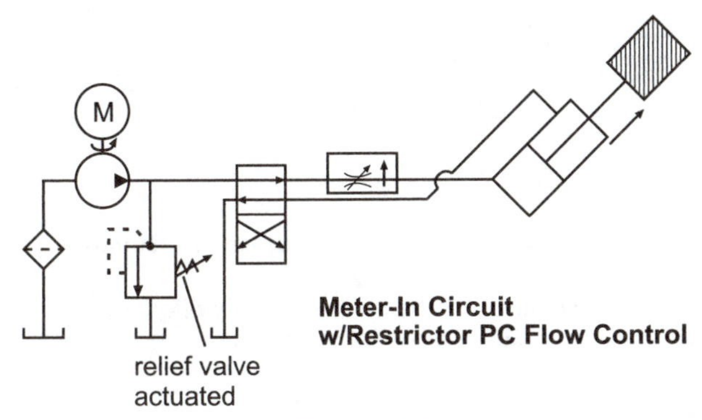

Figure 9-37

Meter-Out Circuit

In some cases, the work load changes direction (load passing over the center point of an arc) or the work load pressure suddenly changes from full to zero pressure (drill breaking through stock). This causes the cylinder to run away.

A flow control valve placed at the outlet port of an actuator controls the rate of flow exiting the actuator. This is a meter-out circuit and gives positive speed control to actuators used in drilling, sawing, boring and dumping operations. A meter-out circuit is a very popular industrial hydraulic flow control circuit. See Appendix B-19.

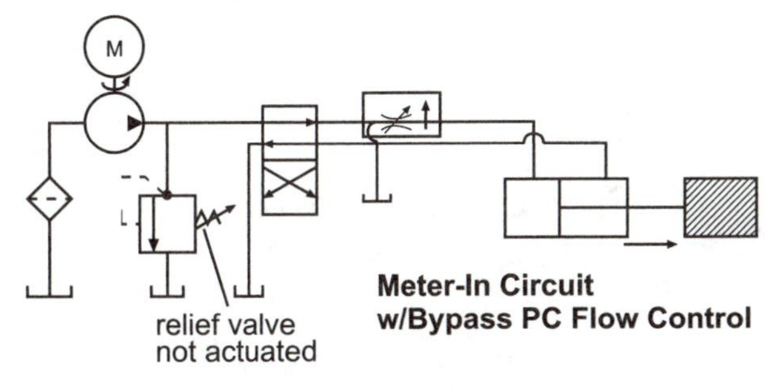

Figure 9-38

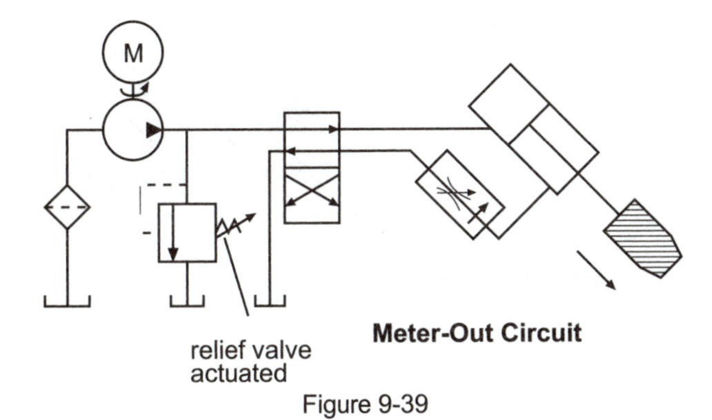

Figure 9-39

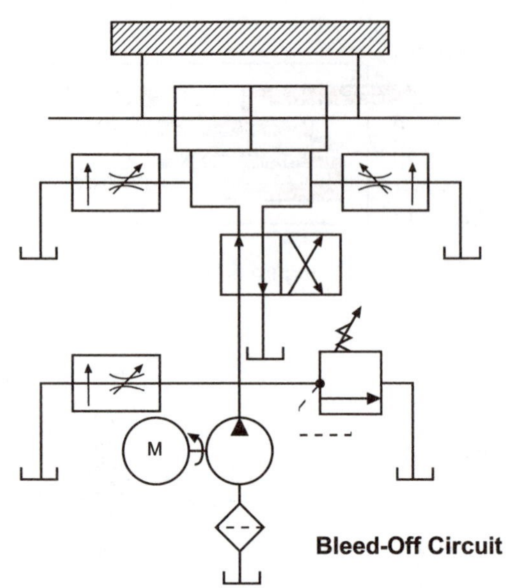

Flow control can be located directly after the pump or in a line to an actuator.

Figure 9-40

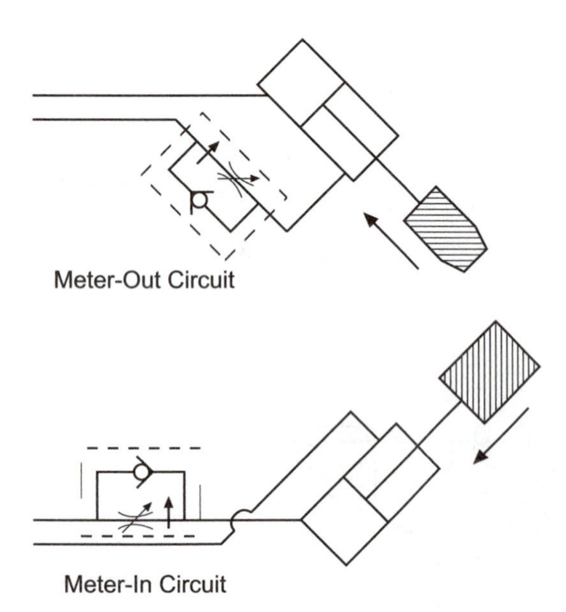

Figure 9-41

Bleed-Off Circuit

Another type of flow control circuit is the bleed-off circuit. In this circuit, the flow control valve does not cause an additional resistance for the pump. It operates by bleeding-off to tank a portion of the pump's flow at the existing system pressure.

Besides generating less heat, a bleed-off circuit can also be more economical than a meter-in or meter-out circuit. For instance, if a flow rate of 100 gpm (379 l/min) had to be reduced to 90 gpm (341.1 l/min), a 90 gpm (341.1 l/min) flow control valve would be needed in a meter-in circuit and, depending on the size of the cylinder, approximately a 70 gpm (265.3 l/min) flow control in a meter-out circuit. Whereas in a bleed-off circuit, a 10 gpm (37.9 l/min) flow control could be used.

Even with these apparent advantages, a bleed-off circuit is not a very popular flow control circuit. This is because a flow control in a bleed-off arrangement only indirectly controls the speed of an actuator. It can precisely meter flow to the tank, but if leakage through various system components increases, actuator speed will decrease.

A bleed-off circuit can be used in any application where precision flow regulation is not required; and where the load offers a constant resistance as in reciprocating grinding tables, honing operations, and vertically lifting a load. See Appendix B-20.

Reverse Flow Through a Flow Control Valve

In the examples seen so far, flow to an actuator was described as being controlled in one direction only. But cylinders and motors usually work in two directions. It is often not required, and even undesirable, to reduce the speed of an actuator in the opposite direction. To bypass a flow control valve when retracting a cylinder or reversing a hydraulic motor, a check valve is used.

Terms And Idioms Associated with Flow Control Valves

NON-JUMP CONTROL - lunge control

PRIORITY FLOW DIVIDER - bypass type pressure compensated flow control valve; secondary flow is to an actuator instead of to tank.

Exercise
Flow Control Valves
40 Points

Instructions: Answer the following questions or solve the following problems as required.

1. Assuming the cylinders to be the same size and the loads to be equal, how does the flow rate through the needle valve differ from situation A to situation B?
   a. flow remains exactly the same
   b. flow decreases
   c. flow increases

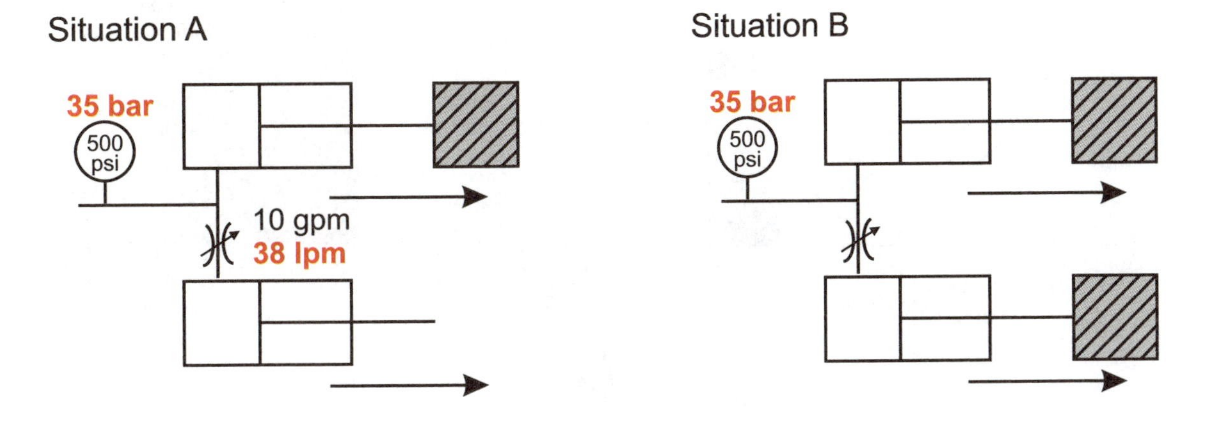

2. Assuming pipe friction to be zero, what do the gages read? (The spring acting on the spool in the pressure compensated flow control has a value of 100 psi/34.5 bar.)

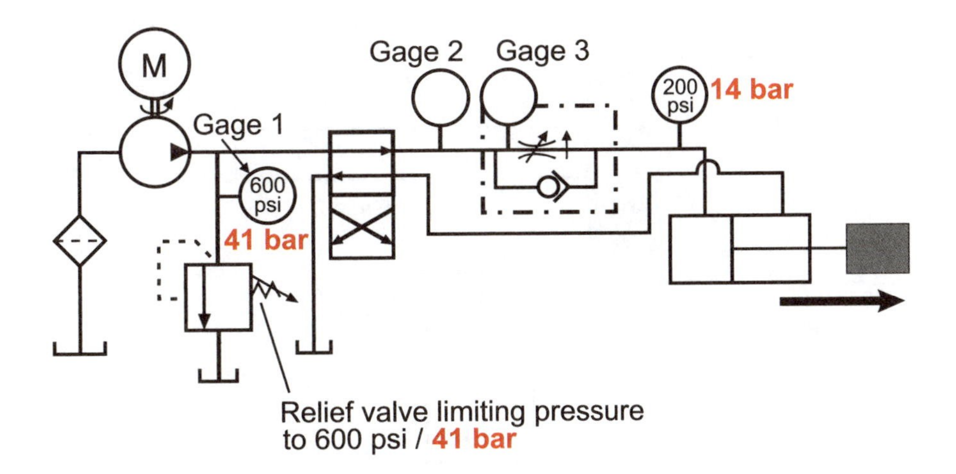

Flow Control Valves (cont.)

3. In situation A, the cylinder rod is pushing out the load. In situation B, there is no load on the cylinder rod.

   The system relief valve is set at 2000 psi (137.9 bar). The flow control valve is metering flow out of the cylinder.

   What do the gages read in each case?

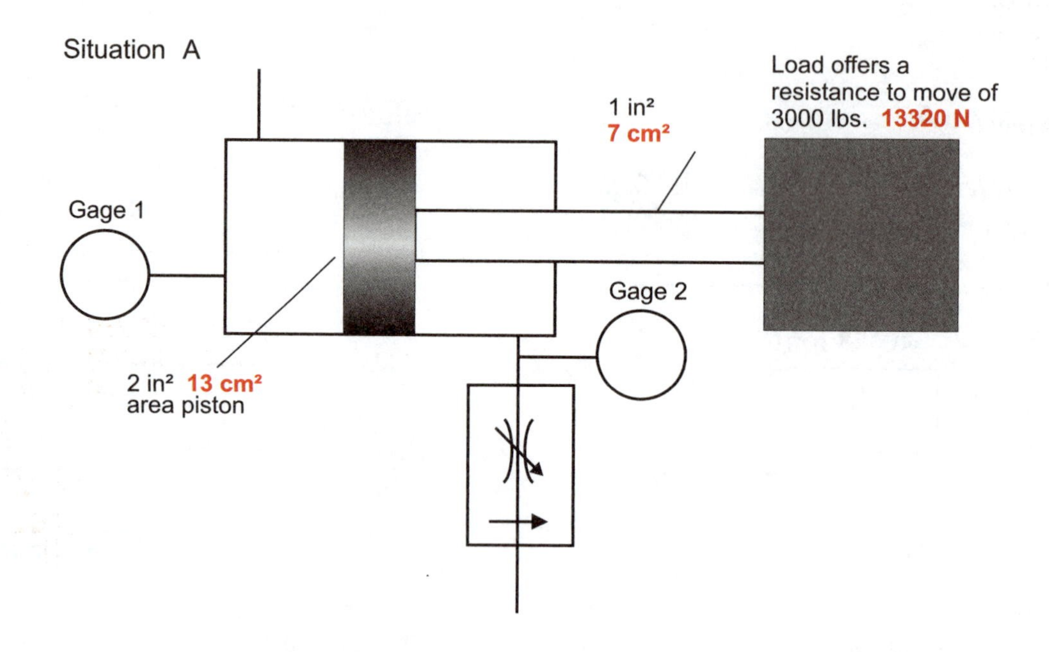

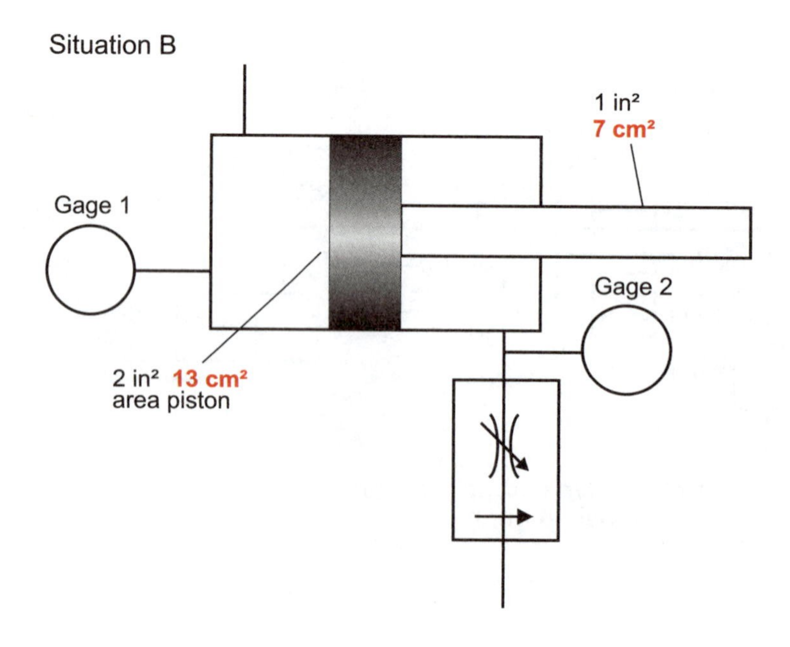

CHAPTER 10

Directional Control Valves

A directional control valve consists of a body with internal passages which are connected and disconnected by a movable part. In directional valves, and in most industrial hydraulic valves we have seen previously, the movable part has been a spool. Spool valves are by far the most common type of directional valve used in industrial hydraulics. For this reason, we will concentrate on these valves - their types and operation.

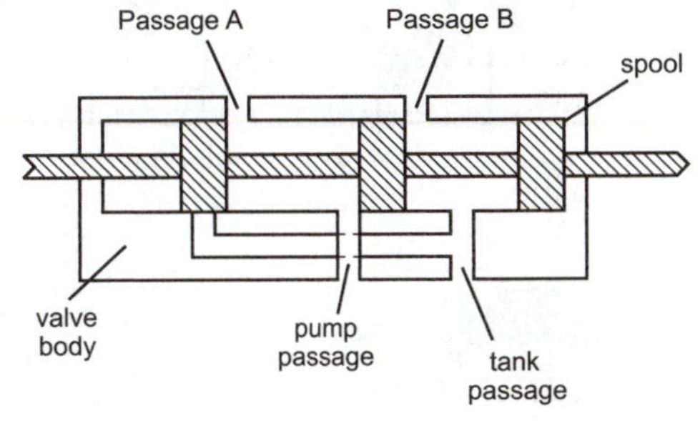

Figure 10-1

4-Way Directional Valve

In the directional valve we have dealt with earlier, the valve consisted of a pump passage, tank passage, and two actuator passages. This valve is known as a 4-way valve because it has four distinct passages within its body.

The function of a 4-way directional valve is to cause the reverse motion of a cylinder or hydraulic motor. To perform this function, the spool directs flow from the pump passage to one actuator passage when it is in one extreme position. At the same time, the spool is positioned so that the other actuator passage is exhausted to tank.

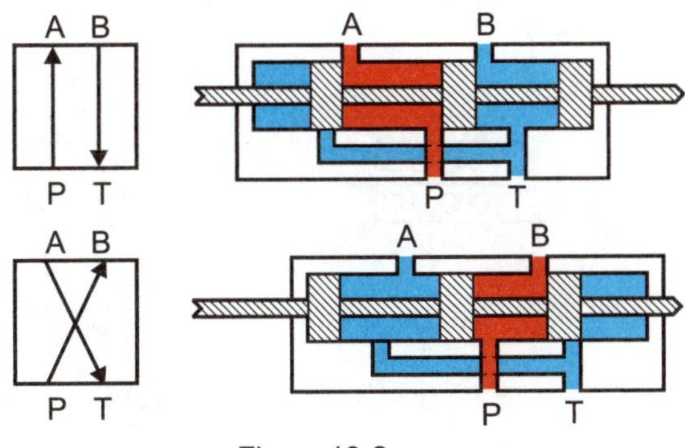

Figure 10-2

4-Way Directional Valves in a Circuit

Since all valves consist of a body and an internal moving part, the moving part of all valves has at least two positions - both extremes. These two positions in a directional valve are depicted by two separate squares. Each square shows by means of arrows how the spool is connecting the passages within the body at that point in time. When the valve is shown symbolically, the two squares are connected together. But, when placed in a circuit, one square, and only one square, is connected in the circuit. With this arrangement, the condition within the valve is shown while an actuator is moving in one direction. To picture the actuator moving in the opposite direction, the other square of the symbol is mentally slid into the circuit.

A directional valve symbol, which shows only two spool conditions within a body, is known as a two-position valve.

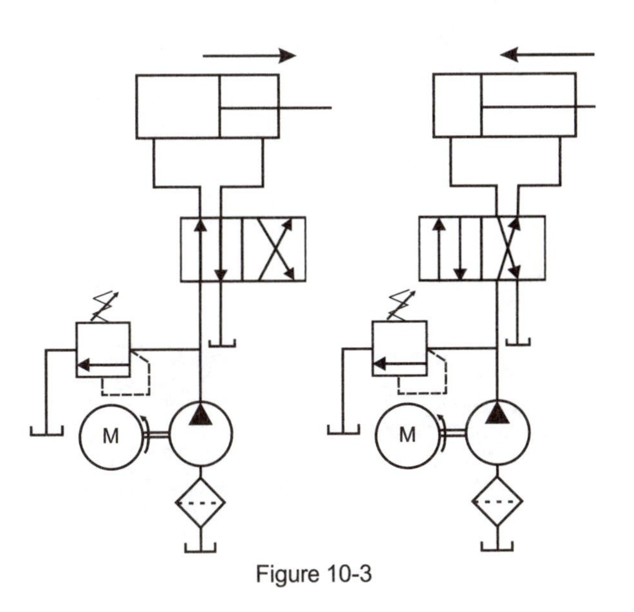

Figure 10-3

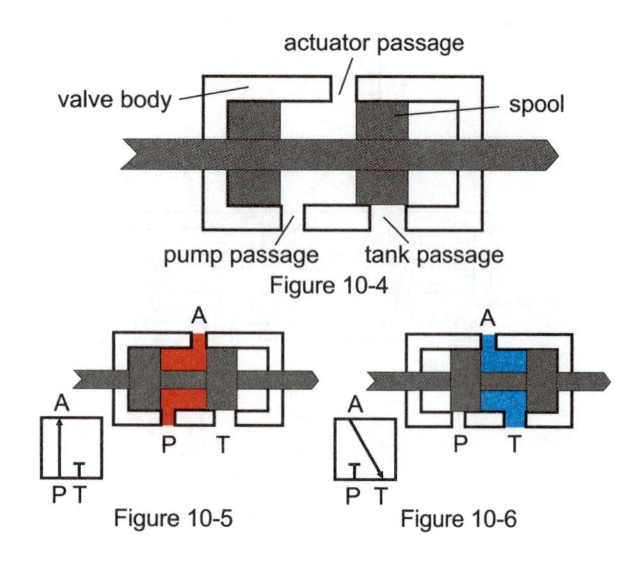

Figure 10-4

Figure 10-5

Figure 10-6

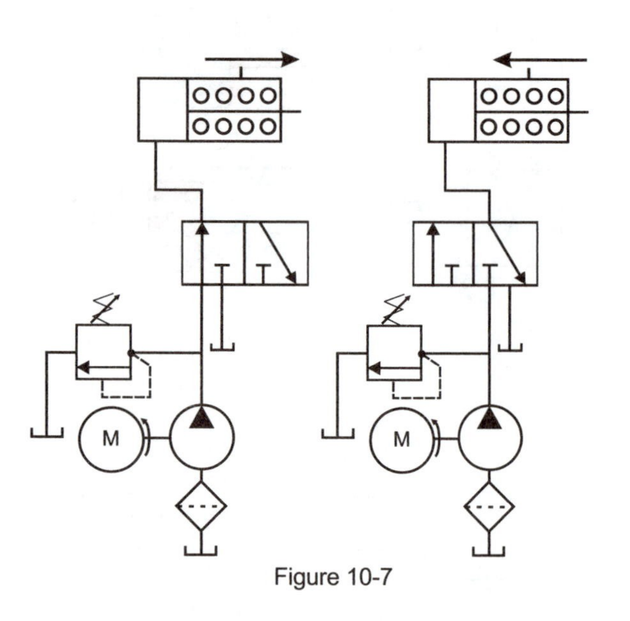

Figure 10-7

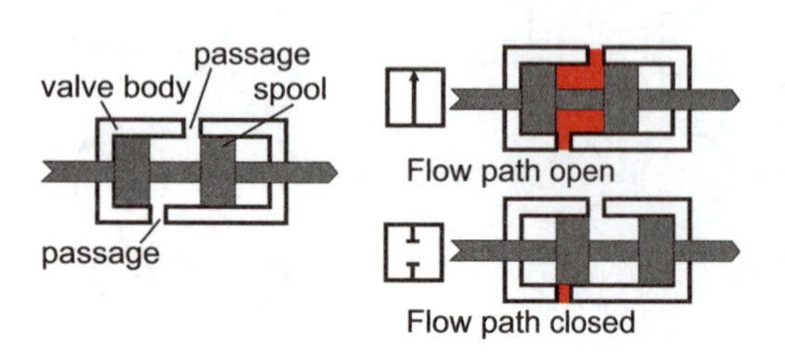

Figure 10-8

3-Way Directional Valve

A 3-way valve consists of three passages within a valve body - pump passage, tank passage, and one actuator passage.

The function of this valve is to pressurize an actuator port when the spool is in one extreme position. When the spool is positioned in the other extreme, the valve exhausts the same actuator port. In other words, the valve alternately pressurizes and exhausts one actuator port.

3-Way Directional Valves in a Circuit

A 3-way directional valve can be used to operate single acting actuators like rams and spring-return cylinders. In these applications, the 3-way valve directs fluid pressure and flow to the cap end side of the cylinder. When the spool is shifted to the other extreme position, flow to the actuator is blocked. At the same time, the actuator passage within the body is connected to tank.

A vertical ram is returned by its own weight or the weight of its load when the actuator passage of a 3-way valve is drained to tank. In a spring return cylinder, the piston rod is returned by a spring within the cylinder body.

In industrial hydraulic applications, 3-way valves are not generally found. If a 3-way function is required, a 4-way valve is usually converted to a 3-way valve by plugging an actuator port.

2-Way Directional Valve

A 2-way directional valve consists of two passages which are connected and disconnected. In one extreme spool position, the flow path through the valve is open. In the other extreme, there is no flow path through the valve.

2-Way Directional Valve in a Circuit

A 2-way directional valve gives an on-off function. This function is used in many systems to serve as a safety interlock and to isolate and connect various system parts.

Directional Valve Sizes and Ratings

Directional control valves used in industrial hydraulic applications come in several basic sizes - 1/4", 3/8", 1/2", 3/4", and 1¼" (6.35 mm, 9.5 mm, 12.7 mm, 19.05 mm, and 31.75 mm). It is common industrial practice to rate the valves on their nominal or average flow rating. Examples of such ratings are respectively at 3-10 gpm (11.37 - 37.9 l/min), 10-20 gpm (37.9 - 45.48 l/min), 40 gpm (75.8 l/min), 80 gpm (303.2 l/min) and 160+ gpm (379 l/min). At this nominal gpm rating the pressure differential from P to A or B to T is approximately 40 psi (2.76 bar).

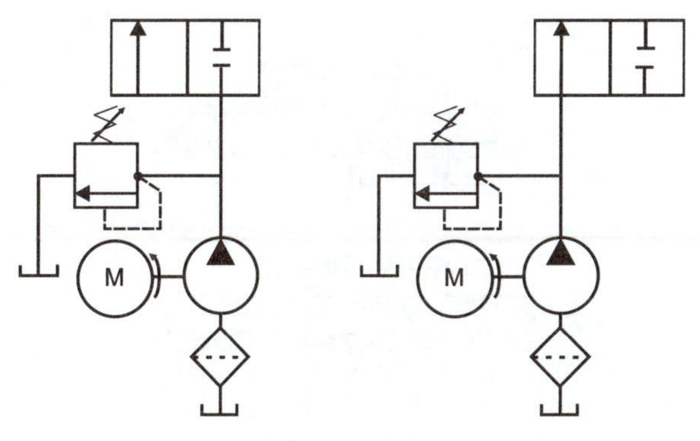

Figure 10-9

Directional Valve Spools

The spool in the 4-way directional valve which we have seen consists of a shaft with three large diameters spaced apart equally. These are known as lands. The spool is identified as a 3-land spool.

Most industrial hydraulic directional valves have either 2-land or 4-land spools. Directional valves with small flow ratings are generally equipped with 2-land spools, while the 4-land spools are most frequently found in valves with large flow ratings.

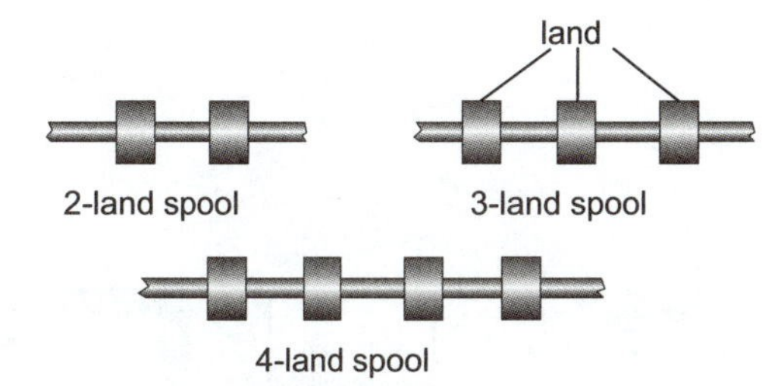

Figure 10-10

Subplate-Mounted 4-Way Valves

The bodies of the 4-way directional valves which have been illustrated had pump and tank passages situated on one side. The two actuator passages were positioned on the opposite side of the body. This arrangement closely followed the symbol for the valve. But, for ease in installation, most industrial hydraulic directional valves are subplate mounted; that is, they are bolted to a plate to which system piping is connected. The ports of subplate-mounted valves are located on the bottom surface of the valve body.

Subplate-mounted, 4-way body with a 4-land spool

Figure 10-11

Directional Valve Actuators

We have seen that a directional valve spool can be positioned in one extreme position or the other. The spool is moved to these positions by mechanical, electrical, hydraulic, pneumatic, or human energy.

Directional valves whose spools are moved by muscle power are known as manually operated or manually actuated valves. Various types of manual actuators include levers, pushbuttons, and pedals.

A very common type of mechanical actuator is a plunger. Equipped with a roller at its top, the plunger is depressed by a cam which is attached to an actuator. Manual actuators are used on directional valves whose operation must be sequenced and controlled at an operator's discretion. Mechanical actuation is used when the shifting of a directional valve must occur at the time an actuator reaches a specific position.

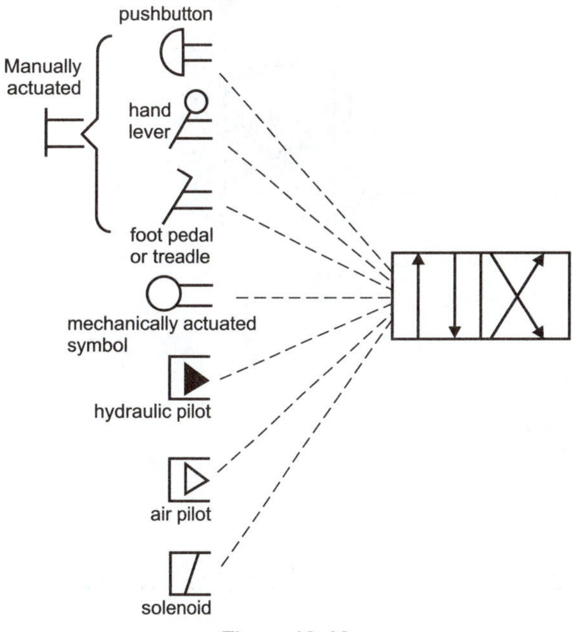

Figure 10-12

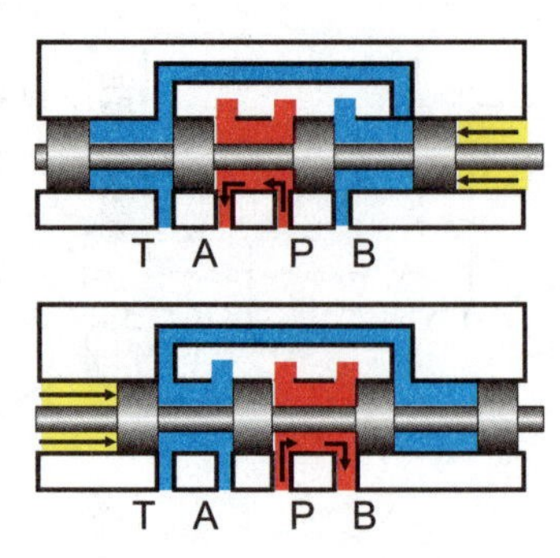

Figure 10-13

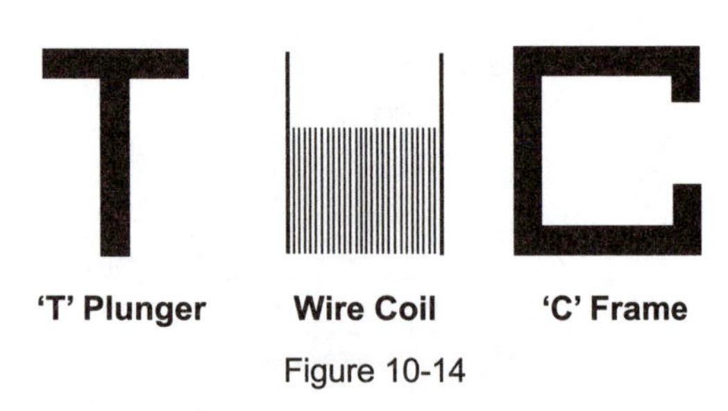

Figure 10-14

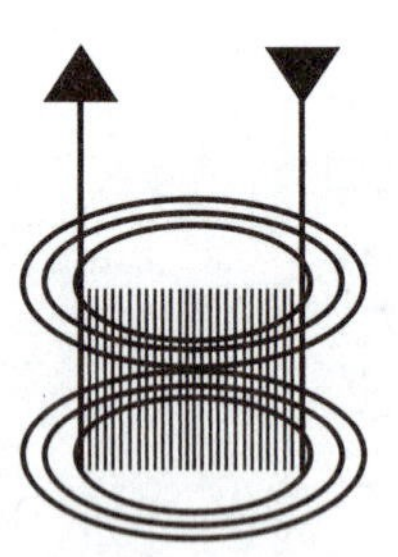

Figure 10-15

Pilot Operation

Directional valve spools can also be shifted with fluid pressure either air or hydraulic. In these valves, pilot pressure is applied to the two extreme spool lands if it is a 4-land spool, or to separate pilot pistons if it is a 2-land spool.

Solenoid Operation

One of the most common ways of operating a directional valve is with a solenoid.

Solenoid

A solenoid is an electromechanical device which converts electric power into linear mechanical force and motion. Its counterpart in a hydraulic system is a cylinder.

Solenoids typically found on industrial hydraulic valves can be divided into "air gap" and "wet armature" types.

Air Gap Solenoid

Of the two types, air gap solenoids are the older design. They are basically an electromagnet made up of a T plunger, wire coil, and C frame. Because of the shape of the plunger and frame around the coil, this is sometimes called a "CT" solenoid.

How an Air Gap Solenoid Works

With an electric current passing through a wire, a magnetic field sets up around the wire. As illustrated earlier, this effect can be seen by sprinkling iron filings on a plastic sheet through which a current-carrying wire is located. The electric current will cause filings to take the shape of its magnetic field.

If the wire were coiled in many turns, the magnetic field would be several times stronger generating around the coil and through its center.

A solenoid depends on this magnetic field to shift a directional valve spool. The more intense the field, the more shifting force will be developed. To intensify a magnetic field, an air gap solenoid is equipped with an iron path, called a C frame, which surrounds the coil. Another iron path, the plunger, is positioned in the center of the coil to concentrate the magnetic field even more; this iron path is movable.

As an air gap solenoid coil receives electric current, the plunger is partially out of the coil. The resultant magnetic field generated from the current, attracts the plunger pulling it in. The directional valve shifts as the plunger hits a pushpin mechanically connected to the valve spool. With the spool shifted, the plunger fully seats within the coil resulting in the magnetic field traveling completely through an iron path.

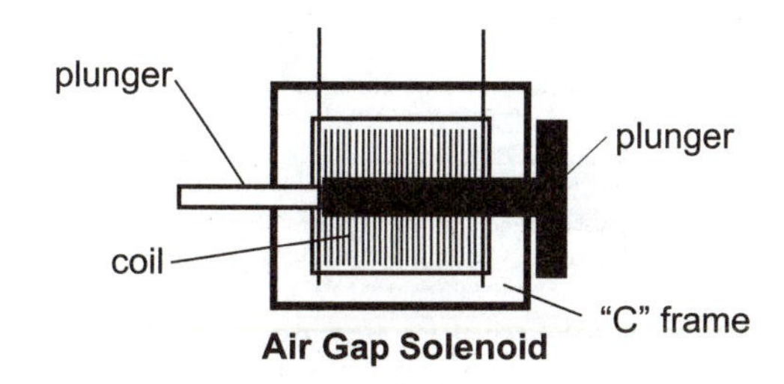

Figure 10-16

Iron is an excellent magnetic conductor; air is a poor magnetic conductor. Air gap solenoid operation depends on the magnetic field pulling in the plunger reducing the high resistance air gap within the coil center. As the plunger moves in, the air gap gradually decreases causing solenoid force to become increasingly stronger. Solenoid force is greater with the plunger pulled in than out.

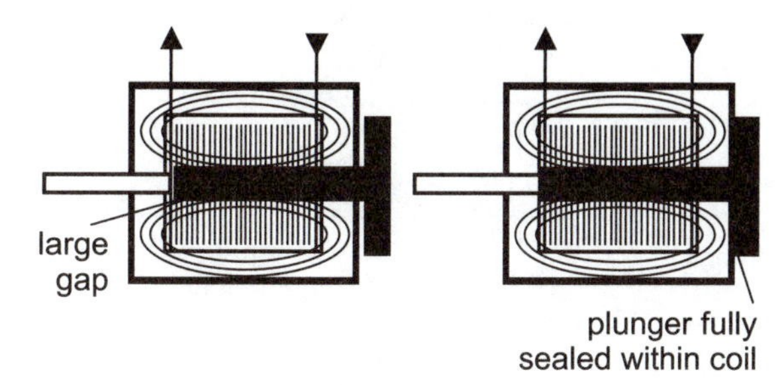

Figure 10-17

Wet Armature Solenoid

Wet armature solenoids are a relatively new arrival on the industrial hydraulic scene. Their acceptance over air gap designs has been due to their increased reliability resulting from better heat transfer characteristics and elimination of pushpin seals which in an air gap solenoid have a tendency to leak.

What a Wet Armature Solenoid Consists Of

A wet armature solenoid consists of a coil, rectangular frame, pushpin, armature (plunger) and tube. The coil is surrounded by the rectangular frame and both are encapsulated in plastic. In the encapsulated unit, a hole runs through the coil center and two sides of the frame. The tube fits within this bore as it is screwed into a directional valve body. Housed within the tube is an armature which is bathed by system fluid through the tank passage within the directional valve. This accounts for the "wet armature" identification.

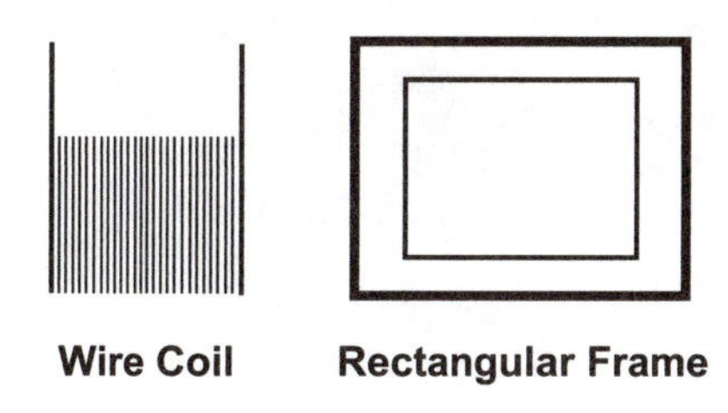

Figure 10-18

How a Wet Armature Solenoid Works

With an electrical current passing through its windings, a magnetic field sets up around the coil. This magnetic field is intensified by the rectangular iron path surrounding the coil and also by the armature in the coil center.

As a wet armature coil receives electric current, the movable armature is partially out of the coil. The magnetic field generated from the current attracts the armature pulling it in. Directional valve shifts as the armature hits a pushpin in contact with the valve spool. With spool shifted, armature is fully centered within the coil resulting in the coil magnetic field traveling completely through an iron path.

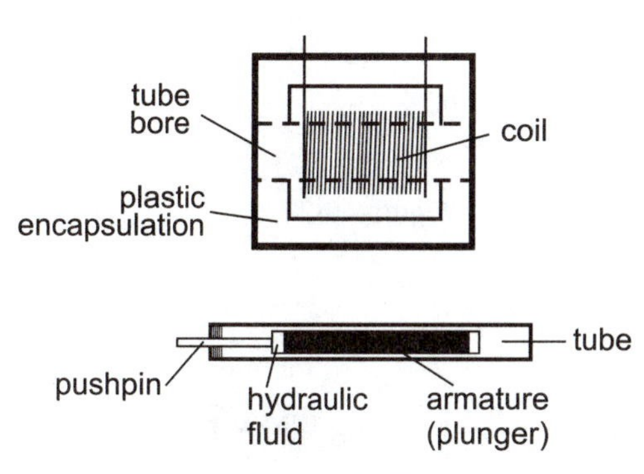

Figure 10-19

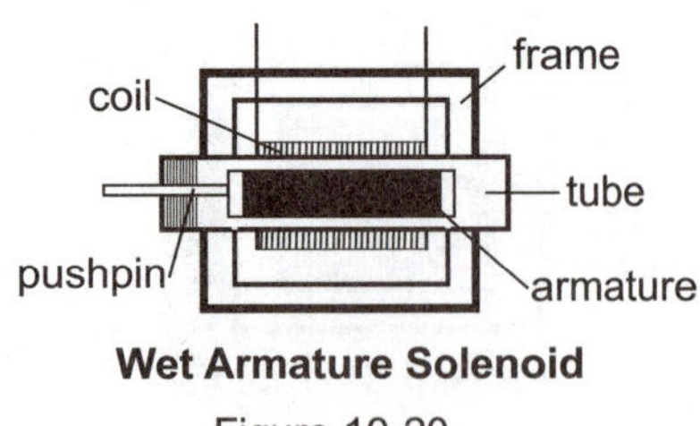

Wet Armature Solenoid

Figure 10-20

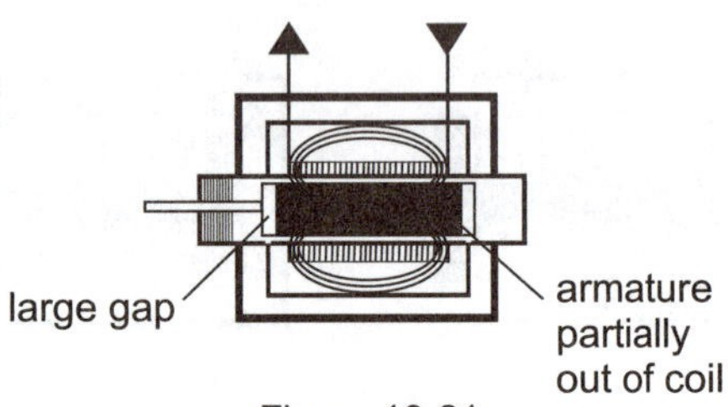

Figure 10-21

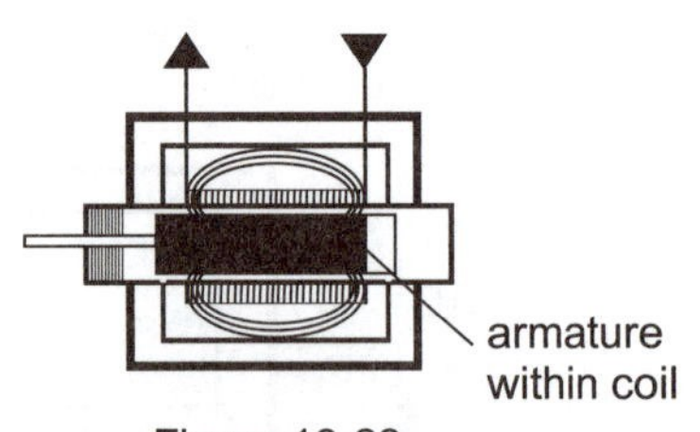

Figure 10-22

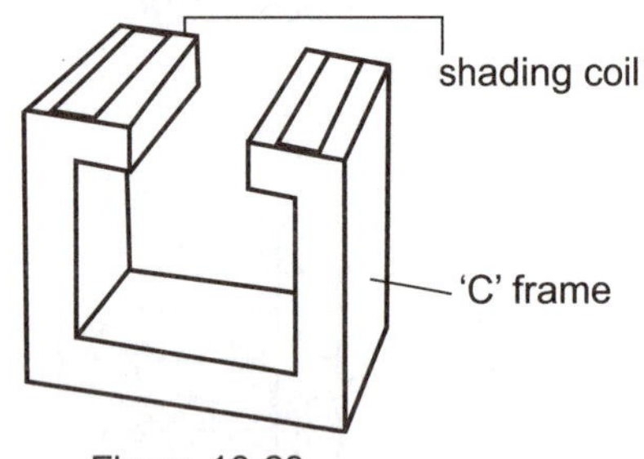

Figure 10-23

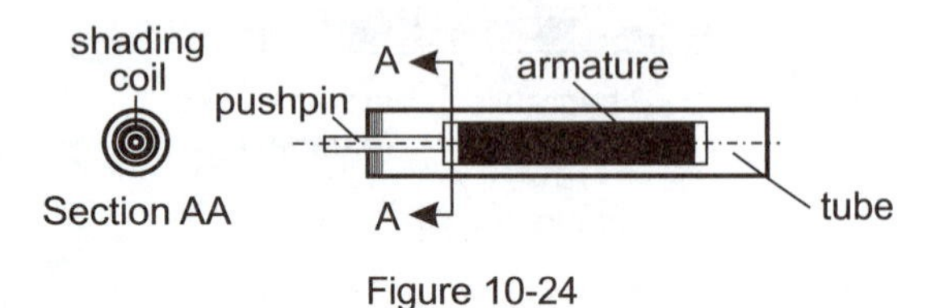

Figure 10-24

Iron is an excellent magnetic conductor; the oil surrounding armature and pushpin is a poor magnetic conductor. Wet armature solenoid operation depends upon the magnetic field pulling in the armature reducing the high resistance gap in the coil center. As the armature moves in, the gap gradually decreases causing solenoid force to become increasingly stronger. Solenoid force is greater with the armature pulled in than out.

AC Hum

Electrical power with an alternating current (AC) is the usual industrial source of control power in the United States. Alternating current in the United States moves from zero to positive, through zero to negative, and back to zero at the rate of 60 times per second or 60 Hz.

Magnetic field and solenoid force are greatest when current is at positive and negative peak values. As current goes through zero, magnetism and solenoid force decrease. This causes the solenoid load (usually a spring-biased spool) to push out the plunger or armature. When magnetism and force build again, the plunger or armature is pulled back in. This motion results in the solenoid humming, buzzing or chattering and is known as AC hum.

Shading Coils

To minimize AC hum and increase solenoid force, shading coils are used. In an air gap solenoid, shading coils are copper loops which are attached to the C frame. In a wet armature solenoid, a shading coil is a copper wire ring at the pushpin end of the tube.

As the solenoid operates, current is generated in a shading coil which lags behind the applied current. Now, as the main magnetic field of the coil has its lowest value, the magnetic field of the shading coil is sufficient to hold the plunger or armature in. As a result, AC hum is greatly reduced.

Eddy Currents

AC magnetic fluctuations also cause small stray currents to develop within the solenoid; these are known as eddy currents. Moving in tiny circles within iron magnetic paths, eddy currents consume power and generate heat reducing solenoid force.

To minimize eddy current effects, the C frame and plunger of air gap solenoids are made up of thin metal sheets which are insulated from one another with an oxide coating. Each metal sheet is known as a lamination. Magnetism can easily flow in its usual path around the coil, but eddy currents cannot flow between laminations because of insulating coatings. Because it is made of laminated sheets stacked together, a C frame is frequently called a "C stack."

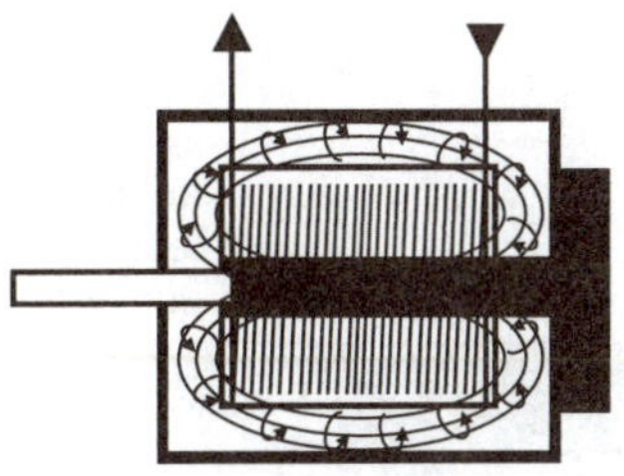

Figure 10-25

In a wet armature solenoid, eddy current effects are minimized by fabricating the rectangular frame surrounding the coil as laminations. However, this is not the case with the armature. Since a wet armature solenoid is more powerful than an air gap type, it is impractical from a durability standpoint to fabricate the armature as laminations. In a wet armature solenoid, the armature is one piece.

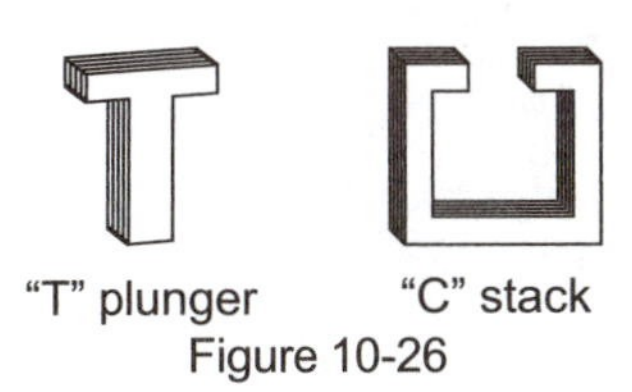

"T" plunger "C" stack

Figure 10-26

By containing eddy currents, heat is reduced and solenoid force is increased.

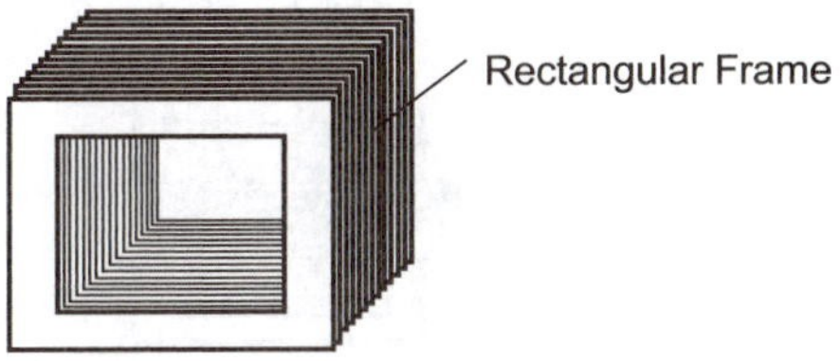

Rectangular Frame

Figure 10-27

Solenoid Inrush Current

Alternating current in a solenoid coil results in solenoid force to shift a directional valve spool. It is also the cause of heat generation which can eventually burn out a solenoid coil. The more current flowing in solenoid windings, the more tendency there is for the solenoid coil to fail.

If the solenoid does not move to close the air gap the current remains high causing excessive heat. Unlike positive displacement, industrial hydraulic systems where fluid flow is relatively constant, current flow in the common electrical system is related to resistance. The greater the resistance to current flow, the less current will develop and the less heat will be generated. Since heat significantly affects solenoid life, the goal, then, is to achieve sufficient shifting and holding force with as little current as possible. This is accomplished by designing into the coil sufficient AC resistance which is known as impedance.

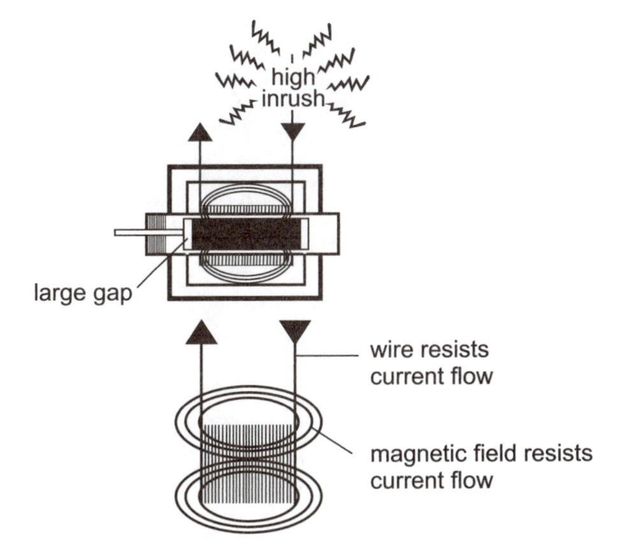

Figure 10-28

Impedance is made up of two elements. One element is the result of pure resistance to electron flow of the conductor material e.g., copper offers less electrical resistance than aluminum. The other element of impedance is the effect generated by the electromagnetic field surrounding the coil. This field tends to hold back or restrict current flow into the coil. The stronger the magnetic field becomes, the less current flows in solenoid windings.

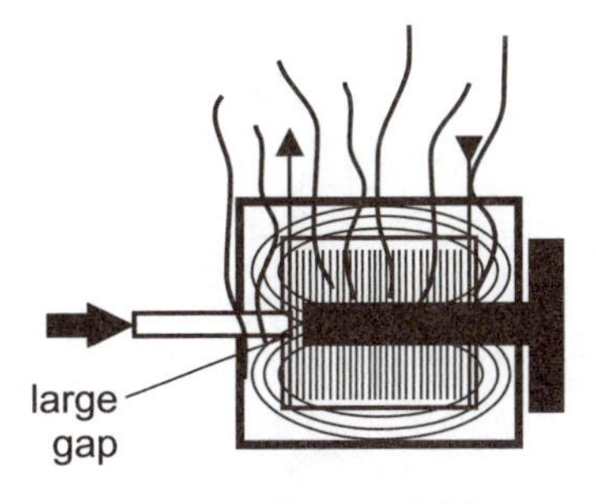

Figure 10-29

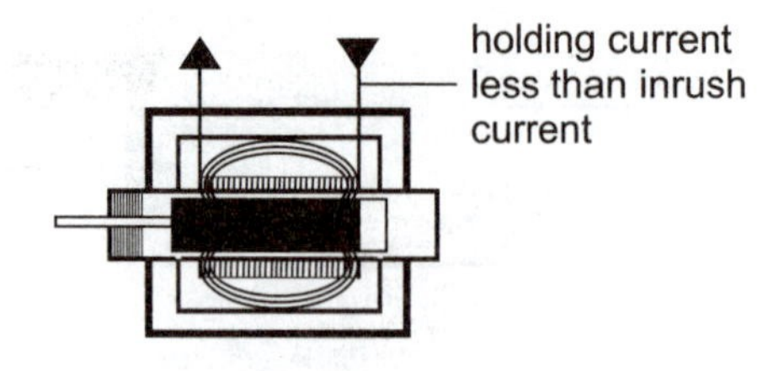

Figure 10-30

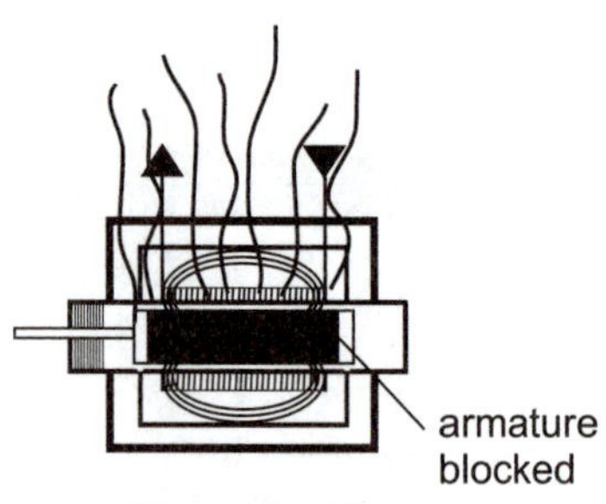

Figure 10-31

As solenoid is energized, its plunger or armature is partially out of the coil. Magnetic field is not at full strength because of the high resistance gap within coil center. At this point, resistance to current flow comes primarily from pure resistance to electron flow of the conductor material. Consequently, a high inrush current is experienced in coil windings. This decreases as plunger or armature moves in completing the magnetic path. With plunger or armature fully seated, impedance is maximum; alternating current within the solenoid is minimum.

Peak inrush current as a solenoid is energized, is several times greater than holding current with plunger or armature fully within the coil. If anything obstructs plunger or armature from fully seating, large amounts of current will rush into the coil generating high temperatures. In a non-encapsulated air gap solenoid, this can cause plastic material on coils ends to melt. With encapsulated air gap or wet armature designs, plastic encapsulation bubbles. And in both solenoid types, wire insulation deteriorates rapidly shorting out the coil in a minute or two. This is the case as a solenoid attempts to shift a spool which is blocked or stuck.

Continuous Duty Solenoid

A continuous duty solenoid is one that can be held energized indefinitely without overheating. Heat dissipating ability of the solenoid is great enough to dissipate most heat generated by the coil's lower holding current.

Solenoids of most industrial hydraulic directional valves are continuous duty solenoids.

Solenoid Limitations

Solenoid operated directional valves have a few limitations. Where a hydraulic system is used in a wet or explosive environment, ordinary solenoids may not be used. Where the cycle life of a directional valve must be extremely long, an electrically controlled solenoid valve is not generally used.

Probably the greatest disadvantage of solenoids is that the force which can be developed by them to shift a directional valve spool is limited. As a matter of fact, the force required to shift a directional valve spool is substantial in the larger sizes. In the larger valves, many times a 1/4" (6.35 mm) or 3/8" (9.5 mm) solenoid operated directional valve is positioned on top of the larger valve. Flow from the small valve is directed to either side of the large valve spool when shifting is required. These valves are designated solenoid controlled, pilot operated directional valves.

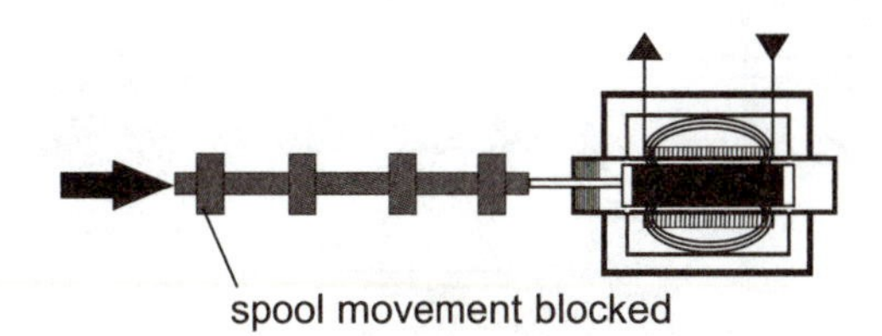

Figure 10-32

Solenoid Failure

Solenoids of direct acting directional valves fail because of heat. This is generally the result of solenoid blockage, high ambient temperatures, or low voltage (insufficient force to seat the plunger).

Solenoid Blockage

A valve spool which has become stuck will block a solenoid plunger or armature from completely closing; this results in a solenoid coil receiving a high inrush current continuously. The solenoid will be incapable of dissipating the developed heat; and the coil will burn out.

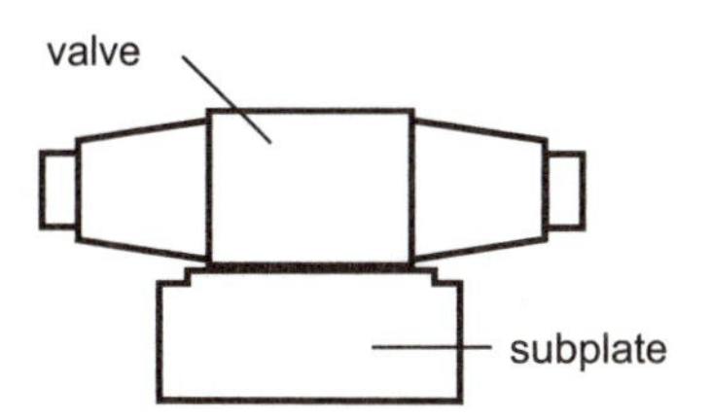

Figure 10-33

Although excessive flow through a valve will block a solenoid, mechanical interference of spool movement is the more frequent cause of blockage. Valve spools can become stuck because of contamination like silt, metal chips, coring sand, and Teflon® tape or because of burrs which build up between spool and valve body. Oxidized oil particles or varnish can also coat a spool eliminating the clearance between spool and body. Varnish can be removed usually by washing with lacquer thinner.

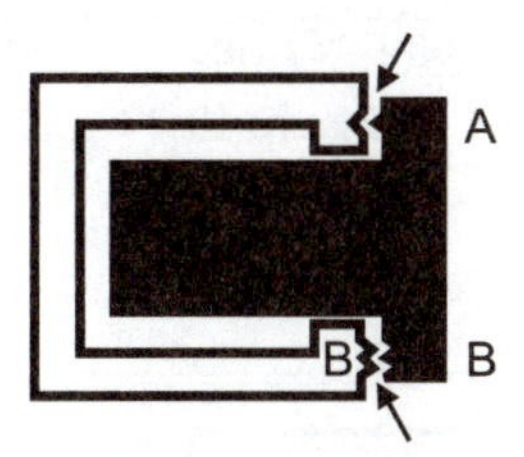

Figure 10-34

A solenoid can also become blocked due to a valve base which is not flat. In this condition, when mounting bolts are tightened, the valve base may warp slightly, restricting spool movement resulting in coil burnout. Generally, the flatness of a valve base is required to be within 0.0003-0.0005 in (.00762 - .0127 mm).

In an air gap solenoid, as a solenoid plunger is drawn into its coil, a wear pattern develops between plunger and C frame. When a solenoid of this type is disassembled, it is recommended that the plunger be replaced in its original position. If not, then different wear patterns on either end of the plunger will not match. The plunger will not fully seat. The solenoid will buzz indicating failure is near.

In some cases, a double solenoid valve may have its solenoids energized at the same time. This usually means one plunger or armature fully seats and the other becomes blocked; one solenoid coil burns out as a result. Simultaneous energizing of two solenoids is frequently caused by a failed or faulty component in the electrical control or an incorrect electrical hookup.

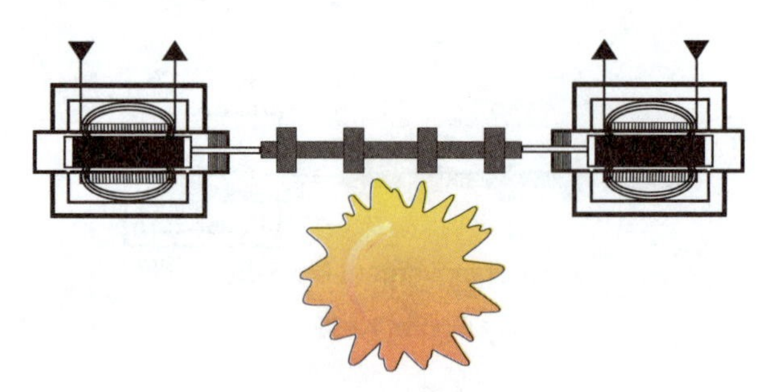

Figure 10-35

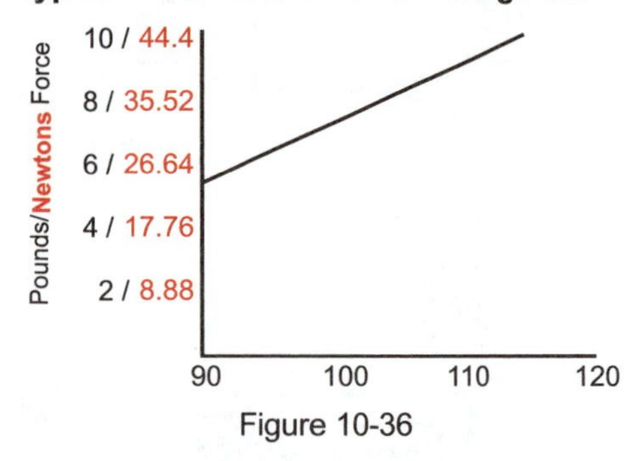

Figure 10-36

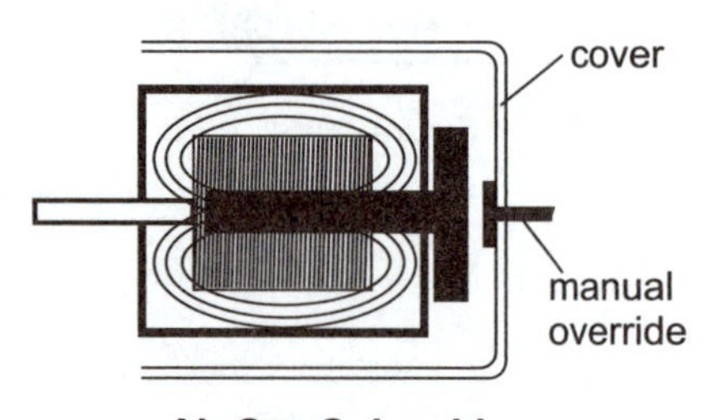

Air Gap Solenoid

Figure 10-37

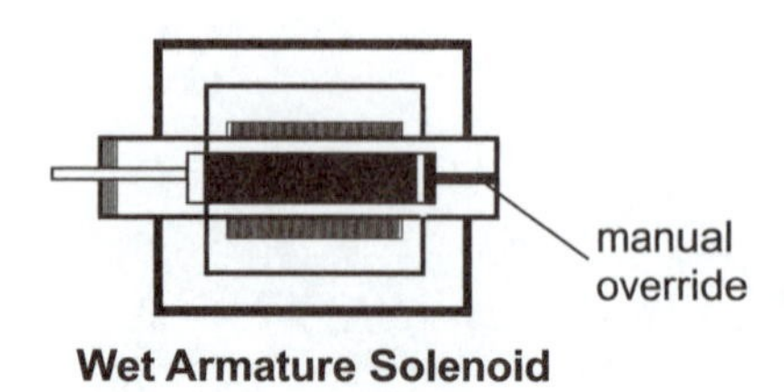

Wet Armature Solenoid

Figure 10-38

High Ambient Temperature

As electrical current pulls in a solenoid plunger or armature, heat must be dissipated from the coil. If the surrounding air temperature is exceedingly high, heat dissipation will be difficult and the coil may burn out. This can be the case if ventilation is poor or the machine is operating near a source of heat.

Low Voltage

When line voltage drops to approximately 100 volts for 115 volt, 60 Hz solenoids, insufficient force is developed to seat plunger or armature within the designed time frame. With inrush current generated for a longer time period, a solenoid operates at a higher temperature eventually failing. In some cases, early stages of coil burnout are indicated by a buzzing or chattering noise. Low voltage problems frequently occur during periods when many utility customers require power. If low voltage is suspected, the power company can install a 24-hour recorder to verify the condition.

Manual Override

Air gap solenoids of industrial hydraulic valves made to industry standards are protected by covers. On cover ends, a small metal pin is located. The pin is positioned directly in line with the solenoid plunger. As the pin is mechanically pushed into the cover, it contacts the plunger hitting the valve pushpin shifting the spool. Solenoid function is overridden. The pin is known as a manual override. The manual override in a wet armature solenoid is located on the end of the tube which houses armature and pushpin.

Manual overrides are used to check movement of a directional valve spool. If a solenoid failed because a spool jammed, spool movement can be checked by pushing in the manual override.

With an air gap solenoid, spool movement frequently cannot be checked with a burnt-out coil in a solenoid cover. When an air gap solenoid fails due to a plunger not fully seating, high temperature causes plastic material on the coil end to melt. Melted plastic obstructs plunger movement within the coil. Under this circumstance, the manual override cannot be pushed into the cover.

Manual overrides are also used to cycle an actuator without energizing the complete electrical control system. If cylinder operation needs checking or flow controls require adjustment, manual overrides can be used to cycle the actuator.

Solenoid Controlled Pilot Operated, Spring Offset

Solenoid operated, pilot operated, two position directional valves are not frequently equipped with two solenoids. This is an unnecessary expense and an additional solenoid to worry about in the system.

More commonly, a two position directional valve uses a solenoid to shift the pilot valve spool to an extreme position. The spool is generally returned to its original position by means of a spring. Two position valves of this nature are known as solenoid controlled pilot operated spring offset directional valves.

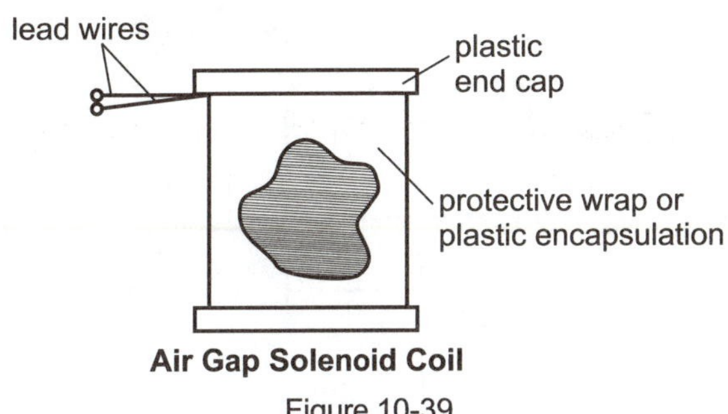

Air Gap Solenoid Coil

Figure 10-39

Normally Passing and Normally Non-Passing Valves

Spring offset 2-way and 3-way valves can be either normally passing or normally non-passing.

When spring offset directional valves are shown symbolically in a circuit, the symbol indicates its normal condition.

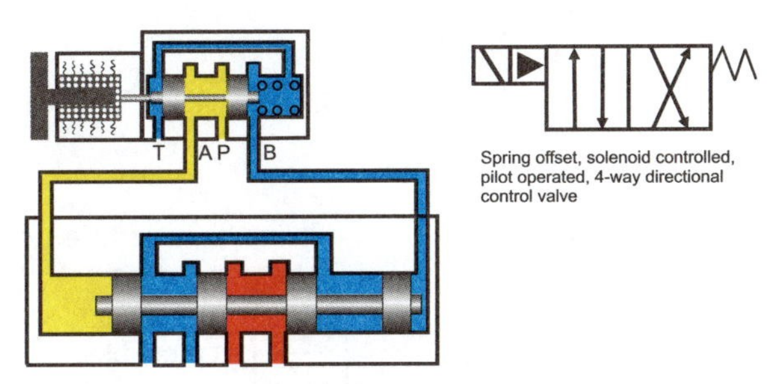

Figure 10-40

In the illustration, two solenoid operated spring offset 2-way valves are positioned in a line. The block of the symbol on which the spring acts determines the valve's unactuated position. In valve A, the spring acts on its non-passing position. As the solenoid is energized, the open block is pushed into the line. Valve A is a normally non-passing 2-way. Valve B is a normally passing 2-way.

With a two-position 3-way valve, there is always an open path through the valve. In one extreme position, P is connected to tank. With a 3-way valve, normally non-passing indicates that the P port is blocked in the normal, unactuated position. In the illustration, valve C is a normally non-passing 3-way; valve D is normally passing 3-way.

A

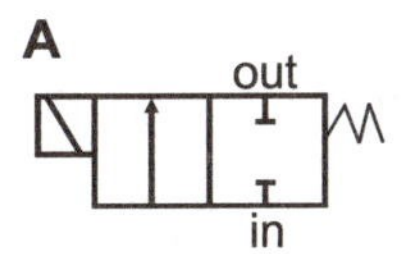

Spring offset - returned, solenoid operated 3-way valve normally closed

B

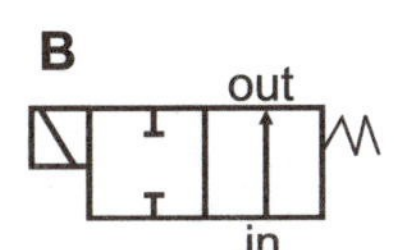

Spring offset - returned, solenoid operated 3-way valve normally open

C

Spring offset - returned, solenoid operated 3-way valve normally closed

D

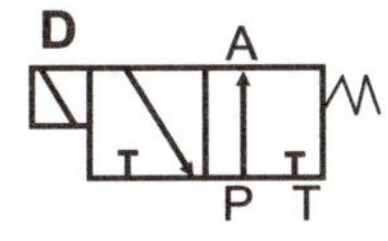

Spring offset - returned, solenoid operated 3-way valve normally open

Figure 10-41

Normally Passing and Normally Non-Passing Valves in a Circuit

Spring offset 2-way and 3-way valves can be used as system interlocks.

In the illustration, a spring offset, solenoid operated, normally non-passing 3-way valve ensures that the load attached to the single-acting cylinder will not rise until an electric signal is transmitted to the solenoid. The signal may indicate that a part is in position, a door is closed, or that the machine is in some manner ready for cycling.

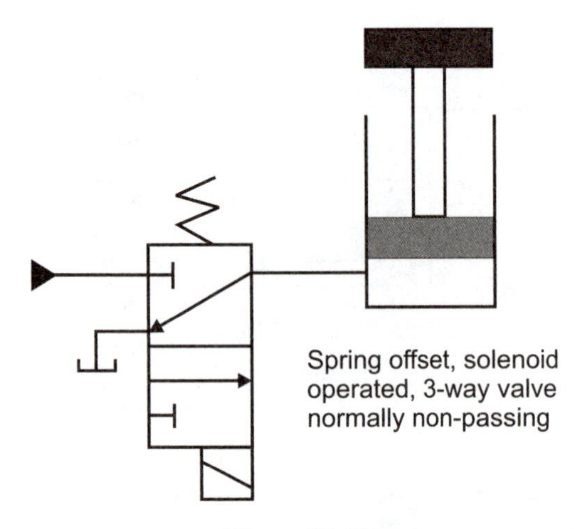

Figure 10-42

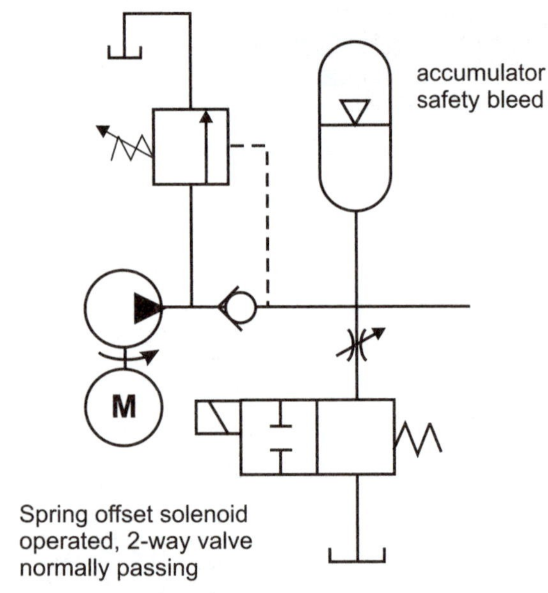

Figure 10-43

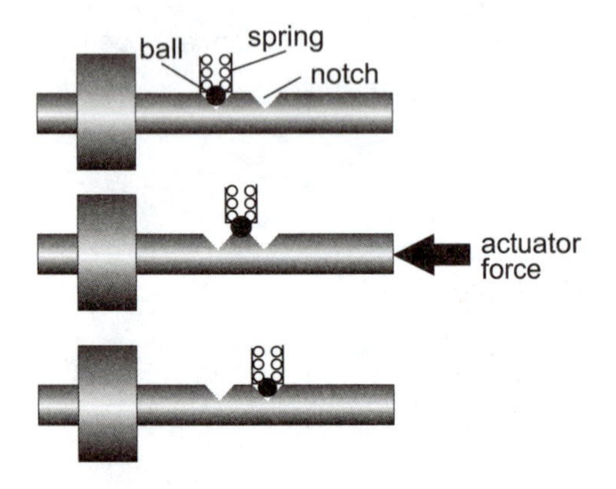

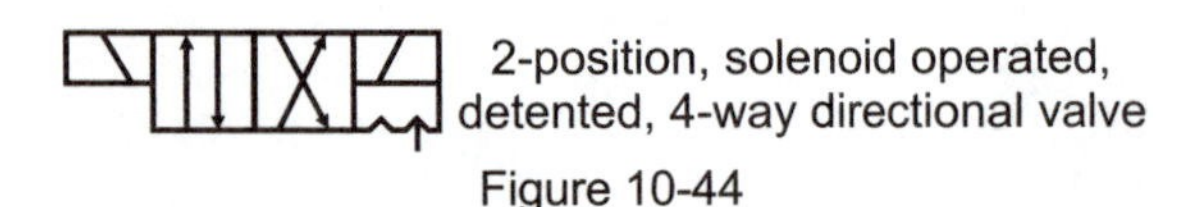

Figure 10-44

In any accumulator circuit, a means should be available of automatically unloading the accumulator(s) once the system is shut down. This can be accomplished with a spring offset, solenoid operated, normally passing 2-way valve. In the example illustrated, the solenoid of a 2-way valve can be energized when the system electric motor is started. This blocks flow through the valve allowing the accumulator to charge. When the system is shutdown, the solenoid is de-energized and a spring pushes the valve spool to its normally passing position. Accumulator bleeds down safely through the needle valve. Therefore, anytime the electric motor is shutdown, the accumulator automatically bleeds off.

Detents

If two actuators are used to shift the spool of a two position valve, detents are sometimes used. A detent is a locking device which keeps a spool in the desired shifted position.

The spool of a detented valve is equipped with notches or grooves. Each notch is a receptacle for a spring-loaded movable part. In the detent illustrated, the movable part is a ball. With the ball in the notch, the spool is held in position. When the spool is shifted, the ball is forced out of one notch and into another notch.

Directional valves which are equipped with detents are not required to keep their actuators energized to achieve a shifted position.

In the following section, we review the center positions of hydraulic 4-way directional valves and then look at their characteristics and application in a system.

Center Conditions

In referring to the possible flow paths through a two position, 4-way directional valve, flow paths as the spool was in either extreme were considered only. But, there are intermediate spool positions. Industrial hydraulic 4-way directional valves are frequently three position valves consisting of two extreme positions and a center position. The two extreme positions of a 4-way directional valve are directly related to actuator motion. They are the power positions of the valve. They control movement of an actuator in one direction and then in the other. The center position of a directional valve is designed to perform logic or to satisfy a need or condition of the system. For this reason, a directional valve center position is commonly referred to as a center condition. There are a variety of center conditions available with 4-way directional valves. Some of the more popular center conditions are the open, closed, tandem, and float centers. These conditions can be achieved within the same valve body simply by using the appropriate spool.

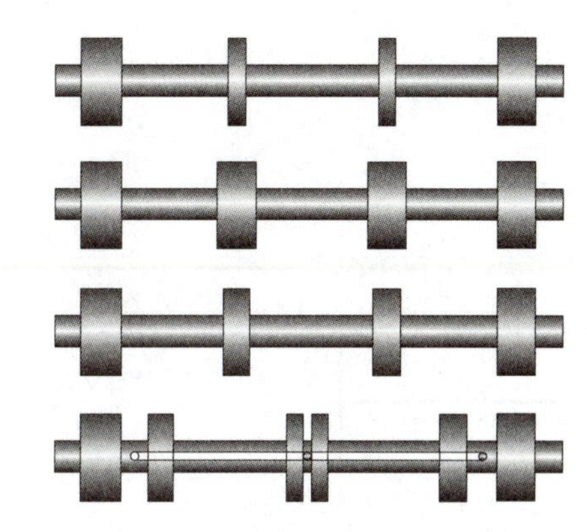
Figure 10-45

Open Center Condition

A directional valve with an open center spool has P, T, A, and B passages all connected to each other in the center position.

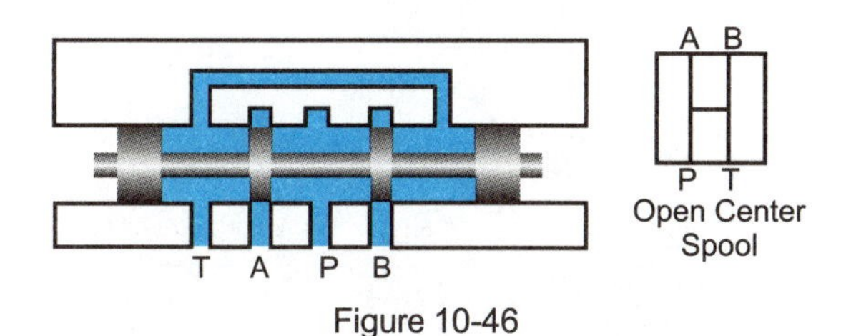

Figure 10-46

Open Center Valve in a Circuit

Open center 4-way valves are many times used in single actuator circuits. In these systems after an actuator has completed its cycle, the directional valve spool is centered and pump flow returns to tank at a low pressure. At the same time, the actuator is free to move. Disadvantages of an open center valve are that no other actuator may be operated while the valve is centered and a cylinder load cannot be held in a mid-stroke position.

In the circuit illustrated, cylinder B extends and retracts as the directional valve is shifted from one extreme position to another. When the valve is centered with cylinder in mid-stroke, cylinder load can continue to move. And, with valve B centered, cylinder A cannot be operated since all pump flow continually dumps to tank.

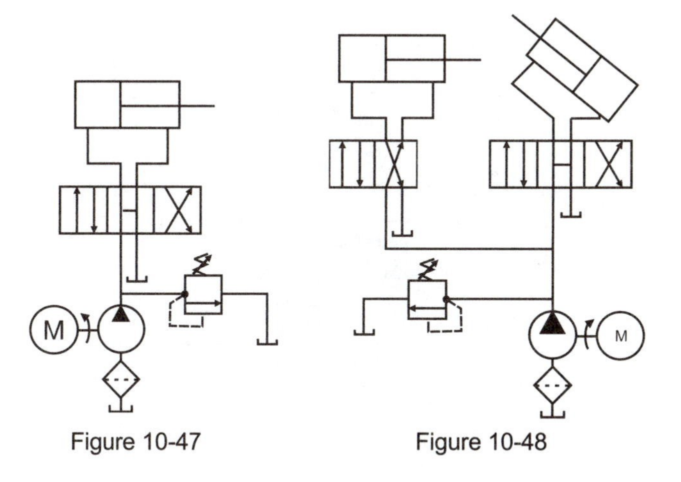

Figure 10-47 Figure 10-48

Closed Center Condition

A directional valve with a closed center spool has P, T, A, and B passages blocked in the center position.

Closed Center Valves in a Circuit

A closed center condition stops the motion of an actuator as well as allowing each actuator in a system to operate independently from one power supply.

Closed center directional valves have some disadvantages. One disadvantage is that pump flow cannot unload to tank through the directional valve during actuator idle time. However, this can be accomplished by venting a pilot operated relief valve.

Another disadvantage of a closed center valve is that the spool leaks just as in any spool valve. This means if the spool is subjected to system pressure for anything longer than a few minutes, pressure will build in the A and B actuator lines. This occurrence can be explained by an example.

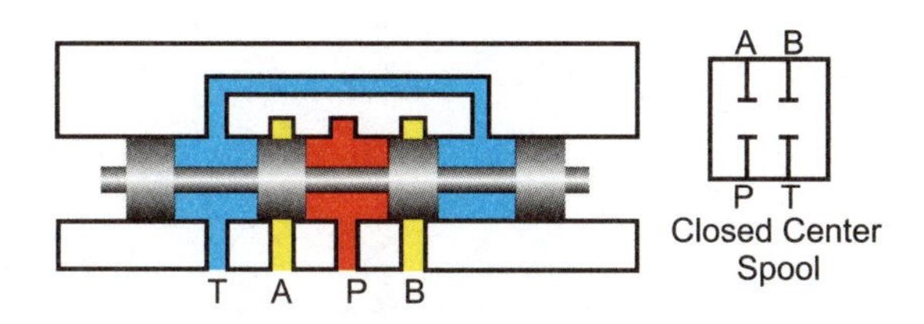

Figure 10-49

In the illustration, two fixed restrictions are placed in a line passing flow to tank. Both restrictions are exactly the same. Tank pressure is assumed to be zero.

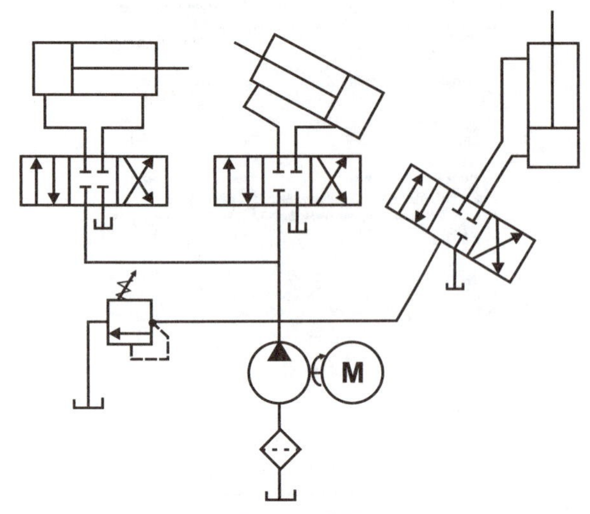

Figure 10-50

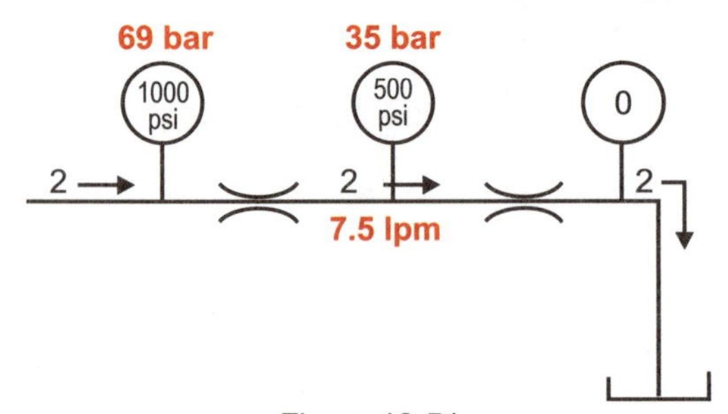

Figure 10-51

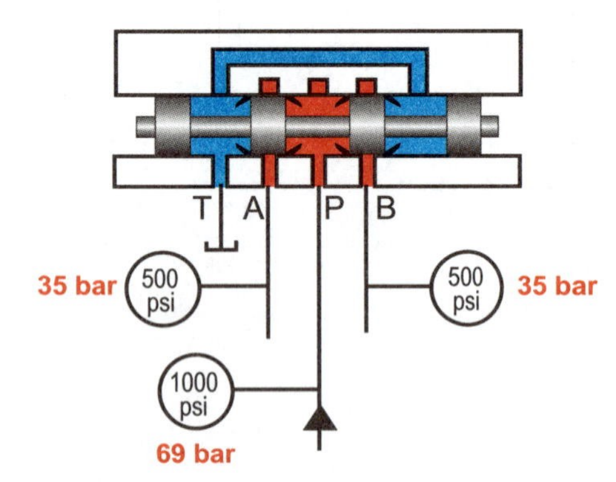

Figure 10-52

Total pressure differential across both restrictions is 1000 psi (68.97 bar). Since restrictions are equal, equal amounts of pressure energy are transformed into heat as 2 gpm (7.58 l/min) passes to tank. Restriction 1 transforms 500 psi (34.48 bar) of the 1000 psi (68.97 bar) into heat. Restriction 2 transforms the remaining 500 psi (34.48 bar) into heat. Pressure differential across each restriction is 500 psi (34.48 bar). Pressure between both restrictions is 500 psi (34.48 bar).

As a closed center directional valve is centered, fluid leaks through the valve. Leakage passes from the P port across a spool land edge into the A port. Since fluid has no place to flow out of the A line, the leakage path continues across the other spool land edge into the tank port.

With the valve spool centered, assume pressure at the P port is 1000 psi (68.96 bar) and tank pressure is zero. Each spool land edge blocking the A port from pressure and tank ports, can be considered a fixed restriction as illustrated in the example. As fluid leaks from the P port into the A port, 500 psi (34.48 bar) is used. As fluid continues to leak into the tank port, it loses another 500 psi (34.48 bar). Pressure in the A port and, therefore, in the A actuator line is 500 psi (34.48 bar).

This action occurs in a similar manner on the other side of spool, building pressure to 500 psi (34.48 bar) in the B port and B actuator line.

In the example circuit, a closed center directional valve is subjected to 1000 psi (68.96 bar) in its P port while in the center position. After a few minutes a pressure of approximately 500 psi (34.48 bar) will be seen in the actuator lines. With 500 psi (34.48 bar) acting on both ends of a single rod cylinder, a force imbalance is generated which tends to extend the rod. If the cylinder does not have a sufficiently large load, it will tend to creep out.

NOTE: In actual practice, pressure at the cylinder rod side would be approximately 500 psi (34.48 bar). Pressure at the cap side would stabilize at something less than 400 psi (27.59 bar). Basically, forces (P x A) across the piston would be practically in balance.

Correcting a rod drift problem of a closed center valve is not accomplished by incorporating a pilot operated check valve at cylinder rod side. Closed center directional valves are not generally used with pilot operated check valves. A PO check can become ineffective as soon as sufficient pressure is built up in its pilot line. PO checks do little to control rod drift of this nature.

Tandem Center Condition

A directional valve with a tandem center spool has P and T passages connected and A and B passages blocked in the center position.

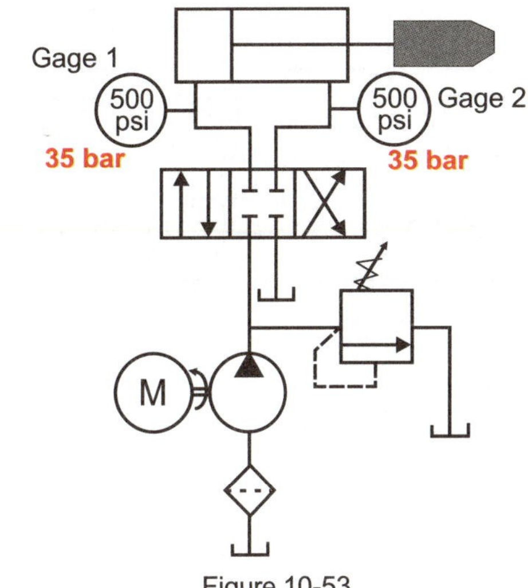

Figure 10-53

Tandem Center Valves in a Circuit

A tandem center condition stops actuator motion, but allows pump flow to return to tank while a system is idling.

Many times tandem center directional valves are connected in series with one valve's T port connected to another valve's P port. With this arrangement, actuators can be operated individually or together. Also, during idle time, pump flow can be unloaded to tank through the directional valve.

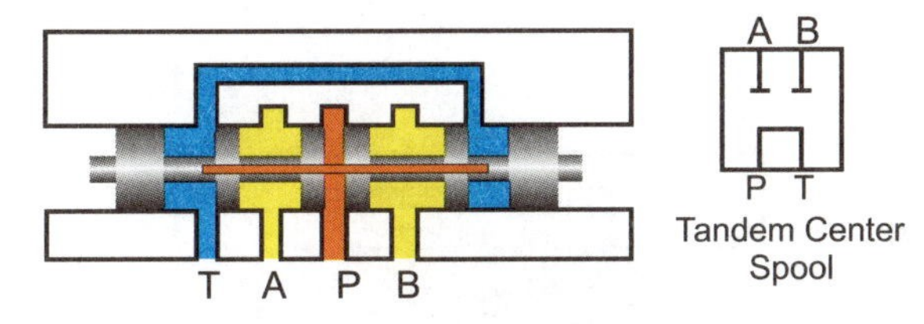

Figure 10-54

In the illustrated circuit, three cylinders have their motion controlled by individual tandem center directional valves. When directional valve A is shifted, parallel arrows are pushed into the circuit; cylinder A extends. Discharge flow passes through valves B and C back to tank. When cylinder A operation is completed, valve A is centered. At this point, either cylinder B or C can be operated. Each cylinder can operate independently; pressure required at each cylinder is determined by load and piston area as is the usual case.

With tandem center directional valves in series, cylinders can be operated simultaneously. If valves A and B are actuated at the same time, cylinder A receives full pump flow; it has priority since valve A is closer to pump. Rod speed of cylinder B is determined by the discharge flow from cylinder A. If valve C is actuated, then, its rod speed would be determined by the discharge from cylinder B. With this arrangement, cylinders can operate at the same time; pressure required at each cylinder will depend upon its load and the load of any downstream cylinder. After work has been completed, all three valves are centered allowing pump flow to dump to tank.

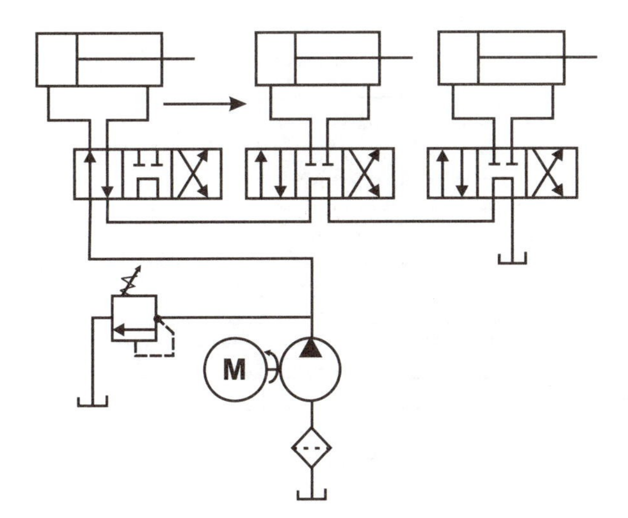

Figure 10-55

It was pointed out previously that various center conditions can be achieved from a 4-way directional valve by inserting the appropriate spool into a valve body. When a tandem center spool is used in a valve body, its flow rating is reduced. And, the center, unloading condition of the spool is not as good as might be expected when looking at a tandem center symbol.

The P and T passages of an industrial hydraulic 4-way valve are not located next to each other. With the P passage in the center and the T passage at both extremes, passages are connected in the cen-

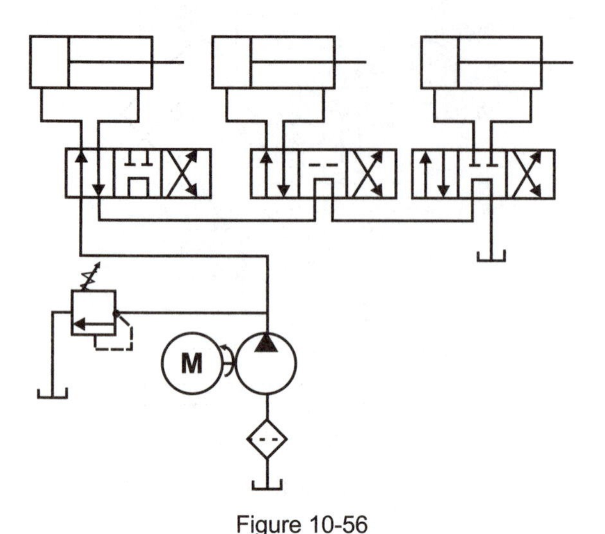

Figure 10-56

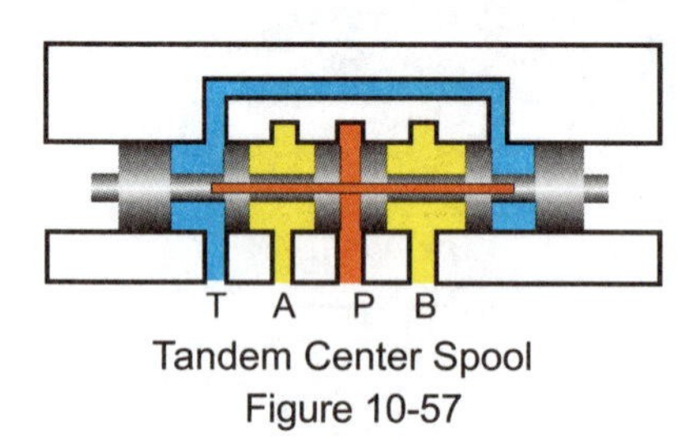

Tandem Center Spool
Figure 10-57

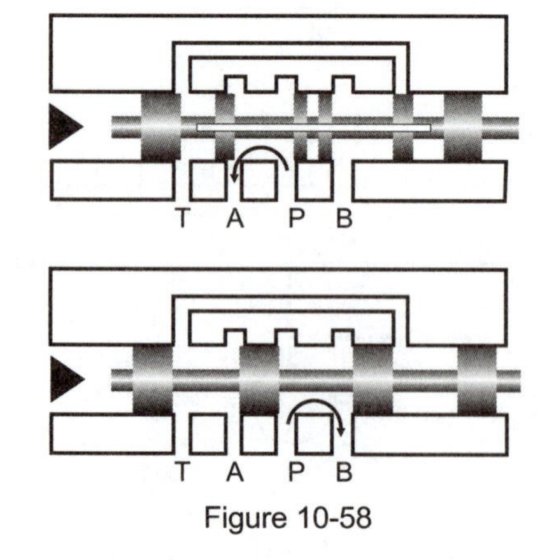

Figure 10-58

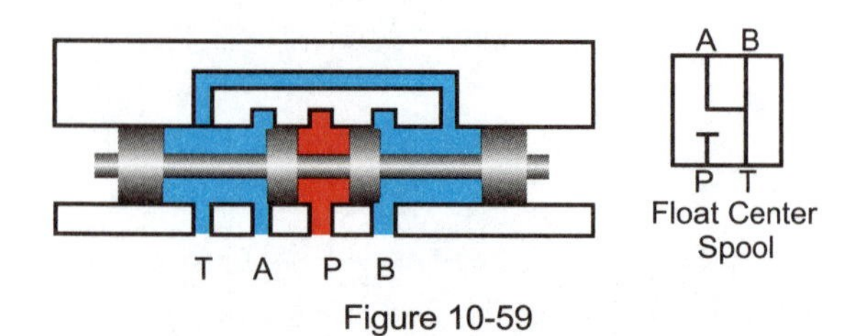

Figure 10-59

ter position by means of a passage through the spool shaft. This is a narrow flow path which can result in a 150 psi (10.35 bar) drop from P to T. If pump flow must pass through several valves, pump/ electric motor may be required to develop substantial power during idle.

In order to allow flow path from P to T in the center position, the spool shaft between the lands is much thicker than in any other spool type. This results in a restricted flow path when the spool is shifted to either extreme.

A tandem center directional valve operates a little differently than directional valves with other spool designs. Because of its construction, when a tandem center spool is shifted toward the right side, flow passes from P to A. With any other spool, such as a closed center spool, flow passes from P to B when shifted to the right.

Consequently, if a tandem valve replaces a valve with a different spool configuration, the actuator controlled by the directional valve will operate backwards if all other things remain the same.

Float Center

A directional valve with a float center spool has the P passage blocked, and A, B, and T passages connected in the center position.

Float Center Valves in a Circuit

A float center condition allows independent operation of actuators tied to the same power source as well as allows free movement of each actuator.

An advantage of a float center is that actuator lines do not have a buildup of pressure when the P passage is blocked as in a closed center valve. This controls piston rod drift.

A disadvantage of this spool is that a load cannot be stopped or held in place. If this is a system requirement, a pilot operated check can be used in conjunction with the valve. Float center spools are sometimes referred to as "PO check" spools for this reason.

If a load controlled by a float center spool must be slowed in its motion when the valve is centered, a float center spool with metering orifices on the A and B lands is used. The orifices restrict the flow to tank through A and B when the spool is centered. This generates a back pressure in an actuator which tends to slow and may even stop motion. Float center spools of this nature are sometimes referred to as "motor spools."

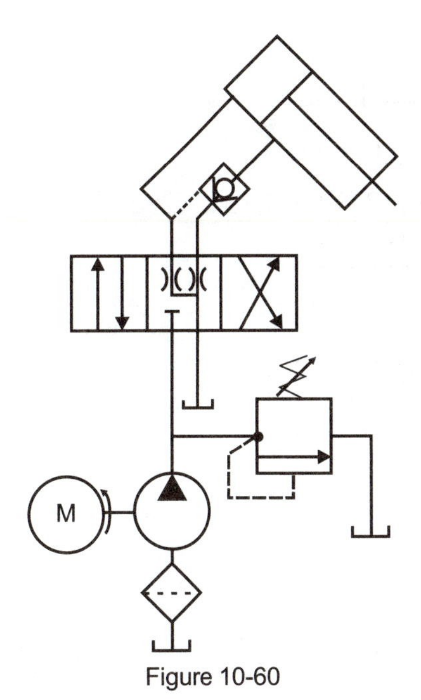

Figure 10-60

Other Center Conditions

Other center conditions besides open, closed, tandem, and float, are available from directional valve manufacturers. These center conditions do not have specific names, but they give even greater flexibility in satisfying system demands.

Crossover Conditions

Two position directional control valves have two extreme positions which usually control actuator direction while doing work. Two position directional valves come equipped with a center position as well. The center position is known as a transition or crossover. This is what an actuator sees for a fraction of a second as a spool shifts from one extreme position to another. Open and closed center conditions are the most frequently used crossovers.

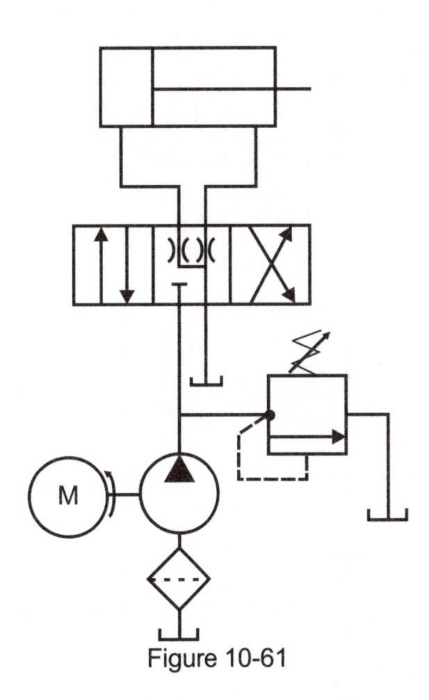

Figure 10-61

A closed center crossover does not allow pump/electric motor pressure to drop during a spool shift.

In the illustrated circuit, a drill cylinder is controlled by a two position, 4-way valve with a closed center crossover. (The crossover condition is illustrated by enclosing the center position with dash lines.) With directional valve shifted so that parallel arrows are in the circuit, the drill cylinder extends. When the piston rod reaches a certain position, directional valve shifts to the crossed arrows. While moving across center, pump pressure is not allowed to fall. This does not permit the piston rod to lunge downward during the shift.

An open center crossover allows actuator lines to bleed slightly before reversal takes place. This is an important consideration in reversing a high inertia load.

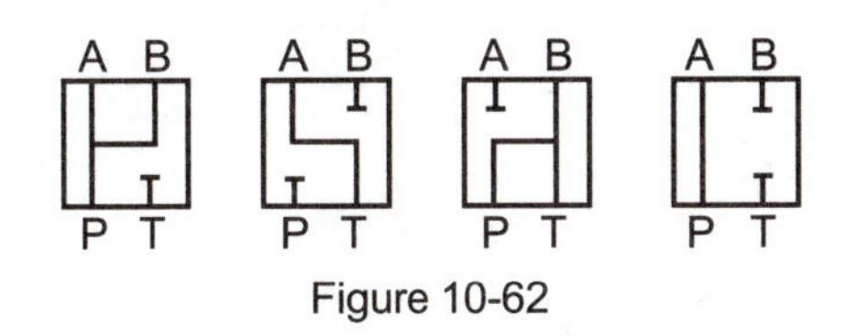

Figure 10-62

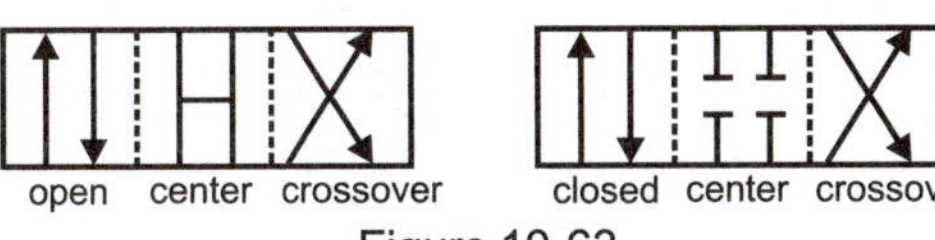

Figure 10-63

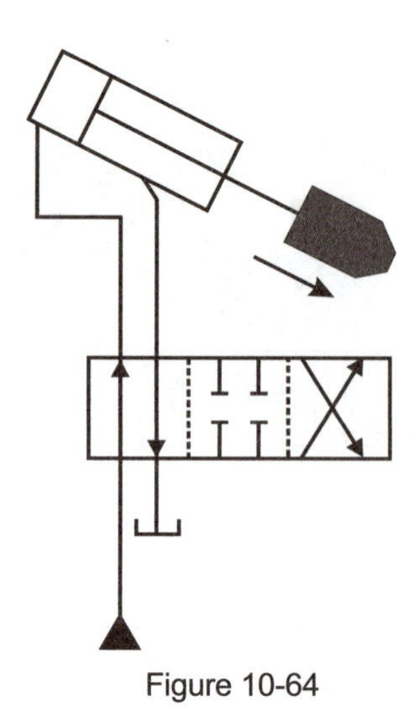
Figure 10-64

In the illustrated circuit, a cylinder is moving a heavy load. When the load reaches a certain position, directional valve is shifted to reverse load direction. As the spool moves to the opposite extreme, an open center crossover bleeds all ports to tank. For a fraction of a second, pressure drops and then builds once again as the spool fully shifts. Bleeding of the ports helps eliminate shock as the load changes direction. If a closed center crossover were used with this type load, load inertia would develop a shock pressure in an actuator line as all ports were blocked while moving to an opposite extreme.

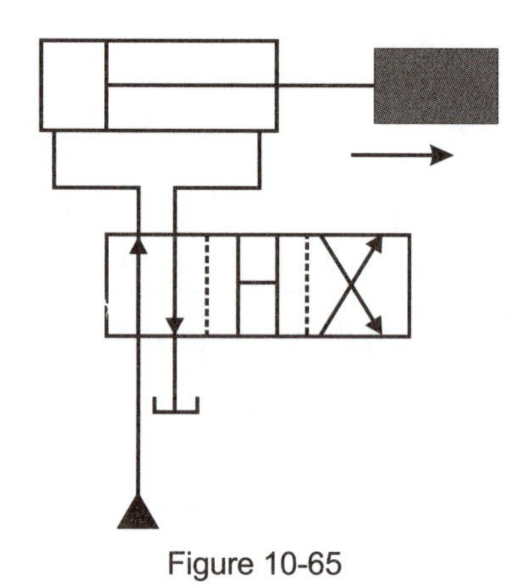
Figure 10-65

Up to this point, we have concentrated on direct solenoid operated directional valves. The remaining text material will deal with solenoid controlled, pilot operated directional valves. We will review their operation and see what accessories are available with these type valves.

Solenoid Controlled Pilot Operated Directional Valve

Industrial hydraulic directional valves come in various sizes. Ranging in port sizes of 1/4" (6.35 mm), 3/8" (9.5 mm), 1/2" (12.7 mm), 3/4" (19.05 mm), 1" (25.4 mm) and 1-1/4" (31.75 mm). As indicated earlier, a very common means of shifting a directional valve spool is with a solenoid. However, this is only practical in 1/4" (6.35 mm), 3/8" (9.5 mm) and some 1/2" (12.7 mm) port sizes. In the larger 3/4" (25.4 mm) and 1¼" (31.75 mm) valves which are designed to handle flow rates of 40 gpm (132.7 l/min) and greater, direct solenoid shifting of a spool is impractical.

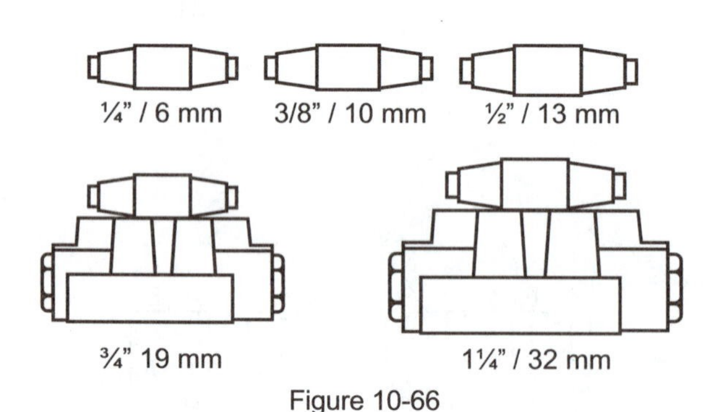

Figure 10-66

In large valves, force required to shift a spool is substantial. A solenoid which could generate the force, would necessarily be quite large. In valves of this type, customarily a 1/4" (6.35 mm) or 3/8" (9.5 mm) solenoid operated directional valve is positioned on top of the main valve body in a piggyback arrangement. Pressurized flow from the small valve is directed to either side of the large valve spool when shifting is required. Instead of controlling the motion of a cylinder or motor, the pilot valve controls the motion of the main valve spool.

A directional valve of this type is known as a solenoid controlled, pilot operated directional valve. Pilot pressure moves the main valve spool, but the solenoid controls the pilot.

Spool Centering in a Pilot Operated Valve

A directional valve with three positions must have an ability to hold its spool in the center position. This can be done with springs or fluid pressure.

Spring centering is the most common means of centering a directional valve spool. A spring centered valve has a spring located at each end of the spool. When the valve is actuated, the spool moves from the center condition to one extreme compressing a spring. When the float center pilot valve is centered, pilot chambers at both spool ends are drained. The spring then returns the main spool to center.

Spring centering can be a problem with valves which are handling flows beyond their intended capacity. Under this condition, the spool may shift to an extreme position. Once shifting force is removed, the spring is intended to push the spool back to center. If pressurized flow is excessive, the fluid will exert enough force to hold the spool shifted.

Centering the spool of the pilot operated directional valve is sometimes accomplished with hydraulic pressure. Pressure centering a valve ensures that the spool will center even if the flow rate through the valve is excessive.

In a pressure centered, pilot operated, 4-way valve, the pilot valve center position has P connected to A and B, T is blocked. In centering the valve, fluid pressure is simply directed to both sides of a spool simultaneously. Since the areas of the spool lands at either end are identical, equal forces are generated when they are exposed to the same pressure. This does not guarantee the spool will center.

In a pressure centered valve, the space between an end spool land and the pilot chamber is sealed by a sleeve which is held against a shoulder in the pilot chamber. This space is also drained of any fluid leakage by means of an external drain passage. When the spool is required to be centered, fluid pressure is directed to the chambers at both spool ends by the pilot valve. Regardless of which extreme position the spool is in, there will be a difference in areas exposed to pressure at the time of centering.

When the spool is centered from the extreme left position, fluid pressure acts on the left end of the spool. Since the right spool end is drained, the spool moves toward the right until it butts up against the sleeve. At this point the spool is centered.

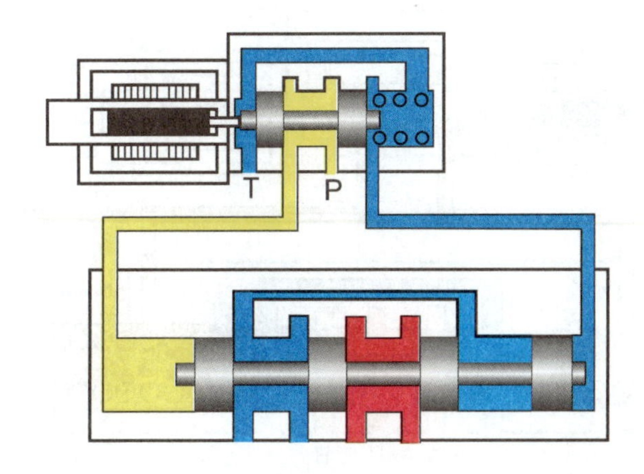

Spring offset, solenoid controlled pilot operated, 4-way directional control valve

Figure 10-67

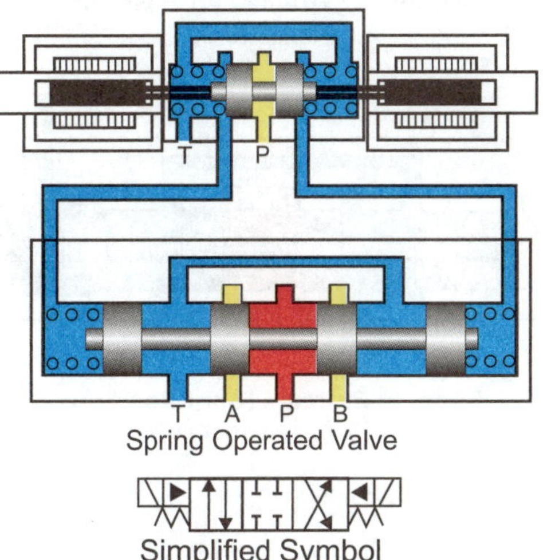

Solenoid-Controlled, Pilot Operated Directional Valve

Figure 10-68

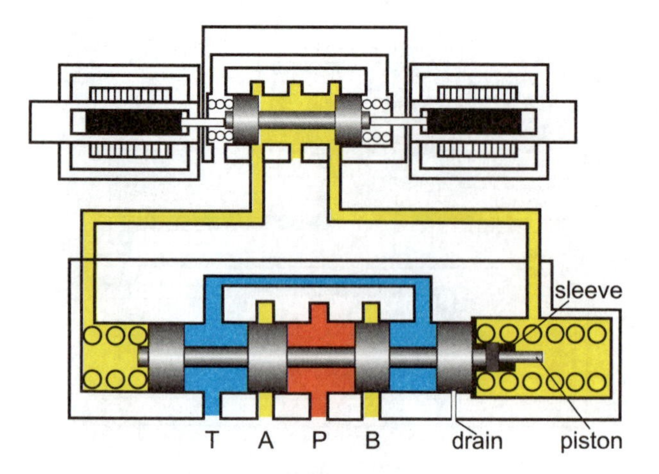

Figure 10-69

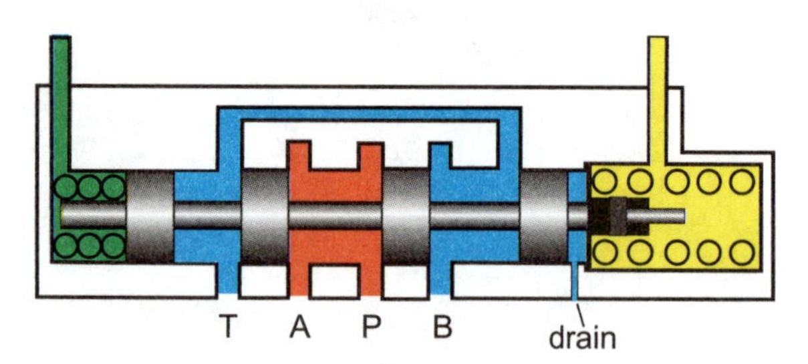

Spool in Extreme Left Position
Figure 10-70

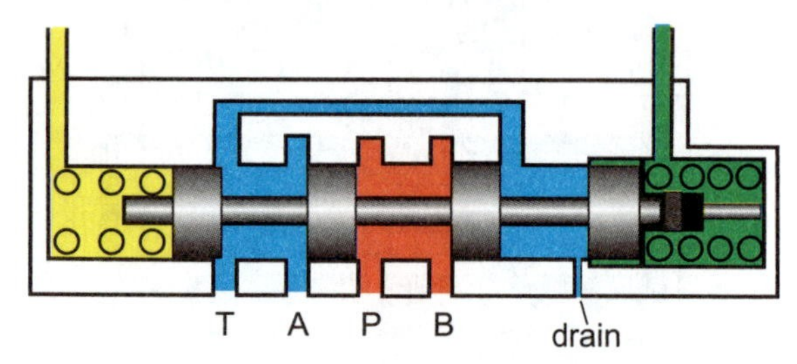

Spool in Extreme Right Position
Figure 10-71

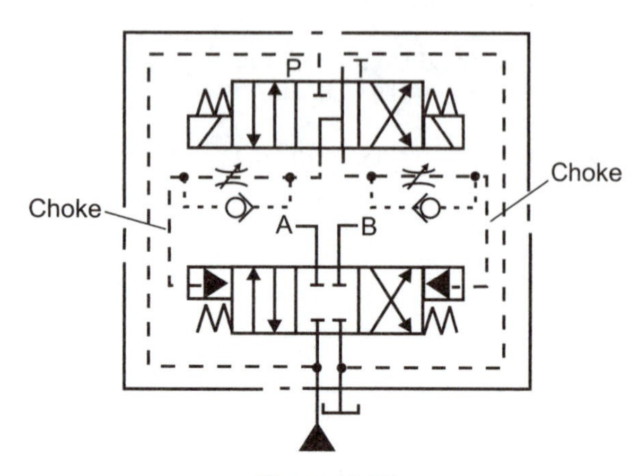

Figure 10-72

When the spool is centered from the extreme right position, fluid pressure acts on the left spool end and the sleeve at the right end. Since the sleeve has more area than the spool land at the left, a larger force is generated to move the spool toward the left. The spool moves left until the sleeve contacts the shoulder in the chamber. At this point, the spool is centered.

When the valve is to be shifted from the sleeve side of the spool, fluid pressure acts on the piston in the sleeve center. he piston then forces the spool to its extreme left position.

Choke Control

As the main spool of the pilot operated directional valve is shifted, shock can be developed as large fluid flows are forced to change direction quickly. A choke control slows the spool shift of the main valve so that shock is reduced.

A choke control is a "sandwich" valve which fits in between a main valve body and pilot valve of a solenoid controlled, pilot operated directional valve. It consists of two needle valves and two bypass check valves. The valve sandwich is arranged so that as the spool shifts in one direction a needle valve meters flow out of the pilot-spring chamber; the other needle valve is bypassed. As the spool shifts in the opposite direction, the other needle valve meters flow out of the other pilot-spring chamber. The more the needle valves are adjusted in, the more flow is restricted or choked. Consequently, the slower is the spool shift.

A choke control does not eliminate shock due to spool shift; it reduces shock. If a crash is experienced each time a large directional valve is shifted, a choke control will reduce the crash to a bang. If the shock is a bang, chokes reduce it to a thump.

Stroke Adjustment

One thing which can be done with the main spool of a pilot operated valve is to limit its travel. This, of course, is not advised with a direct, AC solenoid valve.

A stroke adjustment is a screw adjustment which limits spool travel in the main valve of a solenoid controlled, pilot operated directional valve. Partially limiting spool travel in one direction means that a port is not completely uncovered. This causes a restriction through the valve when the spool is shifted in one direction giving the effect of a coarse needle valve.

If the stroke adjustment is screwed in completely, one extreme position of the valve is blocked out. This in effect transforms a three position valve into a two position valve. The two valve positions are made up of one extreme position and the valve center condition.

Pilot Operated Directional Valve with Stroke Adjustment

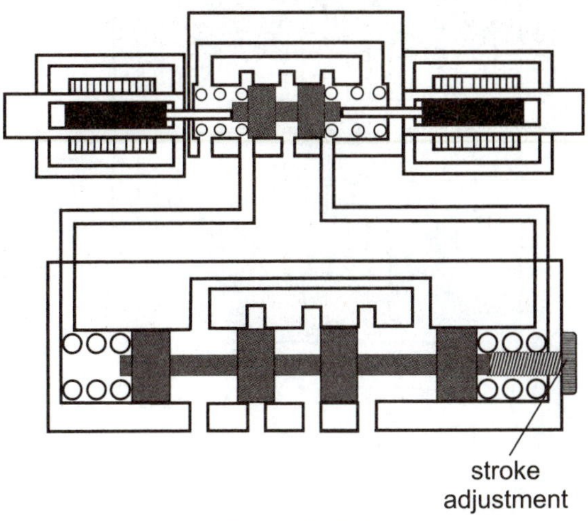

Figure 10-73

Directional Valve Drain

The pilot valve tank passage of a solenoid controlled, pilot operated directional valve is called a drain.

The detailed symbol of a solenoid controlled, pilot operated directional valve shows that the tank port of the pilot valve is connected by means of an internal passage to the tank passage of the main valve. This internal passage connection is known as an internal drain. With an internal drain, pressure in the tank passage must be contended with in shifting the main valve spool.

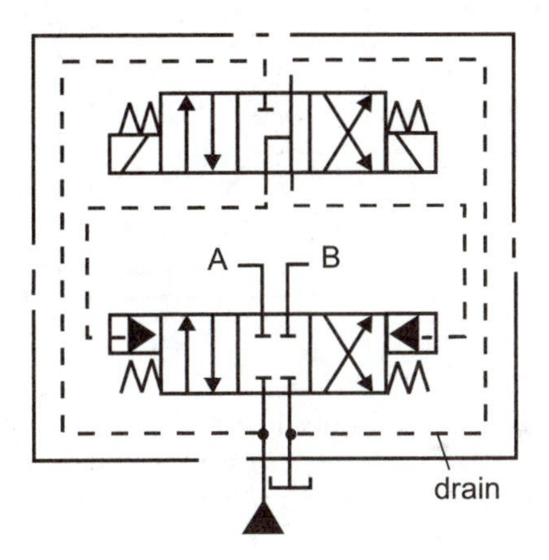

Figure 10-74

From the illustrated symbol, the P port of a closed center directional valve is subjected to 500 psi (34.48 bar) system pressure. This pressure is also supplied to the solenoid pilot valve by means of an internal passage. Backpressure in the tank line is 50 psi (3.45 bar).

With the float center pilot valve centered, each end of the main valve spool is exposed to the backpressure. When the pilot valve is shifted, 500 psi (34.48 bar) is directed to the left spool end. In order to shift the main spool to the right, pilot pressure must overcome 50 psi (3.45 bar) backpressure and the spring which has a value of another 50 psi (3.45 bar). Of course, 500 psi (34.48 bar) can accomplish this quite easily. The main spool shifts.

In some situations, the pilot valve is externally drained. This is the case when shock pressures are experienced in the tank line.

Assume that an internally drained directional valve has its spool held shifted to an extreme position with a pilot pressure of 500 psi (34.48 bar). This is the case in the last example. Keeping in mind that any pressure in the tank line acts on the main spool end drained to tank, assume now that a single acting cylinder discharges in another part of the circuit. Pressure climbs in the tank line to 1000 psi (68.97 bar) for a fraction of a second. Since the tank line is usually common for all valves, this pressure arrives at the main spool end drained to tank. With 500 psi (34.48 bar) on one side and 1000 psi (68.97 bar) on the other, the spool shifts in an undesired direction for a fraction of a second. This results in erratic motion at an actuator.

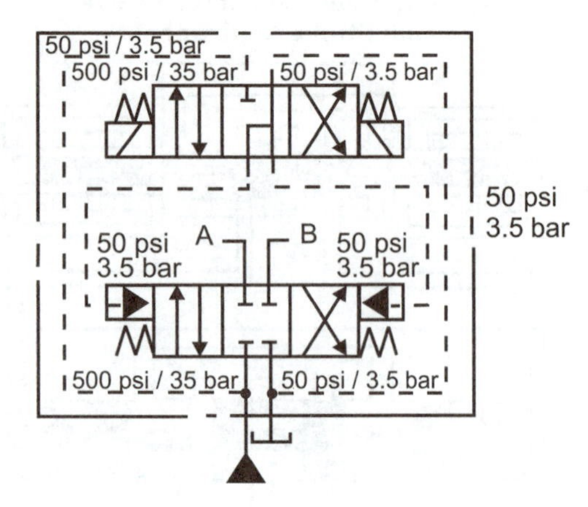

Figure 10-75

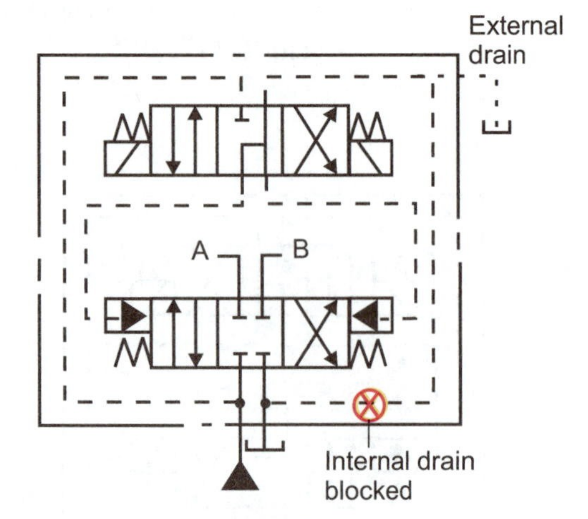

Figure 10-76

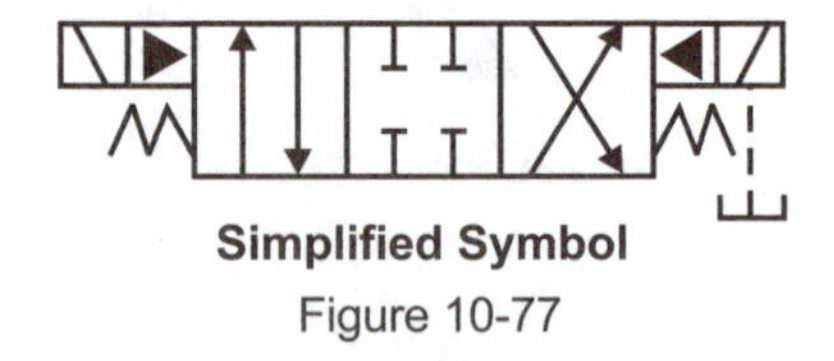

Simplified Symbol

Figure 10-77

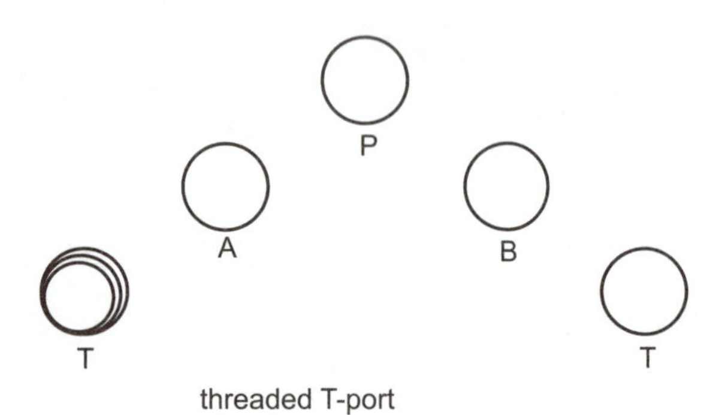

Figure 10-78

To remedy the situation, the valve should be externally drained. This is accomplished by removing the pilot valve from the main valve. This exposes the pilot valve porting on top of the main valve. The port pattern will consists of a P port, A and B ports, and two T ports. One of the T ports will be threaded. A pipe plug is inserted into this port blocking the internal drain. The pilot valve is replaced and a separate drain line is connected to the external drain port (Y port) of the main valve subplate. (NFPA. DO2 foot pattern shown.)

With a pilot operated directional valve externally drained, back pressure on a spool land is very low during shifting.

Externally draining a pilot operated directional valve is also required with tank line back pressure check valves. This will be illustrated below.

Directional Valve Pilot Pressure

Pressure supplied to the pilot valve of a solenoid controlled, pilot operated directional valve is known as pilot pressure. This is used to shift the main valve spool.

The composite symbol of a solenoid controlled, pilot operated directional valve shows that the P port of the pilot valve is connected by means of an internal passage to the P passage of the main valve. This is an internal pilot connection.

With an internal pilot, pilot pressure to shift the main valve spool has the same value as system pressure. From a previous example, a system pressure of 500 psi (34.48 bar) was supplied to the main valve spool. This was used in overcoming spring and back pressure at a spool end during shifting. As long as the pilot pressure is sufficient to overcome spring and back pressure, the spool will shift.

In some situations, the valve is required to be externally piloted. This is the case when internal pilot pressure is either too low or too high.

Assume that the system in which an internally piloted directional valve is operating, has erratic system pressure. In one instant system pressure might be 500 psi (34.48 bar); the next instant it's 50 psi (3.45 bar). This situation cannot be depended on to shift the valve. To solve the problem, the valve can be externally piloted with an external, dependable source of pilot pressure.

Externally piloting a directional valve is also desirable if system pressure is quite high. Assume that a system pressure of 2500 psi (172.4 bar) is supplied internally to a pilot valve. When the main valve spool is required to be shifted from one extreme position to another, the pilot valve is shifted. 2500 psi (172.4 bar) acts on a spool end, accelerating the spool quickly to an extreme. This can generate considerable shock.

Besides using choke controls, the pilot valve can be externally supplied with a pilot source of lower pressure.

To change from internal pilot to external pilot, an internal pipe plug must be added. On the side of the main valve body, a pipe plug or SAE plug covers a port which communicates with the internal pilot passage. This plug is removed. Inside the port, the pilot passage and another threaded port can be seen. The internal port is plugged with a very small pipe plug (1/16 in./1.6 mm). This blocks the internal pilot. The outer pipe plug, or SAE plug, is then replaced and the external pilot port (X port) on the main valve subplate is connected to the source of pilot pressure.

NOTE: Before such changes are attempted, the components manufacturer should be consulted.

Externally piloting a pilot operated directional valve is sometimes required when a backpressure check valve is positioned in the pressure line ahead of the valve.

Backpressure Check Valve

Pilot pressure to shift the main spool of a solenoid controlled, pilot operated directional valve frequently comes through the main valve body from the main system pressure.

In the illustrated symbol, a pilot operated directional valve with a closed center is shown. With 500 (34.48 bar) psi present at the valve P port, 500 psi (34.48 bar) is supplied to the pilot valve through an internal passage. To shift the main spool, the solenoid pilot valve is shifted. 500 psi (34.48 bar) is directed to one end of the main spool and the spool shifts.

Solenoid controlled, pilot operated directional valves which have P connected to tank in the center position, do not have a readily available supply of pressure for the pilot valve. This is the case with open and tandem center directional valves.

In the illustration, pump flow returns directly to tank through system piping. Gages in the line indicate that system pressure at gage 1 is 60 psi (4.1 bar); and at gage 2 pressure is 20 psi (1.4 bar).

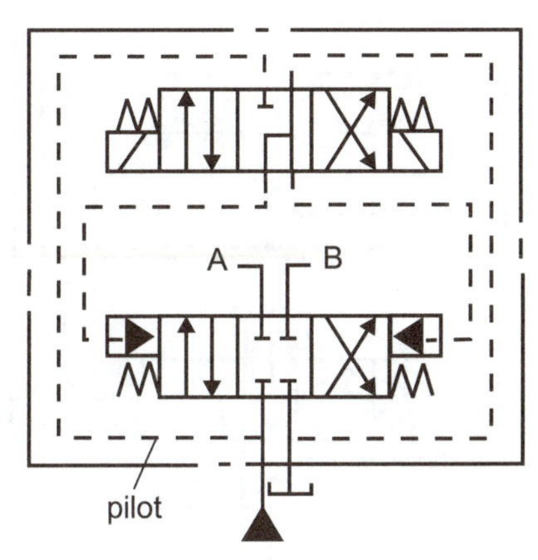

Figure 10-79

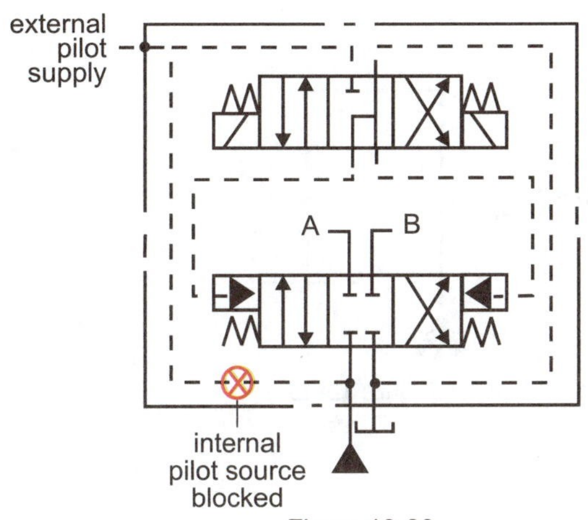

Figure 10-80

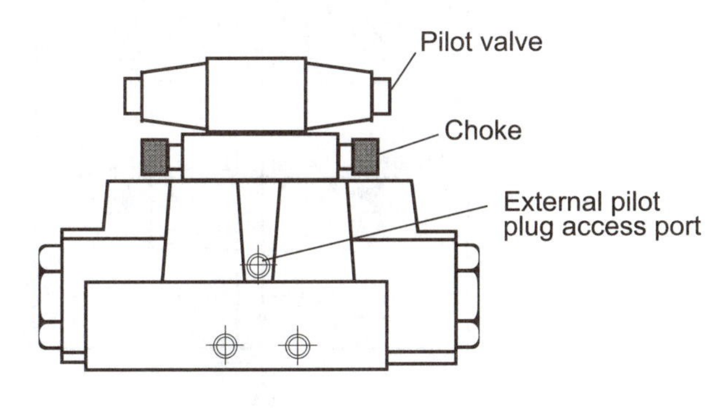

Figure 10-81

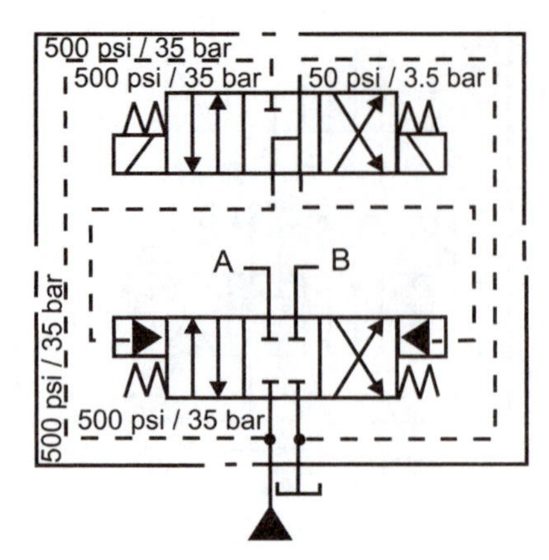

Figure 10-82

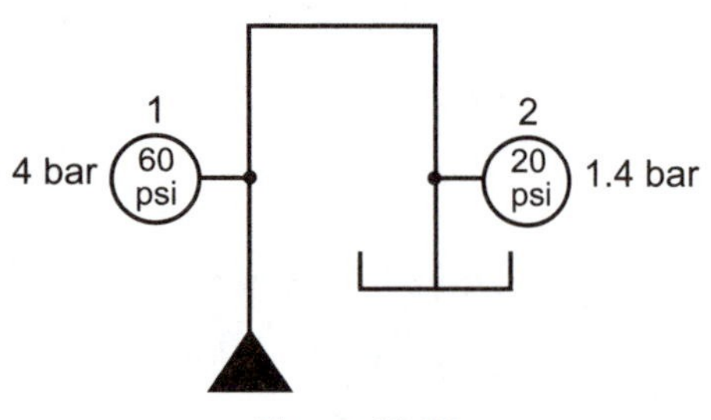

Figure 10-83

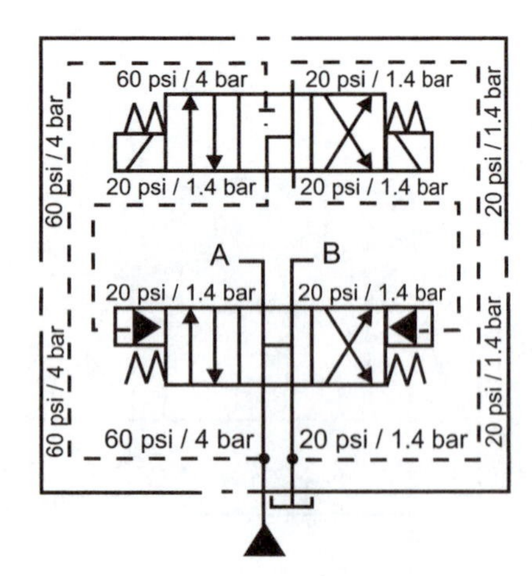

Figure 10-84

This exact situation exists with a tandem and open center directional valve while centered. Gage 1 would indicate the point within a valve where the internal pilot pressure is picked up. Gage 2 would indicate the point where the pilot drain is connected internally.

With this situation, 60 psi (4.1 bar) is supplied to the pilot valve to shift the main valve spool. This pressure must overcome 20 psi (1.4 bar) backpressure and a 50 psi (3.45 bar) spring. Of course, this cannot be done.

To remedy the situation, the valve can be externally supplied with pilot pressure from a remote source. The pilot valve could also be externally drained, but in this instance eliminating 20 psi (1.4 bar) backpressure may not ensure that the spool will shift.

Frequently, with tandem and open center pilot operated valves, back pressure check valves are used. A backpressure check has a stiffer spring than a normal check valve. This spring requires that more pressure be present at the valve inlet to push the poppet off its seat. The additional pressure is then supplied to the pilot valve.

Backpressure check valves can be positioned in either pressure or tank lines. They can be located within the valve or external to the valve.

Some manufacturers offer a 65 psi (4.48 bar) backpressure check valve within the P port of the main valve. Its effect can be illustrated by considering a previous example.

In the illustration, pump flow dumps directly to tank through system piping. Gage 1 indicates 60 psi (4.1 bar); gage 2 indicates 20 psi (1.4 bar). Putting a 65 psi (4.48 bar) check valve after the gage 1 point, requires pressure, in order to flow back to tank, be an additional 65 psi (4.48 bar) for a 125 psi (8.6 bar) total. If gage 1 indicated the point where internal pilot pressure is picked up, then 125 psi (8.6 bar) pilot pressure would be supplied to the pilot valve. When shifting is required, 125 psi (8.6 bar) is directed to either spool end overcoming 20 psi (1.4 bar) back pressure and a 50 psi (3.45 bar) spring.

Some valve manufacturers offer a 65 psi (4.48 bar) backpressure check valve in the tank port of the main valve. Its effect can be illustrated by the following example:

Once again, pump flow dumps directly to tank through system piping. Gage 1 indicates 60 psi (4.1 bar); gage 2 indicates 20 psi (1.4 bar). With a 65 psi (4.48 bar) back pressure check placed after gage 2, fluid pressure must be an additional 65 psi (4.48 bar) for an 85 psi total (5.8 bar) in order to get to tank. This also means gage 1 indicates an additional 65 psi (4.48 bar) or 125 psi (8.6 bar).

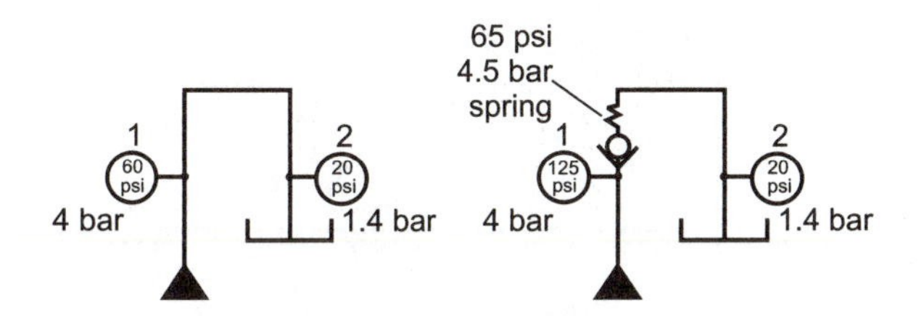

Figure 10-85

Assume that gage point 1 is where pilot pressure is picked up and gage 2 is the place where the internal drain is connected. With this situation, the pilot valve is supplied with 125 psi (8.6 bar) pilot pressure. In order to shift the valve, this pressure must overcome a back pressure of 85 psi (5.8 bar) and 50 psi (3.45 bar) spring. The spool will not shift. To remedy the situation, the pilot valve is externally drained eliminating the 85 psi (5.86 bar) back pressure.

A back pressure check valve may also be located externally in the tank line.

Anytime a tank line back pressure check is used with tandem and open center valves, the valve is externally drained.

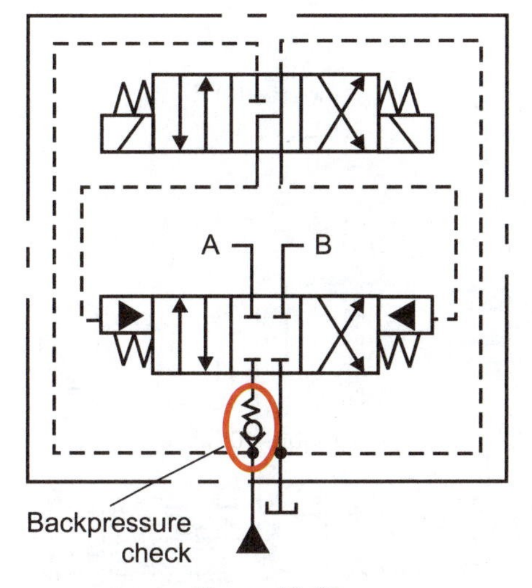

Figure 10-86

Electrohydraulic Directional Flow Control Valves

Over the years an increasing need for higher response, stiffer systems, better flow characteristics and an efficient means of interfacing electronic control systems with hydraulic control systems has arisen within the industrial hydraulic industry. The use of electrohydraulic directional flow control valves has met this need.

In general, we will discuss two types of electrohydraulic valves; namely, proportional and servo valves. Which type is used in a particular hydraulic system depends upon the sophistication of performance required by the system.

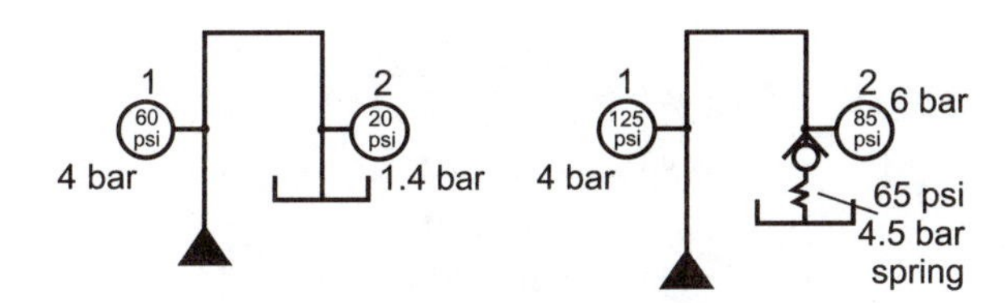

Figure 10-87

Proportional vs. Servo

There are several areas that distinguish the electrohydraulic proportional valve from a typical electrohydraulic servo valve. These areas are in the overall response of the valve, the spool center condition, the hysteresis, repeatability, and threshold of the valve, and the filtration requirements of the valves.

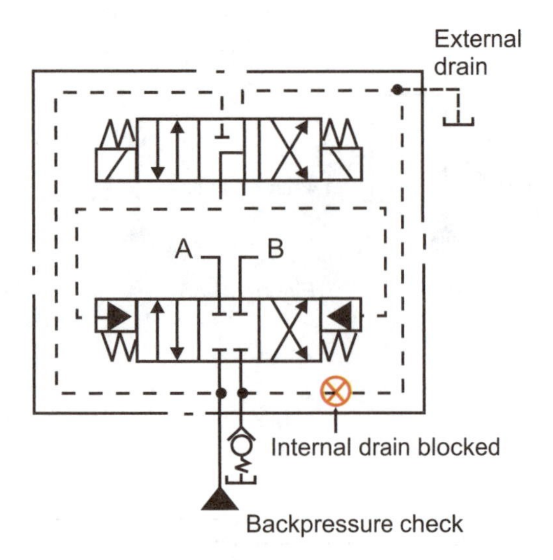

Figure 10-88

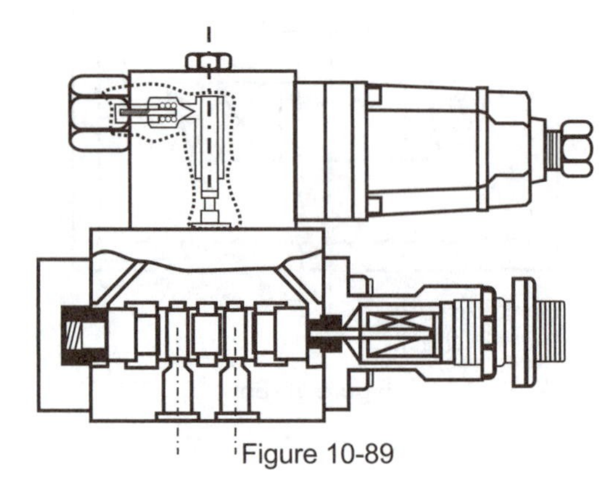

Figure 10-89

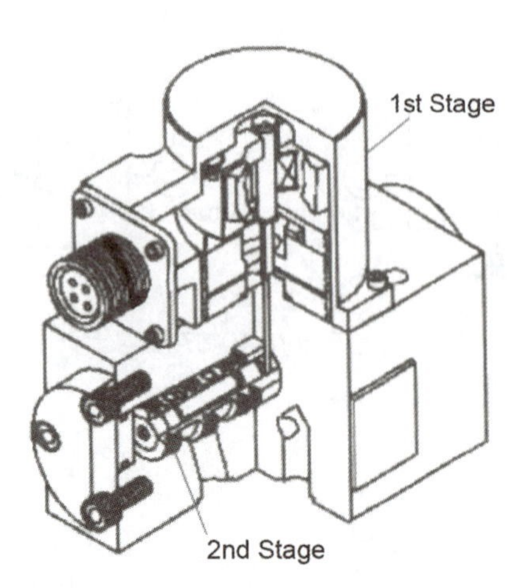

Figure 10-90

Response

Response is generally defined for proportional valves as the time required for the valve to achieve maximum rated flow due to an electrical step input command signal.

However, the response of a servo valve is generally expressed in terms of its "frequency response." The frequency response is expressed in hertz and is the frequency of a small sine wave command signal that causes the flow amplitude to be -3db when compared with the flow amplitude at a low frequency reference level (.1Hz). The -3db point occurs when the output flow amplitude is 70.7% of the flow amplitude at the low frequency.

The response for the proportional valve vs. servo valve is 2-10 Hz vs. 10-300 Hz respectively.

Spool Center Condition

The spool center condition or crossover characteristics of the main spool between the proportional valve and servo valve is quite noticeable when related to cost of valve and valve stability.

Servo valves are critically lapped by carefully matching both the width and the position of the spool lands to the metering recesses within the valve body. In other words, the spool and valve body or sleeve are matched to produce a line-line contact by hand fitting (this adds greatly to their cost).

Proportional valves are designed so that the spools and valve bodies are interchangeable. This typically results in a crossover overlap on the order of 10-30% of the total spool stroke. The overlap creates a flow condition known as deadband. Though not a problem in circuits dealing with velocity control, deadband can cause instability and loss of resolution in a closed loop feedback positioning system.

Hysteresis, Repeatability, Threshold

Hysteresis is an accepted measurement of the difference in electrical command signal as percentage of rated electrical command for a given flow output level, when the flow output setting is alternately approached from above and below the desired flow output.

Repeatability is an indication of the ability of the valve to repeat a given flow when a given electrical command signal is repeatedly applied to the valve.

Threshold is the smallest discreet change in electrical command signal that will produce a corresponding change in the output flow.

"Open loop" proportional valves may exhibit levels as high as 10%. Typical levels for "closed loop" proportional and servo valves are 3% and less.

Filtration Requirements

Particulate contamination is the enemy of all hydraulic systems and especially servo valves. Because of their close tolerances, filtration requirements of 3 micrometres are specified. Proportional valves are a little more tolerant of contamination and require filtration of 10 micrometres.

NOTE: Some proportional valves use small servo valves as the pilot head, thus, requiring additional filtration for the flow of fluid being supplied to the pilot head.

Proportional Valve Construction

A typical proportional valve consists of the torque motor pilot valve, adapter block, wire mesh pilot filter, internal pilot pressure regulator, main spool and body, and LVDT (Linear Variable Differential Transducer).

Another style of proportional valve uses proportional solenoids to operate the main valve spool direct with a positional transducer attached to the end of the valve spool to provide a feedback signal. Still another type of proportional solenoid control pilot operated valve design is used.

How a Direct Operated Proportional Solenoid Directional Valve Works

The main spool is held in the center condition by springs. Ports P, T, A and B are all blocked by the lands of the spool.

When solenoid "A" is energized with a positive voltage, the main spool is moved to the right, proportional to the input voltage directing flow from port P to B.

The positional transducer or LVDT is attached directly to the main spool which measures the precise movement (position) of the spool and feeds this back to the electronics as a voltage signal.

Briefly in the electronics, the feedback signal and the solenoid input signal are compared generating what is called an "error." The electronics will then supply either a + voltage or - voltage input signal to the proportional solenoid "A" or "B" that is energized, moving the spool either right or left respectively until the "error" equals zero.

At the same time, flow from the valve is increased or decreased.

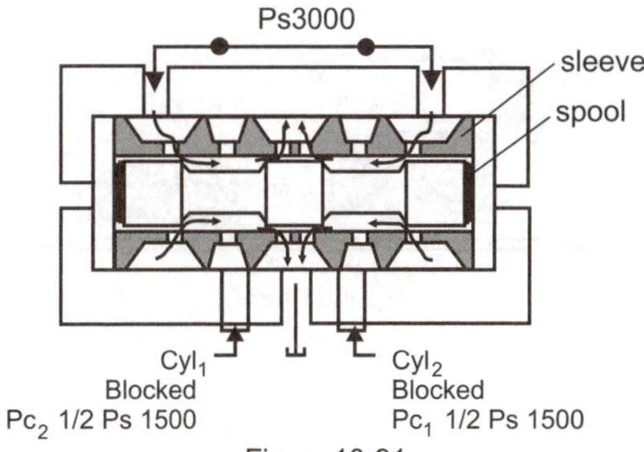

Figure 10-91

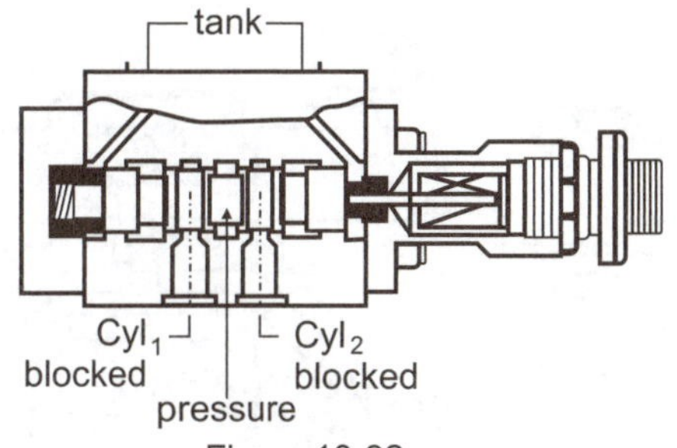

Figure 10-92

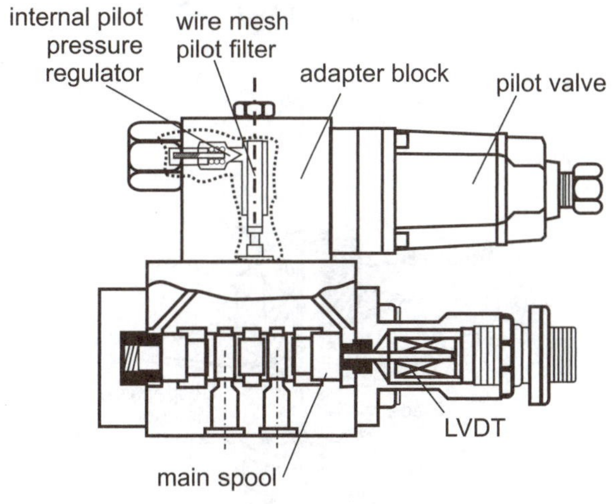

Figure 10-93

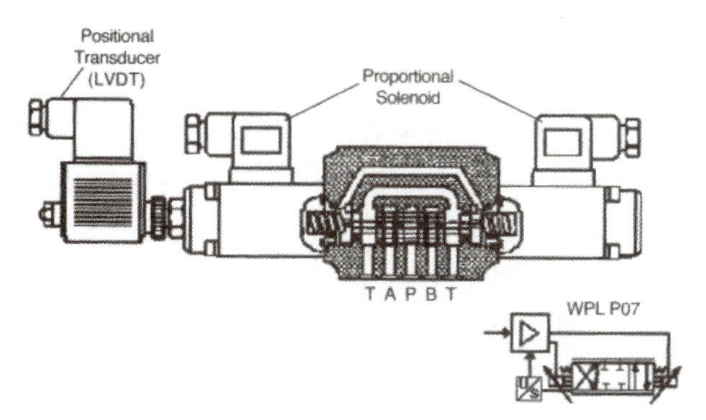

Figure 10-94

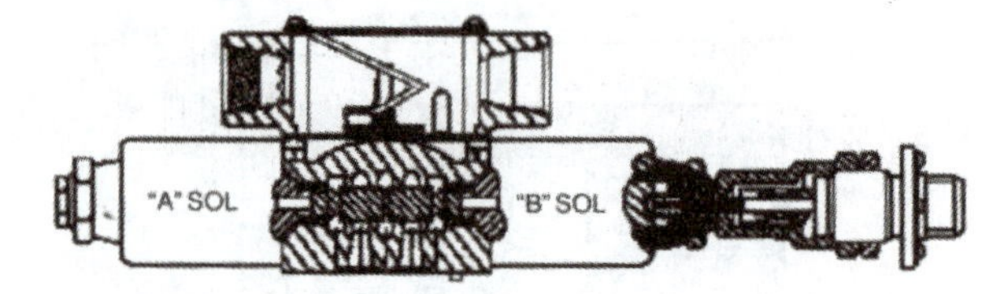

Figure 10-95

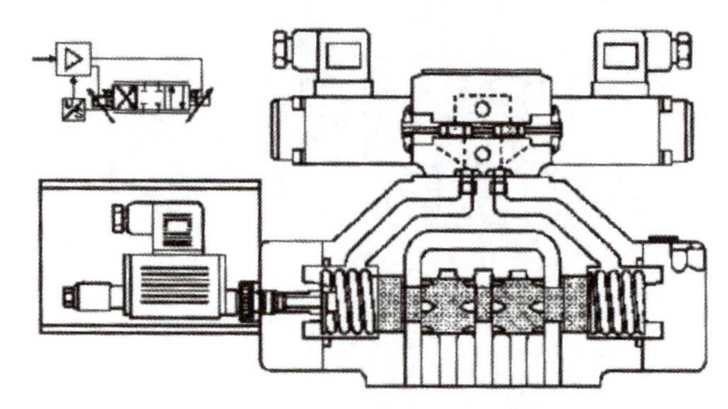
Figure 10-96

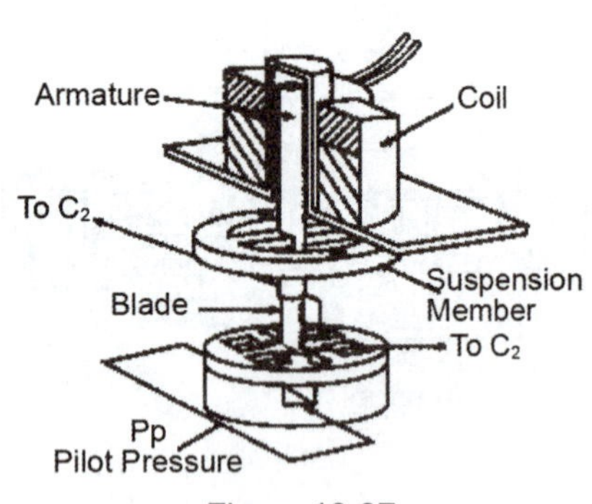

Figure 10-97

How a Pilot Operated Proportional Solenoid Controlled Directional Valve Works

The main spool is held in the center condition by springs. Ports P, A, B and T are blocked by the land areas of the spool. With neither proportional solenoid energized, the pilot spool blocks flow from the pilot supply port.

Energizing proportional solenoid "A" forces the pilot spool to the right which converts the electrical signal into a pressure signal. Pilot oil is directed to the right spring cavity of the main spool. The metering slots on the main spool open progressively based upon the amount of pressure supplied by the pilot valve.

Attached to the main spool is a positional transducer or LVDT. It functions in the same way as previously discussed for direct operated valves.

A Pressure Differential Proportional Directional Valve (Torque Motor Pilot)

A pressure differential type proportional directional valve is a two stage unit. The pilot valve sometimes called the 1st stage and the main spool or 2nd stage valve.

1st Stage Pilot Valve Consists Of

The pilot valve basically consists of coil, armature, suspension member, diverter plate, and blade. Some pilot valves incorporate a built-in relief valve and filter to limit pilot pressure and prevent contamination of the orifice(s) in the blade.

The combination of magnet, coil and armature assembly is commonly called a torque motor.

2nd Stage Main Slave Consists Of

The 2nd stage or main valve consists of a spool, return springs, and an LVDT (Linear Variable Differential Transducer). This stage is quite similar to a standard directional valve.

How the Pilot Valve Works

Current from the electronics passing through the coil will create a north or south pole on top of the armature. The polarity of the current will determine the direction of movement of the armature between the poles of the magnet. As the armature is attracted to one of the magnet poles, it pivots on the suspension member and thus moves the diverter blade.

This movement will be proportional to the magnitude of the command. The blade in the center position partially interrupts each jet in a manner to provide equal pressure in both receivers (C1 & C2).

When the blade moves from center, it will increase the interruption of the other jet stream. This will decrease the pressure in one receiver and increase the pressure in the other receiver. The resulting differential pressure between C1 and C2 will cause the second stage spool to shift proportionally.

How the Main Valve Works

As was explained, the valve spool is moved back and forth by the differential pressure generated between C1 and C2. However, these are not the only forces acting on the spool. Additional forces due to flow forces, dirt, friction and pressure loading can cause this type of proportional valve main spool to change position. To counteract these problems, an LVDT is attached to the spool which generates an accurate electrical signal which is feedback to the electronics to indicate the spool position, known as feedback.

The LVDT feedback is compared to the input command signal within the electronics. If the two are not equal, the electronics increases or decreases the electrical power to the pilot coil, thereby adjusting the delta p across P1 to P2. This repositions the spool where the command inputs indicate it should be.

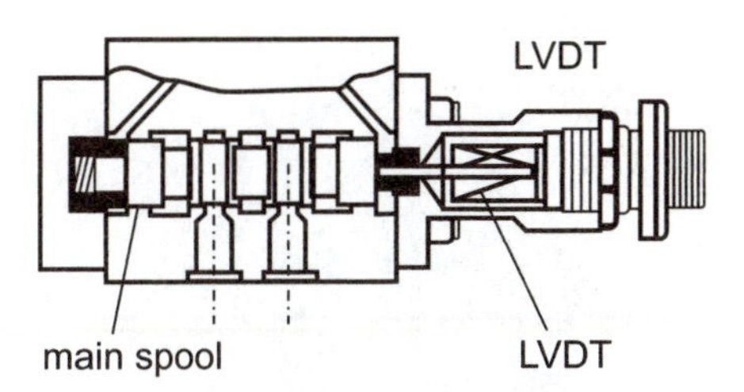

Figure 10-98

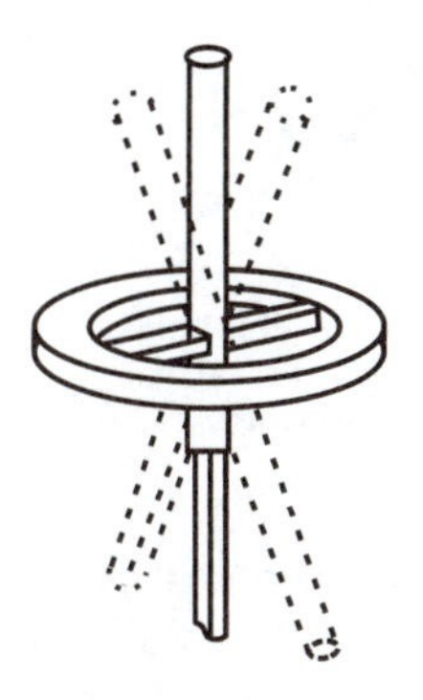

Armature Motion

Figure 10-99

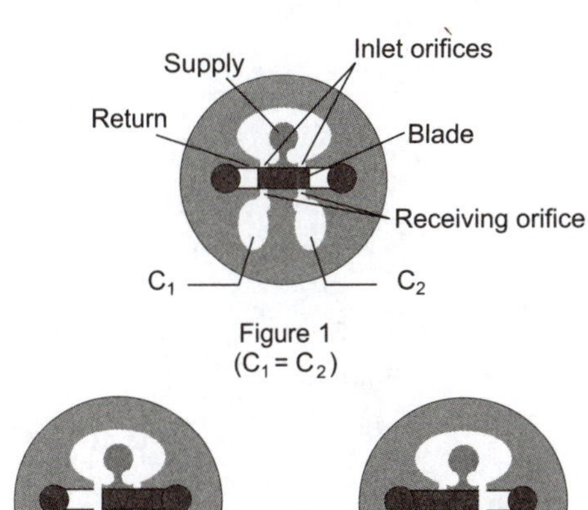

Figure 1
($C_1 = C_2$)

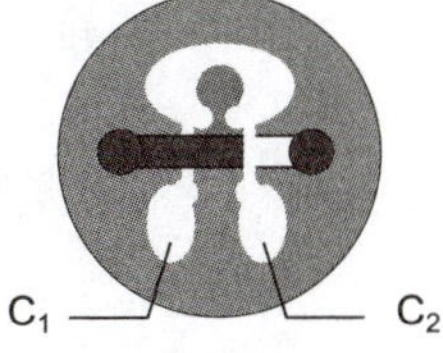

Figure 2
(C_1 pressure greater than C_2)

Figure 3
(C_2 pressure greater than C_1)

Figure 10-100

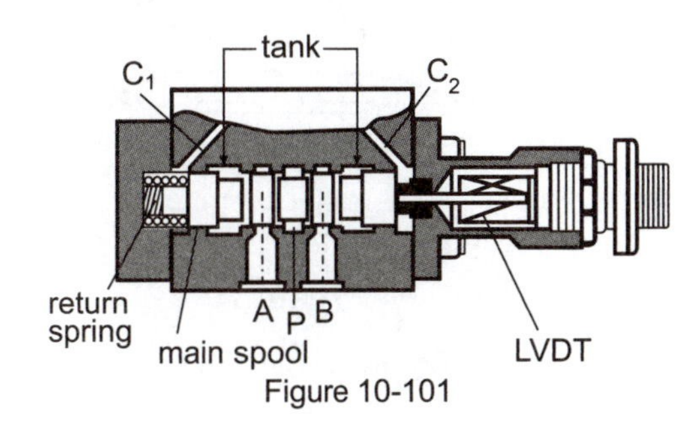

Figure 10-101

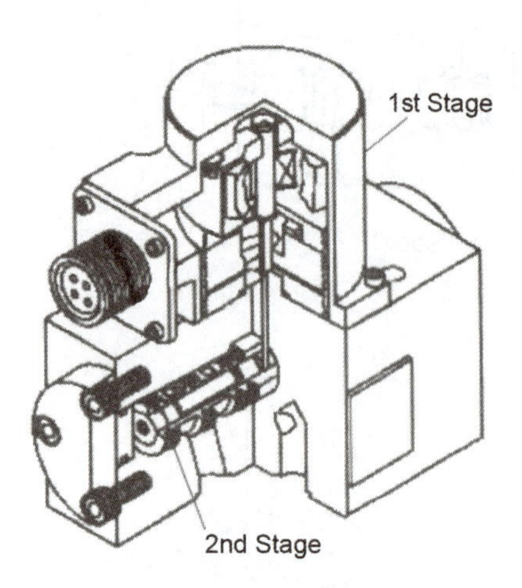

Figure 10-102

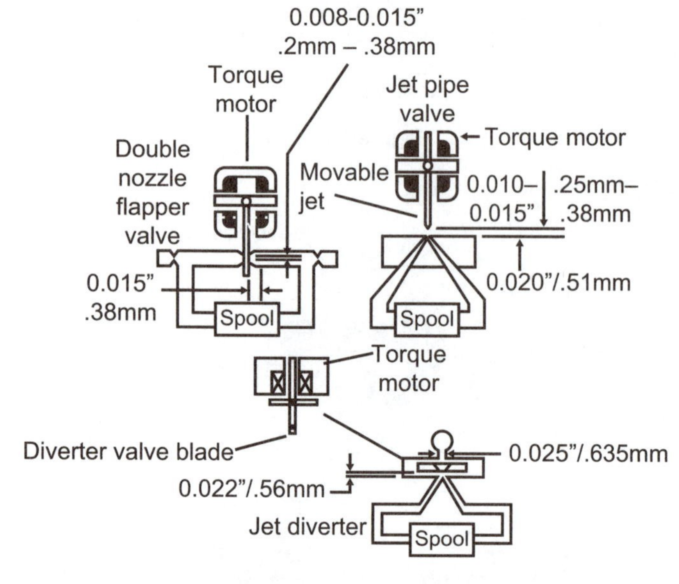

Figure 10-103

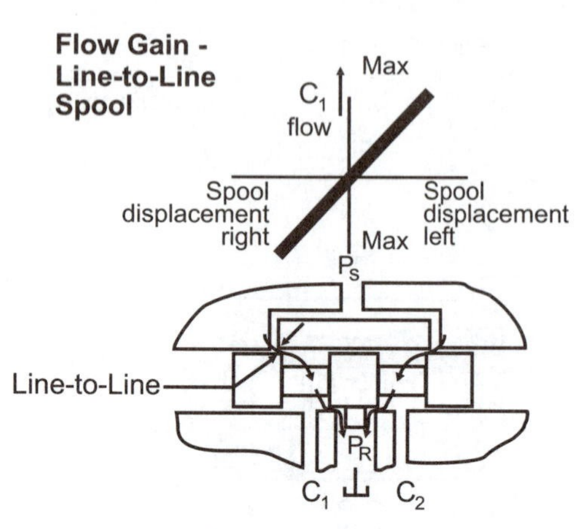

Figure 10-104

What a Servo Valve Consists Of

A typical servo valve consists of a first stage (pilot valve) mounted onto the second stage (spool valve) with a mechanical feedback of spool position to the torque motor assembly.

Types of First Stages

There are three most common first stage designs, flapper nozzle, jet pipe and jet diverter. The minimum orifice for a flapper nozzle valve is the 0.0015 (0.38mm) clearance between the flapper and the 0.010/0.015 (0.25mm/0.38mm) diameter nozzle.

A jet pipe valve has a typical nozzle diameter of 0.008/0.010 (0.2mm/0.25mm) for the minimum orifice.

A jet diverter valve has a minimum orifice of 0.020 (0.51mm).

It is not within the context of this textbook to suggest which first stage is best suited for an application, but it should be noted that the larger the orifice, the more contamination tolerant is the valve.

Types of 2nd Stage Spool Designs

The condition or matching of the second stage spool lands at the center position can and does vary depending upon the requirements of the system and/or tolerances during manufacturing.

The most common requirement dictates an edge condition of the spool lands to valve body ports to be as close to "line to line" as possible. Optional spool land matching includes "underlapped" and "overlapped" conditions. Each one of these creates unique flow characteristics within the valve.

Line to Line Condition

This line to line condition results in an ideal flow gain plot where the output flow to the cylinder ports is zero with the spool in the center position and increases immediately with spool travel.

Underlapped Condition

An underlapped condition has more clearance between the spool land edges and metering notches or ports in the spool sleeve or valve body. This results in a higher leakage flow in the center spool position. Note that with this condition, the spool must travel through the underlap before proportional flow begins to the cylinder port.

Overlapped Condition

An overlapped condition will reduce leakage flow to a minimum and no cylinder port flow will occur until the spool has traveled through the overlap. It should be noted that this configuration creates what is known as "deadband" in the valve operation.

Deadband is a zone of valve movement in either direction from center where no actuator (cylinder or motor) response to input signal occurs.

NOTE: This condition also occurs in proportional valves previously discussed.

How a Servo Valve Works

A typical first stage section operation was explained in the proportional valve operating discussion and it is the same for servo valves.

The differential pressure created by the first stage is applied across the ends of the second stage spool and will cause it to move. In order to locate the spool into a position that is proportional to the electrical input command a feedback spring connects the first stage armature and second stage spool together.

This spring can be considered as a cantilever beam that is sized to provide a linear resisting force that is equal to the torque motor force for every spool position. When there is no electrical command, the diverter blade/armature assembly, the feedback spring, and the center position of the spool have a center line relationship.

Applying an input command signal to the torque motor causes the diverter blade/ armature assembly to move generating a differential pressure between PC1 and PC2. Again the spool is moved and at the same time deflecting the feedback spring developing a torque in opposition to the motor torque. When the spool has moved to the point where these two torques are equal, the diverter blade/armature assembly, the feedback spring and the spool are essentially recentered and spool movement stops at this new position related to input command signal.

As the input command signal is reduced to zero, the torques go to zero and the spool returns to its center position.

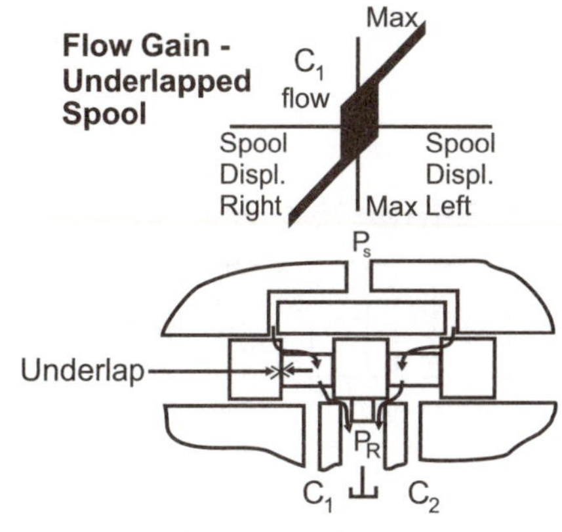

Figure 10-105

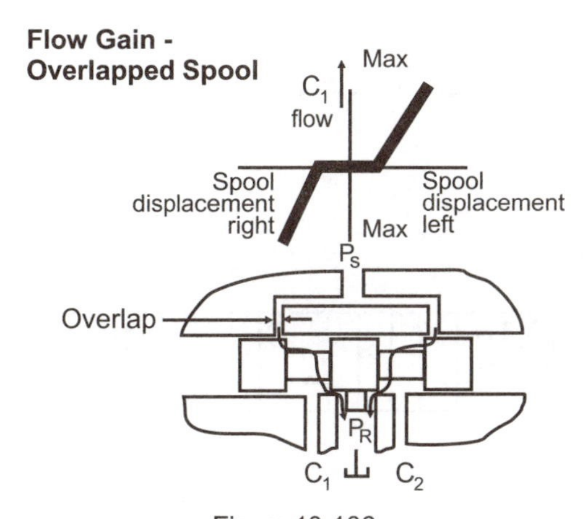

Figure 10-106

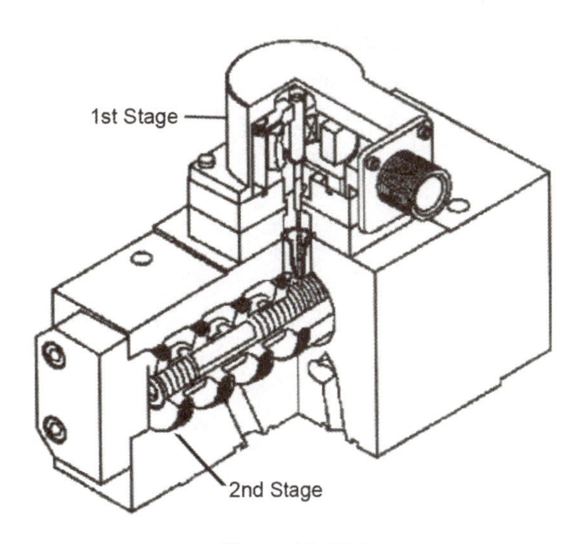

Figure 10-107

Exercise
Directional Control Valves
50 Points

1. Describe in detail the following symbols:

A. 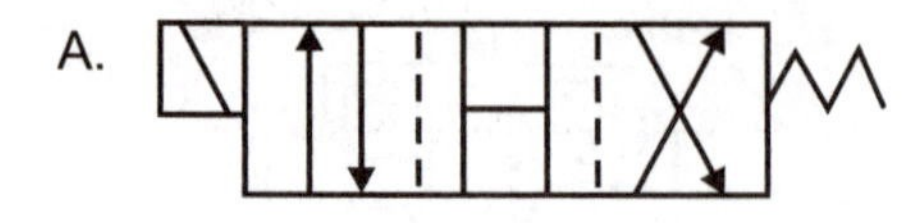

B. 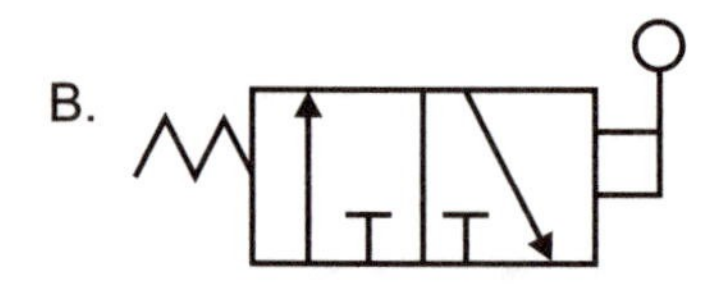

C. 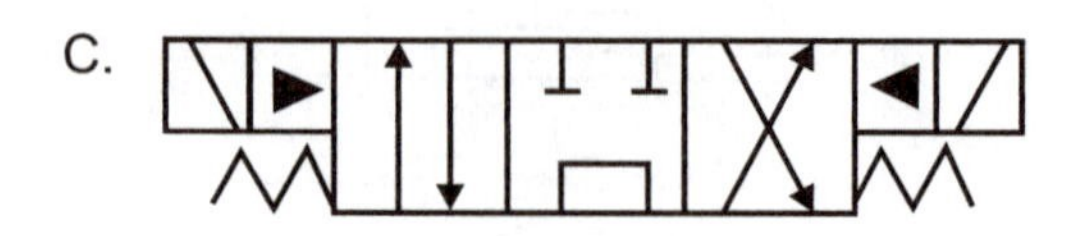

D. 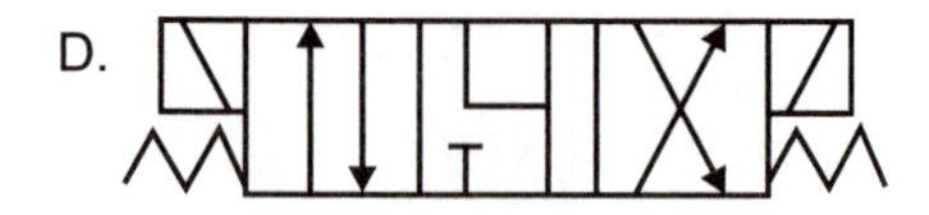

Directional Control Valves (cont.)

2. **SITUATION:** A cylinder receives a flow rate of 100 gpm (37.9 l/min). The cylinder is required to slow down in mid-stroke and go into a feed rate. The return stroke is at the full 100 gpm (379 l/min) pump flow.

   PROBLEM: Design the system with the components listed below using the appropriate ANSI or ISO standard symbol.

   1 - pump

   1 - electric motor

   1 - filter

   1 - reservoir

   1 - relief valve

   1 - directional valve

   1 - cylinder

   1 - deceleration valve

   1 - check valve

   1 - pressure compensated flow control

Directional Control Valves (cont.)

3. In the circuit illustrated, assume that the directional valve has been centered for several minutes.

   What do the gages in the cylinder lines read?

   Gage 1_______________ Gage 2_______________

   How is the cylinder affected?

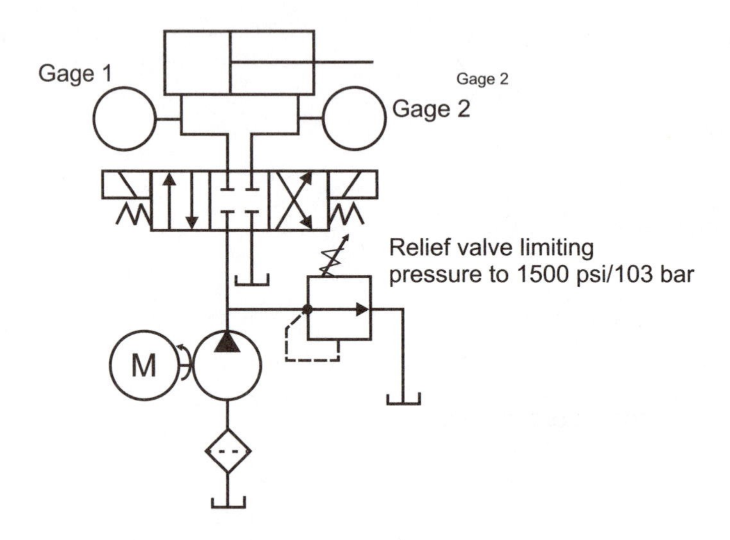

Directional Control Valves (cont.)

4. It was shown earlier that cylinder rod speed could be increased by regeneration. However, the maximum force developed by the cylinder was diminished because of it.

   In the circuit illustrated, select a directional valve so that in one position regeneration is achieved; in a second position maximum force is received; in a third position, the cylinder is retracted.

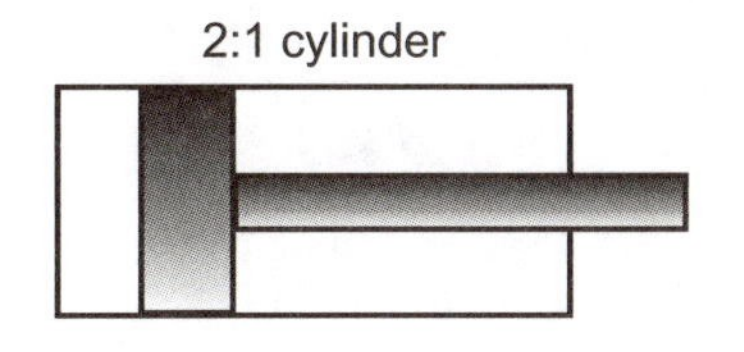

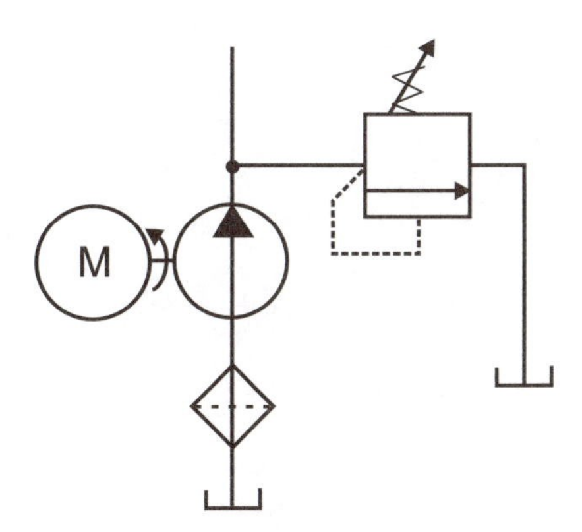

CHAPTER 11

Pressure Control Valves

As we have seen, maximum system pressure can be controlled with the use of a normally non-passing pressure valve. With the primary port of the valve connected to system pressure and the secondary port connected to tank, the spool in the valve body is actuated by a predetermined pressure level at which point primary and secondary passages are connected and flow is diverted to tank. This type of normally non-passing pressure control is known as a relief valve.

Pressure Adjustment

In a pressure control valve, spring pressure is usually varied with a screw adjustment which compresses or decompresses the springs.

Uses of a Non-Passing Pressure Valve

Normally non-passing pressure control valves have many uses in a hydraulic system. Besides using the valve as a system relief, a non-passing pressure control can be used to cause one operation to occur before another. It can also be used to counteract external mechanical forces in the system.

Sequence Valve

A normally non-passing pressure control valve which causes one operation to occur before another is referred to as a sequence valve.

Sequence Valve in a Circuit

In a clamp and drill circuit, the clamp cylinder must extend before the drill cylinder. To accomplish this, a sequence valve is positioned in the leg of the circuit just ahead of the drill cylinder. The spring in the sequence valve will not allow the spool to connect primary and secondary passages until pressure is high enough to overcome it.

Flow to the drill cylinder is blocked. Therefore, the clamp cylinder will extend first. When the clamp contacts the work piece, the pump applies more pressure to overcome the resistance. This rise in pressure actuates the spool in the sequence valve. Primary and secondary passages connect. Flow goes to the drill cylinder.

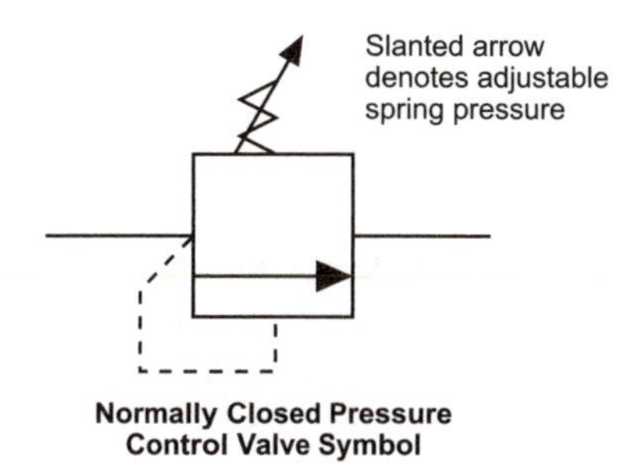

Normally Closed Pressure Control Valve Symbol

Figure 11-1

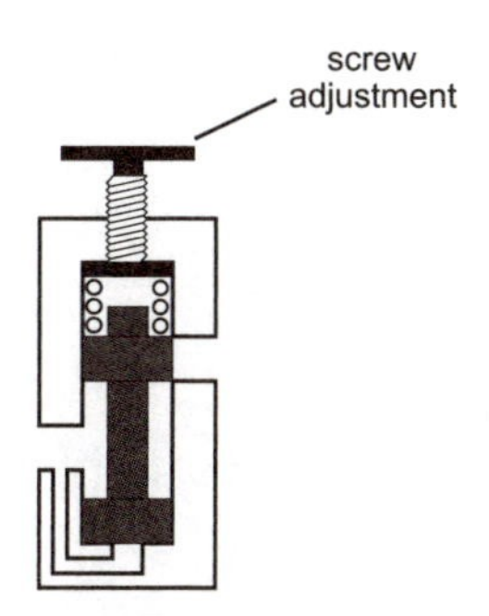

Figure 11-2

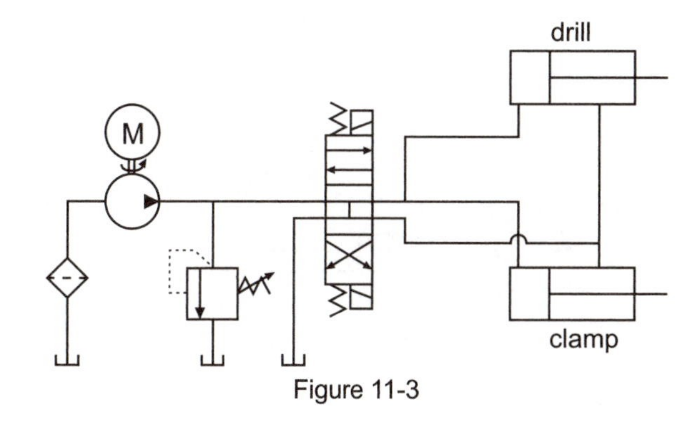

Figure 11-3

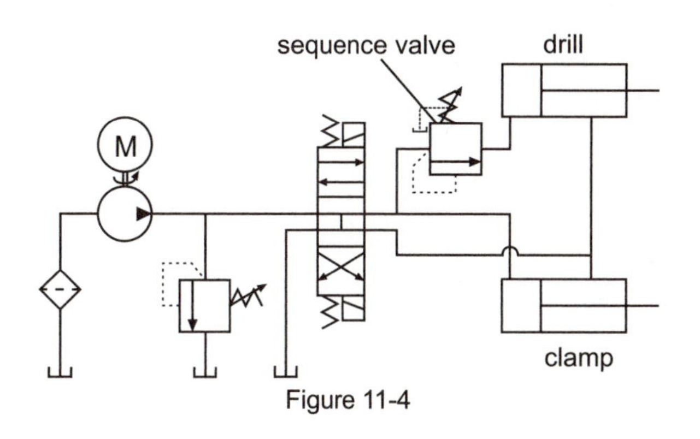

Figure 11-4

Counterbalance Valve

A normally non-passing pressure control valve can be used to balance or counteract a weight such as the platen of a press. This valve is called a counterbalance valve.

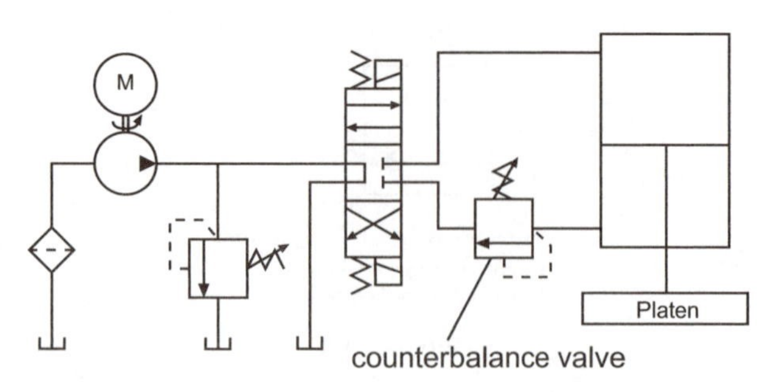

Figure 11-5

Counterbalance Valve in a Circuit

In a press circuit, when the directional valve directs flow to the cap end of the cylinder, the weight of the platen attached to the cylinder rod will fall uncontrollably. Pump flow will not be able to keep up.

To avoid this situation, a normally non-passing pressure valve is located downstream from the press cylinder. The spool in the valve will not connect primary and secondary passages until a pressure, which is sensed at the bottom of the spool, is greater than the pressure developed by the weight of the platen. (In other words, when fluid pressure is present at the cap end of the piston.) In this way the weight of the platen is counterbalanced throughout its downward stroke.

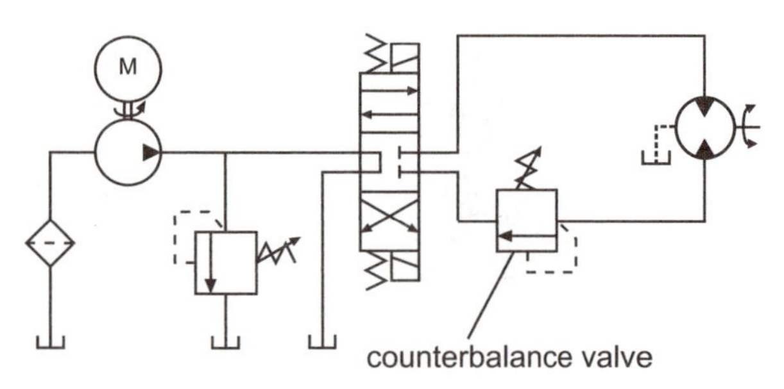

Figure 11-6

A counterbalance valve can also be used to retard the spinning motion of a weight attached to a motor shaft.

A hydraulic motor, which is turning a heavy wheel, may run away once the momentum of the wheel has built up. A counterbalance valve, positioned at the outlet of the motor, will not open until pressure is present at the motor outlet. This pressure counteracts the force of the spinning weight.

Simple Relief Valve

A simple relief valve basically consists of valve body with a spool which is biased by a heavy spring. When pilot pressure at the spool end opposite the spring is high enough, the spool moves up opening a path to tank for pump flow.

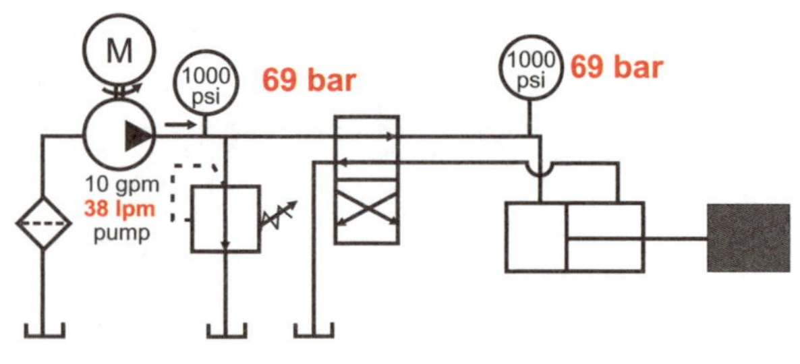

Figure 11-7

In the circuit illustrated, the simple relief valve is set to relieve 10 gpm (37.9 lpm) when pressure at pump outlet reaches 1000 psi (68.97 bar). This does not mean that at a pressure level of 1000 psi (68.97 bar), the valve suddenly opens dumping flow to tank. The valve opens at a point below 1000 psi (68.97 bar) bleeding off fluid to tank and progressively bleeds more off as 1000 psi (68.97 bar) is approached.

In the previous lesson, we saw that clearances between moving parts of a pumping mechanism acted like restrictions allowing flow to slip back to pump inlet. As wear became excessive, the size of the restriction increased so much that pump/electric motor could have its total flow leak back to pump inlet. A relief valve can be thought of somewhat in the same manner.

The moving part inside a relief valve forms a restriction back to tank as the valve is operating. The restriction begins to appear in the system when a predetermined pressure level is reached. As system pressure and pilot pressure at the bottom of the spool increases, the size of the restriction also increases allowing more flow back to tank. At the pressure setting of the valve, the size of the restriction is large enough to accept all pump flow.

When pressure in a system drops below valve cracking pressure, the opening to tank through the valve disappears from the circuit.

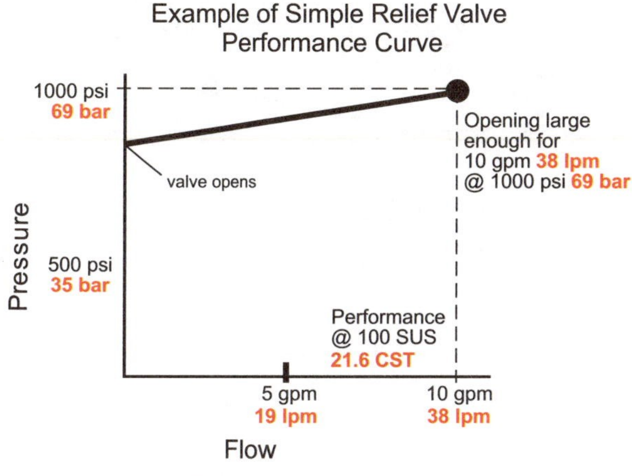

Figure 11-8

Relief Valve Setting Affected by Temperature

Since the opening through a relief valve is actually a restriction, flow through it is affected by temperature just as with any other restriction.

With a relief valve in a system set to relieve 10 gpm (37.9 lpm) at 1000 psi (68.97 bar), 10 gpm (37.9 lpm) can be pushed through the relief valve opening when 1000 psi (68.97 bar) is present at the valve inlet. This assumes that the liquid viscosity remains constant.

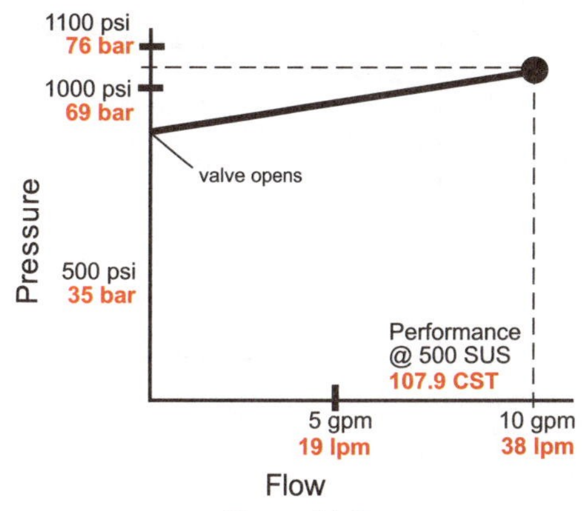

Figure 11-9

With fluid in a machine reservoir at room temperature, its viscosity might be 400-500 SUS (86.3 - 107.9 CST). If the system relief valve were adjusted to handle 10 gpm (37.9 lpm) at 1000 psi (68.97 bar) while at this viscosity, pump pressure might be limited to only 900 psi (62 bar) or less once the fluid heated up. With fluid at operating temperature and its viscosity decreased to 100 SUS (21.6 CST), 10 gpm (37.9 lpm) could be pushed back to tank through the valve with less pressure.

The opposite also occurs when the relief valve is set at operating viscosity. At start up, this will mean that pump pressure could approach 1100 psi (75.9 bar).

In a system where maximum pump pressure is rather critical, the system should be allowed to warm up before relief valve adjustments are made.

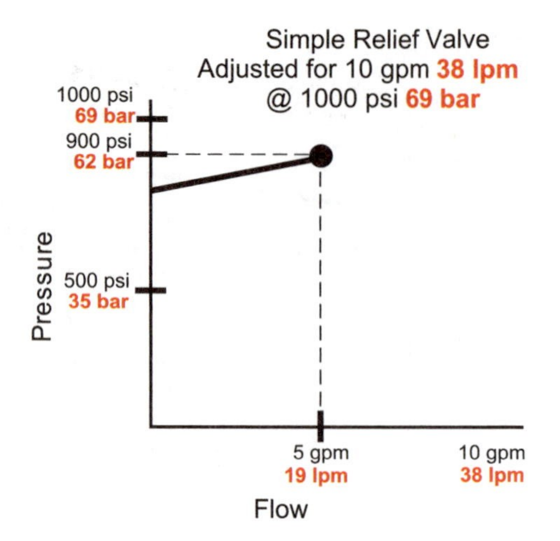

Figure 11-10

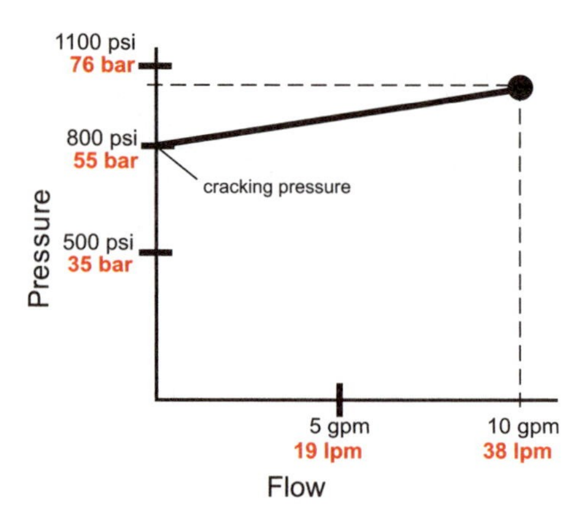

Figure 11-11

Relief Valve Setting Affected by Pump Wear

Setting of a relief valve is affected by pump wear.

With a relief valve set to relieve 10 gpm (37.9 lpm) at 1000 psi (68.97 bar), a pump flow of 10 gpm (37.9 lpm) can pass through the relief valve opening once 1000 psi (68.97 bar) is present at valve inlet. As the pump wears, discharge flow decreases and less flow passes through the relief valve. Pump/ electric motor pressure will decrease as its flow dumps over the relief valve.

With a relief valve adjusted for 10 gpm (37.9 lpm) at 1000 psi (68.97 bar), system relief valve pressure might be only 900 psi (62 bar) as a reduced flow of 5 gpm (18.95 lpm) passes through the valve. It would appear that the relief valve setting was changed.

Cracking Pressure – Simple Relief Valve

Cracking pressure is the point at which a relief valve begins to open a flow path back to tank. This point for a simple relief valve is somewhat below the relief valve setting. Simple relief valves have characteristically early cracking pressures.

In the illustrated performance curve for a simple relief valve, the valve is set to relieve a pump flow of 10 gpm (37.9 lpm) when pressure reaches 1000 psi (68.97 bar). The curve points out that the valve cracks open at 800 psi (55.2 bar) increasing in size as pump pressure approaches 1000 psi (68.97 bar). Finally, at 1000 psi (68.97 bar) the size of the restriction through the valve is large enough to accept 10 gpm (37.9 bar).

System Affected by Early Cracking Pressure

A relief valve with an early cracking pressure can be a disadvantage for a system. Assume that a pump/ electric motor develops 750 psi (51.7 bar) to push a flow of 10 gpm (37.9 lpm) out to an actuator. 550 psi (37.9 bar) is required for the load; 200 psi (13.8 bar) is used to overcome liquid resistances. With a simple relief valve set for 1000 psi (68.97 bar), the valve remains closed; all pressurized liquid flow is directed toward the work load.

If work load pressure increases to 700 psi, (48.28 bar) then pump/electric motor must increase its output pressure to 900 psi (62.1 bar). (Discharge flow assumed constant.) With a cracking pressure of 800 psi (55.2 bar), the valve at the current pump pressure is bleeding off 5 gpm (18.95 l/min) of pump flow back to tank. This means actuators are not filled as quickly and work is performed at a slower rate. It also means heat is unnecessarily being generated as fluid passes through the relief valve restriction.

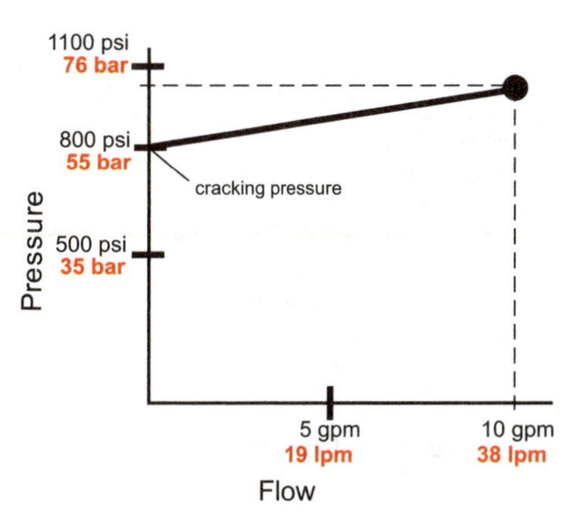

Figure 11-12

If work load pressure were 750 psi (51.7 bar), pump/electric motor might develop 950 psi (65.5 bar) to push its flow out to the system. With this being higher up on the relief valve curve, more flow passes to tank. At this point, rod speed decreases even more since a greater amount of flow is dumping back to tank; and, more heat is generated unnecessarily.

Normally Passing Pressure Valve

A normally non-passing pressure control valve has primary and secondary passages disconnected, and pressure at the bottom of the spool is sensed from the primary port.

A normally passing pressure valve has primary and secondary passages connected, and pressure at the bottom of the spool is sensed from the secondary port.

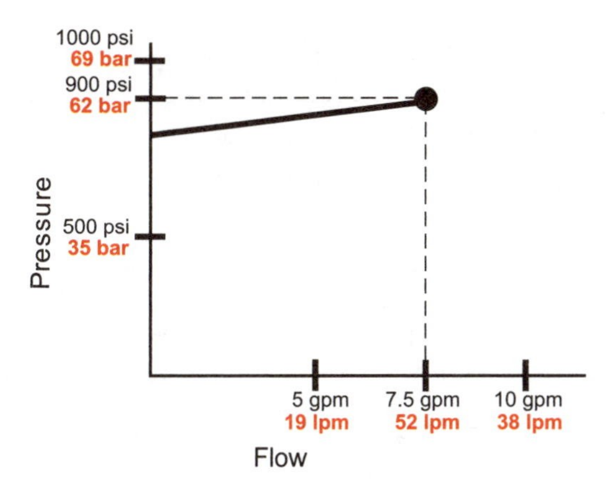

Figure 11-13

Pressure Reducing Valve

A pressure reducing valve usually is a normally passing pressure control valve.

How a Pressure Reducing Valve Works

A pressure reducing valve operates by sensing fluid pressure after it has passed through the valve. As pressure downstream equals the setting of the valve, the spool is partially closed causing a restricted flow path. This restriction turns any excess pressure energy ahead of the valve into heat.

If pressure after the valve drops off, the spool will open and allow pressure to build once again.

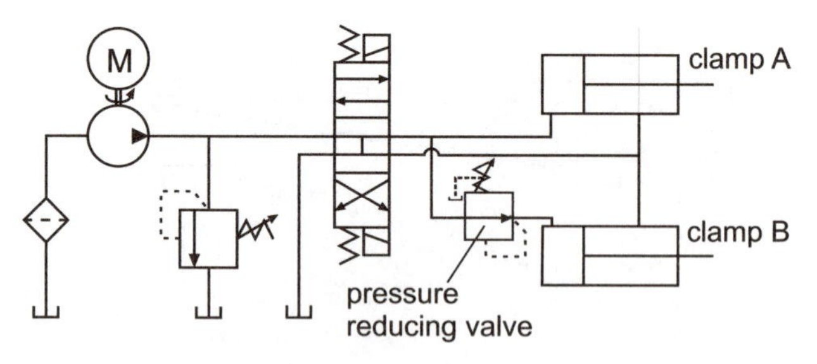

Figure 11-14

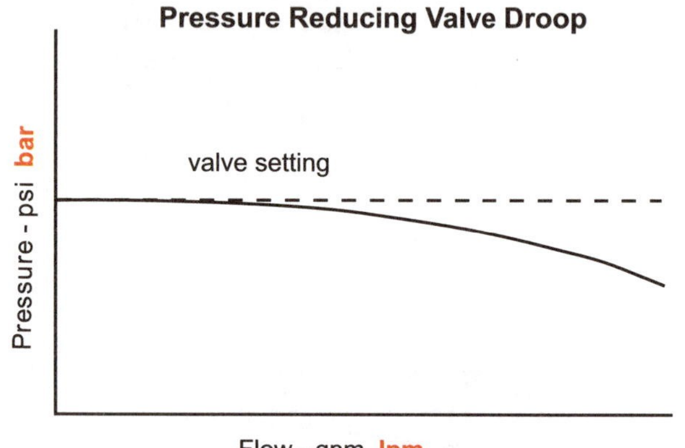

Figure 11-15

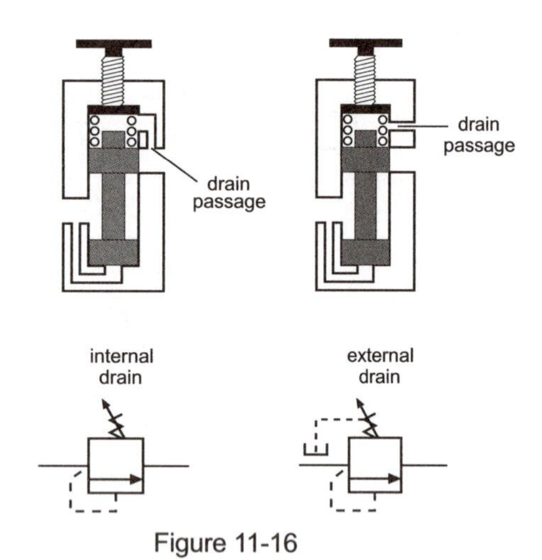

Figure 11-16

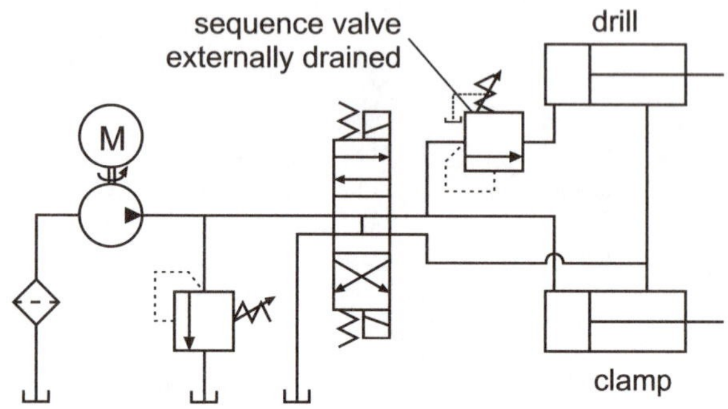

Figure 11-17

Pressure Reducing Valve in a Circuit

The illustrated clamp circuit requires that clamp cylinder B apply a lesser force than clamp A. A pressure reducing valve placed just ahead of clamp cylinder B will allow flow to go to the cylinder until pressure reaches the setting of the valve. At this point, the valve spool is actuated causing a restriction to that leg of the circuit. Excess pressure ahead of the valve is turned into heat. Cylinder B clamps at a reduced pressure.

Pressure Reducing Valve Droop

With the same valve setting, the reduced pressure downstream from a pressure reducing valve will be lower when the valve is handling its rated flow than when it is operating under deadheading conditions as in a clamp circuit. This difference in reduced pressures is known as pressure reducing valve droop. Droop is a characteristic of all reducing valves and becomes more pronounced as system pressure rises and flow increases.

A 15 gpm (56.85 lpm) pressure reducing valve could droop 50 psi (3.45 bar) at its rated flow and rated operating pressure. A 100 gpm (37.9 lpm) pressure reducing valve may droop as much as 150 psi (10.3 bar).

Drains

The spool in a pressure control valve moves within a passage. There is some leakage of fluid into the passage above the spool. This is a normal occurrence which serves to lubricate the spool.

In order for a pressure valve to operate properly, the area above the spool must be continuously drained so that the liquid does not impair the movement of the spool. This is accomplished with a passage within the valve body which is connected to the reservoir.

Internal Drain

If the secondary passage of a pressure valve is connected to the reservoir, as in relief valve and counterbalance valve applications, the drain passage is internally connected to the valve's secondary or tank passage. This is known as an internal drain.

External Drain

If the secondary passage of a pressure valve is a pressure line (or in other words does work) as in sequence valve and pressure reducing valve applications, the drain passage is connected to tank by means of a separate line. This is an external drain.

Sequence valves and pressure reducing valves are always externally drained.

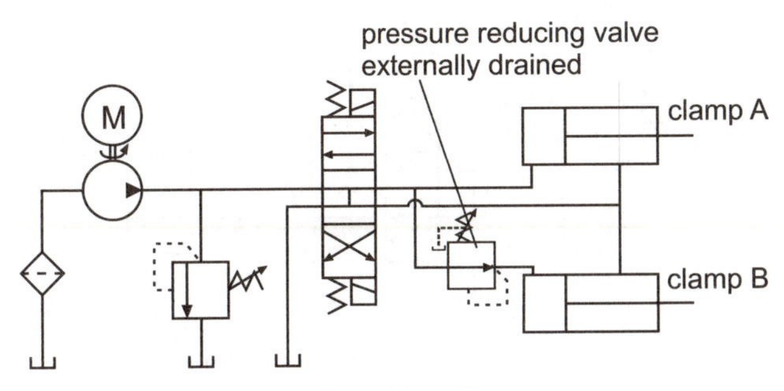

Figure 11-18

Direct and Remote Operation

Up to this point, we have seen that pressure controls sense pressure from a passage within the valve body. In normally non-passing valves, pressure is sensed from the primary passage. In a pressure reducing valve, pressure is sensed from the secondary passage. This type of pressure sensing is identified as direct operation.

Pressure control valves can also sense pressure in another part of a system by means of an external line. This is remote operation.

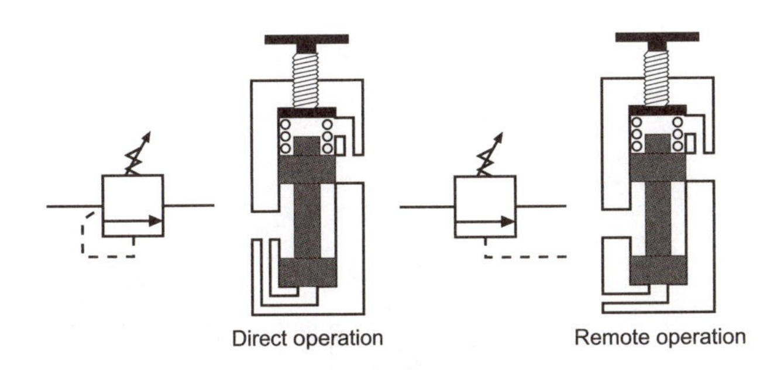

Figure 11-19

Unloading Valve

An unloading valve is a remotely operated normally non-passing pressure control valve which directs flow to tank when pressure in a remote part of a system reaches a predetermined level.

Unloading Valve in a Circuit

A directly operated relief valve used in an accumulator circuit means that once the accumulator is charged, the pump's flow returns to tank at the relief valve setting. This is a waste of horsepower and an unnecessary generation of heat.

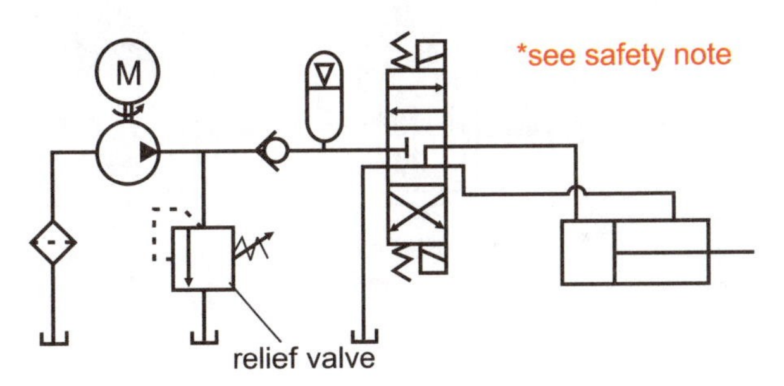

Figure 11-20

A remotely operated unloading valve, with its pilot line connected downstream from the check valve, will allow pump flow to return to tank at a minimum pressure when the accumulator is charged to the valve setting.

The pump is not required to apply a high pressure to operate the unloading valve because the valve is operated from pressure in another part of the system.

Since pressure applied by the pump is negligible, so is the horsepower.

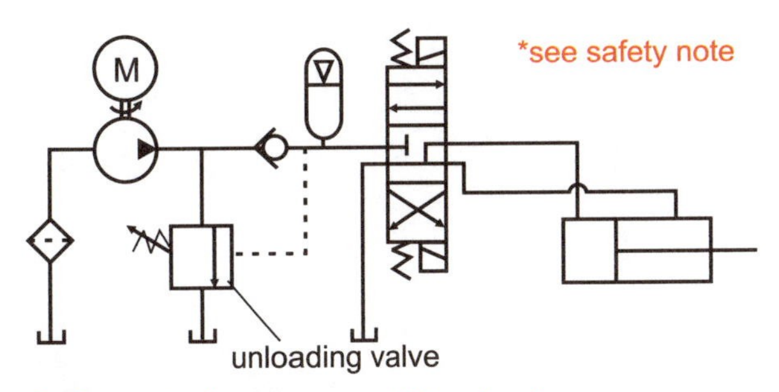

Safety note: In any accumulator circuit a means should be available of automatically unloading the accumulator when the machine is shut down.

Figure 11-21

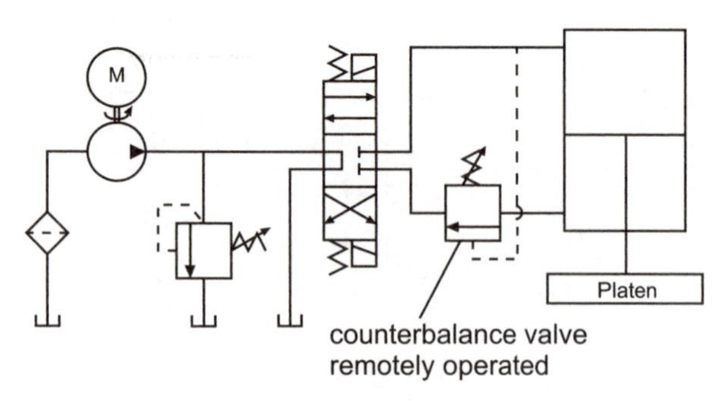

Figure 11-22

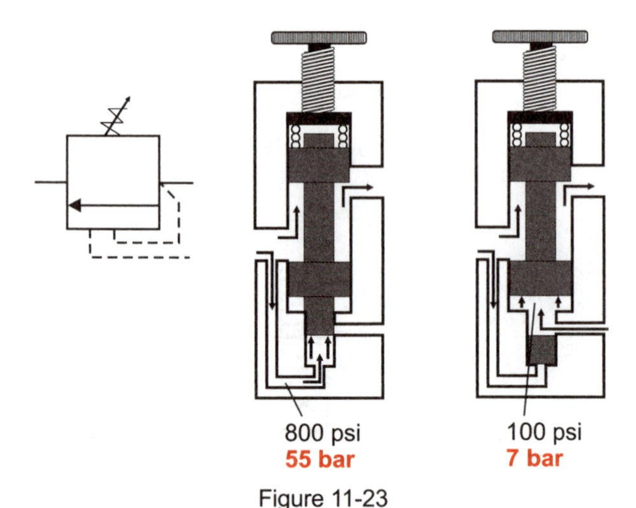

Figure 11-23

Remotely Operated Counterbalance Valve

A directly operated counterbalance valve, positioned downstream from a cylinder supporting a heavy platen, effectively balances or cancels out the weight of the platen. If the platen is required to move through the material during the pressing process, the weight of the platen does not add to the total pressing force. If this is undesirable, the pilot line of the valve is remotely connected to the other cylinder line.

Remotely Operated Counterbalance Valve in a Circuit

With remote operation, the platen is still balanced on its downward stroke and use can be made of the platen's weight in pressing. If the platen attempts to pull away from fluid flow, pressure will drop off in the upstream cylinder line as well as in the pilot line. The valve will close and allow the flow to catch up. During the pressing operation, the valve will be wide open. No back pressure will act on the rod end side of the piston. The platen's weight can be added to the pressing force.

NOTE: This simplified circuit may need refining to achieve a smooth operation.

Directly Operated Counterbalance Valve in a Motor Circuit

A motor circuit is illustrated which uses a directly operated counterbalance valve to control the runaway tendency of a spinning load. With the valve set for 800 psi (55.2 bar), a back pressure is always present while the load is spinning. This pressure keeps the load from running away from pump flow, but it also means that pressure at motor inlet must be 800 psi (55.2 bar) more than the work load pressure. This is a disadvantage which is overcome by a brake valve.

Brake Valve

A brake valve is a normally non-passing pressure control valve with both direct and remote pilots connected simultaneously for its operation. This valve is frequently used with hydraulic motors instead of a directly operated counterbalance valve.

What a Brake Valve Consists Of

A brake valve consists of a valve body with primary and secondary passages, internal and remote pilot passages, spool, piston, bias spring, and spring adjustment.

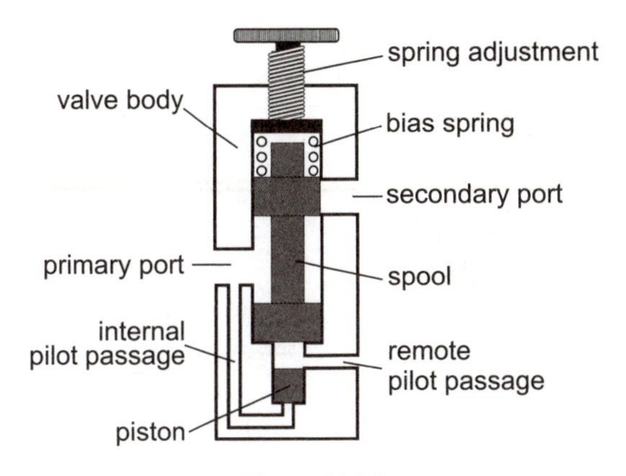

Figure 11-24

How a Brake Valve Works

A brake valve is a normally non-passing valve. Assume that the spring biasing the spool is adjusted for 800 psi (55.2 bar) direct operation. When pressure in the internal pilot passage reaches 800 psi (55.2 bar), the piston moves up pushing the spool and opens a passage through the valve.

If pressure falls below 800 psi (55.2 bar), the valve closes. This operates as the directly operated counterbalance valve which we saw earlier.

The piston on which the internal pilot pressure acts, has much less cross sectional area than the spool. The area ratio is frequently 8:1. With the remote pilot connected to the opposite motor line, a pressure of only 100 psi (6.89 bar) is needed to open the valve since it acts on the bottom of the spool with eight times more area than the piston.

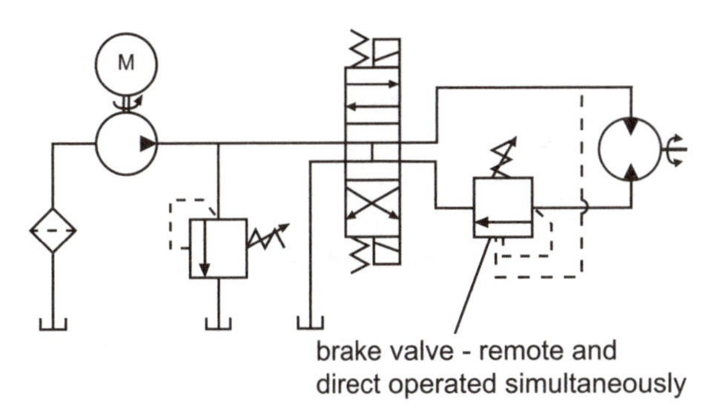

Figure 11-25

Brake Valve in a Circuit

With a brake valve set for 800 psi (55.2 bar), the valve will be open when 100 psi (16.89 bar) is present in the motor inlet line. Pressure at motor inlet will be whatever it takes to turn the load only (assuming that this pressure is above 100 psi/68.97 bar) If the load attempts to run away pressure drops off in the motor inlet line. The brake valve closes and does not reopen until a back pressure of 800 psi (55.2 bar) is generated to slow down the load.

A brake valve is a normally non-passing pressure control valve whose operation is directly tied to the needs of a motor load.

Reverse Flow

A normal requirement of all pressure valves, except relief and unloading valves, is that reverse flow must be able to pass through the valve.

Since normally non-passing pressure valves sense pressure from the primary passage, as soon as flow is reversed, pressure in the primary passage falls off. The spool is de-actuated. Primary and secondary passages are disconnected. Flow through the valve is blocked. Since we cannot go through the valve, we go around the valve by using a check valve.

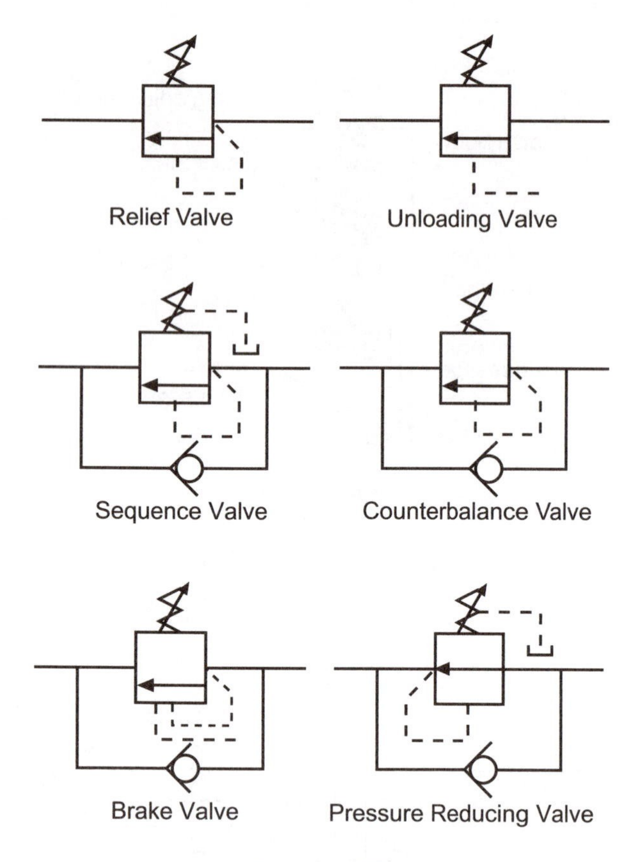

Figure 11-26

Normally passing pressure valves sense pressure from the secondary passage. It would appear that as long as reverse flow pressure ahead of the valve remains below valve setting, passage through the valve will remain open and no check valve is required. This is true. However, any rise in pressure above the setting will result in the spool being slammed shut. As a precautionary measure, many times a check valve is used with a pressure reducing valve for reverse flow.

Generalizations About Pressure Control Valves

Some generalizations can be made about pressure control valves:

A. Pressure control valves, whose secondary ports are pressurized, have external drains. (Sequence and pressure reducing valves)

B. Pressure control valves, whose secondary ports are connected to tank, generally have internal drains. (Relief, unloading, counterbalance, and brake valves)

C. To pass reverse flow through a pressure control valve, a check valve is used.

Pressure Control Valve Symbols

From the beginning of the lesson we have been building the symbols for the various types of pressure control valves. The complete symbol for each valve is illustrated.

Terms and Idioms Associated with Pressure Control Valves

DIRECT OPERATED - We have been referring to “direct operated” as meaning that the valve’s spool is piloted or actuated by an internal pilot line within the valve body. “Direct operated” is also commonly used to denote that the spool in the valve is held biased by spring pressure.

DUMPING OVER THE RELIEF VALVE - flow passing through the relief valve.

Exercise
Pressure Control Valves
40 Points

1. In the diagram, the directly operated counterbalance valve is set at 550 psi (37.9 bar).

   A. What does the gage read when the platen is not moving and is being suspended in mid-stroke?

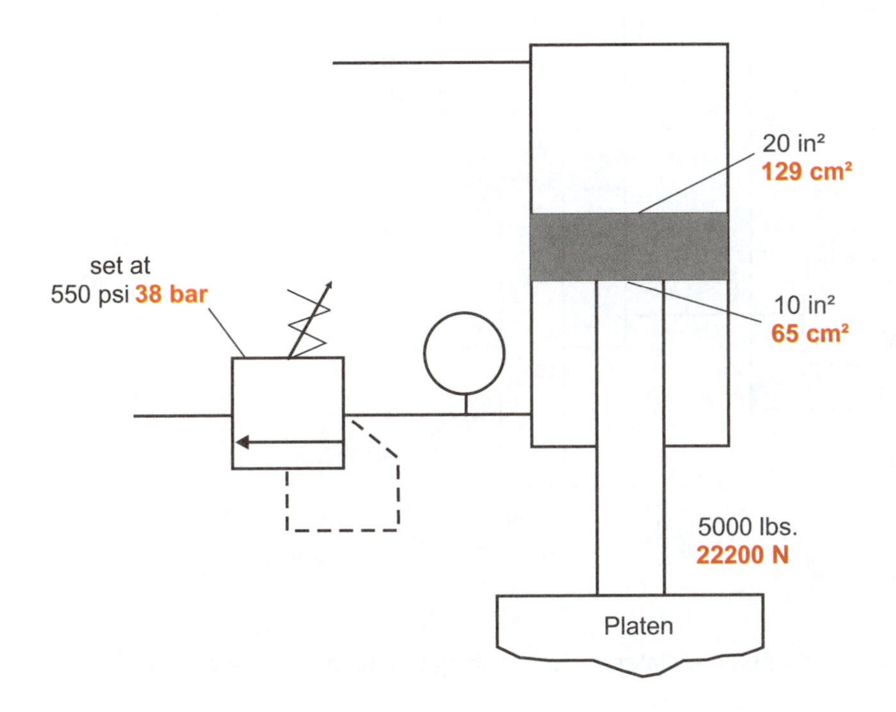

   B. What does the gage read when the platen is approaching the material to be pressed?

   ____

   C. Assume that the system relief valve is set for 1000 psi (68.97 bar) and that the movement of the cylinder rod is severely restricted once the platen contacts the material to be crushed. What is the maximum pressing force if the platen moves through the material during the pressing operation?

   ____

Pressure Control Valves (cont.)

2. In the diagram, the remotely operated counterbalance valve is set at 100 psi (6.8 bar).

   A. What does the gage read when the platen is not moving and is being suspended in mid-stroke?

   ____

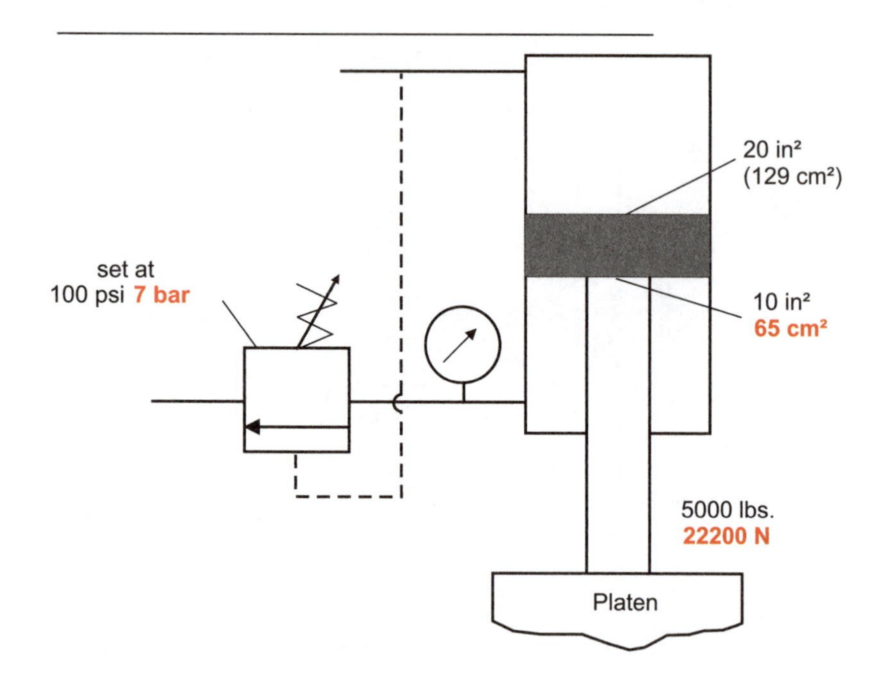

   B. What does the gage read when the platen is approaching the material to be pressed?

   ____

   C. Assume that the system relief valve is set for 1000 psi (68.96 bar) and that the movement of the cylinder rod is severely restricted once the platen contacts the material to be crushed. What is the maximum pressing force if the platen moves through the material during the pressing operation?

   ____

Pressure Control Valves (cont.)

3. **SITUATION:** The clamp cylinder must extend first and clamp at a pressure of 500 psi (34.48 bar).

   PROBLEM: Add the appropriate valves.

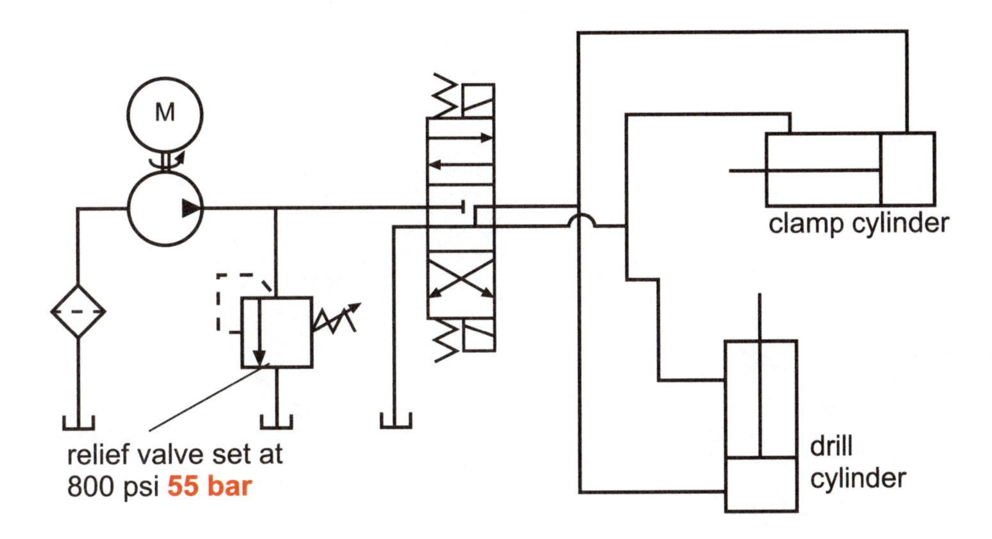

CHAPTER 12

Pilot Operated Pressure Control Valves

Unlike a simple or direct operated pressure control valve, where a spool is held biased by spring pressure only, a pilot operated valve has its spool biased by both fluid and spring pressure. This combination eliminates the high override commonly found in direct operated pressure valves.

Override Characteristic of Direct Operated Valves

A pressure valve's override characteristic may be best described by an example:

The graph shows the functioning of a direct operated relief valve operating in a typical system. The valve is required to relieve a system flow of 10 gpm (37.9 l/min) at 1000 psi (68.97 bar). In order to perform this function, the valve begins to open or "crack" at a lower pressure. This causes a small portion of system flow to return to tank. As pressure increases, the spool spring is continuously compressed in order to form a larger opening for the increasing flow returning to tank. Finally at 1000 psi (68.97 bar), a total flow of 10 gpm (37.9 l/min) passes through the valve. If for some reason system flow is increased, pressure will rise above or "override" the 1000 psi (68.97 bar) level. A direct operating valve acts in this manner because of compression of the spool spring.

Override Characteristic of Pilot Operated Valves

A pilot operated relief valve avoids an early cracking pressure and severe override by eliminating the heavy spool spring. Fluid pressure and a light spring bias the valve spool. When a certain system pressure is reached, the spool is actuated. Any slight override which results with increased flow is primarily due to the compression of the light spring. To understand the operation of a pilot operated pressure control valve, initially, we will concentrate on the operation of a pilot operated relief valve.

What a Pilot Operated Relief Valve Consists Of

A pilot operated relief valve consists of two valves — a main valve and a pilot valve. The main valve is made up of a body with inlet and outlet ports, a spool with an orifice, and a spool bias spring. The pilot valve consists of a dart, dart bias spring, and screw adjustment.

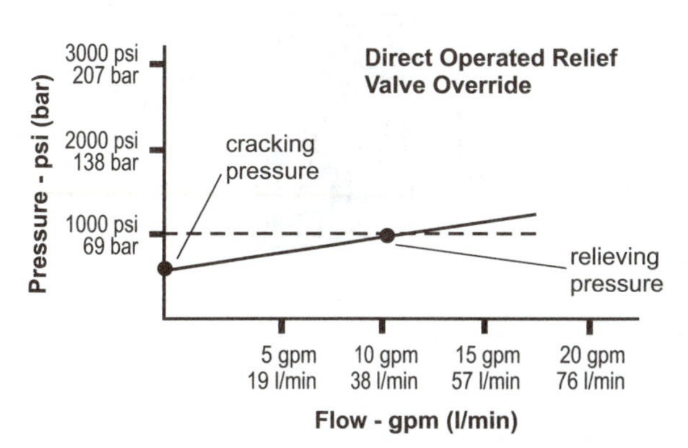

Figure 12-1

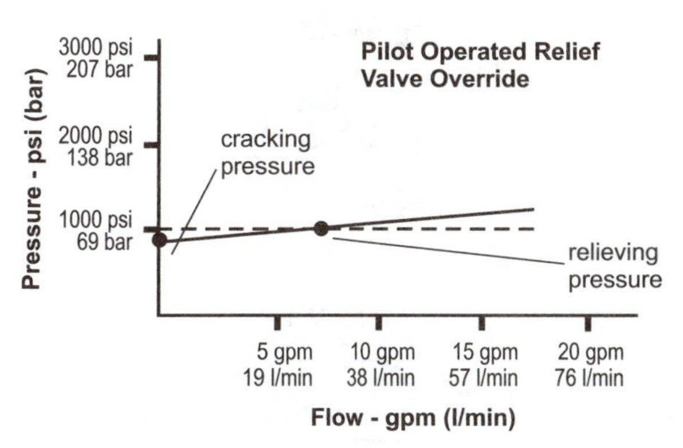

Figure 12-2

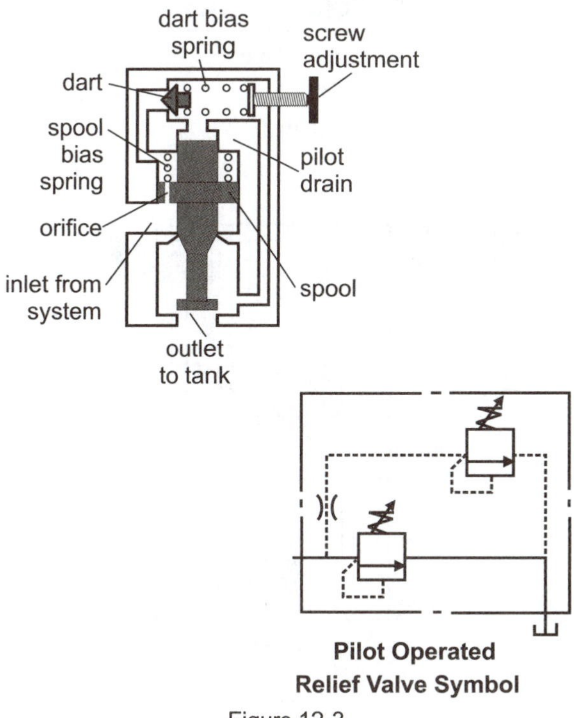

Figure 12-3

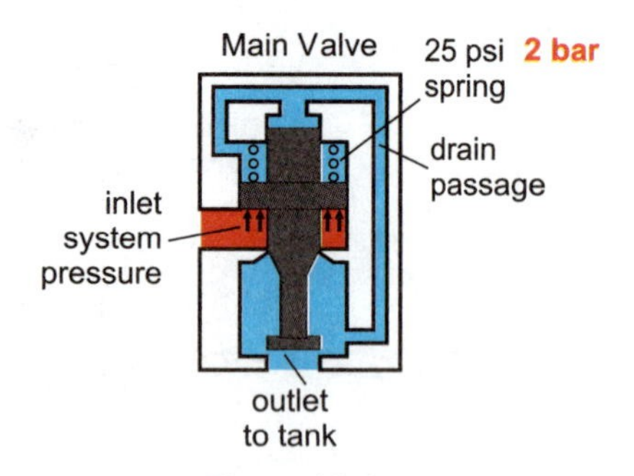

Figure 12-4

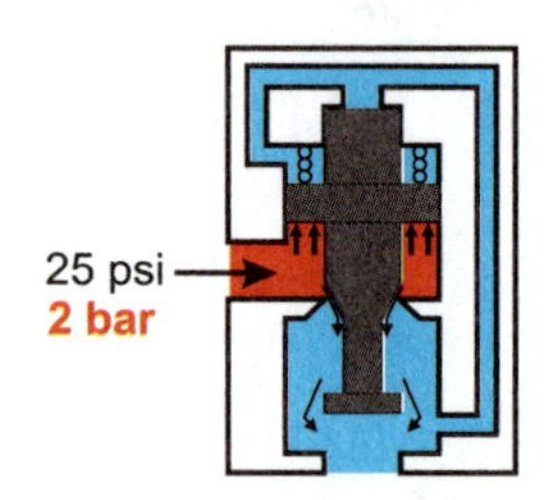

Figure 12-5

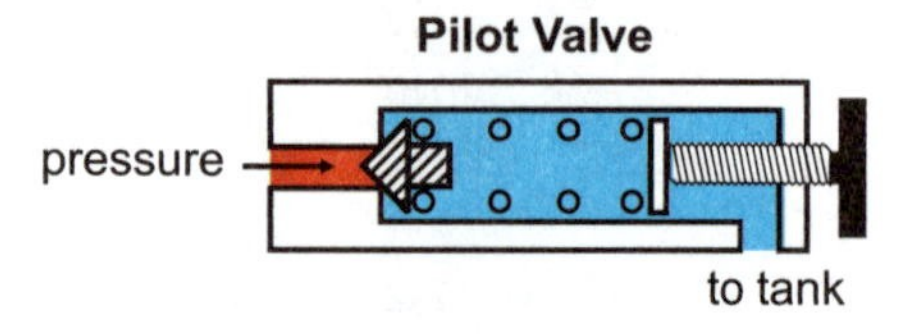

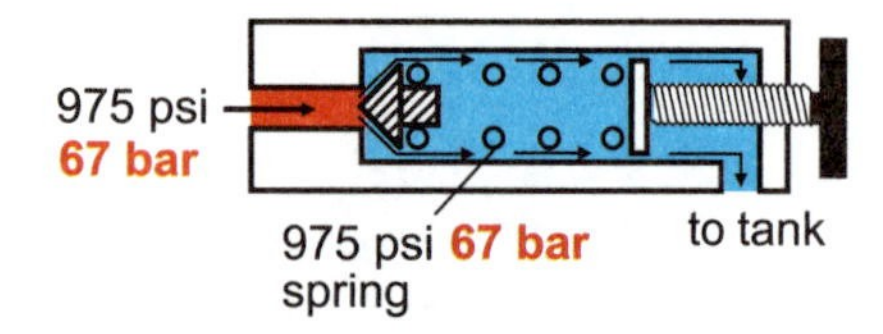

Figure 12-6

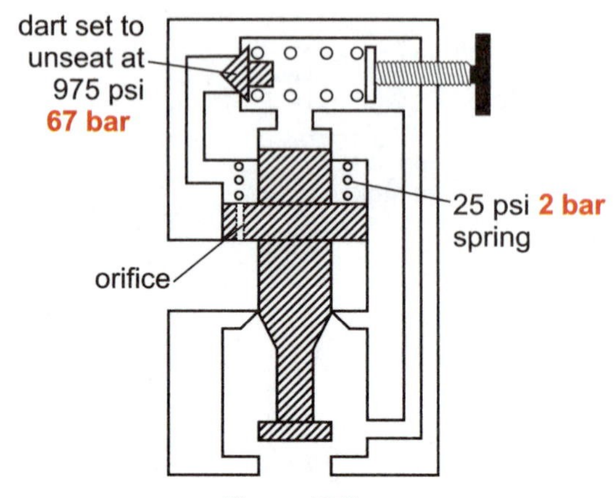

Figure 12-7

How a Pilot Operated Relief Valve Works

To understand the operation of a pilot operated relief valve, we will look at the independent operation of the main valve and the pilot valve. The main valve spool is biased by a light spring. The stem of the main valve spool plugs the outlet to tank. System pressure acts on the area under the spool skirt. Any leakage past the spool is internally drained back to tank through a passage in the valve body.

If the spring biasing the main valve spool has a value of 25 psi (1.7 bar), the spool will be pushed up and system flow will pass to tank when system pressure reaches 25 psi (1.7 bar). In this way, the valve functions as any of the spring biased pressure control valves which we have seen up to this time.

The movable part of the pilot valve is a dart. The area of the dart exposed to hydraulic pressure is relatively small. The spring which biases the dart on its seat is rather stiff. The combination of small area and stiff spring means that the dart will remain seated until a high pressure is reached.

If the spring biasing the dart has a value of 975 psi (67.2 bar), the dart will remain seated until this pressure is reached. At this time, the dart will unseat and flow will pass to tank. Consequently, pressure is limited to 975 psi (67.2 bar). In this manner, the pilot valve also acts like any of the spring biased pressure control valves we have seen previously.

The pilot valve is a simple, spring-biased pressure control which handles small flows at high pressures. The main valve is also a simple, spring-biased pressure control which handles larger flows at low pressure. By using the two together, large flows can be handled at high pressures without the consequence of an early cracking pressure or severe override.

In a pilot operated relief valve, the main valve spool is biased by light spring pressure and fluid (pilot) pressure in the spring chamber. The maximum fluid pressure which is allowed to bias the spool is determined by the setting of the pilot valve.

To allow pressure to accumulate in the spring chamber, an orifice or hole is drilled through the main valve spool skirt.

To illustrate the operation of a pilot operated relief valve, assume that the spring biasing the main valve spool has a value of 25 psi (1.7 bar) and that the pilot valve will limit the pilot pressure in the spring chamber to 975 psi (67.2 bar).

With a system pressure of 800 psi (55.17 bar), 800 psi (55.17 bar) is acting to push the spool up. 800 psi (55.17 bar) is transmitted through the orifice to the spring chamber and acts to hold the spool down. The areas exposed to pressure on either side of the spool skirt are equal. Therefore, the spool is balanced except for the 25 psi (1.7 bar) spring. Consequently, there is a hydraulic pressure of 800 psi (55.17 bar) trying to unseat the spool and a total hydraulic and mechanical pressure of 825 psi (56.89 bar) keeping the spool seated.

With a system pressure of 900 psi (62.1 bar), 900 psi (62.1 bar) at the bottom of the spool acts to push the spool up. A total mechanical and hydraulic pressure of 925 psi (63.79 bar) acts to keep the spool down.

When system pressure rises to 990 psi (68.28 bar), 990 psi (68.28 bar) will act to push the spool up. Since the pilot valve is set to limit the fluid pressure in the spring chamber to 975 psi (67.2 bar), the pilot valve dart is unseated and the pilot pressure above the spool is 975 psi (67.2 bar). This is a total hydraulic and mechanical pressure of 1000 psi (68.97 bar) acting to hold the spool down. The total pressure acting down is still more than the pressure acting up. The maximum pressure which can bias the spool in the down position is 1000 psi (68.97 bar). If pressure below the spool attempts to rise above 1000 psi (68.97 bar), the spool will be pushed up and flow will pass to tank.

In our example, up to a pressure of 975 psi (67.2 bar), the total mechanical and fluid pressure biasing the spool will be 25 psi (1.7 bar) greater than system pressure. Between 975 psi (67.2 bar) and 1000 psi (68.97 bar), the difference becomes less until at any pressure over 1000 psi (68.97 bar) the main valve spool is unseated.

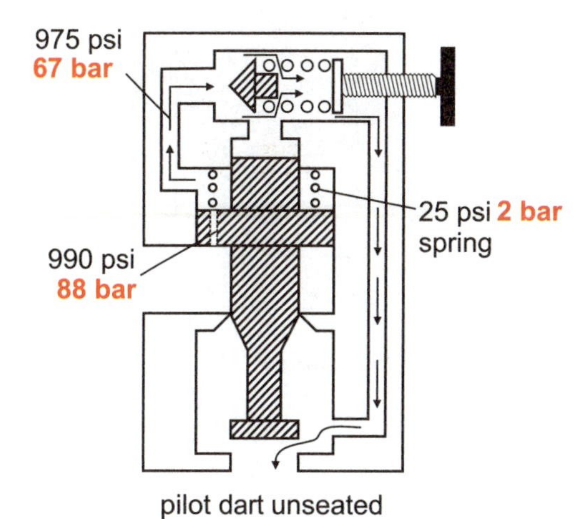

pilot dart unseated

Figure 12-8

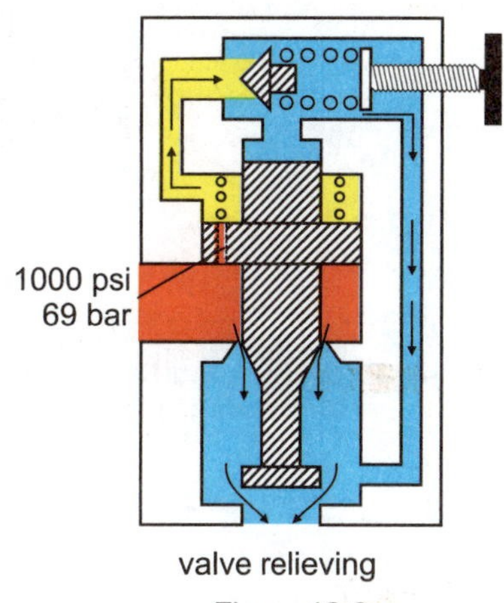

valve relieving

Figure 12-9

Other Pilot Operated Pressure Control Valves

In addition to relief valves, sequence, counterbalance, unloading, and pressure reducing valves can also be pilot operated. Just as relief valves, the other pilot operated pressure controls consists of a pilot valve and main valve spool. The spools in these valves are different from a relief valve spool, but pilot pressure is still sensed through a passage in the main valve spool.

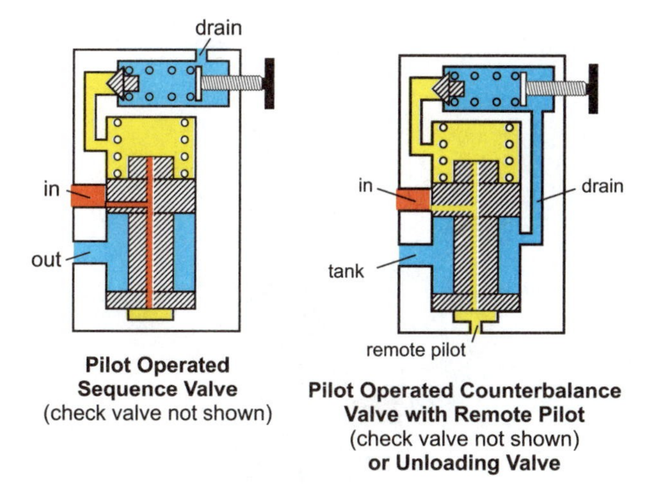

Pilot Operated Sequence Valve (check valve not shown)

Pilot Operated Counterbalance Valve with Remote Pilot (check valve not shown) **or Unloading Valve**

Figure 12-10

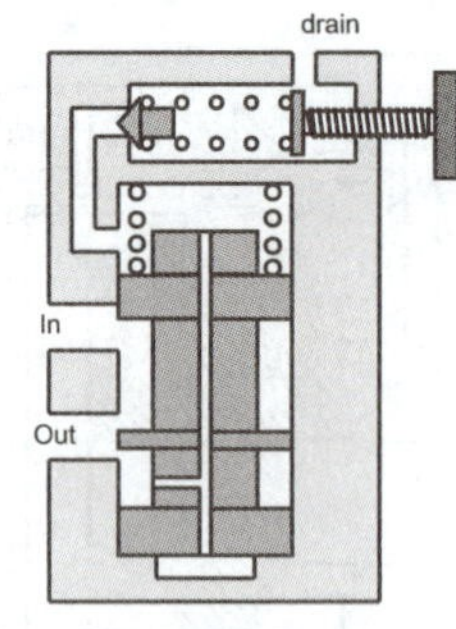

Pilot Operated Pressure Reducing Valve

Figure 12-11

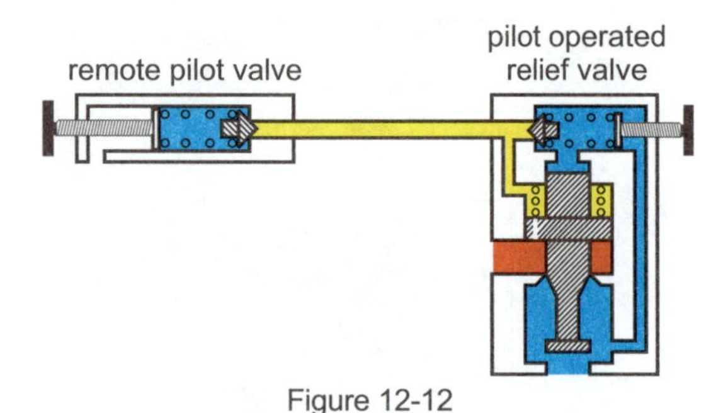

Figure 12-12

Example Pilot Operated Relief Valve Performance Curve

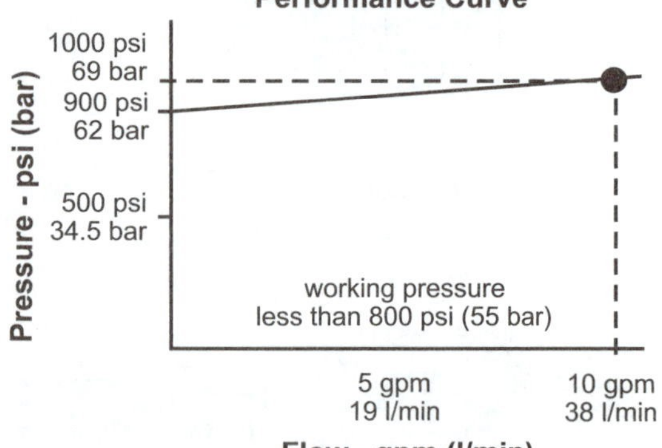

Figure 12-13

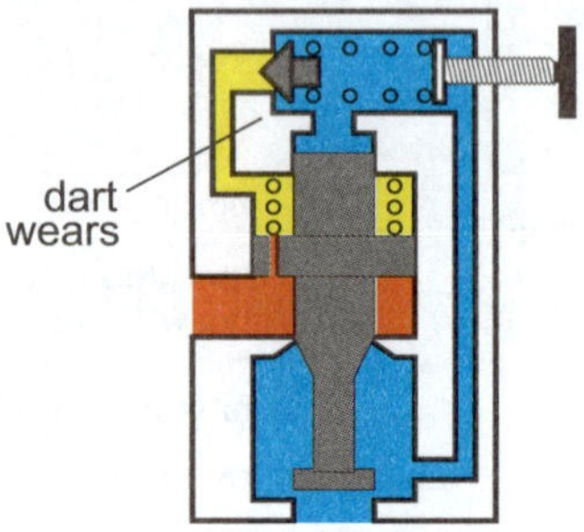

dart wears

Figure 12-14

Remote Pilot Adjustment

Since fluid pressure is used to bias the main valve spool, a pilot operated pressure control valve can be adapted for remote adjustment. With an additional pilot valve connected to the spring chamber of a pilot operated valve, the maximum pressure in the spring chamber of the main spool will be limited to the setting of the remote pilot valve if that setting is lower than the other pilot valve. With this arrangement, the remote pilot valve can be mounted on a panel for ease in adjustment by a machine operator.

In the illustration of remote adjustment, a remote pilot valve is used in conjunction with a pilot operated relief valve. This is a very common application. Pilot operated unloading, counterbalance, sequence and pressure reducing valves can also be remotely adjusted.

Cracking Pressure - Pilot Operated Relief Valve

Since the main spring of a pilot operated relief valve is relatively light, the valve cracking pressure is closer to full flow pressure than in a simple relief valve.

The illustrated performance curve for a pilot operated relief valve points out that the valve cracks open and reaches full flow within 100 psi (6.89 bar) of its maximum pressure setting of 1000 psi (68.92 bar). Whereas, with the simple relief valve seen earlier, the valve cracked open and reached full flow within 200 psi (13.8 bar) while handling the same flow.

Operating Problems with a Pilot Operated Relief Valve

Certain problems can arise while using a pilot operated relief valve. Many problems can be traced to dart wear, orifice plugging, and excessive line lengths between remote and main pilot valves.

As a pilot operated relief valve operates, the dart in the pilot valve can wear as it moves on and off its seat. This causes leakage from the main spring chamber back to tank. If wear, and consequently leakage become excessive, sufficient pilot pressure above the spool skirt will not be maintained. As a result, valve and system will become erratic.

In cases where systems are not properly protected with filters, dirt can plug the orifice in the spool. This means pressure can no longer pass to the pilot-spring chamber on top of the spool. When this occurs, the valve will actuate one time. The spool will not close or will close very slowly since fluid pressure has a difficult time transmitting through the orifice.

In some instances, a problem occurs where excessive line lengths exist between main valve and a remote pilot adjustment. It appears that the farther a remote pilot valve is removed from its main valve, the more chance of pressure pulsations, fluid compressibility, and valve springs interacting to generate irritable noises and excessive vibration. For this reason, it is usually recommended that line lengths be kept to a minimum.

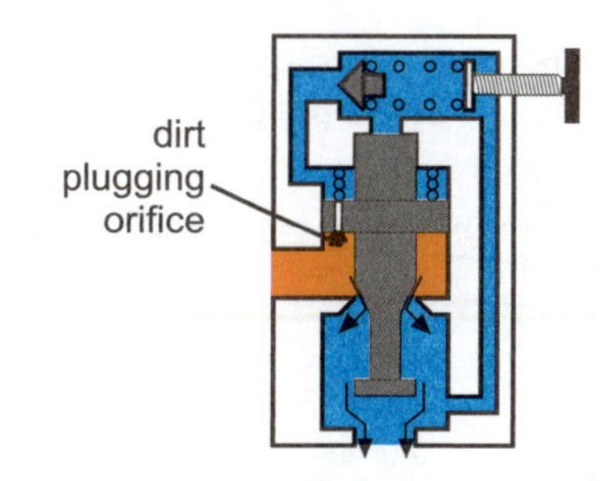

Figure 12-15

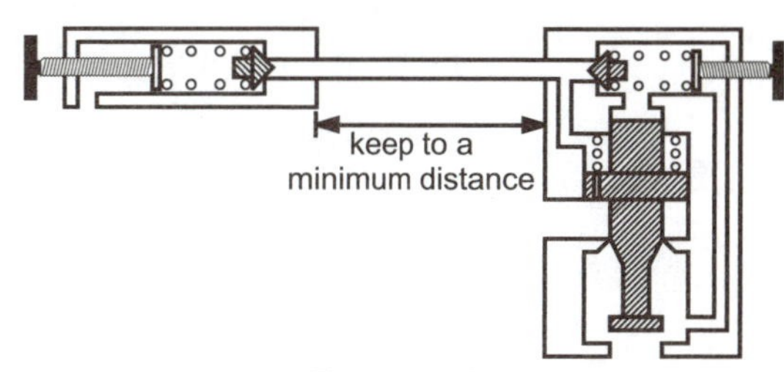

Figure 12-16

Adjusting a Relief Valve

With pressure gage positioned close to pump outlet or plugged into the relief valve body, and with reservoir fluid at its operating temperature, a system relief valve can be adjusted.

With the relief valve set sufficiently high so that no flow passes through the valve, the gage will indicate the amount of pressure the pump/electric motor must develop to overcome resistances of load and liquid as the system is doing useful work. This is pump/electric motor working pressure.

A relief valve should be adjusted so that the valve cracking pressure is above pump/electric motor working pressure. Valve cracking pressure can be determined from a manufacturer's catalog.

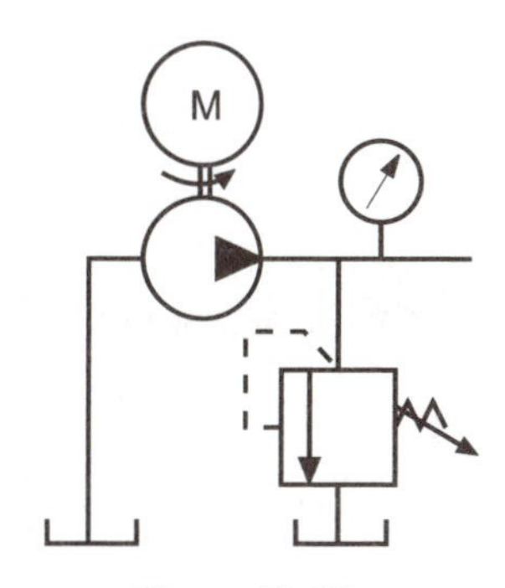

Figure 12-17

Adjusting a simple relief valve may mean the valve is set 200 psi (13.8 bar) over working pressure. This causes no harm or system inefficiency when the system is doing work. But, once the system becomes solid, pump pressure will have to climb 200 psi (13.8 bar) before the valve will accept its full flow.

With a pilot operated relief valve in the same circuit, the valve could be set at a lower pressure; its cracking to full flow range is less than a simple relief valve. Full pump flow could then pass through the valve at perhaps 100 psi (6.89 bar) over working pressure. This generates less heat.

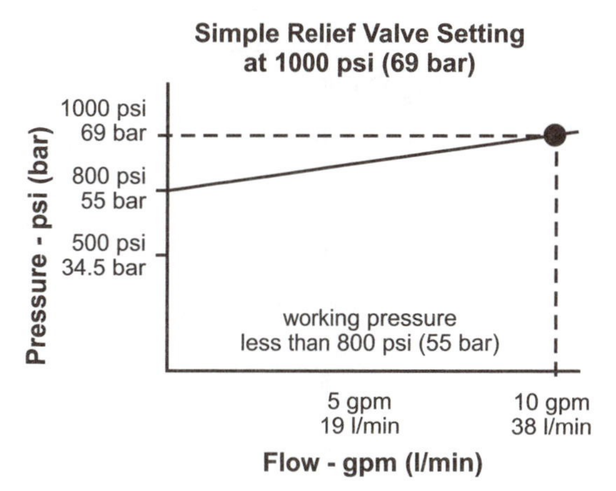

Figure 12-18

Using 100 psi (6.89 bar) less each time the relief valve is actuated does not seem like much, but with a flow rate of 10 gpm (37.9 l/min) this saves .6 hp (447.6 watts). With 25 gpm (94.75 l/min), it saves 1.5 hp (1119 watts). And, with 50 gpm (189.5 l/min), 2.9 hp (2163.4 watts) is saved.

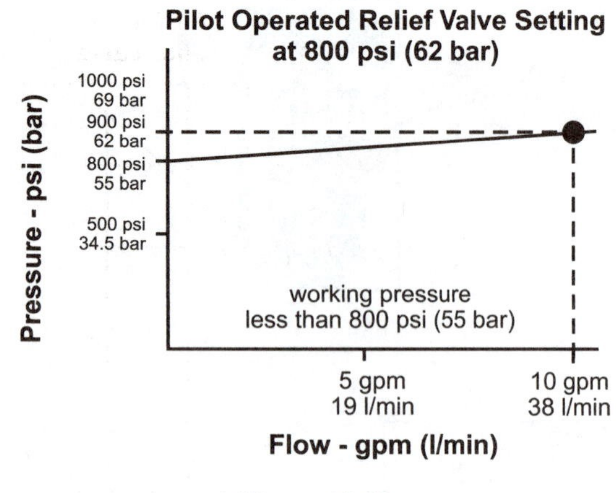

Figure 12-19

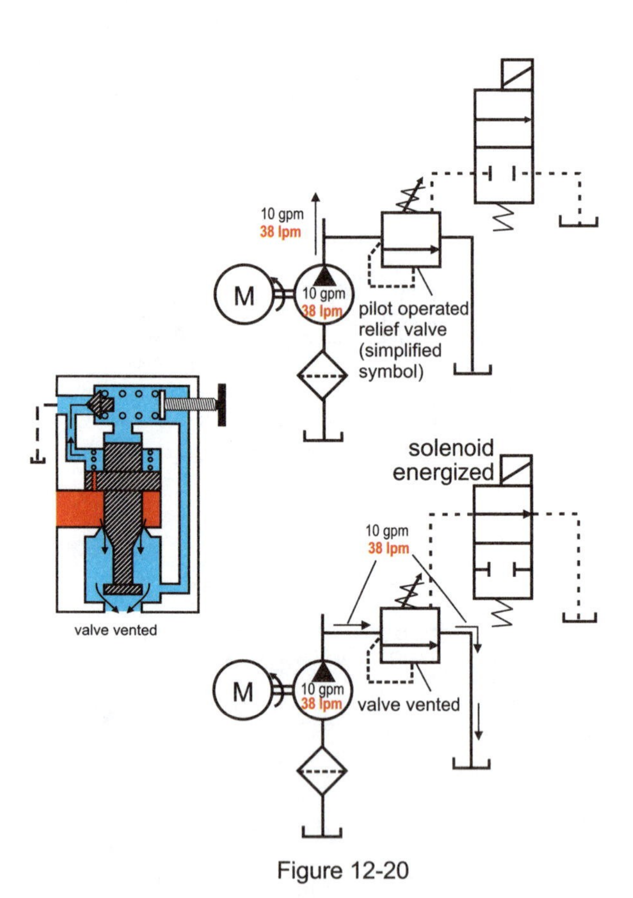

Figure 12-20

Dumping flow over a relief valve is only designed to occur for a short period of time. If actuators are stalled for anything above a few seconds, the pump/electric motor should be unloaded. In the following sections, we learn what high and low vent are and we see that unloading a pump doesn't necessarily mean the electric motor is unloaded.

Pump Unloading

During idle time, when actuators are stalled, flow should not be dumped over a relief valve for anything more than a few seconds. Relief valves are designed to be used for short periods of time. If this rule is not followed, considerable heat will be added to the system unnecessarily.

Anytime actuators are stalled for an extended period, horsepower generation by pump/electric motor should be as close as possible to zero. In some instances, this is accomplished by shutting off the electric motor, but in most cases this is impractical.

A common way of reducing power generation by pump/electric motor is unloading pump flow back to tank through a valve. Unloading pump flow drops discharge pressure and consequently power generation.

Unloading Through a Pilot Operated Relief Valve

A common way of unloading pump flow during idle time is "venting." Venting a relief valve refers to releasing the fluid pressure biasing the main spool of a pilot operated relief valve. By releasing this pilot pressure, the only pressure holding the spool closed is the relatively light pressure of the spring. This results in the pump applying a relatively low pressure to return its flow to tank.

In the circuit illustrated, a solenoid operated directional valve is connected to the vent line of a pilot operated relief valve. With the solenoid de-energized, maximum pump pressure is determined by the relief valve setting. When the solenoid is energized during idle time, pump flow returns to tank at whatever pressure it takes to overcome the relatively light spring biasing the spool.

High and Low Vent

Pilot operated relief valves can be equipped with either a high or low vent spring.

A low vent spring exerts minimal pressure on the main spool of a pilot operated relief valve. This allows pump flow to return to tank through the valve with the least possible resistance during venting. A low vent spring is the standard main spring of a pilot operated relief valve.

A high vent spring is a stiffer spring biasing the main spool. It is a greater resistance to tank during venting.

If venting is done to allow pump flow to return to tank at a low pressure, the use of a stiff spring may be questioned. Even though a high vent spring is a greater resistance to tank, it is found in systems where better valve response time during deventing is required. It also ensures that the valve will close especially if any back pressure is present in the tank line.

A comparison of high and low vent spring operation is given in the illustrated graph. We have seen how pressure and consequently power, is limited by a valve when actuators are not doing any work. Pressure is also limited with a valve when actuators are working. In the next section, we see how this is accomplished in a hi-lo system and in flow control circuits.

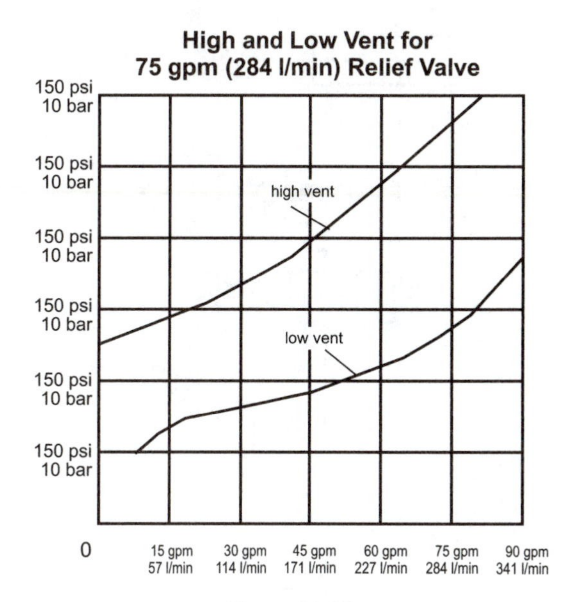

Figure 12-21

Limiting Pump Pressure While Working

In an effort to match hydraulic power generation with mechanical power output, valves are used to limit pump pressure. This is the case when actuators are working as well as not working.

In a simple hydraulic system, as pump/electric motor develops its flow, it applies within its capabilities whatever pressure is necessary to overcome resistance of load and liquid. The maximum pressure is determined by the single setting of a relief valve. In some systems, it is advantageous to have pump/electric motor flow discharge at different pressures depending on the circuit cycle.

Limiting the pressure developed by pump/electric motor as a system cycles is found in a hi-lo system and in some flow control valve circuits.

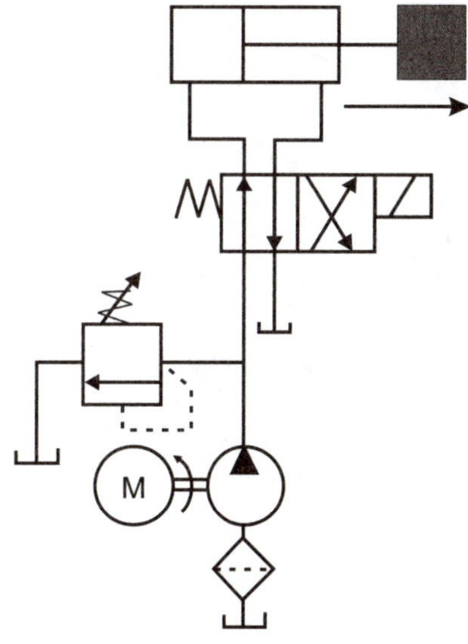

Figure 12-22

Hi-Lo System

A hi-lo system consists of two pumps — one high volume, the other low volume. Both pump flows combine under low pressure to give a large rate of flow. Yet, when system pressure climbs above a certain value, the high volume pump is unloaded while the low volume pump does the work.

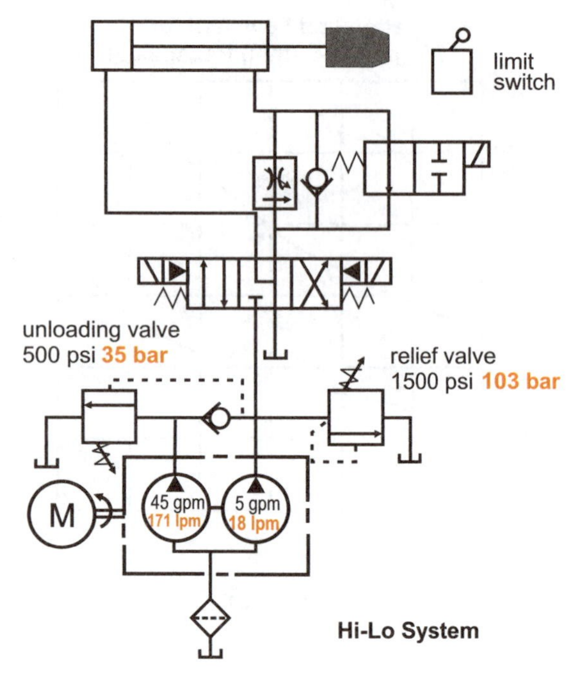

Figure 12-23

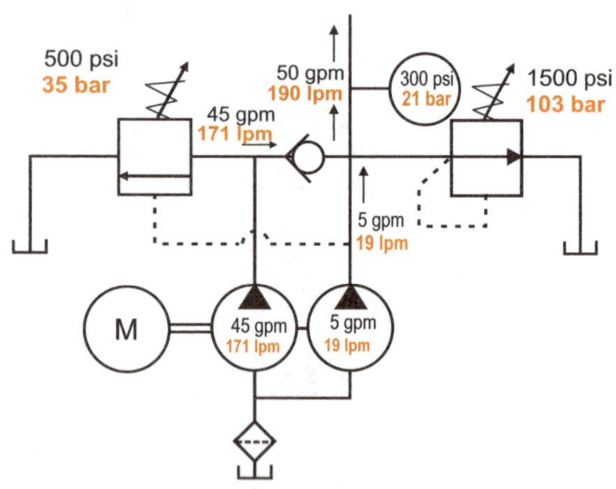

Figure 12-24

In the illustrated hi-lo system, the cylinder is not required to work through its entire stroke. Work is performed through the last few inches of stroke only.

A flow rate of 4.5 gpm (17.05 l/min) flowing into the cylinder is required while the cylinder is doing its work. Yet, to reduce the amount of time getting to and from the work, 50 gpm (189.5 l/min) is desired. If a 50 gpm (189.5 l/min) pump were used in this system, 45 gpm (170.6 l/min) would dump back to tank over the relief valve while work is performed. This would be a gross mismatch of power generated to power used.

A hi-lo system satisfies the system demand by combining a 45 gpm (170.6 l/min) and 5 gpm (18.95 l/min) pump flow. When the electric motor is turned on, the 45 gpm (170.6 l/min) passes through the check valve adding to the 5 gpm (18.95 l/min) flow; 50 gpm (189.5 l/min) passes out into the system extending the cylinder at a relatively low pressure. When the work load is contacted and work pressure is desired, pump/electric motor pressure starts climbing toward the relief valve setting of 1500 psi (103.4 bar). As it passes through the 500 psi (34.5 bar) pressure level, the normally non-passing unloading valve opens allowing the 45 gpm (170.6 l/min) pump to unload while the 5 gpm (18.95 l/min) pump continues to work. This action eliminates unnecessary power generation by the 45 gpm (170.6 l/min) pump when it is not needed.

After work has been completed, the directional valve is shifted to retract the cylinder and pressure drops to a low level once again. This closes the unloading valve. The 45 gpm (170.6 l/min) adds to the 5 gpm (18.95 l/min) retracting the cylinder quickly.

A hi-lo system gives high volume at low pressure and low volume at high pressure. In this way, power generation is more evenly matched to actuator output.

A hi-lo system is one means of limiting pump pressure by a valve when a system is working. In the next section, we shall see how maximum pump pressure can be controlled in a flow control circuit by a solenoid operated relief valve.

Flow Control Circuit

A flow control valve is a restriction. If it is used in a meter-in or meter-out application, pump/electric motor attempts to overcome the restriction by applying a higher pressure. This actuates the relief valve. The desired flow goes through the flow control; excess flow dumps over the relief valve.

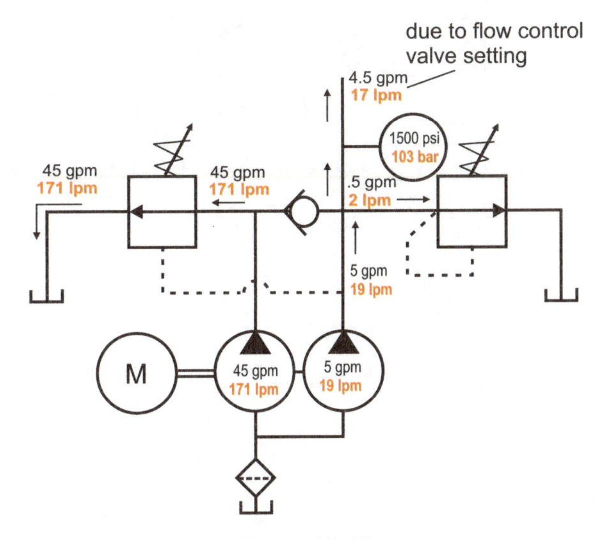

Figure 12-25

As far as a pump/electric motor is concerned, with a meter-in or meter-out circuit, flow is discharged at the relief valve setting even though the load requires much less pressure. Pump/electric motor generate more power than required; and, the excess turns into heat.

Excess power generation is a common occurrence with flow control circuits. Flow control valves are not the most efficient means of reducing pump flow, but they are many times very accurate and economical. Once again, the best that can be hoped for when metering-in or metering-out is to match as close as practical power generated with actuator output.

In the circuit illustrated, two cylinders are connected to the same source of hydraulic power. Cylinder A requires 10 gpm (37.9 l/min) when it extends. Cylinder B's extension flow rate must be 6 gpm (22.74 l/min). For this reason, a flow control is positioned in the supply line end of valve B and adjusted so that 6 gpm (22.74 l/min) is metered into the cylinder.

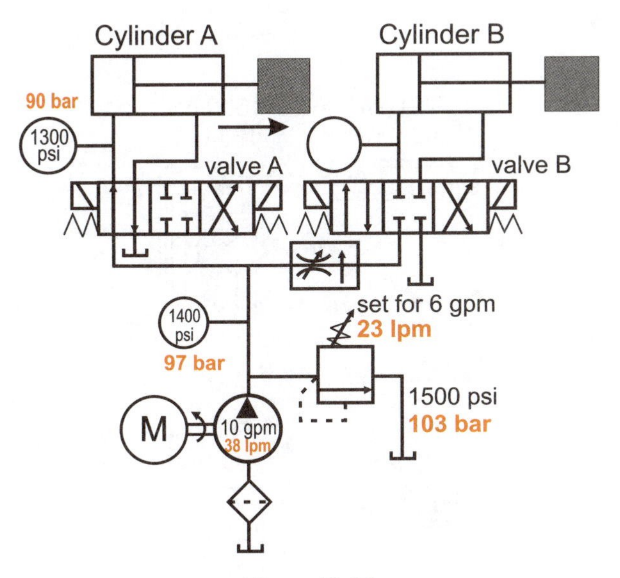

Figure 12-26

The relief valve for the system is set at 1500 psi (103.4 bar). Cylinder A requires 1300 psi (89.7 bar) to move its load, yet cylinder B needs only 700 psi (48.3 bar). This means a considerable amount of unnecessary power will be generated as cylinder B extends and 4 gpm (15.16 l/min) dumps over the relief valve.

With pump/electric motor operating and directional valve A shifted, 10 gpm (37.9 l/min) discharges from pump outlet at a pressure of 1400 psi (96.6 bar). 1300 psi (89.7 bar) is used to equal the load at cylinder A; 100 psi (6.9 bar) is used to overcome liquid resistances. At this point, cylinder A is using 7.6 HP (5.7 Kwatts) (10 gpm x 1300 psi x .000583) to do work and the pump/electric motor is generating 8.2 HP (6.1 Kwatts). This is a good match.

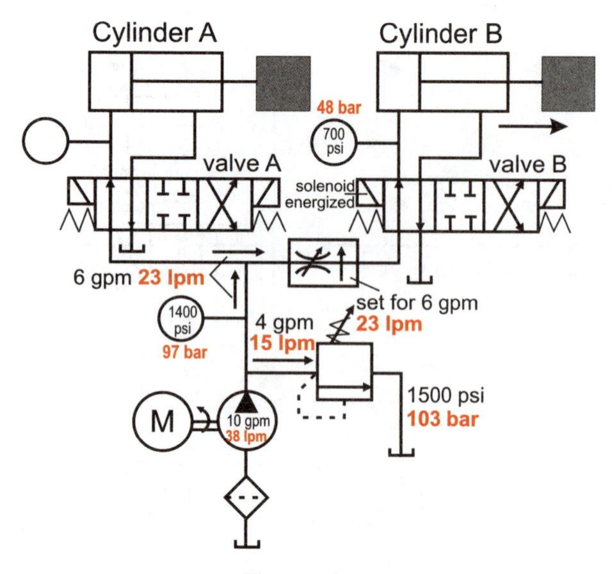

Figure 12-27

After cylinder A has performed its work, directional valve A is de-energized and directional valve B is shifted. With the pressure compensated flow control valve set for 6 gpm (22.74 l/min), pump/electric motor cannot discharge its 10 gpm (37.9 l/min) flow into cylinder B causing pressure at the pump outlet to rise to 1500 psi (103.4 bar) even though cylinder B only requires 700 psi (48.3 bar). At 1500 psi (103.4 bar), the relief valve is actuated and allows the excess volume of 4 gpm (15.2 l/min) to return to tank. At this point, pump/electric motor is generating 8.7 hp (6.5 Kwatts) (10 gpm x 1500 psi x .000583); cylinder B is using only 2.4 hp (1.8 Kwatts) (6 gpm x 700 psi x .000583). This is a gross mismatch, 4.9 hp (3.2 Kwatts) is turned into heat.

To more closely match input and output power, the system described can be equipped with a solenoid operated relief valve which automatically lowers its setting when cylinder B is working.

Solenoid Operated Relief Valve

In the description of a pilot operated relief valve, it was indicated that the pilot valve setting determined at which point the valve would limit pump pressure. It was also shown that a remote pilot valve connected to the main spool spring chamber would control relief valve operation as long as its setting were lower than the main pilot valve. A solenoid operated relief valve takes advantage of this arrangement.

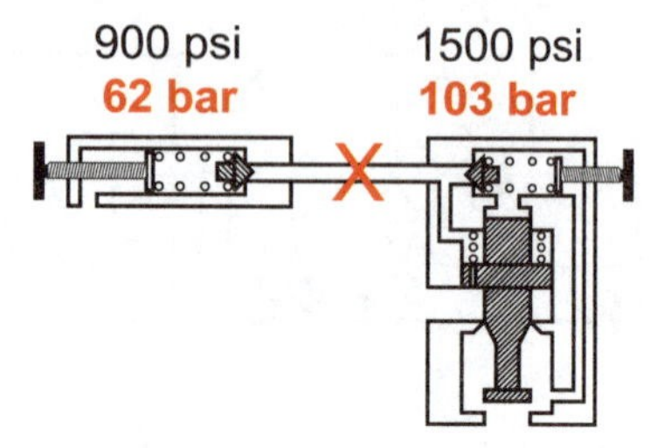

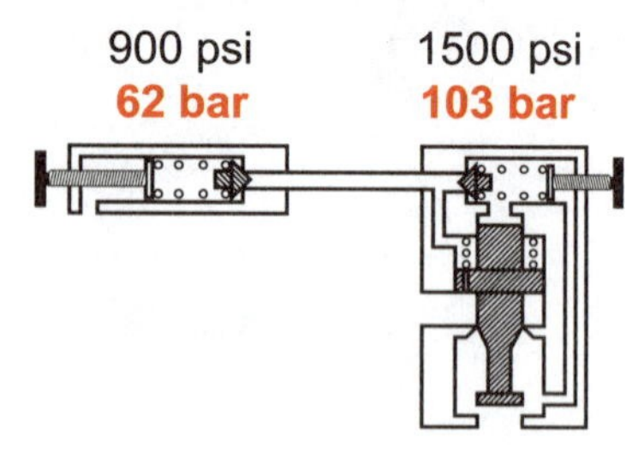

Figure 12-28

What a Solenoid Operated Relief Valve Consists Of

A solenoid operated relief valve consists of a pilot operated relief valve, directional valve, and remote pilot valve. The directional valve is mounted on top of the relief valve body. The remote pilot valve can be mounted on a fixture or panel near the main valve.

How a Solenoid Operated Relief Valve Works

A solenoid operated relief valve changes setting as the directional valve mounted on top of its housing is shifted.

In the illustration, a remote pilot valve is connected to a pilot operated relief valve. With the remote valve set for 900 psi (62.1 bar) and the main pilot valve adjusted to 1500 psi (103.4 bar), the remote pilot valve controls the relief valve operation. However, if the remote valve were disconnected, the main pilot valve would determine relief pressure. The action of connecting and disconnecting the remote pilot valve in a solenoid operated relief valve is performed by the directional valve.

Solenoid Operated Relief Valve in a Circuit

With a solenoid operated relief valve in our two-cylinder circuit, power generation can be more evenly matched when cylinder B is working.

The main pilot valve is set to limit pump/electric motor pressure to 1500 psi (103.4 bar); the remote pilot valve is set for 900 psi (62.1 bar).

When cylinder A is working, electric circuitry controlling the solenoid operated relief valve is arranged so that its directional valve is de-energized. Maximum pump pressure is limited to 1500 psi (103.4 bar). Pump/electric motor discharges 10 gpm at 1400 psi (96.6 bar). Cylinder A uses 7.6 hp (5.7 Kwatts) and pump/electric motor generates 8.2 hp (6.1 Kwatts).

When cylinder B is working, the directional valve of the solenoid operated relief valve is energized. This action causes the remote pilot valve and main valve to connect limiting pump/electric motor pressure to 900 psi (62.1 bar). At this point, 5.3 (3.9 Kwatts) hydraulic horsepower is being generated and 4.1 HP (3.0 Kwatts) is used by cylinder B. This is a closer match than if the relief valve setting were kept at 1500 psi (103.4 bar).

In the circuit just described, the solenoid actuated, spring offset valve connected and disconnected a remote pilot valve to the main spring chamber of a pilot operated relief valve. If the remote pilot port of the valve were connected to tank instead of to a remote pilot valve, the relief valve would be vented when the directional valve was shifted. This is commonly done.

Solenoid operated relief valves are frequently equipped with 3-position directional valves. This allows three different relief valve settings — one for each directional valve position.

Referring to our two cylinder circuits once again, the illustration shows that the solenoid operated relief valve now has a directional valve with three positions. The directional valve has a tandem center.

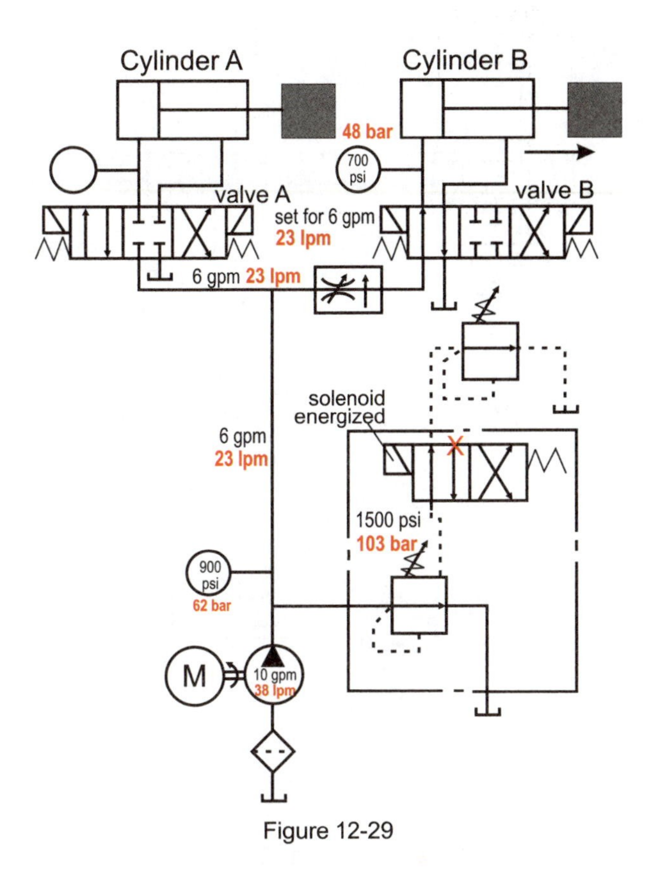

Figure 12-29

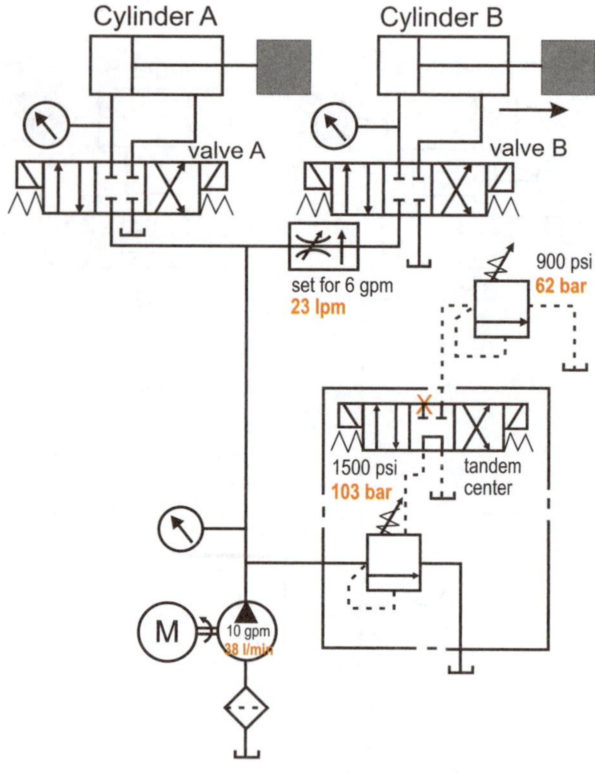

Figure 12-30

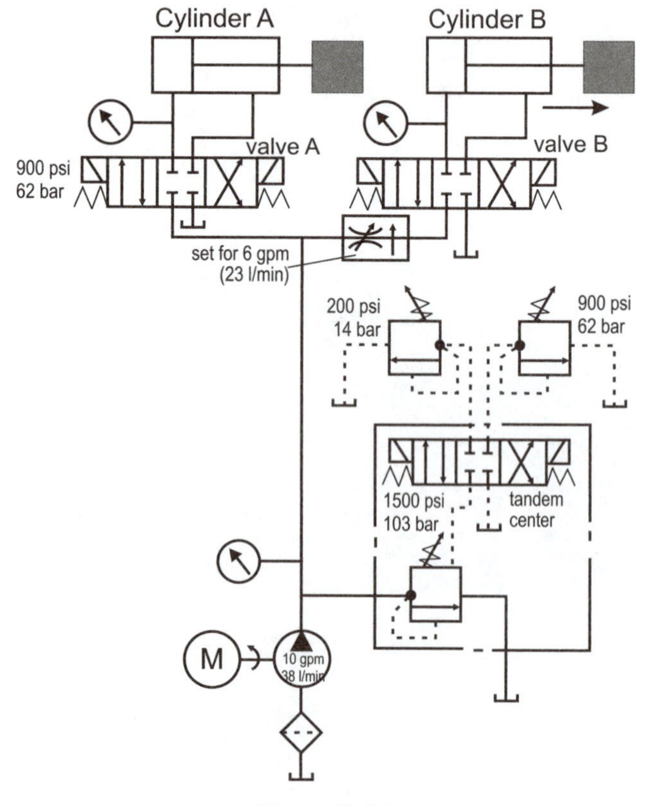

Figure 12-31

Electrical circuitry is arranged so that when cylinder A is working, the directional valve is not connecting the main relief spring chamber with anything. Maximum pump/electric motor pressure is therefore limited to 1500 psi (103.4 bar).

When cylinder B works, the relief directional valve joins remote pilot valve to spring chamber limiting pump/electric motor pressure to 900 psi (62.1 bar).

When work is completed, the directional valve is centered connecting main spring chamber to tank. The relief valve vents unloading the pump.

If the relief directional valve were equipped with a closed center, three maximum working pressures could be selected.

Referring to our two cylinder circuit, assume that both cylinders retract separately and that the flow rate entering the rod side of cylinder B is required to be no more than 6 gpm (22.74 l/min). With flow control adjusted to meter 6 gpm (22.74 l/min) into the cylinder, excess pump flow would have to be dumped over the relief valve at the relatively high setting of 900 psi (62.1 bar) when retract pressure is 50 psi (3.4 bar). This generates unnecessary power and heat.

Incorporating another remote pilot valve adjusted for 200 psi (13.8 bar), the relief directional valve could connect main spring chamber with this low pressure setting during cylinder B retraction. Power generation could then more evenly be matched with actuator output during this portion of the cycle.

We saw that as pump flow was unloaded to tank through various valves, the generated power at pump outlet dropped to a low level. However, this does not mean that the electric motor coupled to the pump is completely unloaded. Pump overall efficiency must be considered.

Overall Pump Efficiency

To reduce the amount of power and heat generated by a pump/electric motor when it is not required to do work, pump flow should be directed to tank at a low pressure. As has been illustrated, this can be accomplished through the center position of a directional valve, an unloading valve of a hi-lo system, or by venting a pilot operated relief valve.

When pump flow is unloaded to tank, pressure and therefore generated hydraulic power, drop to a very low level. However, it should not be thought that the electric motor no longer has to work. A pump's overall efficiency has to be considered.

Overall efficiency takes into account a pump's mechanical efficiency as well as its volumetric efficiency. Pump overall efficiency can be determined by dividing hydraulic horsepower delivered to the system by the input horsepower of the electric motor.

An expression which describes overall efficiency is:

$$\text{Overall efficiency (\%)} = \frac{\text{hydraulic hp output} \times 100}{\text{input hp of electric motor}}$$

If a particular pump/electric motor delivered 10 gpm (37.9 l/min) to a system at 1000 psi (68.97 bar), this would be 5.8 HP (4.3 Kwatts) (10 gpm x 1000 psi x .000583). If the electric motor driving the pump had to develop 7 hp (5.2 Kwatts), then pump overall efficiency would be 83%.

Industrial hydraulic pumps are generally designed to be operated at pressures above 200-400 psi (13.8-27.5 bar). Consequently, the overall efficiency of a pump is greatly diminished below this low pressure range.

An electric motor developing 7 hp (5.2 Kwatts) turns a pump which develops 10 gpm (37.9 l/min) at 1000 psi (68.97 bar) (5.8 HP/4.3 kWatts). When work is no longer required and pump flow is unloaded to tank, hydraulic power drops to a very low level. However, the electric motor may still have to develop 1 hp (.74 kWatts) because of the pump's overall efficiency at low pressure.

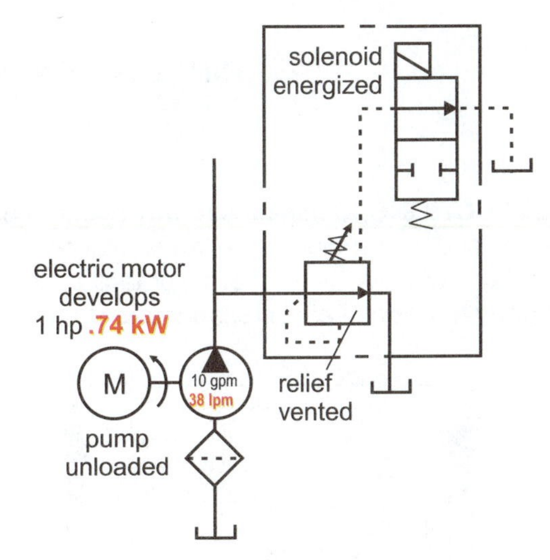

Figure 12-32

Exercise
Pilot Operated Pressure Control Valves
35 Points

1. **INSTRUCTIONS:** Color the cutaway diagram of the pilot operated relief valve. Red is system pressure; blue is drain and tank pressure; yellow is pilot pressure. Assume no leakage passes themain valve spool or the pilot dart. No fluid will be in drain or tank passages unless the pilot dart or main valve spool is unseated. Color by function. If pilot pressure and system are the same psi, pilot is yellow and system is red.

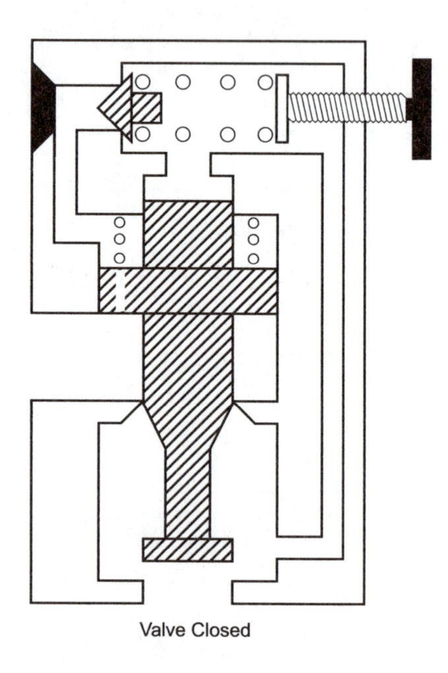

Valve Closed

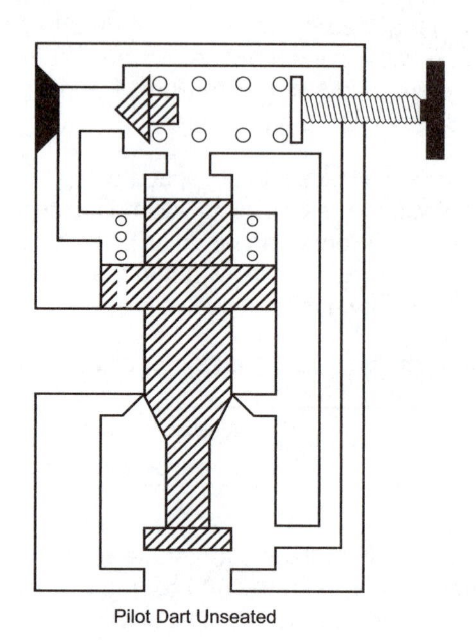

Pilot Dart Unseated

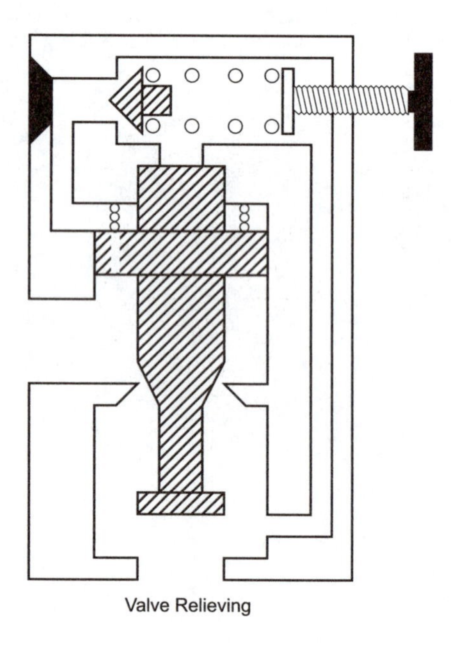

Valve Relieving

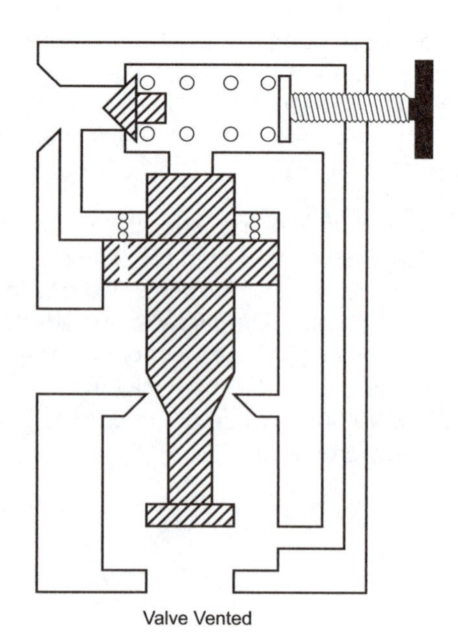

Valve Vented

CHAPTER 13

Hydraulic Pumps

Hydraulic pumps convert the mechanical energy transmitted by its prime mover (electric motor, internal combustion engine, etc.) into hydraulic working energy. Pumping action is the same for every pump. All pumps generate an increasing volume at the suction side and a decreasing volume at the pressure side. However, the elements which perform the pumping action are not the same in all pumps. The type of pump used in an industrial hydraulic system is a positive displacement pump. There are many types of positive displacement pumps. For this reason, we must be selective and concentrate on the most popular. These are vane, gear, and piston pumps. The first group of pumps discussed will be fixed displacement pumps.

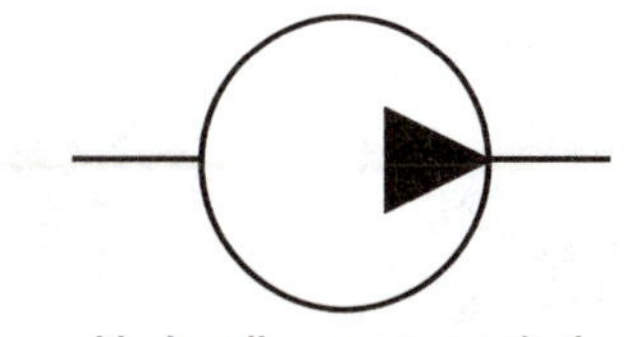

Hydraulic pump symbol

Figure 13-1

Vane Pumps

Vane pumps generate a pumping action by causing vanes to track along a ring.

What a Vane Pump Consists Of

A pumping mechanism of a vane pump basically consists of rotor, vanes, ring, and a port plate with kidney-shaped inlet and outlet ports.

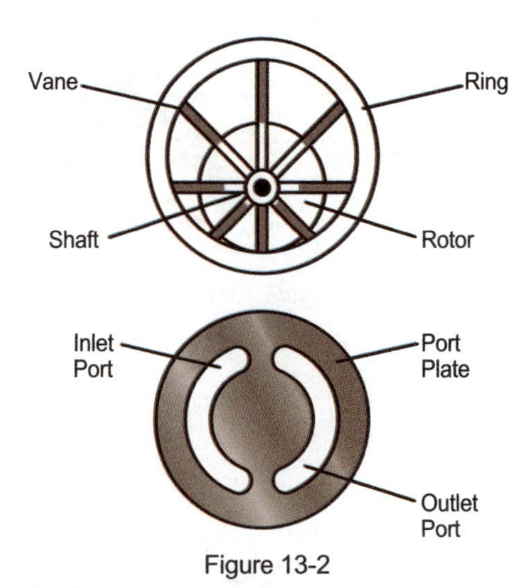

Figure 13-2

How a Vane Pump Works

The rotor of a vane pump houses the vanes and it is driven by a shaft which is connected to a prime mover. As the rotor is turned, vanes are thrown out by centrifugal force and track along a ring. (The ring does not rotate.) As the vanes make contact, a positive seal is formed between vane tip and ring.

The rotor is positioned off-center to the ring. As the rotor is turned, an increasing and decreasing volume is formed within the ring.

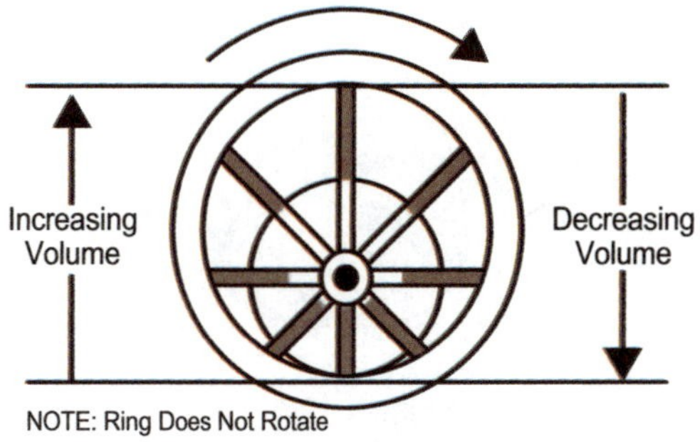

Increasing Volume

Decreasing Volume

NOTE: Ring Does Not Rotate

Figure 13-3

Since there are no ports in the ring, a port plate is used to separate incoming fluid from outgoing fluid. The port plate fits over the ring, rotor, and vanes. The inlet port of the port plate is located where the increasing volume is formed. The port plate's outlet port is located where the decreasing volume is generated. All fluid enters and exits the pumping mechanism through the port plate. (The inlet and outlet ports in the port plate are, of course, connected respectively to the inlet and outlet ports in the pump housing.)

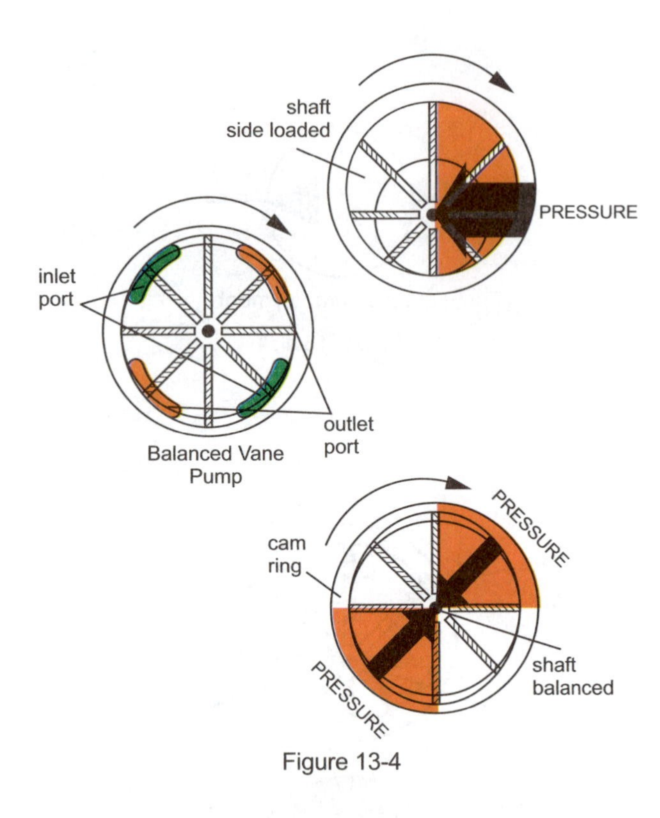

Figure 13-4

Balanced Vane Pump Design

In a pump, two very different pressures are involved — working pressure of a system and less-than-atmospheric pressure. In the vane pump which has been described, one half of the pumping mechanism is at less-than- atmospheric pressure. The other half is subjected to full system pressure. This results in side loading the shaft which could be severe when high system pressures are encountered. To compensate for this condition, this ring is changed from circular to cam-shaped. With this arrangement, the two pressure quadrants oppose each other and the forces acting on the shaft are balanced. Shaft side loading is eliminated.

Therefore, a balanced vane pump, consists of a cam ring, rotor, vanes, and a port plate with inlet and outlet ports opposing each other. (Both inlet ports are connected together, as are the outlet ports, so that each can be served by one inlet or one outlet port in the pump housing).

Constant volume, positive displacement vane pumps, used in industrial systems, are generally of the balanced design.

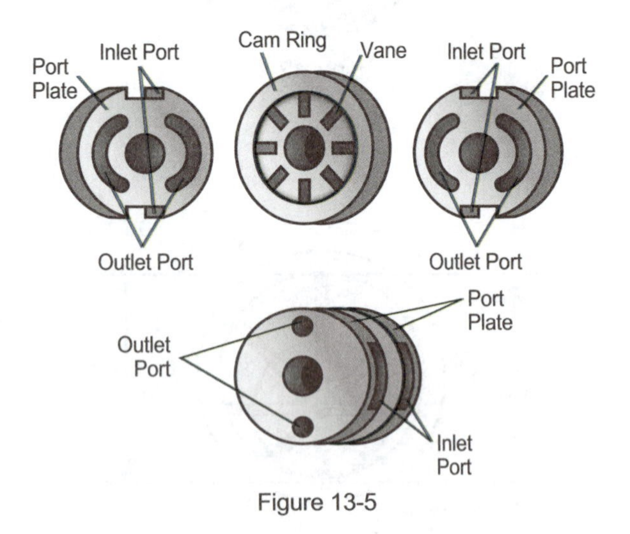

Figure 13-5

Cartridge Assembly

The pumping mechanism of industrial vane pumps is often an integral unit called a cartridge assembly. A cartridge assembly consists of vanes, rotor, and a cam ring sandwiched between two port plates. (Note that the port plates of the cartridge assembly are somewhat different in design than the port plates previously illustrated).

An advantage of using a cartridge assembly is easy pump servicing. After a period of time when pump parts naturally wear, the pumping mechanism can be easily removed and replaced with a new cartridge assembly. Also, if for some reason the pump's volume must be increased or decreased, a cartridge assembly with the same outside dimension, but with the appropriate volume, can be quickly substituted for the original pumping mechanism.

Vane Loading

Before a vane pump can operate properly, a positive seal must exist between vane tip and cam ring. When a vane pump is started, centrifugal force is relied on to throw-out the vanes and achieve a seal.

(This is the reason that the minimum operating speed for most vane pumps is 600 rpm).

Once the pump is primed and system pressure begins to rise, a tighter seal must exist at the vane so that leakage does not increase across the vane tip. To generate a better seal at high pressures, industrial vane pumps direct system pressure to the underside of the vane. With this arrangement, the higher system pressure becomes, the more force is developed to push the vane out against the cam ring.

Hydraulically loading a vane in this manner develops a very tight seal at the vane tip. But, if the force loading the vane is too great, vanes and cam ring would wear excessively and the vanes would be a source of drag.

As a compromise between achieving the best seal and causing the least drag and wear, manufacturers design their pumps so that the vanes are only partially loaded.

The use of vanes with a chamfer or beveled edge is one way in which high vane loading is eliminated. With these vanes, the complete underside vane area is exposed to system pressure as well as a large portion of the area at the top of the vane. This results in a balance of most of the vane. The pressure which acts on the unbalanced area is the force which loads the vane.

In high pressure systems, the use of a vane with a beveled edge still results in too much wear and too much drag. The use of this type vane in a high pressure pump is not satisfactory. Another arrangement is used.

Common vane construction of high pressure vane pumps consists of dual vanes, intra-vanes, spring-loaded vanes, pin-vanes, and angled vanes. The dual vane construction consists of two vanes in each vane slot. Each vane is almost completely balanced. And, a good seal is achieved because two vanes are used.

The intra-vane is another type of vane construction which consists of a small vane within a large vane with a beveled edge. System pressure is directed to the area above the small vane. This again results in less vane loading.

Very similar to the intra-vane construction is the pin-vane construction. In pin-vane construction pressure is directed to the underside of a pin. The pin then forces the vane out against the cam ring.

With a spring-loaded vane, spring pressure at the bottom of the vane is primarily what loads the vane.

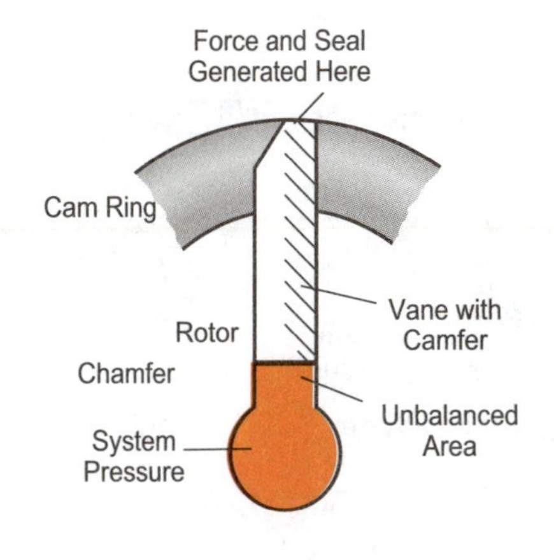

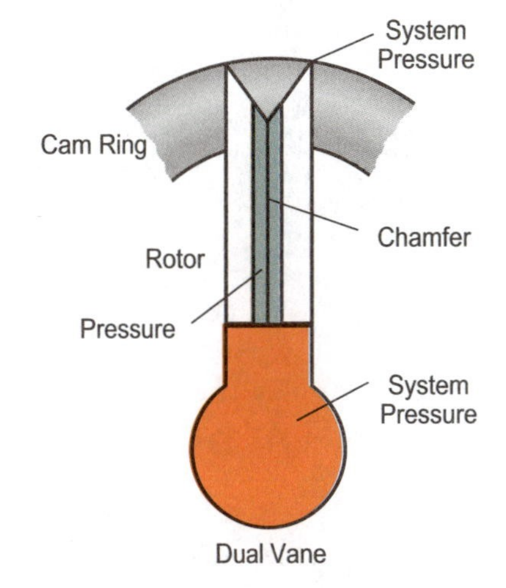

Figure 13-6

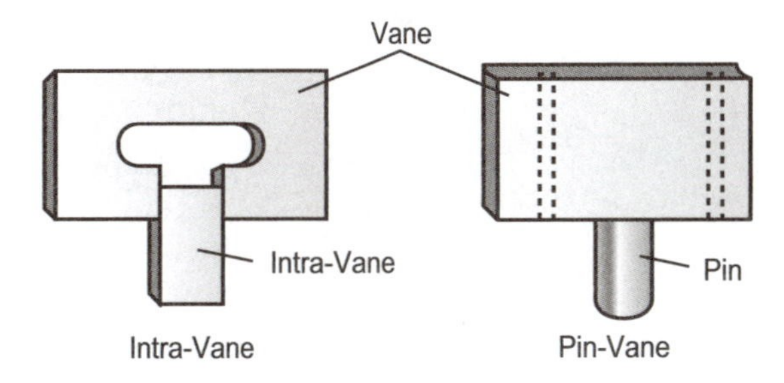

Figure 13-7

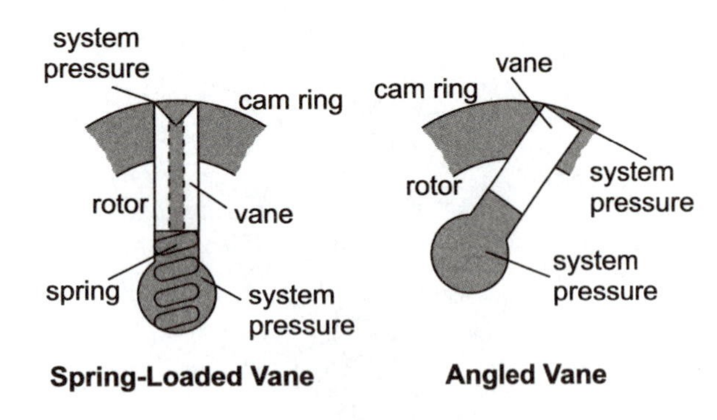

Spring-Loaded Vane **Angled Vane**

Figure 13-8

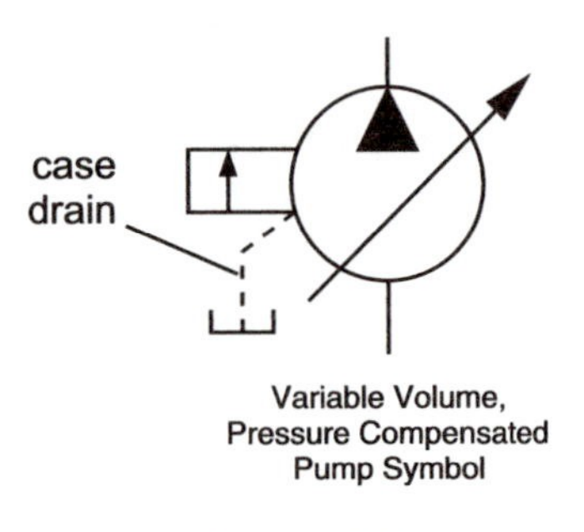

Variable Volume, Pressure Compensated Pump Symbol

Figure 13-9

Another means of reducing vane loading is positioning the vanes in the rotor at an angle. This results in a slight loading of the vane without using any other mechanical device.

Double Pumps

The vane pump which has been described is referred to as a single pump; that is, it consists of one inlet, one outlet and a single cartridge assembly. Vane pumps are also available as a double pump.

A double vane pump consists of housing with two cartridge assemblies, one or two inlets, and two separate outlets. In other words, a double pump consists of two pumps in one housing. A double pump can discharge two different flow rates from each outlet. Since both pump cartridges are connected to a common shaft, one prime mover is used to drive the whole unit.

Double pumps are many times used in hi-lo circuits and where two different flow rates are supplied from the same power unit.

Double pumps give up to twice the flow of a single pump without an appreciably larger unit.

Case Drain

All variable volume, pressure compensated pumps must have their housings externally drained. The pumping mechanisms in these pumps move extremely fast when pressure compensation is required. Any buildup of fluid within the housing would hinder their movement.

Also, any leakage which accumulates in a pump housing is generally directed back to the pump's inlet side. The leakage from a variable volume pump, while it is compensating, is generally hot. If it were diverted to the inlet side, the fluid would get progressively hotter. Externally draining the housing alleviates the problem.

The external drain of a pump housing is commonly referred to as a case drain.

Gear Pumps

Gear pumps generate a pumping action by causing gears to mesh and unmesh.

What a Gear Pump Consists Of

A gear pump basically consists of a housing with inlet and outlet ports, and a pumping mechanism made up of two gears. One gear, the drive gear, is attached to a shaft which is connected to a prime mover. The other gear is the driven gear.

How a Gear Pump Works

As the drive gear is turned by a prime mover, it meshes with and rotates the drive gear. The action of teeth meshing and unmeshing generates an increasing and decreasing volume. At the inlet where gear teeth unmesh (increasing volume), fluid enters the housing. The fluid is then trapped between the gear teeth and housing, and carried to the other side of the gear.

At this point, the gear teeth mesh (decreasing volume) and force the fluid out into the system.

A positive seal in this type pump is achieved between the teeth and the housing, and between the meshing teeth themselves.

Gear pumps are generally an unbalanced design.

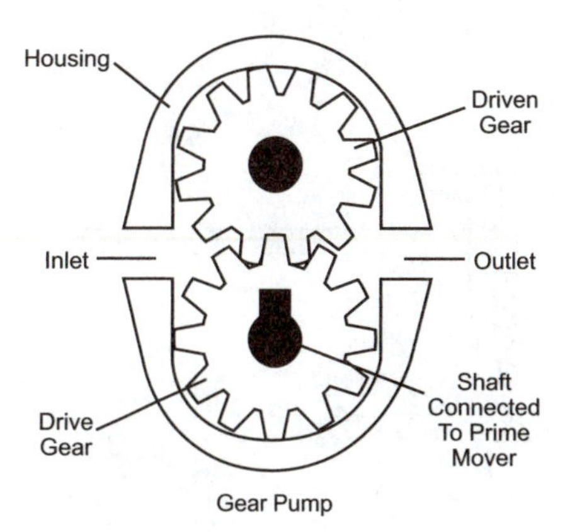

Figure 13-10

External Gear Pumps

The gear pump that has been described above is an external gear pump; that is, both meshing gears have teeth on their outer circumferences. These pumps are sometimes referred to as gear-on-gear pumps.

There are basically three types of gears used in external gear pumps — spur, helical, and herringbone. Since the spur gear is the easiest to manufacture, this type pump is the most common and the least expensive of the three.

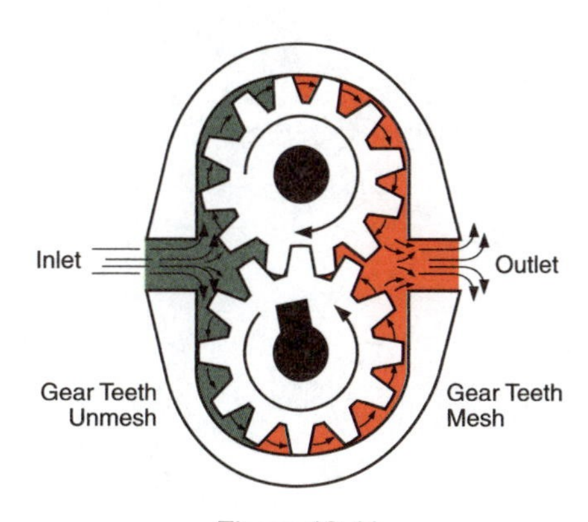

Figure 13-11

Internal Gear Pump

An internal gear pump consists of one external gear which meshes with the teeth on the inside circumference of a larger gear. This type pump is sometimes referred to as gear-within-gear pump. The most common type of internal gear pump in industrial systems is the gerotor pump.

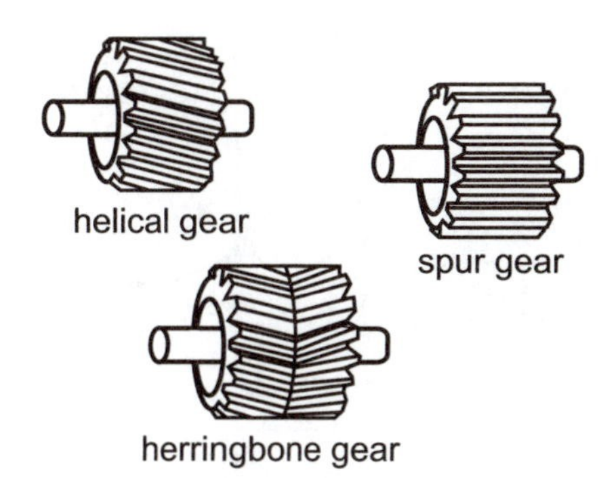

Figure 13-12

Gerotor Pump

A gerotor pump is an internal gear pump with an inner drive gear and an outer drive gear. The inner gear has one less tooth than the outer gear.

As the inner gear is turned by a prime mover, it rotates the larger outer gear. On one side of the pumping mechanism, an increasing volume is formed as gear teeth unmesh. On the other half of the pump, a decreasing volume is formed. A gerotor pump has an unbalanced design.

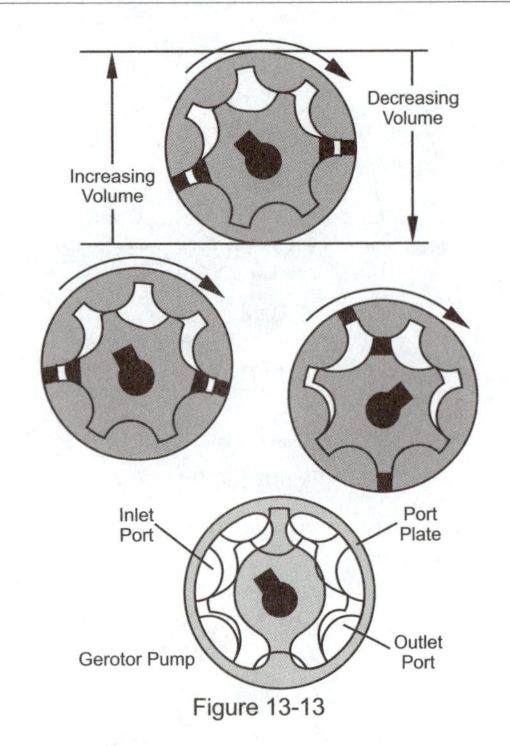

Figure 13-13

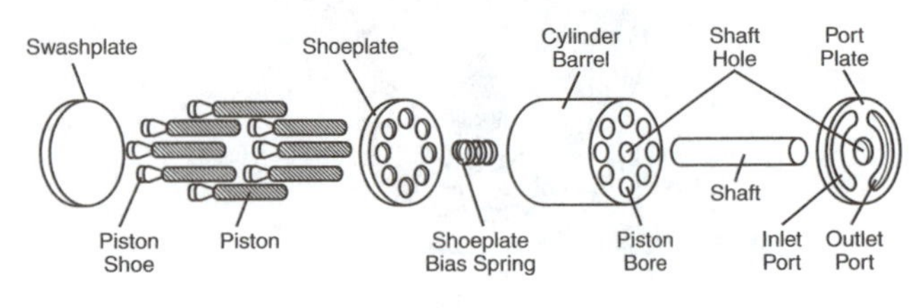

Figure 13-14

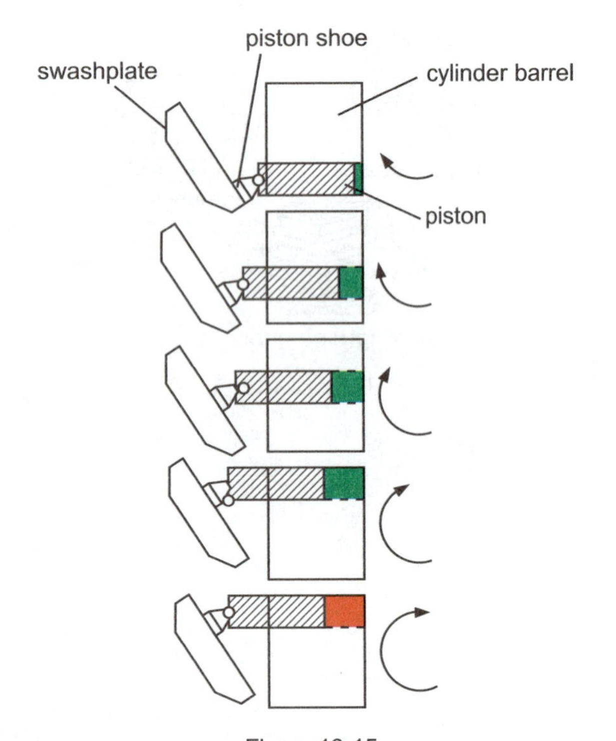

Figure 13-15

Fluid entering the pumping mechanism is separated from the discharge fluid by means of a port plate as in a vane pump.

While fluid is carried from inlet to outlet, a positive seal is maintained as the inner gear teeth follow the contour of crests and valleys of the outer gear.

Piston Pumps

Piston pumps generate a pumping action by causing pistons to reciprocate within a piston bore.

What a Piston Pump Consists Of

The pumping mechanism of a piston pump basically consists of a cylinder barrel; pistons with shoes; swashplate; shoeplate; shoeplate bias spring; and port plate.

How a Piston Pump Works

Earlier, we have seen one example of a piston pump. This pump generates an increasing and decreasing volume by means of a plunger being pulled and pushed in and out of a cylinder body. It was pointed out that the disadvantages of this type of pump were that the pump developed a pulsating flow and that it could not be easily operated by an electric motor or internal combustion engine.

However, a piston can be made to reciprocate easily by the turning motion of a prime mover as well as developing a smooth flow.

In the example illustrated, a cylinder barrel with one piston bore is fitted with one piston. A swashplate is positioned at an angle. The shoe of the piston rides on the surface of the swashplate.

As the cylinder barrel is rotated, the piston shoe follows the surface of the swashplate. (The swashplate does not rotate.) Since the swashplate is at an angle, this results in the piston reciprocating within the bore. In one half of the circle of rotation, the piston moves out of the cylinder barrel and generates an increasing volume.

In the other half of the circle of rotation, this piston moves into the cylinder barrel and generates a decreasing volume. In actual practice, the cylinder barrel is fitted with many pistons. The shoes of the pistons are forced against the swashplate surface by a shoeplate and bias spring. To separate the incoming fluid from the discharge fluid, a port plate is positioned at the end of the cylinder barrel opposite the swashplate.

A shaft is attached to the cylinder barrel which connects it with the prime mover. This shaft can be located at the end of the barrel where the porting is taking place. Or, more commonly, it can be positioned at the swashplate end. In this case, the swashplate and shoeplate have a hole in their centers to accept the shaft. If the shaft is positioned at the other end, just the port plate has a shaft hole.

The piston pump which has been described above is known as an axial or in-line piston pump; that is, the pistons are rotated about the same axis as the pump shaft.

Axial piston pumps are the most popular piston pumps in industrial applications. Other types of piston pumps include the bent-axis and radial piston pumps.

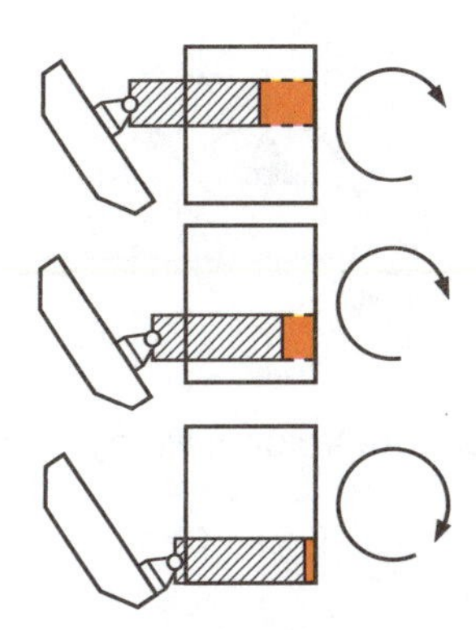

Figure 13-16

Bent-Axis Piston Pump

The pumping mechanism of a bent-axis piston pump consists of a cylinder barrel, pistons, port plate, flange, piston linkages, and a drive shaft. In this pump the cylinder barrel is not in-line with the drive shaft, but is at an angle to the shaft. Because of this arrangement, as the shaft is turned the pistons are pulled out of the barrel during one half of the barrel rotation. This generates an increasing volume. On the other half of the barrel rotation, the pistons are pushed in and a decreasing volume is formed. In this pump, incoming fluid is separated from discharge fluid by means of a port plate.

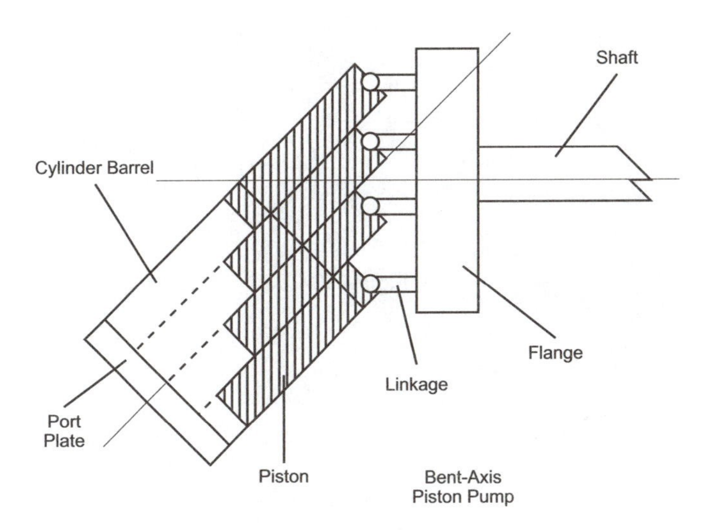

Figure 13-17

Radial Piston Pump

The pumping mechanism of a radial piston pump basically consists of a cylinder barrel, pistons with shoes, a ring, and a valve block.

The action of a radial piston pump is quite similar to a vane pump. But, instead of using vanes to track along a ring, the pump uses pistons.

The cylinder barrel, which houses the pistons, is positioned off-center to the ring. As the cylinder barrel is rotated, an increasing volume is formed within the cylinder barrel during one half of the barrel rotation. During the other half, a decreasing volume is formed.

Fluid enters and is discharged from the pump through the valve block in the center of the pump.

Figure 13-18

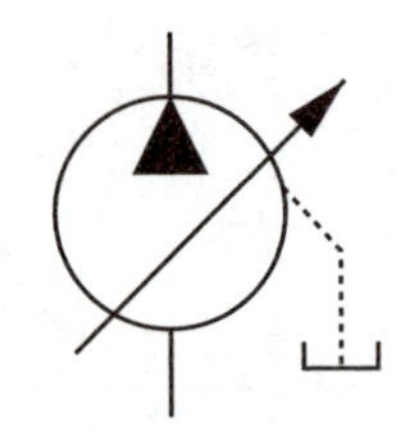

Variable Volume Pump Symbol

Figure 13-19

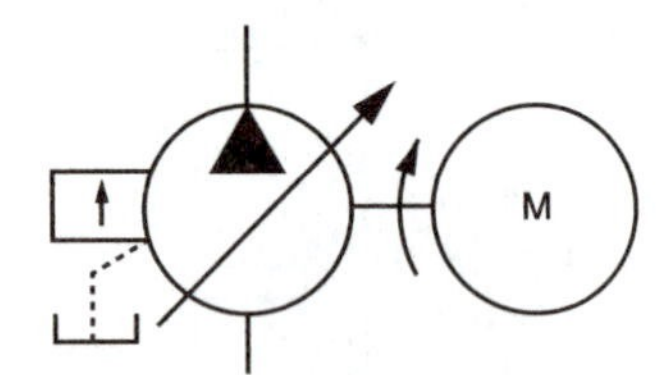

Figure 13-20

Variable Volume Vane Pumps

A positive displacement vane pump delivers the same volume of fluid for each revolution. Industrial pumps are generally operated at 1200 or 1800 rpm. This infers that the pump flow rate remains constant.

In some cases, it is desirable that a pump's flow rate be variable. One way of accomplishing this is by varying the speed of the prime mover. This is usually economically impractical. The only other way, then, to vary the output of a pump is to change its displacement.

The amount of fluid which a vane pump displaces is determined by the difference between the maximum and minimum distance the vanes are extended and the width of the vanes. While the pump is operating, nothing can be done to change the width of a vane. But, a vane pump can be designed so that the distance the vanes are extended can be changed. This is known as a variable volume vane pump.

Regulating Hydraulic Power with Variable Volume Pumps

Since hydraulic power consists of gpm (l/min) and psi (bar), power generation by pump/electric motor can be controlled by reducing the generation of flow as well as limiting pressure. This is accomplished with the use of variable volume vane and piston pumps which are pressure compensated. These pumps have the ability to reduce their displacement once system pressure reaches a certain level.

The intent of this lesson is to describe how pressure compensated, variable volume pumps can more easily match hydraulic power generated with mechanical power output and how wear affects their operation.

To begin the lesson, we look at the operation of variable volume vane and piston pumps.

What a Variable Volume Vane Pump Consists Of

The pumping mechanism of a variable volume vane pump basically consists of a rotor, vanes, a cam ring which is free to move from side to side, a port plate, a thrust bearing to guide the cam ring, and something to vary the position of the cam ring. In our illustration, a screw adjustment is used.

Variable volume vane pumps are unbalanced pumps. Their rings are circular and not cam-shaped. However, they are still referred to as cam rings.

Since the cam ring in this type pump must be free to move, the pumping mechanism does not come as a cartridge assembly.

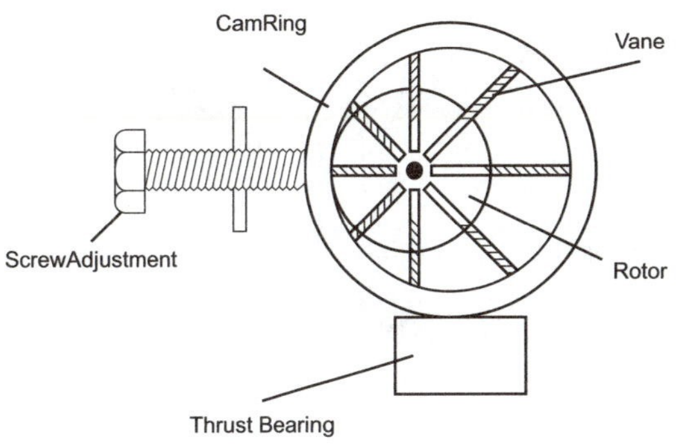

Figure 13-21

How a Variable Volume Vane Pump Works

With the screw adjusted in, the rotor is held off center with regard to the cam ring.

When the rotor is turned, an increasing and decreasing volume is generated. Pumping occurs.

With the screw adjustment turned out slightly, the cam ring is not as off center to the rotor as before. An increasing and decreasing volume is still being generated, but not as much flow is being delivered by the pump. The exposed length of the vanes at full extension has decreased.

With the screw adjustment backed completely out, the cam ring naturally centers with the rotor. No increasing and decreasing volume is generated. No pumping occurs. With this arrangement a vane pump can change its output flow anywhere from full flow to zero flow by means of the screw adjustment.

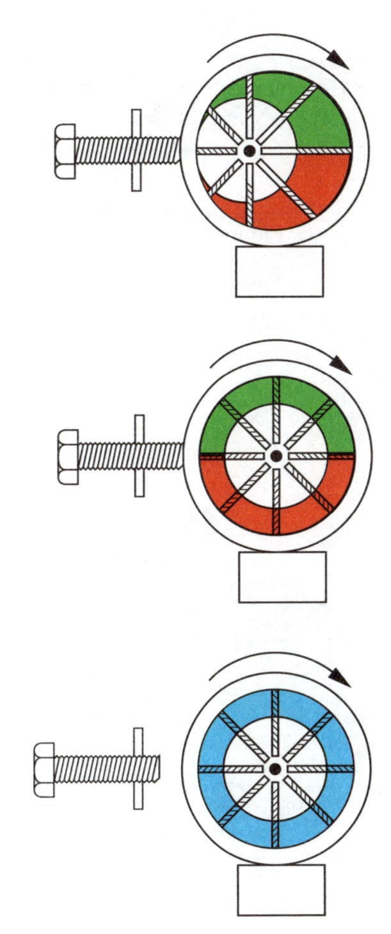

Figure 13-22

How the Pressure Compensator of a Variable Volume Vane Pump Works

The pressure compensator of a variable volume vane pump is the large adjustable spring which holds the cam ring off center against the volume adjustment. This causes an increasing and decreasing volume to form and pumping to occur. One half of the ring is under less-than-atmospheric conditions and the remaining half is pressurized.

As pumping occurs, pressure generates a force on the inner surface of the cam ring. Because of the design of the kidney-shaped ports in the port plate, the generated force tends to move the ring in the main direction of, but not exactly at, the thrust bearing. A portion of the force is taken up by the compensator spring.

When pressure on the ring and force become large enough, the ring begins to move in the direction of the spring reducing discharge flow. The ring cannot move through the thrust bearing because it is a mechanical stop. Once the ring centers with the rotor, it stops and discharge flow ceases. The cam ring will not cross to the other side of the rotor because this would cause the pressure half of the pumping mechanism to form an increasing volume which is suction operated. Pressure would immediately drop resulting in the spring pushing the ring off center once again.

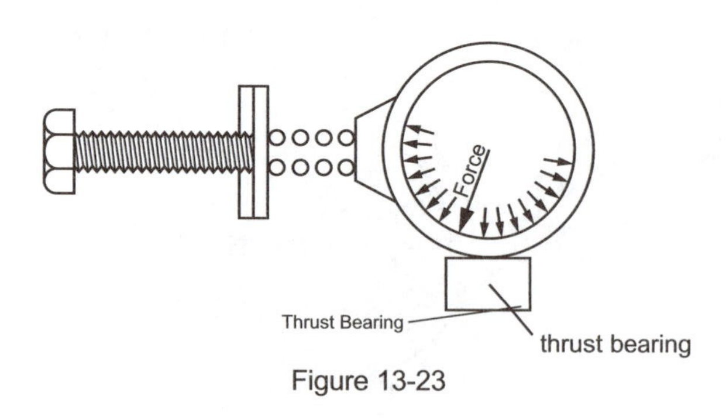

Figure 13-23

Figure 13-24

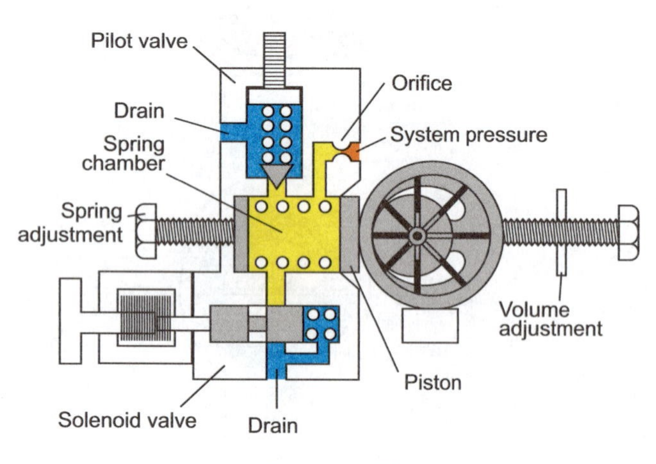

Figure 13-25

The more the pressure compensator adjustment is screwed in, the more the compensator spring is compressed and the greater the force holding the ring off center. To protect the pump from too high of a pressure adjustment, pressure compensated vane pumps are commonly equipped with over pressure stops.

In the next section, the operation of a variable volume axial piston pump is described. We find that the compensator in this pump basically functions as a pilot operated relief valve.

Vane Pump Dual Compensation

In a vane pump, a dual compensator consists of a solenoid valve, pilot valve, and orifice which senses system pressure.

With the solenoid valve closed, fluid pressure is allowed to accumulate in the spring chamber through the sensing orifice. This adds to the spring pressure biasing the cam ring against the volume adjustment. The maximum amount of fluid pressure is determined by the pilot valve setting. Compensator setting is a function of spring and fluid pressure.

If the spring were adjusted for 600 psi (41.4 bar) and the pilot valve were adjusted for 900 psi (62 bar), the compensator setting would be 1500 psi (103.4 bar).

A second compensator setting is achieved when the solenoid valve is energized. This action vents the spring chamber of fluid pressure. Spring pressure only holds the cam ring against the volume adjustment. Since this has a value of 600 psi (41.4 bar) in this example, the pump now compensates at 600 psi (41.4 bar).

Variable Volume from a Gear Pump

The output volume of a gear pump is determined by the volume of fluid each gear displaces, and by the rpmat which the gears are turning. Consequently, the output volume of gear pumps can be altered by replacing the original gears with gears of different dimensions or by varying rpm.

Gear pumps, whether of the internal or external variety, do not lend themselves to a change in displacement while they are operating. There is nothing that can be done to vary the physical dimensions of a gear while it is turning.

One practical way, then, to vary the output flow from a gear pump is to vary the speed of its prime mover. This can be easily done when the pump is being driven by a variable speed internal combustion engine. It can also be accomplished electrically by using a variable speed electric motor.

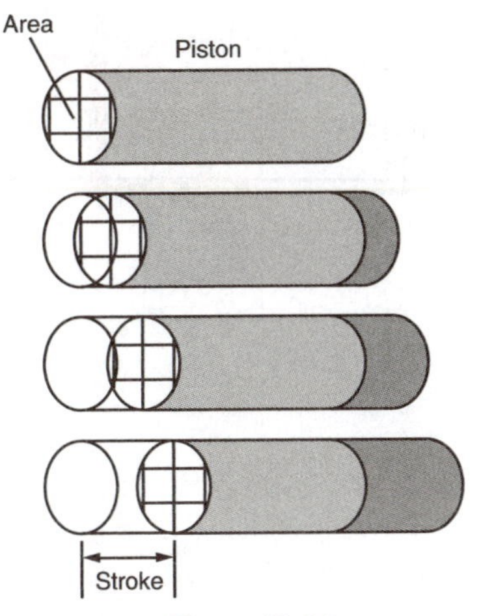

Figure 13-26

Variable Volume Axial Piston Pump

The amount of fluid which an axial piston pump displaces per revolution is determined by the cross sectional area of the piston and the distance pistons are stroked within their cylinder bores. While pumping takes place, nothing can be done to change the cross sectional area of the pistons. But, a variable volume axial piston pump is designed so that the distance pistons are stroked can be varied.

What a Variable Volume Axial Piston Pump Consists Of

The pumping mechanism of a variable volume axial piston pump basically consists of a barrel, port plate, pistons with shoes, an adjustable swashplate, shoeplate, maximum volume adjustment, control piston, orifice, compensator spool, pilot valve, and various bias springs.

Pistons are housed in the bores of the barrel. Attached to the barrel is a shaft which is coupled to an electric motor. The shaft can be connected at either the port plate or swashplate side.

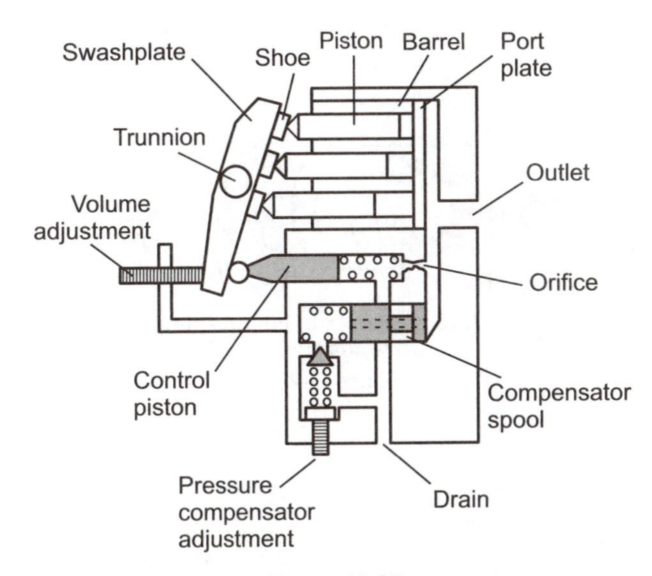

Figure 13-27

How the Volume Adjustment of a Variable Volume Axial Piston Pump Works

The volume adjustment of a variable volume axial piston pump is a threaded rod which mechanically limits the swashplate angle.

With the volume adjustment backed completely out, the spring biasing the control piston positions the swashplate at an extreme angle. As the barrel is rotated by a prime mover, piston shoes are forced by mechanical bias to contact the swashplate surface resulting in the pistons reciprocating within the bore. In one half of the circle of rotation, pistons move out of the cylinder barrel generating an increasing volume; fluid fills the increasing volume through the inlet port. In the other half of the circle of rotation, pistons are pushed into the cylinder barrel generating a decreasing volume; fluid is pushed out to the system through the outlet port.

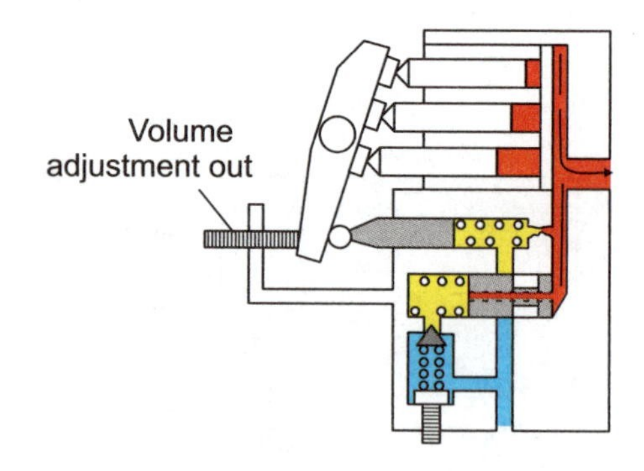

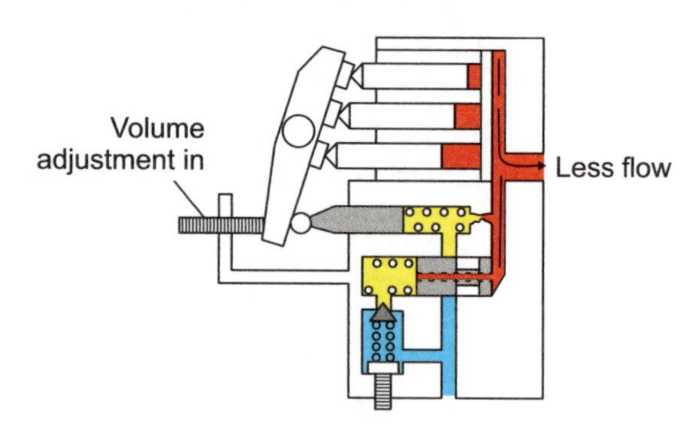

Figure 13-28

With the swashplate at its extreme angle, maximum volume discharges from the pump. If the threaded rod were screwed in slightly so that the swashplate had less of an angle, pistons would not reciprocate to such a great degree. This results in less fluid discharging into the system.

The more the volume adjustment of a variable volume axial piston pump is screwed in, the less flow is discharged.

Overcenter Axial Piston Pumps

As was illustrated, the displacement of an axial piston pump, and therefore its output volume, can be varied by changing the angle of the swashplate. It was also shown that the pump will develop no flow when the swashplate is centered with the cylinder barrel.

Some swashplates of axial piston pumps have the capability of crossing over center. This results in the increasing and decreasing volumes being generated at opposite ports. Flow through the pump reverses.

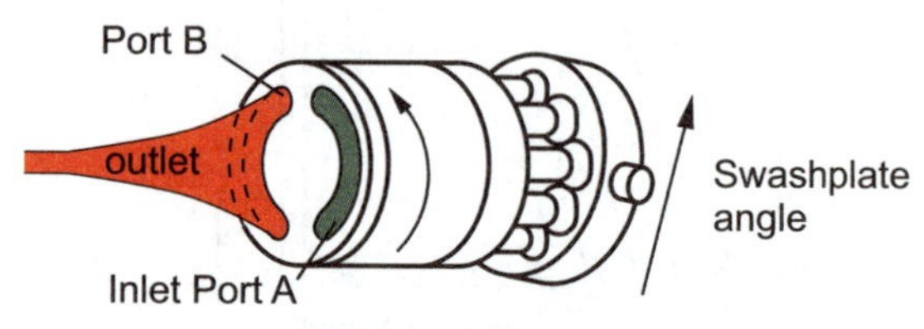

Overcenter Axial Piston Pump
(Drive shaft not shown)
Figure 13-29

From the overcenter axial piston pump illustrated, it can be seen that ports A and B can be either an inlet or outlet port depending upon the angle of the swashplate. This takes place with the cylinder barrel rotating in the same direction.

Overcenter axial piston pumps are often used in hydrostatic transmissions which will be covered in a later section.

Axial piston pumps can be variable displacement, pressure compensated and variable displacement, or variable displacement and overcenter. These combinations are also available with the bent-axis and radial design piston pumps.

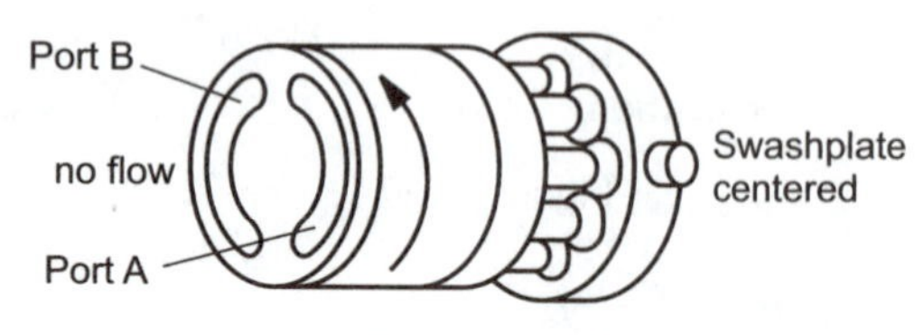

Figure 13-30

How the Pressure Compensator of a Variable Volume Axial Piston Pump Works

The pressure compensator of a variable volume axial piston pump uses a combination of spring and fluid pressure.

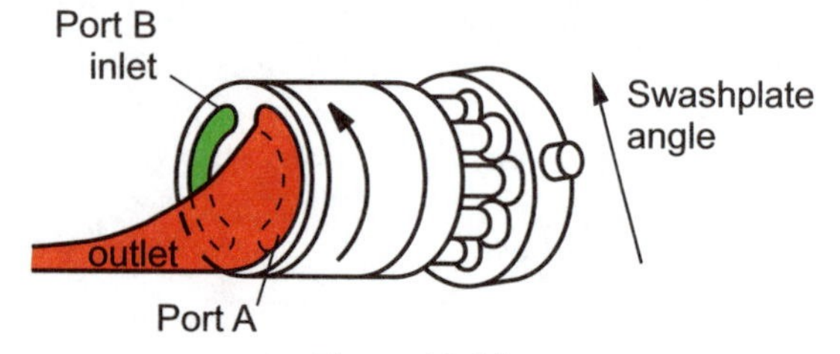

Figure 13-31

In the pressure compensated vane pump seen previously, a single spring acted as the compensator. This was possible because the force moving the ring did not act directly opposite the spring, but at an angle. Most of the force was taken up by the thrust bearing. In a pressure compensated axial piston pump, this is not the case.

As a pressure compensated axial piston pump is operating, pressure acts on the ends of the pistons doing the pumping. With the centerline of the swashplate trunnion offset from the centerline of the barrel, the pistons under pressure attempt to push the swashplate to an upright position. This force can be considerable and is counteracted directly by spring and fluid pressure acting on the control piston.

A pressure compensator of a typical axial piston pump consists of a control piston, compensator spool, adjustable pilot valve, and bias springs. When pressure at pump outlet is high enough to overcome the setting of the pilot valve and the bias springs, the swashplate is pushed back to an upright position by the pressurized pistons.

Once the swashplate achieves a zero angle, it stops. The swashplate cannot be pushed to the other side of center since this would cause the pressure half of the pumping mechanism to form an increasing volume which is suction operation. Pressure would immediately drop resulting in the spring offsetting the swashplate once again.

The adjustable pilot valve, compensator spool and its bias spring are very similar in operation to a pilot operated relief valve. Pump discharge pressure is sensed through an orifice in the compensator spool. It is allowed to accumulate in the bias spring chamber of the spool and the amount of accumulated pressure is limited by the pilot valve. When pressure at the end of the compensator spool is sufficient to overcome the pilot valve setting and the bias spring, the compensator spool will move venting the control piston spring chamber.

Assume that the spring biasing the compensator spool has a value of 100 psi (6.9 bar) and that the pilot valve will limit pilot pressure in the spring chamber to 700 psi (48.3 bar).

With a system pressure of 500 psi (34.5 bar), 500 psi (34.5 bar) acts to push the spool to the left. 500 psi (34.5 bar) is transmitted through the orifice to the spring chamber acting to hold the spool in place. Areas exposed to pressure on either side of the spool are equal; therefore, the spool is balanced except for the 100 psi (6.9 bar) spring. At this point, 500 psi (34.5 bar) attempts to push the spool to the left and a total hydraulic and mechanical pressure of 600 psi (41.4 bar) holds the spool in place.

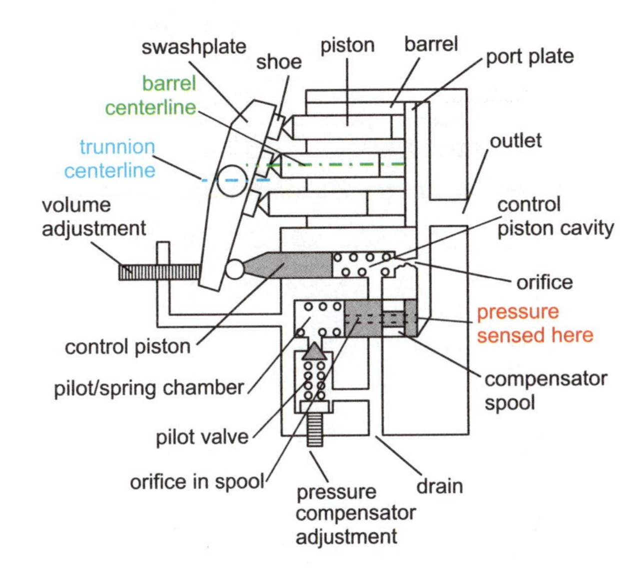

Figure 13-32

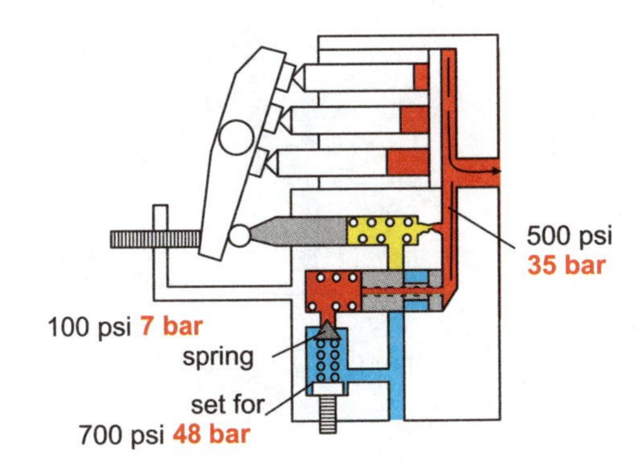

Figure 13-33

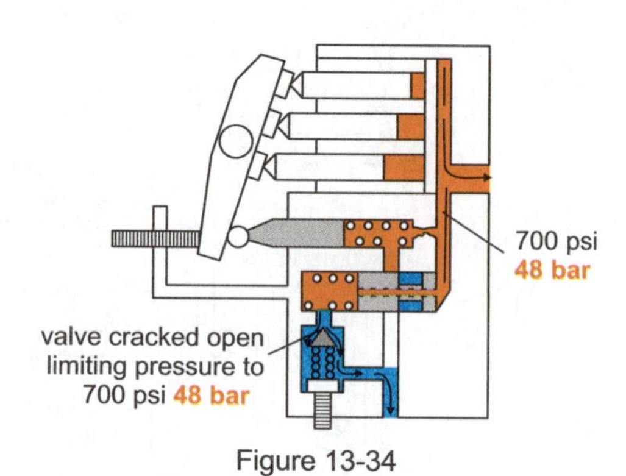

Figure 13-34

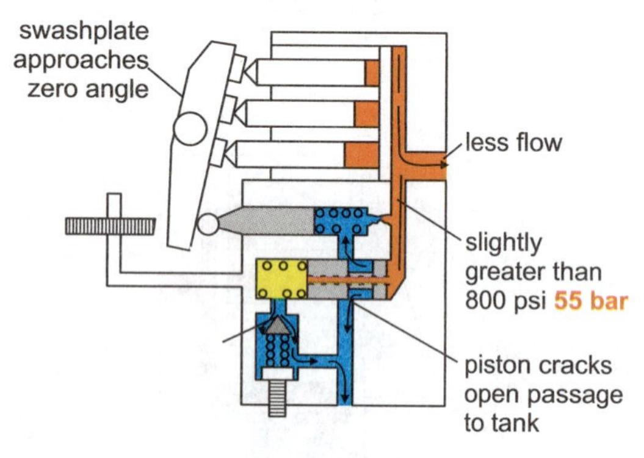

Figure 13-35

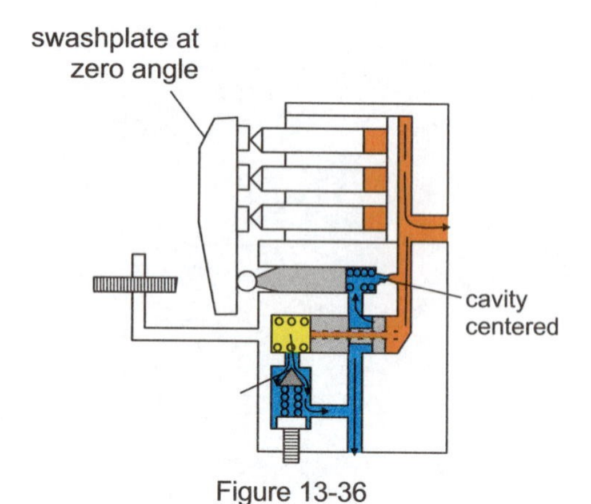

Figure 13-36

With a system pressure of 600 psi (41.4 bar), 600 psi (41.4 bar) at the right end of the spool acts to push the spool to the left. A total mechanical and hydraulic pressure of 700 psi (48.3 bar) acts to hold the spool.

When system pressure climbs to 700 psi (48.3 bar), the pilot valve opens limiting the pilot/spring chamber to 700 psi (48.3 bar). This means the maximum pressure biasing the compensator spool is 800 psi (55.2 bar).

As pump discharge pressure climbs above 800 psi (55.2 bar), the spool moves cracking open the control piston cavity. Pressure in this cavity begins to drop resulting in a smaller generated force at the control piston and an inability to keep the swashplate fully displaced. Consequently, discharge flow decreases. As pressure at pump outlet increases, the compensator spool moves more, resulting in less discharge flow. In this example, let's assume that as pressure climbs an additional 100 psi (6.9 bar) above 800 psi (55.2 bar), the spring biasing the compensator spool will have been compressed, and the spool will have moved sufficiently to vent the control piston cavity, de-stroking or compensating the pump. The pump is fully compensated at 900 psi (62 bar).

(The spring biasing the control piston is relatively light and can be neglected in calculating compensating pressure.)

When pressure drops at pump outlet, the compensator spool closes off the control cavity de-venting the chamber. Pumping action resumes.

The more the pressure compensator pilot valve adjustment is screwed in, the more pilot pressure it allows to bias the compensator spool before compensation occurs.

With the operation of pressure compensated, variable volume vane and axial piston pumps in mind, we compare in the next section their operation with a fixed pump.

Piston Pump Dual Compensation

A dual compensator of a piston pump incorporates a solenoid valve, and another pilot valve.

With the solenoid valve closed, the pump compensates in its normal manner. Fluid pressure is allowed to accumulate in the compensator spool spring chamber. The amount of pressure is determined by the main pilot valve. This pressure adds to the compensator spring pressure biasing the compensator spool.

If the pilot valve were adjusted for 1300 psi (89.66 bar) and the compensator spring allowed pressure to climb an additional 200 psi (13.8 bar) before full compensation, the pump would be fully compensated at 1500 psi (103.4 bar).

Assume now that the auxiliary pilot valve is adjusted for 400 psi (27.6 bar). When the solenoid valve is energized, the auxiliary pilot valve connects with the compensator spring chamber. Maximum pressure is allowed to accumulate in the chamber is now 400 psi (27.6 bar). With the spring having a full compensation value of 200 psi (13.8 bar), the pump now compensates at 600 psi (41.4 bar).

With a dual compensator accessory, generated power can more evenly match output power at actuators with different requirements.

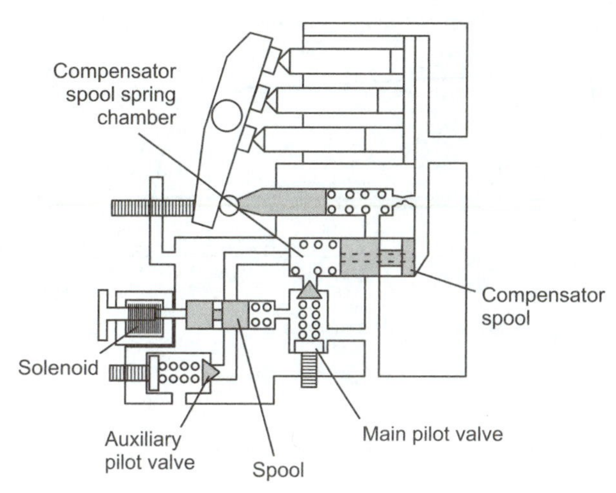

Figure 13-37

Comparing Fixed and Pressure Compensated Variable Volume Operation

Illustrated are two curves. One curve shows the discharge flow from a 10 gpm (37.9 l/min) fixed displacement pump as it operates between zero and 800 psi (55.2 bar). The other curve shows the discharge flow from a 10 gpm (37.9 l/min) variable volume pump as it operates between the same pressures. Both curves are the same and they point out that the discharge flow rate from each unit at 800 psi (55.2 bar) is approximately 9.25 gpm (35.05 l/min); this is 4.3 hp (3.7 kW).

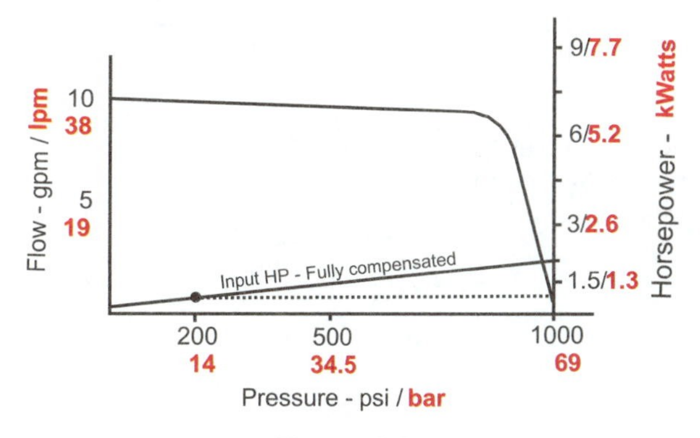

Figure 13-38

Assume that the fixed displacement pump/ electric motor pressure is limited by a relief valve setting of 1000 psi (68.97 bar). Referring to the fixed displacement curve between zero and 1000 psi (68.97 bar), we see that at 1000 psi (68.97 bar) pump/electric motor is developing 9 gpm (34.11 l/min) which is 5.2 (3.8 kW) hydraulic horsepower. This is the maximum power that the unit is allowed to develop; and as it passes through a relief valve, it is transformed into heat.

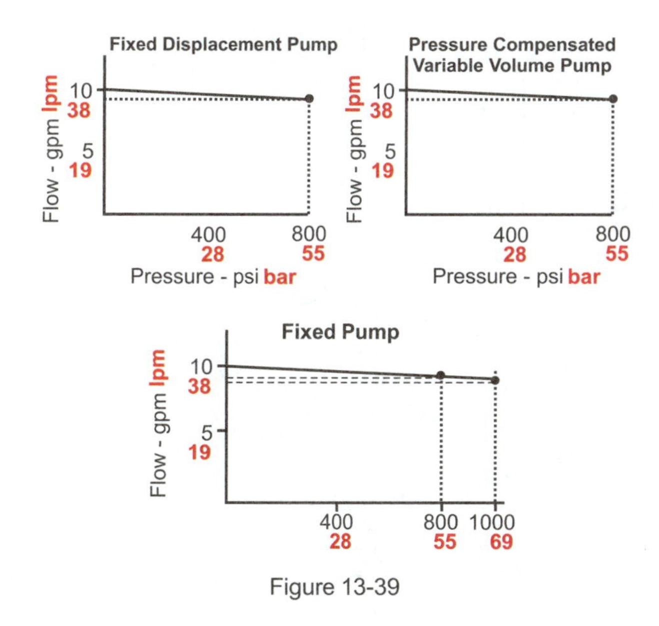

Figure 13-39

Pressure compensated vane pumps are characteristically fully compensated within 200 psi (13.8 bar). Piston pumps compensate fully within 100 psi (6.9 bar) or less. Since a vane pump has a wider compensating band, the curves of pressure compensated, variable volume vane pumps will be shown. This will more easily illustrate a pump's operating characteristic during compensation.

Assume now that the pressure compensated variable volume pump has a compensator setting of 1000 psi (68.97 bar). Referring to the pressure compensated pump curve between zero and 1000 psi (68.97 bar), we find that at 800 psi (55.2 bar) and

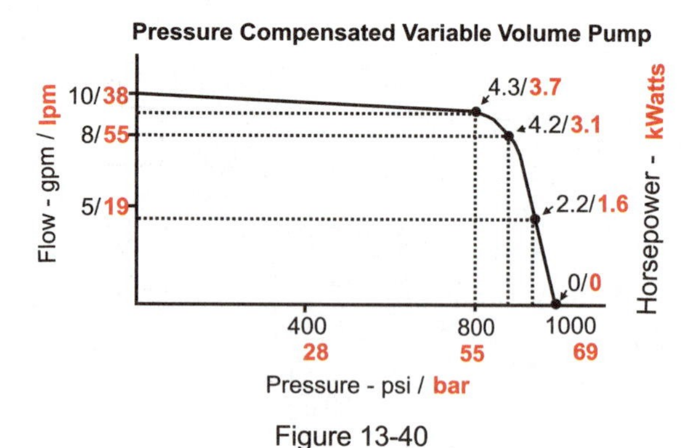

Figure 13-40

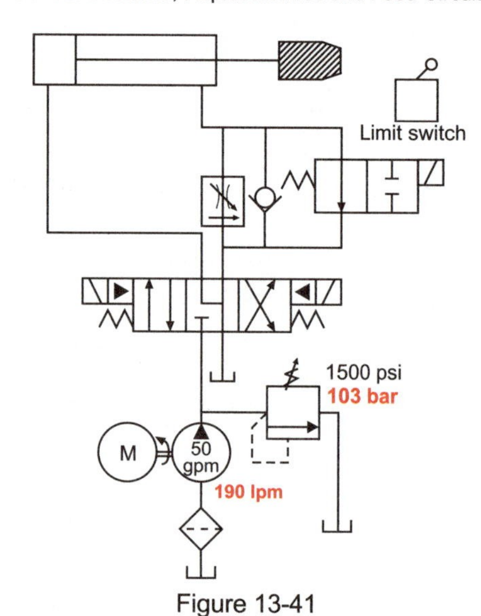

Figure 13-41

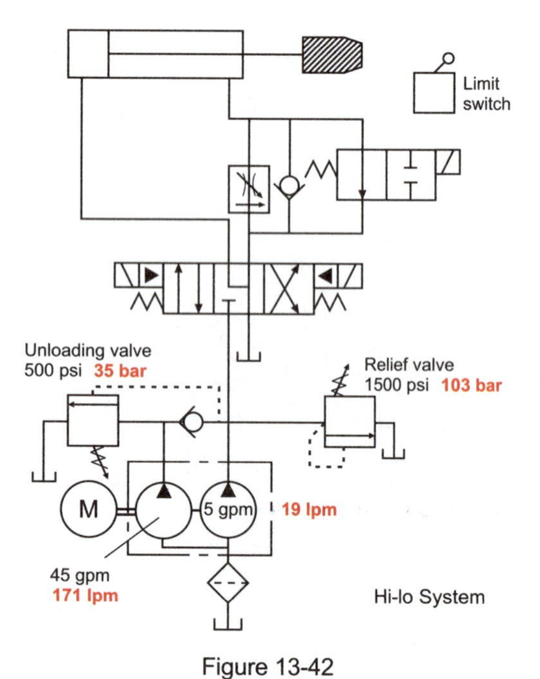

Figure 13-42

9.25 gpm (34.86 l/min) (4.3 hp/3.7 kW), discharge flow begins to decrease. Consequently, generated power decreases as system pressure approaches 1000 psi (68.97 bar). At a pressure of 900 psi (62.1 bar) discharge flow is 8 gpm (30.3 l/min) (4.2 hp/3.1 kW). At 950 psi (67.85 bar), discharge flow is 4 gpm (15.16 l/min) (2.2 hp/1.6 kW). At 1000 psi (68.97 bar), discharge flow rate is zero gpm (0 hp).

With a pressure compensated pump coupled to a prime mover, generated hydraulic power is not only controlled by limiting pressure, but it is reduced by regulating discharge flow. This means less heat is generated in a system since generated power is not wasted over a relief valve.

We have seen that when a pressure compensated pump is fully compensated, the generated hydraulic power to the system is zero. However, the prime mover coupled to the pump must still develop power because of internal leakage and mechanical inefficiencies of the pump.

Hydraulic Pumps in a Circuit

The systems which have been described to this point were primarily constant volume systems. In these systems, the pump would apply whatever pressure was required to get its full flow out into the system.

In the circuit illustrated, the 50 gpm (189.5 l/min) pump delivers its flow to the work cylinder to achieve a rapid advance. When the cylinder rod reaches the limit switch, the solenoid of the two-way valve is energized and the cylinder rod goes into a feed rate. At this point, the pump is delivering to the system 43.7 hp (32.6 kW) (50 gpm x 1500 psi x .000583). If a hi-lo system were used in the same system, less horsepower would be used.

Let us assume that the feed rate required is 3.7 gpm (14.02 l/min). With a hi-lo system, we could use a 45 gpm (170.6 l/min) pump and 5 gpm (18.95 l/min) pump to achieve the rapid advance. During the feed rate, the 45 gpm (170.6 l/min) is dumped to tank through the unloading valve. The 5 gpm (18.95 l/min) pump develops the required flow for the feed rate. At this point, 4.4 hp (3.3 kW) is delivered to the system. (5 gpm x 1500 psi x .000583).

A pressure compensated, variable volume pump can make the system even more efficient. With this type pump, the relief valve is eliminated. When the cylinder goes into its feed rate, the compensator only allows the pump to deliver 3.7 gpm (14.02 l/min) at 1500 psi (103.4 bar). This is equal to 3.2 hp (2.3 kW).

Systems which use pressure compensated, variable volume pumps are quite efficient.

Standby Pressure

It was illustrated earlier that pressure compensated pumps may or may not be compensated during machine idle time. This was due to the input power requirements at various compensator settings.

From the illustrated curve, it can be seen that input power is quite low when the pump is compensating at 200 psi (13.8 bar) approximately .5 horsepower (.37 kW) as compared to compensating at 1000 psi (68.97 bar) where the input is 1.5 horsepower (1.1 kW). This can be taken advantage of during machine idle time.

A pressure compensated vane pump is illustrated. The pump is equipped with a dual compensator.

Assume that the spring is adjusted for 200 psi (13.8 bar) and the pilot valve is adjusted for 800 psi (55.2 bar). The pump will be fully compensated at 1000 psi (68.97 bar).

When the solenoid valve is energized during machine idle, the spring chamber is vented compensating the pump at 200 psi (13.8 bar). The pump stands by at this pressure until the work cycle resumes.

A pressure compensated piston pump is also illustrated. This unit is equipped with a solenoid valve.

Assume the compensator spool spring has a value of 200 psi (13.8 bar) at full compensation. The pilot valve is adjusted for 800 psi (55.2 bar). The pump will be fully compensated at 1000 psi (68.97 bar).

During machine idle, the solenoid valve is energized venting the compensator spool spring chamber. The pump compensates at 200 psi (13.8 bar) and stands by until work is required.

By causing a pressure compensated pump to compensate at a low pressure during idle time, minimum power is used.

Horsepower Limiting

Many variable volume, vane and piston pumps have a horsepower limiting option. This enables a pump/electric motor not to exceed a given horsepower regardless of load resistance.

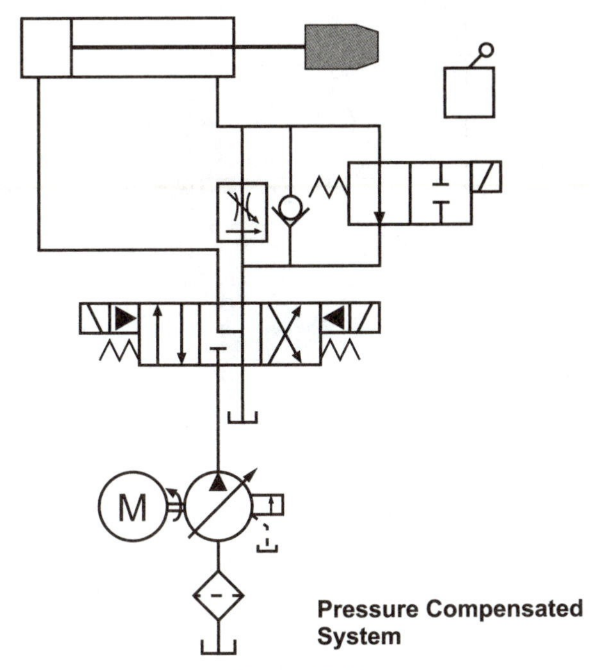

Figure 13-43

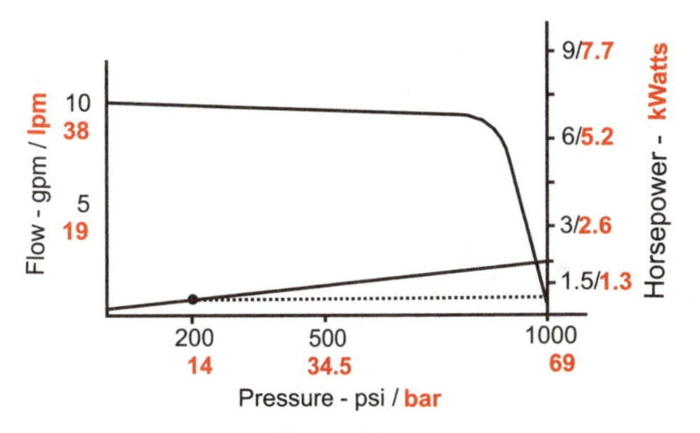

Figure 13-44

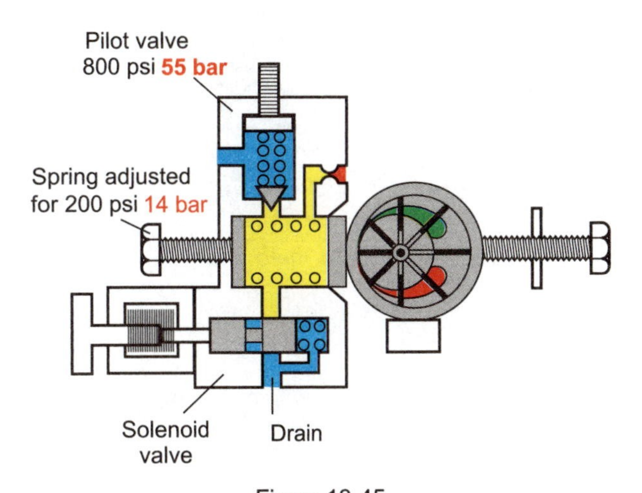

Figure 13-45

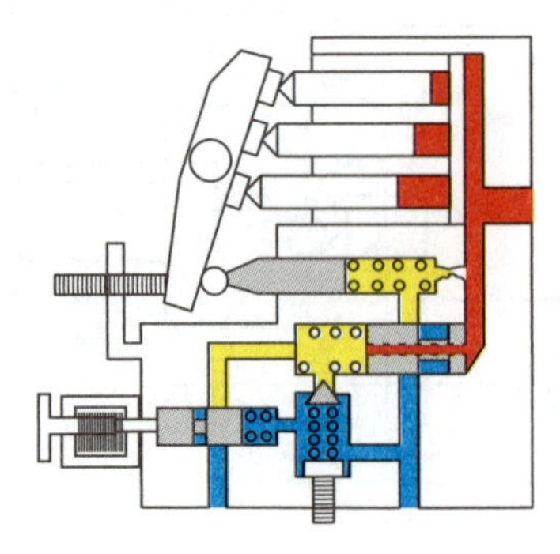

Figure 13-46

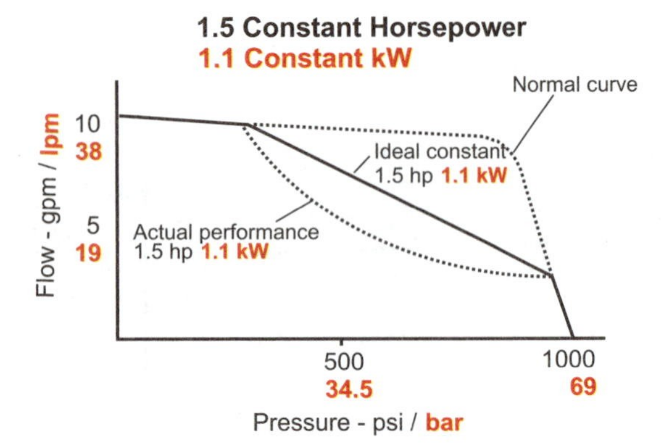

Figure 13-47

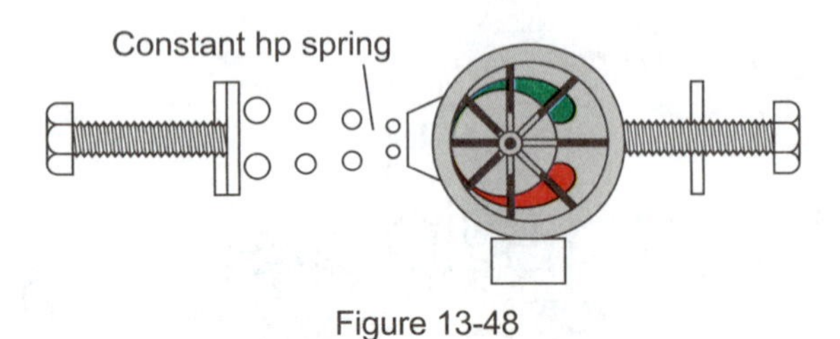

Figure 13-48

Horsepower is work performed at a rate of 550 ft.lb./sec (746 W). A 775 lb. loading moving 1 ft. in 1 second is 1.5 (1.1 kW) horsepower. Likewise a 1550 lb. load moving 1 ft. in 2 seconds, or a 2325 lb. load moving 1 ft. in 3 seconds, is 1.5 (1.1 kW) horsepower.

Consequently, an electric motor coupled to a variable volume pump with a horsepower limiting option could move a 775 lb. (3441 N), 1550 lb. (6882 N), and 2325 lb. (10323 N) load using the same horsepower (kW). With a fixed displacement or ordinary pressure compensated unit, this is not the case; horsepower would increase as load resistance increased. (This assumes the pressure compensated pump is not compensating.) A horsepower limiting option is achieved in a vane pump by using a compensator spring with certain compression characteristics. As the cam ring moves closer to rotor center, spring pressure increases and flow decreases. In this manner horsepower is kept relatively constant.

In a piston pump, constant horsepower is achieved with the addition of a pilot valve and control piston of a specific contour. As swashplate approaches a zero angle, the control piston contour increases the pilot setting of the valve. Compensator pressure increases as flow decreases keeping horsepower relatively constant. Since actuator speed decreases as load increases, a constant horsepower feature is found in applications where electric motor size is more important than cycle time. Examples of such systems would be commercial garbage compactors and lift trucks.

Volumetric Efficiency

While turning at a constant rpm, we generally think of a positive displacement pump as delivering a constant rate of flow regardless of system pressure. This is not entirely true. As system pressure increases, the internal leakage of various pumping mechanisms increase. This results in a decreased output flow. The degree to which this happens is known as volumetric efficiency.

The expression which describes volumetric efficiency is:

$$\text{Volumetric efficiency (\%)} = \frac{\text{actual output} \times 100}{\text{theoretical output}}$$

For example, if a particular pump had a theoretical output at 1200 rpm of 10 gpm (37.9 l/min), but the actual delivery at 1000 psi (68.97 bar) were 9 gpm (34.1 l/min), the volumetric efficiency would be 90%.

Typically, piston pumps have an initial volumetric efficiency in the high 90's. Gear and vane equipment range from the high 80's to the mid 90's in volumetric efficiency.

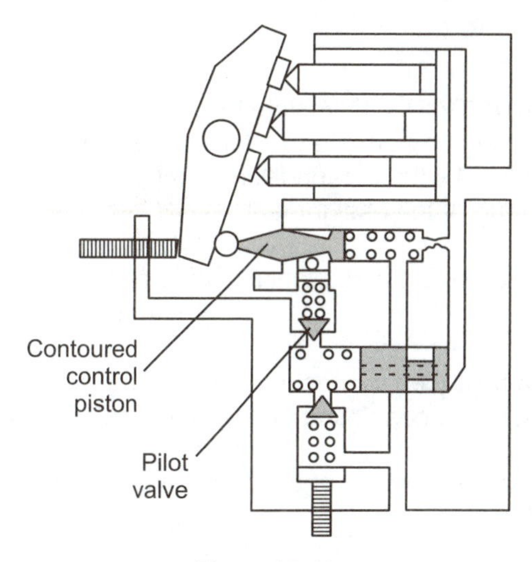

Figure 13-49

Overall Efficiency

The overall efficiency of a hydraulic pump takes into account a pump's mechanical efficiency as well as its volumetric efficiency. This can be determined by dividing the hydraulic horsepower delivered to the system by the pump, by the input horsepower of its prime mover. An expression which describes overall efficiency is:

$$\text{Overall efficiency (\%)} = \frac{\text{hydraulic hp output} \times 100}{\text{input hp of prime motor}}$$

For instance, if a particular pump delivered 10 gpm (37.9 l/min) to a system at 1000 psi (68.97 bar), this would equal 5.8 hp (4.3 kW) (hp = gpm x psi x .000583).

If an electric motor driving the pump had to develop 7 hp (5.2 kW), then the overall efficiency would be 83%.

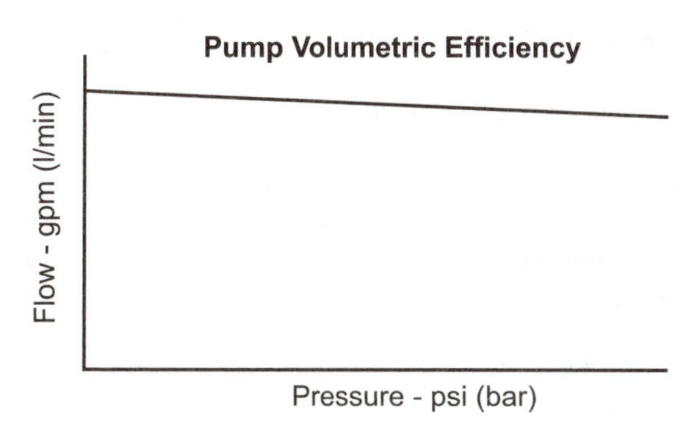

Figure 13-50

The overall efficiencies of industrial hydraulic gear, vane, and piston pumps operating at a 1000 psi (68.97 bar) is approximately 85%. Conversely, the overall efficiencies of these same pumps at 200 psi (13.8 bar) is about 60-70%.

Industrial hydraulic pumps are generally designed to operate at pressures above 200 psi (13.8 bar). For this reason, the overall efficiency of a pump is diminished at this low pressure.

Terms and Idioms Associated with Hydraulic Pumps

COMBINATION PUMP – a double pump which is equipped with accessory valving to perform relief and unloading functions.

SLIPPAGE – internal leakage of a pumping mechanism.

TWO-STAGE PUMP – a pump with two pumping mechanisms (stages). The units are connected in series; that is, one feeds the other. With this arrangement, two pumping mechanisms rated at 1000 psi (68.97 bar) individually, could be used to a maximum pressure of 2000 psi (137.9 bar). Two-stage pumps were popular in the days of hydraulics when high pressure pumps were not available.

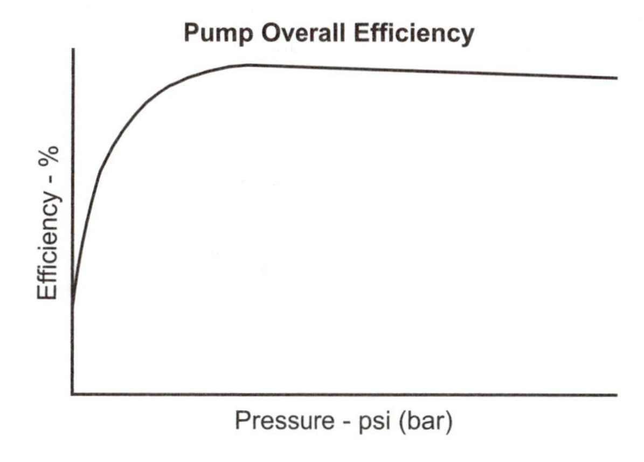

Figure 13-51

Exercise Hydraulic Pumps

INSTRUCTIONS: Complete the sentences below by filling in the blanks with the appropriate word. In each blank, the number of letters for the correct word is shown by dashes. One dash for each word is circled. After all the blanks have been filled in, take all of the circled letters and form two words which answer the question at the end of the assignment.

1. The pumping mechanism of a constant volume vane pump is usually an integral unit known as a(n) _ _ _ _(_)_ _ _ _ assembly.
2. In a balanced vane pump design, the two pressure quadrants in the _ _(_) ring oppose each other so that the forces acting on the shaft are balanced.

   _ _ _
3. The inner contour of the ring in a pressure compensated, variable volume pump has a(n) _(_)_ _ _ _ _ _ shape.
4. All pressure compensated, variable volume pumps have a(n) _ _ _(_) drain.
5. At the inlet of a gear pump, gear teeth _ _(_)_ _ _.
6. A spur gear pump is a(n) _ _ _(_)_ _ _ _ gear pump.
7. In industrial applications, the _ _ _(_)_ _ _ is a popular internal gear pump.
8. The three most commonly used pumps in industrial hydraulic systems are vane, piston, and _ _ _(_).
9. The pistons in an axial piston pump are held against the swashplate by a(n) _ _ _ _(_)_ _ _ _ and bias spring.
10. If the swashplate of an axial piston pump can cross over center, the flow through the pump can be _ _(_)_ _ _ _ _.

QUESTION: What is a required pump partner?

ANSWER: ( ) ( ) ( ) ( ) ( )   ( ) ( ) ( ) ( ) ( )

CHAPTER 14

Hydraulic Motors

Hydraulic motors convert the working energy of a hydraulic system into rotary mechanical energy.

Hydraulic motors operate by causing an imbalance which results in the rotation of a shaft. This imbalance is generated in different ways depending upon the motor type.

Hydraulic motors are positive displacement devices; that is, as it receives a constant flow of fluid, the motor speed will remain relatively constant regardless of the pressure.

Hydraulic motors used in an industrial system can be divided into vane, gear, and piston types.

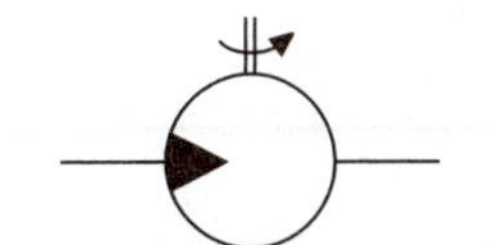

Hydraulic motor symbol (unidirectional)

Figure 14-1

Motor Drains

Motors used in industrial hydraulic systems are almost exclusively designed to be bidirectional (operate in both directions). Even motors which operate in a system in only one direction (unidirectional) are probably bidirectional motors in design.

To protect its shaft seal, vane, gear, and piston, bidirectional motors are generally externally drained.

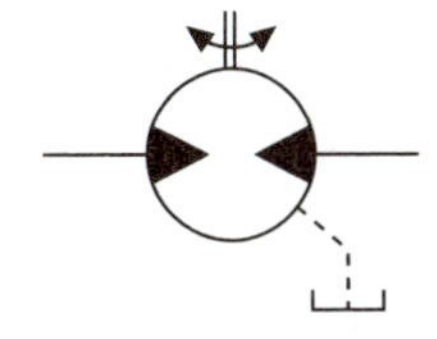

Figure 14-2

Vane Motors

A vane motor is a positive displacement motor which develops an output torque at its shaft by allowing hydraulic pressure to act on vanes which are extended.

What a Vane Motor Consists Of

The rotating group of a vane motor basically consists of vanes, rotor, ring, shaft and a port plate with kidney-shaped inlet and outlet ports.

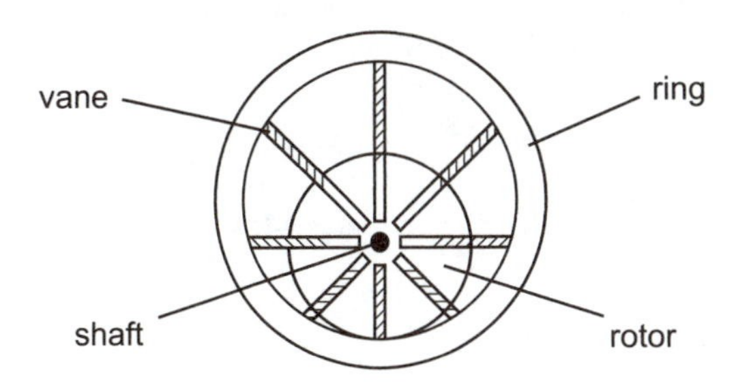

Figure 14-3

How a Vane Motor Works

All hydraulic motors operate by causing an imbalance which results in the rotation of a shaft. In a vane motor, this imbalance is caused by the difference in vane area exposed to hydraulic pressure. In our illustration, with the rotor positioned off-center with respect to the ring, the area of the vanes exposed to pressure increases toward the top and decreases at the bottom. When pressurized fluid enters the inlet port, the unequal areas of the vanes result in a torque being developed at the motor shaft.

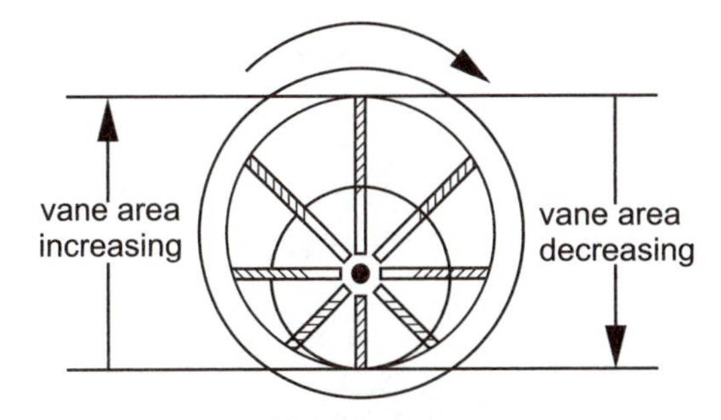

Figure 14-4

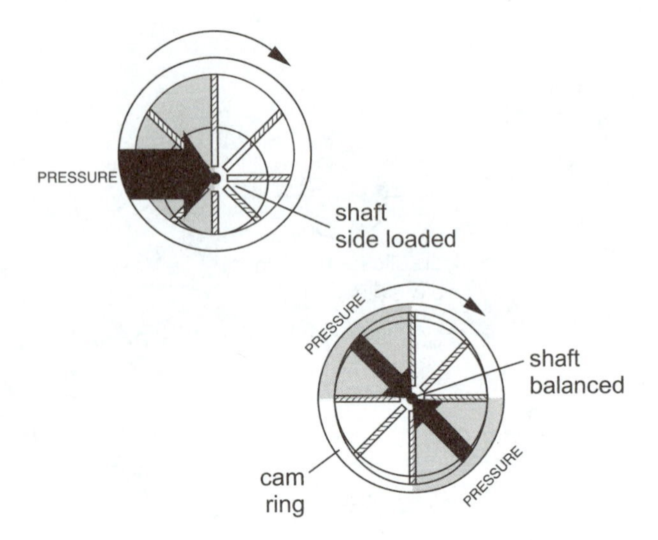

Figure 14-5

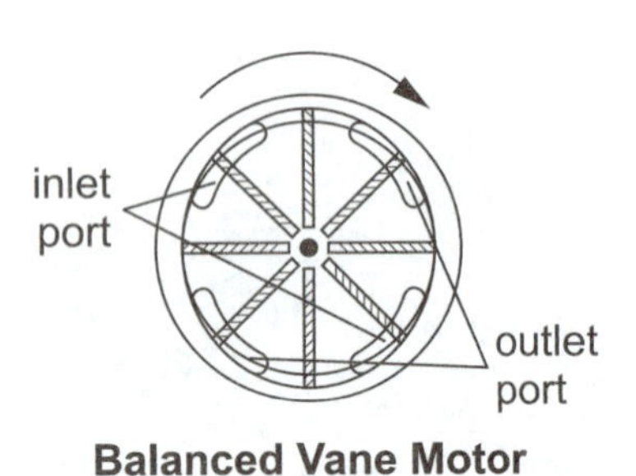

Balanced Vane Motor

Figure 14-6

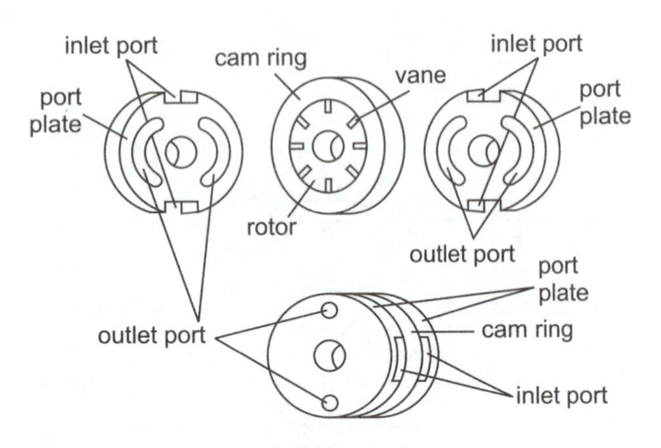

Cartridge Assembly

Figure 14-7

The larger the exposed area of the vanes, or the higher the pressure, the more torque will be developed at the shaft. If the torque developed is large enough, the rotor and shaft will turn.

Balanced Vane Motor Design

In a hydraulic motor, two different pressures are involved — system working pressure at the inlet and tank line pressure at the outlet. This results in side loading the shaft which could be severe at high system pressures. To avoid shaft side loading, the inner contour of the ring is changed from circular to cam-shaped. With this arrangement, the two pressure quadrants oppose each other and the forces acting on the shaft are balanced. Shaft side loading is eliminated.

A balanced vane motor consists of a cam ring, rotor, vanes, and a port plate with inlet and outlet ports opposing each other. (Both inlet ports are connected together, as are the outlet ports, so that each can be served by one inlet or one outlet port in the pump housing).

Vane motors used in industrial hydraulic systems are generally of the balanced design.

Cartridge Assembly

The rotating group of industrial vane motors is usually an integral cartridge assembly. The cartridge assembly consists of vanes, rotor, and a cam ring sandwiched between two port plates.

An advantage of using a cartridge assembly is easy motor servicing. After a period of time when motor parts naturally wear, the rotating group can be easily removed and replaced with a new cartridge assembly. Also, if the same motor is required to develop more torque at the same system pressure, a cartridge assembly with the same outside dimensions, but with a larger exposed vane area, can be quickly substituted for the original.

Extending a Motor's Vanes

Before a vane motor will operate, its vanes must be extended. Unlike a vane pump, centrifugal force cannot be depended on to throw-out the vanes and create a positive seal between cam ring and vane tip. Some other way must be found.

There are two common means of extending the vanes in a vane motor. One method is spring loading the vanes so that they are extended continuously. The other method is directing hydraulic pressure to the underside of the vanes.

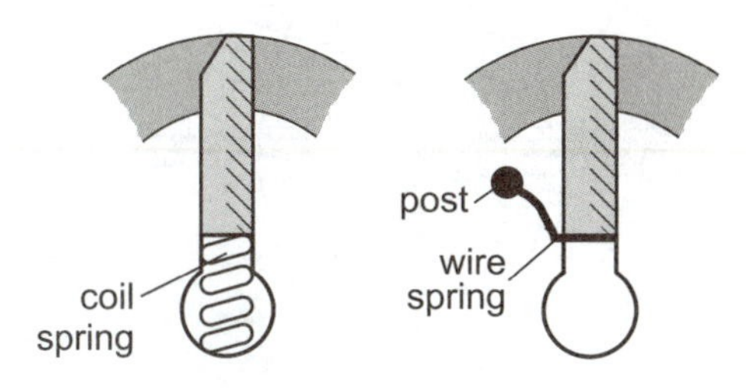

Figure 14-8

Spring loading is accomplished in some vane motors by positioning a coil spring in the vane chamber.

Another way of loading a vane is with the use of a small wire spring. The spring is attached to a post and moves with the vane as it travels in and out of the slot.

In both types of spring loading, fluid pressure is directed to the underside of the vane as soon as torque is developed.

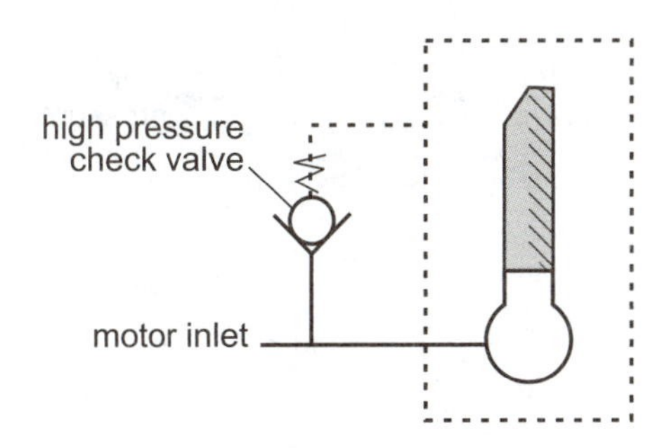

Figure 14-9

Another means of extending a motor's vanes is with the use of fluid pressure. In this method, fluid is not allowed to enter the vane chamber area until the vane is fully extended and a positive seal exists at the vane tip. At this time, pressure is present under the vane. When fluid pressure is high enough to overcome the spring force biasing the internal check valve, fluid will enter the vane chamber and develop a torque at the motor shaft. The internal check valve in this instance performs a sequencing function.

When the load attached to a motor's shaft is allowed to freewheel, the load is allowed to coast to a stop. A motor, which uses hydraulic pressure to extend its vanes, requires a 65 psi (4.1 bar) to 120 psi (8.3 bar) check valve in the tank line if the load is allowed to freewheel. The backpressure, which is generated because of the tank line check valve, keeps the vanes from retracting. This slows down the load more quickly.

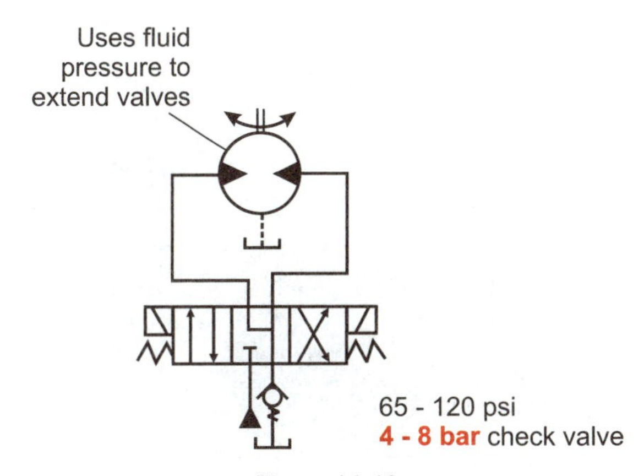

Figure 14-10

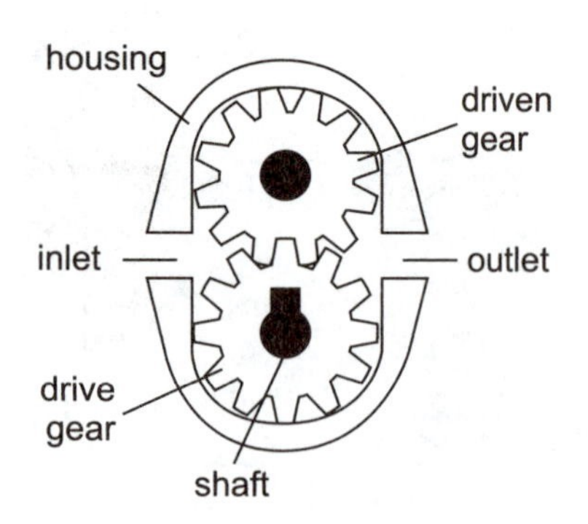

Figure 14-11

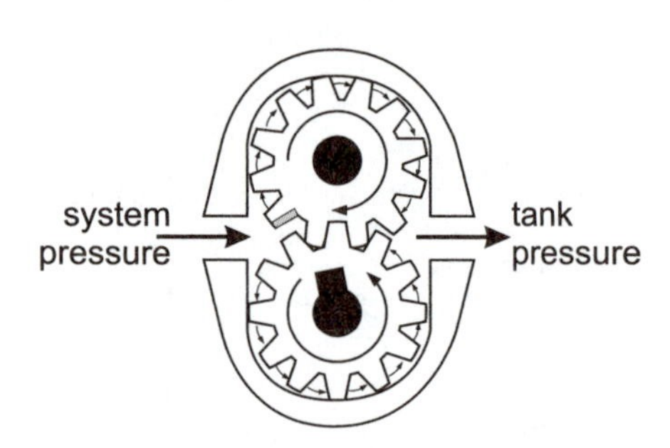

Figure 14-12

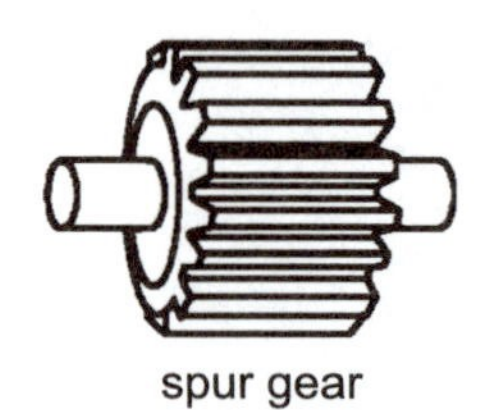

Figure 14-13

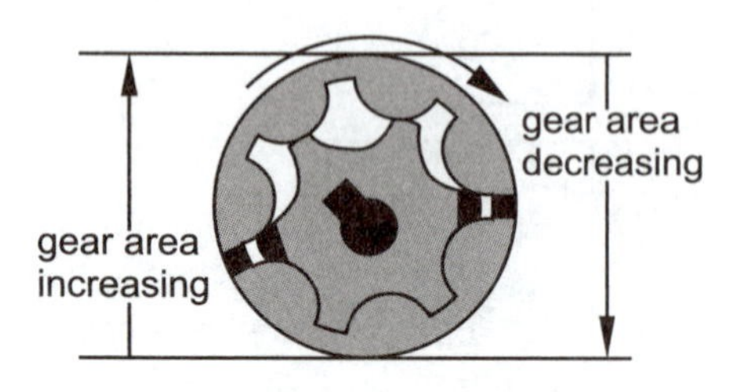

Figure 14-14

Gear Motors

A gear motor is a positive displacement motor which develops an output torque at its shaft by allowing hydraulic pressure to act on gear teeth.

What a Gear Motor Consists Of

A gear motor basically consists of a housing with inlet and outlet ports, and a rotating group made up of two gears. One gear, the drive gear, is attached to a shaft which is connected to a load. The other gear is the driven gear.

How a Gear Motor Works

Hydraulic motors operate by causing an imbalance which results in the rotation of a shaft. In a gear motor, this imbalance is caused by gear teeth unmeshing.

In the illustration of the gear motor, the inlet is subjected to system pressure. The outlet is under tank pressure. As gear teeth unmesh, it can be seen that all teeth subjected to system pressure are hydraulically balanced except for one side of one tooth on one gear (shaded area). This is the point where torque is developed. Consequently, the torque developed by a gear motor of this type is a function of one side of one gear tooth. The larger the gear tooth or the higher the pressure, the more torque is produced.

You may wonder why the gear teeth do not turn in the opposite direction. To rotate in the opposite direction, gear teeth would have to mesh instead of unmesh. Gears which mesh generate a decreasing volume which pushes fluid out of the housing. The gears have no other choice but to unmesh.

External Gear Motor

The gear motor which has been described above is an external gear motor; that is, both meshing gears have teeth on their outer circumferences. The type of gear used in this motor is a spur gear.

Internal Gear Motor

An internal gear motor consists of one external gear which meshes with the teeth on the inside circumference of a larger gear. A popular type of internal gear motor in industrial systems is the gerotor motor.

Gerotor Motor

A gerotor motor is an internal gear motor with an inner drive gear and an outer drive gear which has one more tooth than the inner gear. The inner gear is attached to a shaft which is connected to a load.

The imbalance in a gerotor motor is caused by the difference in gear area exposed to hydraulic pressure at the motor inlet.

In the gerotor motor illustration, the exposed area of the inner gear increases at the inlet.

Fluid pressure acting on these unequally exposed teeth, results in a torque at the motor shaft. The larger the gear, or the higher the pressure, the more torque will be developed at the shaft.

Fluid entering the rotating group of a gerotor motor is separated from the fluid exiting the motor by means of a port plate with kidney-shaped inlet and outlet ports.

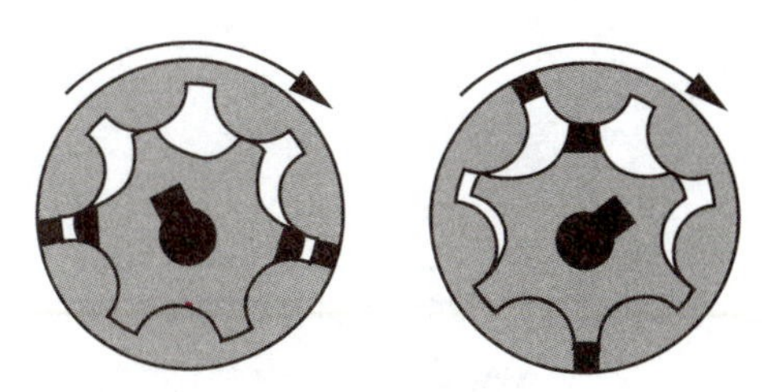

Figure 14-15

Gerotor Pump

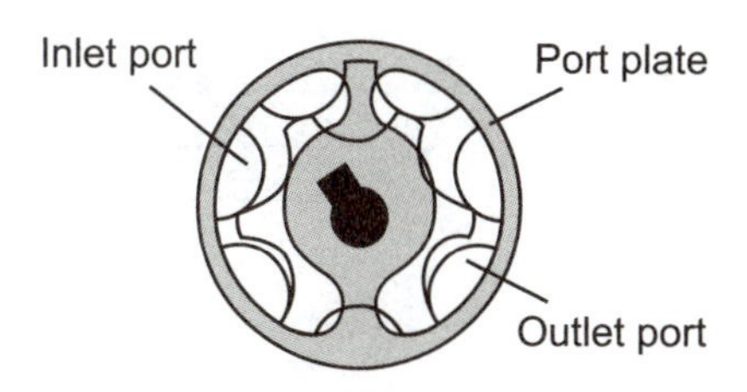

Figure 14-16

Piston Motors

A piston motor is a positive displacement motor which develops an output torque at its shaft by allowing hydraulic pressure to act on pistons.

What a Piston Motor Consists Of

The rotating group of a piston motor basically consists of swashplate, cylinder barrel, pistons, shoeplate, shoeplate bias spring, port plate, and shaft.

The pistons fit inside the cylinder barrel. The swashplate is positioned at an angle and acts as a surface on which the shoe side of the piston travels. The piston shoes are held in contact with the swashplate by the shoeplate and bias spring. A port plate separates incoming fluid from the discharge fluid. A shaft is connected to the cylinder barrel. In our example, it is attached at the port plate end.

How a Piston Motor Works

To illustrate how a piston motor works, let us observe the operation of one piston in a cylinder barrel of an axial piston motor.

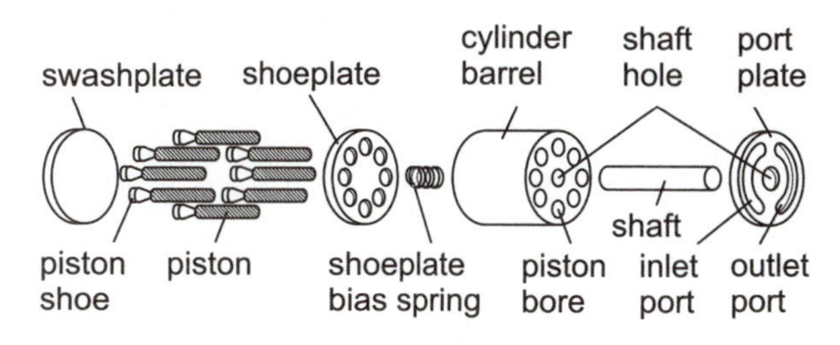

Axial Piston Pump

Figure 14-17

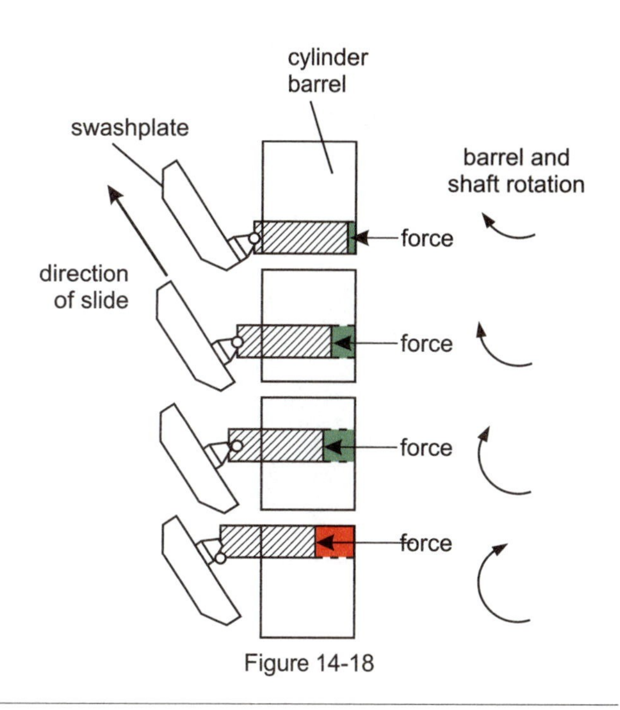

Figure 14-18

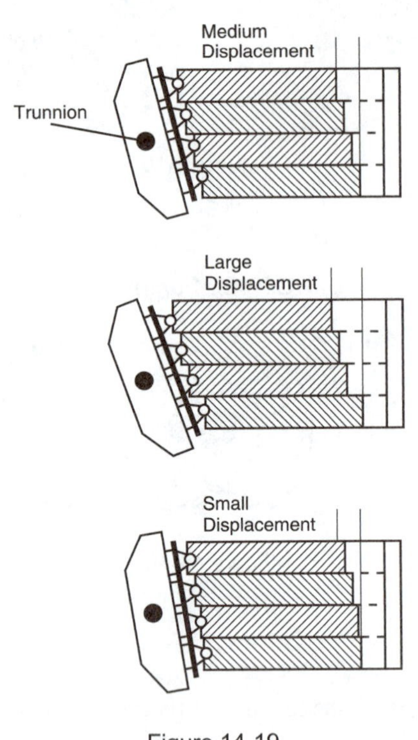

Figure 14-19

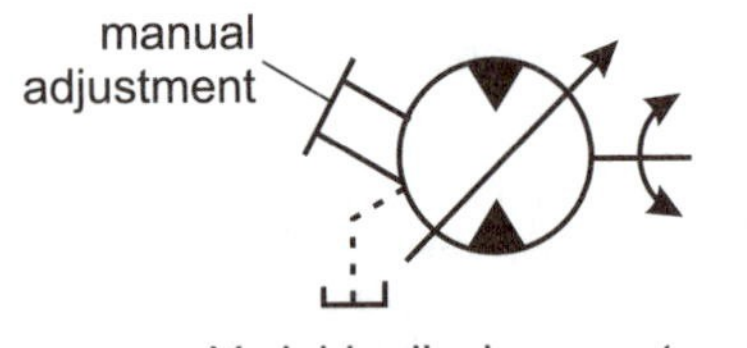

Variable displacement motor

Figure 14-20

With the swashplate positioned at an angle, the piston shoe does not have a very stable surface on which to position itself. When fluid pressure acts on the piston, a force is developed which pushes the piston out and causes the piston shoe to slide across the swashplate surface. As the piston shoe slides, it develops a torque at the shaft attached to the barrel. The amount of torque depends on the angle of slide caused by the swashplate and the pressure in the system. If the torque is large enough, the shaft will turn.

Torque continues to be developed by the piston as long as it is pushed out of the cylinder barrel by fluid pressure. Once the piston passes over the center of the circle, it is pushed back into the cylinder barrel by the swashplate. At this point, the piston bore will be open to the outlet port of the port plate.

A single piston in a piston motor develops torque for only half of the full circle of rotation of the cylinder barrel and shaft. In actual practice, a cylinder barrel of a piston motor is fitted with many pistons. This allows the motor shaft to continuously rotate as well as obtain maximum torque.

Of the vane, gear, and piston motors which have been described, only piston motors are available as variable displacement.

Variable Displacement Axial Piston Motors

The displacement of an axial piston motor, or any piston motor, is determined by the distance the pistons are reciprocated in the cylinder barrel.

Since the swashplate angle controls this distance in an axial piston motor, we need only to change the angle of the swashplate to alter the piston stroke and motor displacement.

With a large swashplate angle, the pistons have a long stroke within the cylinder barrel.

With a small swashplate angle, the pistons have a short stroke within the cylinder barrel.

By varying the angle of the swashplate then, the motor's displacement and consequently its shaft speed, and torque output can be changed.

Overcenter Axial Piston Motors

Some swashplates of axial piston motors have the capability of crossing overcenter. A motor of this type is able to reverse its shaft rotation without changing the direction of flow through the motor.

In the overcenter axial piston motor illustrated, the motor shaft is not shown. But, you can imagine it as being attached to the cylinder barrel at the port plate end through the swashplate side. We can see from the illustration that changing the angle of the swashplate by crossing overcenter results in a different direction of slide for the pistons. Consequently, cylinder barrel and motor shaft rotate in the reverse direction. This takes place with fluid flow passing through the motor in the same direction.

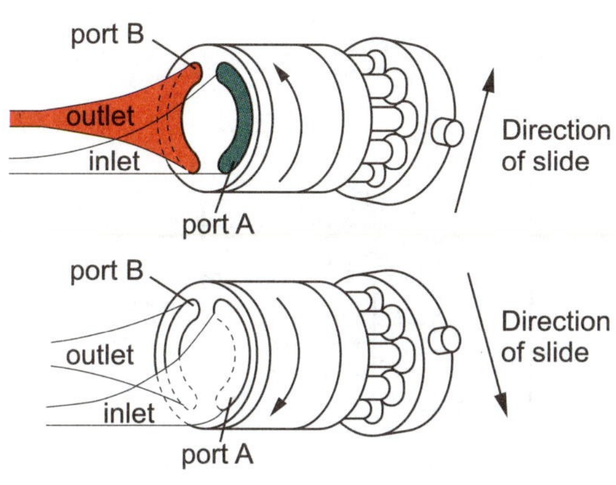

Figure 14-21

Hydraulic Motors in a Circuit

We have seen that the torque developed by a hydraulic motor is the result of hydraulic pressure acting on the motor's rotation group. In these situations, we made the assumption that no hydraulic pressure was present after the motor. Even though this side of the motor is generally drained to tank, tank line pressure, or back pressure, can be as high as 100 psi (6.9 bar) in some systems. The force generated by the back pressure on the rotating group must be overcome before a load can be turned.

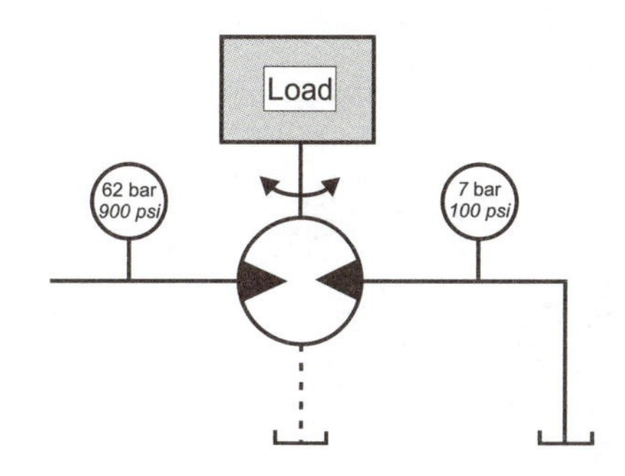

Figure 14-22

In the illustration, the load attached to the motor's shaft can be turned, theoretically, with a pressure of 800 psi (55.2 bar) at the motor inlet. The back pressure in the tank line is 100 psi (6.9 bar). With this condition, the load will not turn. An additional 100 psi (6.9 bar) must be present at the motor inlet to equal or offset the 100 psi (6.9 bar) backpressure. With 900 psi (62.1 bar) at the inlet, 900 psi (62.1 bar) is used to develop and turn the load. 100 psi (6.9 bar) offsets the back pressure. The 800 psi (55.2 bar) which was calculated to turn the load actually indicated the required pressure differential.

To accurately control the speed of a hydraulic motor, a meter-out circuit is used.

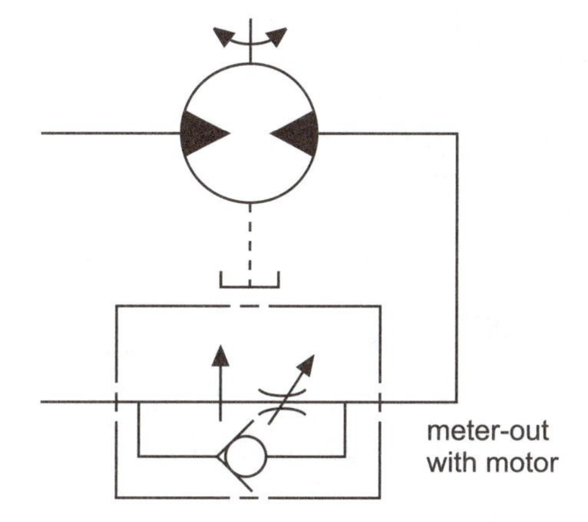

Figure 14-23

Hydraulic motors are generally externally drained. This means a portion of the flow entering the motor ends up as leakage. As the torque requirement and the pressure at the motor increases, more flow runs out the drain. As a result, motor shaft speed decreases.

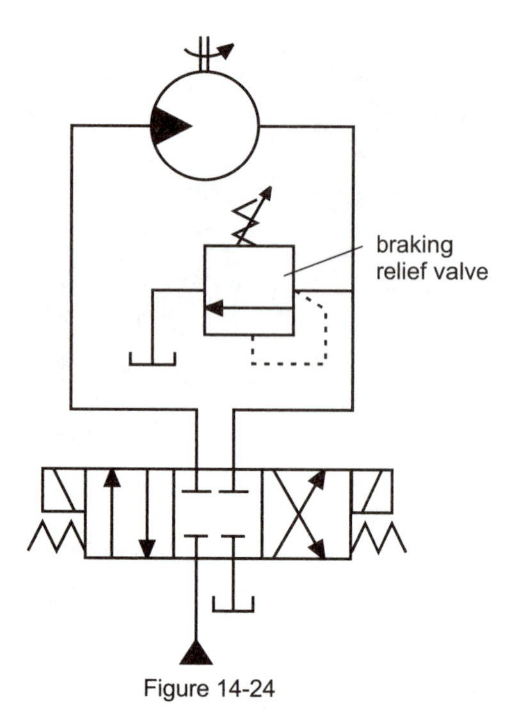

Figure 14-24

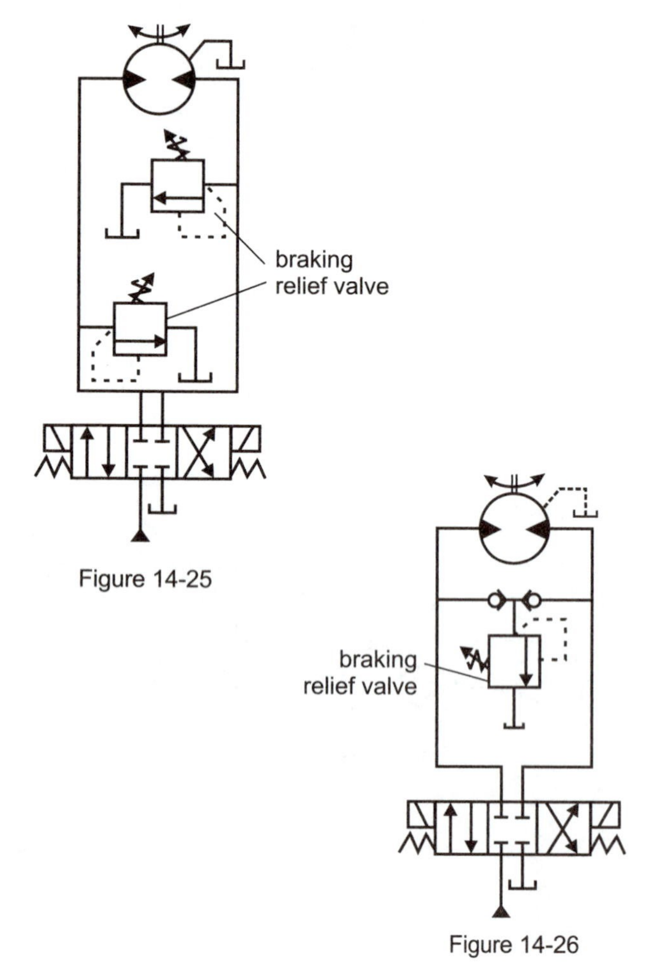

Figure 14-25

Figure 14-26

A meter-out circuit controls the flow as it discharges from the motor and is not concerned with leakage. This is the only circuit which can control a motor's shaft speed accurately regardless of load.

One of the major concerns in motor circuits is the control of the load attached to the motor shaft. We have seen previously that a brake valve will keep a load from running away as well as allow the motor to develop full torque.

A brake valve senses the load. It automatically responds to the load's demand. Many times, the braking function is required to be a matter of choice rather than automatic. For example, in a conveyor system, which has no overrunning load and requires braking only periodically, a directional valve is used to select the braking function.

Braking is performed by shifting the directional valve, usually to its center position, and blocking the flow out of the motor. When the pressure at the motor outlet increases to the braking relief valve setting, the valve opens and brakes the motor.

If the motor requires braking in both directions, a braking relief valve can be connected through check valves to both motor lines. No matter which way the motor is rotated, braking is performed by the same valve.

In some applications, two braking pressures are necessary. For example, a conveyor, which is loaded in one direction and unloaded in the opposite direction, would require two different braking pressures to make most efficient use of its cycle time.

When two different braking pressures are required, two braking relief valves are connected in the motor lines. Each valve handles flow in different directions. Braking relief valves applied in this manner can also be used to achieve approximate starting and stopping positions with dissimilar loads in opposite directions.

A braking relief valve is a common, ordinary relief valve placed in a motor line. It is not a special valve.

The setting of a braking relief valve is higher than the setting of the system relief valve.

Motor Cavitation

All of the motor circuits shown so far do not take into account that hydraulic motors can cavitate. They will cavitate just as a pump if a sufficient supply of fluid is not received at its inlet port while turning. This means that anytime a motor is braking, the motor inlet must not be blocked.

In a unidirectional motor circuit, this requirement can be met by allowing the motor inlet to be connected to tank through the center position of the directional valve.

When braking occurs, any less-than- atmospheric pressure at the motor inlet will result in fluid being drawn from the reservoir.

Make-Up Checks

In a bidirectional motor circuit, supplying liquid to the motor inlet during braking is usually done with low pressure (5 psi .34 bar or less) check valves positioned in each line. These are known as make-up check valves.

Crossover Relief Valves

A bidirectional motor circuit, using braking relief valves in both directions, can be designed so that the discharge from the relief valves is connected to the opposite motor lines. From first glance, it may appear that these "crossover" relief valves would keep the motor's inlet well supplied since the motor's discharge fluid is re-directed to motor inlet. However, make-up checks are still required since some fluid is lost through the motor drain and leakage across the directional valve. This arrangement of crossover relief valves and make-up checks is a very common bidirectional motor circuit.

Hydrostatic Drive

Hydraulic motors used in combination with various pumps is termed a hydrostatic drive. A hydrostatic drive can be either "open loop" or "closed loop."

Open Loop

An open loop hydrostatic drive has the motor inlet connected to pump outlet and the motor outlet connected to tank. The motor rotation is stopped or reversed with a directional valve. The speed of the motor depends u[on pump flow rate and motor displacement.

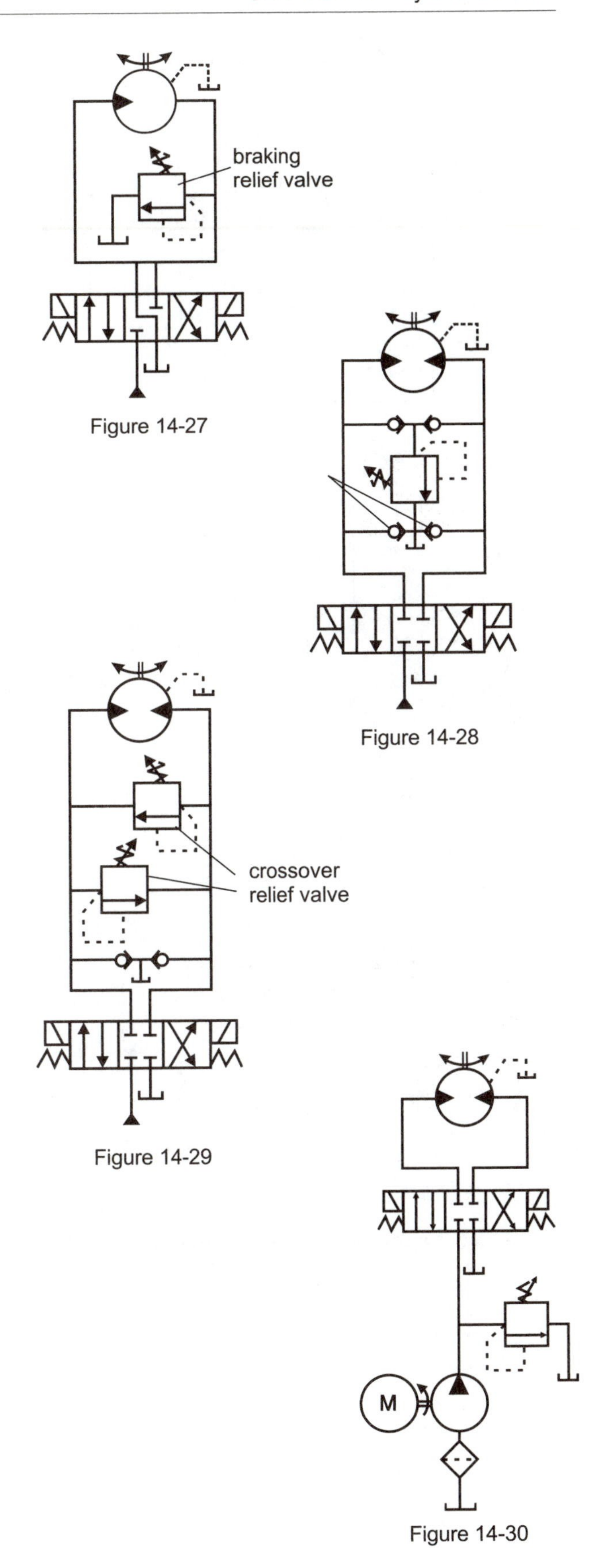

Figure 14-27

Figure 14-28

Figure 14-29

Figure 14-30

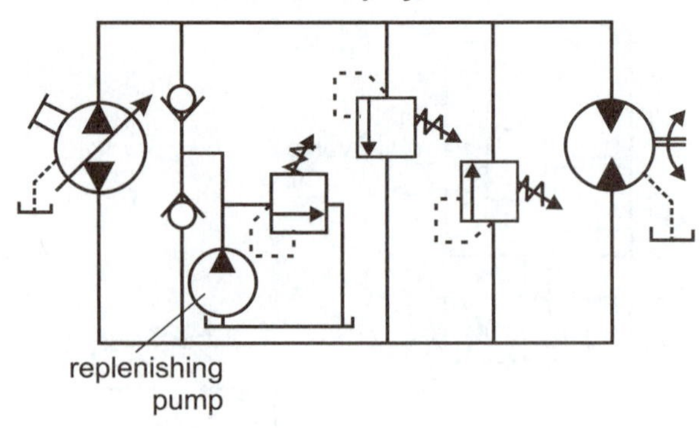

Figure 14-31

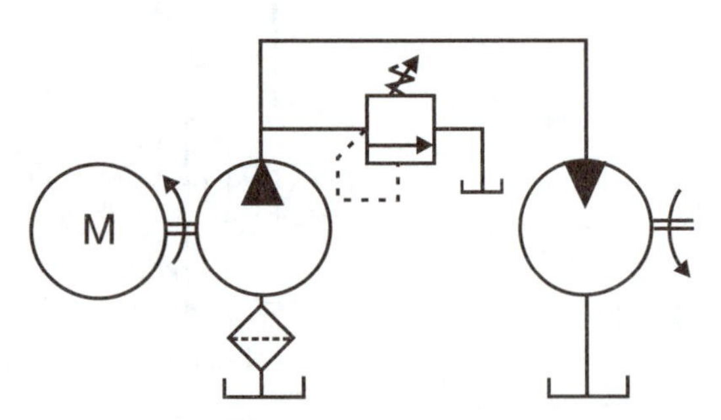

Fixed horsepower, torque and speed

Figure 14-32

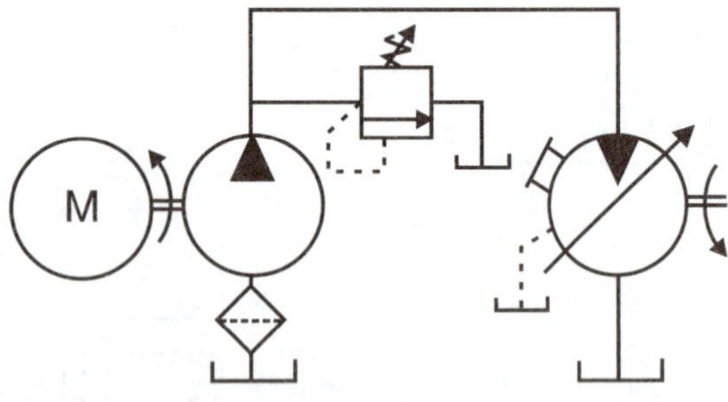

Fixed horsepower, variable speed and torque

Figure 14-33

Closed Loop

A closed loop hydrostatic drive has motor inlet connected to pump outlet and motor outlet connected to pump inlet. The closed loop schematic shows motor rotation in either direction with variable pump input which will vary motor speed and direction. Any leakage in the system is made up by the replenishing pump. A small reservoir is used in this system since most of the system fluid is carried and stored in the system piping. Closed loop hydrostatic drives are compact.

Pump-Motor Combinations

Various types of pumps and motors can be combined to achieve different system requirements.

A constant displacement pump used with a fixed displacement motor results in a fixed hydraulic horsepower being developed by the pump. Torque and shaft speed are constant at the motor. (Assume input rpm is constant.)

A constant displacement pump combined with a variable displacement motor results in a fixed hydraulic horsepower being delivered to a motor. But, shaft speed and torque are variable at the motor.

A variable displacement pump used with a fixed displacement motor results in a constant torque at the motor. Since the pump's flow rate can be changed, the horsepower delivered to the motor and the motor shaft speed can be varied.

A system using both a variable displacement pump and a variable displacement motor has the flexibility of variable speed, torque, and horsepower.

Hydrostatic Transmission

In common terminology, anytime a variable displacement pump or motor is used in a pump-motor circuit, the system is labelled a hydrostatic transmission.

In the closed loop hydrostatic transmission illustrated, the variable displacement, overcenter pump can vary the speed of the motor shaft as well as reverse shaft rotation. In closed loop systems of this nature, a small pump, called a replenishing pump, is used to make up for any leakage which occurs in the system.

Closed loop hydrostatic transmissions are compact systems. The reason being the reservoir is small and flow controls and directional valves are not needed to reverse or control the speed of shaft rotation.

Hydraulic Motors Vs. Electric Motors

Hydraulic motors have certain advantages over electric motors. Some of these advantages are:

1. instant reversing of a motor's shaft
2. stalling for indefinite periods without damage
3. torque control throughout its operating speed
4. dynamic braking easily accomplished
5. a weight to horsepower ratio of .5 lb./hp (0.2668 kg) compared to 10 lb/hp (4.536 kg) for electric motors.

Motor Torque

The amount of torque developed by a specific motor is a function of system pressure acting on the imbalance of the motor displacement. This is expressed by the formula:

$$\text{Torque (lb. in.)} = \frac{\text{pressure (lb/in}^2\text{) x motor disp. (in}^3\text{)}}{2\pi}$$

$$\text{Torque (N-m)} = \frac{\text{pressure (N/m}^2\text{) x motor disp. (m}^3\text{)}}{2\pi}$$

The formula indicates that the higher the pressure or the greater the motor displacement, the more torque will be developed. The equation for torque is very similar to the equation for cylinder force.

$$\text{Force (lbs.)} = \text{pressure (lb/in}^2\text{) x area (in}^2\text{)}$$

$$\text{Force (N)} = \text{pressure (bar) x area (m}^2\text{)}$$

Instead of using force (lbs./N), the motor formula uses torque (lb. in./N-m). psi [(lb/in^2) (N/m^2)] denotes fluid pressure in both equations. The cylinder force formula indicates area [(in^2) (m^2)]; the motor torque formula indicates motor displacement [(in^3) (m^3)]. But by dividing displacement [(in^3) (m^3)] by 2π, in^3 (m^3) displacement can be treated as the unbalanced area within the motor.

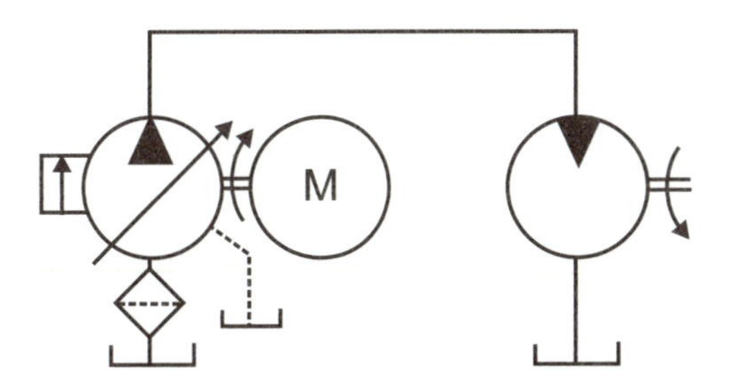

Variable horsepower and speed, constant torque

Figure 14-34

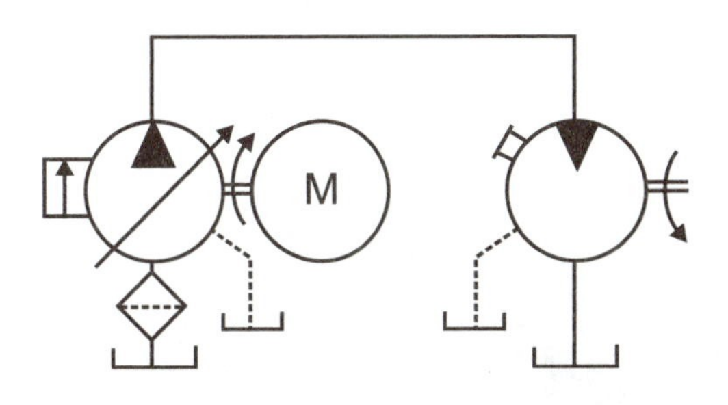

Variable horsepower, speed and torque

Figure 14-35

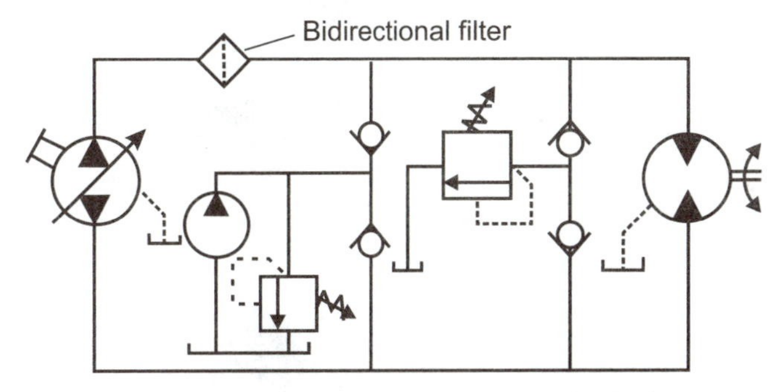

Closed loop hydrostatic transmission

Figure 14-36

Figure 14-37

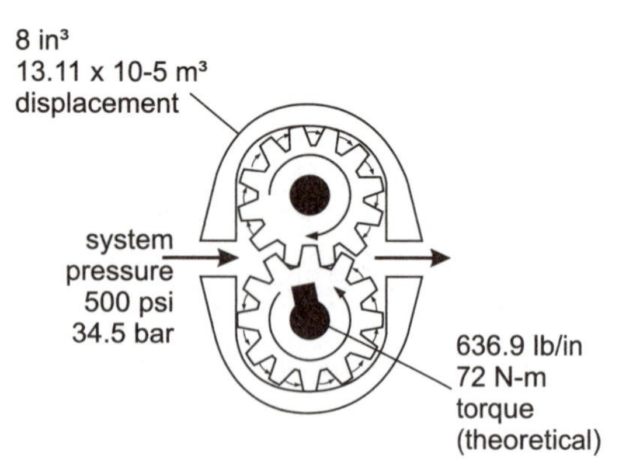

Figure 14-38

Assume that a motor with an 8 in³ (13.11 x 10^{-5} m³) displacement is subjected to 500 psi (34.48 bar). To determine the output torque in terms of lb. in., the above information is plugged into the motor torque formula.

$$\text{Torque (lb. in.)} = \frac{\text{pressure (lb/in}^2\text{) x motor disp. (in}^3\text{)}}{2\pi}$$
$$= \frac{500 \times 8}{2 \times 3.14}$$
$$= \frac{4000}{6.28}$$
$$= 636.9 \text{ lb. in. (71.9 N-m)}$$

With 500 psi (34.48 bar) present at the inlet of an 8 in³ (13.11 x 105m³) motor, the equation indicates that 636.9 lb. in. (71.9 N-m) of torque is developed at the motor shaft.

Figure 14-39

Running, Breakaway and Starting Torque

The equation for motor torque refers to theoretical torque; it does not take into account motor inefficiency.

Running torque indicates the amount of torque a motor produces to keep a load turning. Running torque may also refer to the requirement of the load to keep it turning.

When it refers to a motor, running torque indicates the actual torque that a hydraulic motor can develop to keep a load turning. It takes into account motor inefficiency and is expressed as a percentage of theoretical torque. The running torque of common gear, vane, and piston motors is approximately 90% of theoretical.

Breakaway torque refers to a demand of a load which must be satisfied before it will rotate.

Breakaway torque is the torque required of a hydraulic motor to get a load turning. More torque is required to start a load moving than to keep it moving. If a hydraulic motor is not capable of developing sufficient torque to breakaway a load, the load will not move.

Starting torque refers to a capability of a hydraulic motor; it indicates the amount of torque which a motor can develop to start a load turning. In some cases, this is much less than running torque.

Starting torque is expressed as a percentage of theoretical torque. It ranges between 60-90% of theoretical. Assume a load needs 500 lb-in (56.5

N-m) of torque to keep it turning, but the breakaway torque required is 600 lb-in (67.8 N-m). A specific motor may be capable of developing a running torque of 500 lb-in (56.5 N-m), but its starting torque capability at a maximum system pressure may only be 450 lb-in (50.85 N-m). The starting torque capability of a hydraulic motor must be equal to or greater than the breakaway requirement of the load. In this example, the motor could not turn the load.

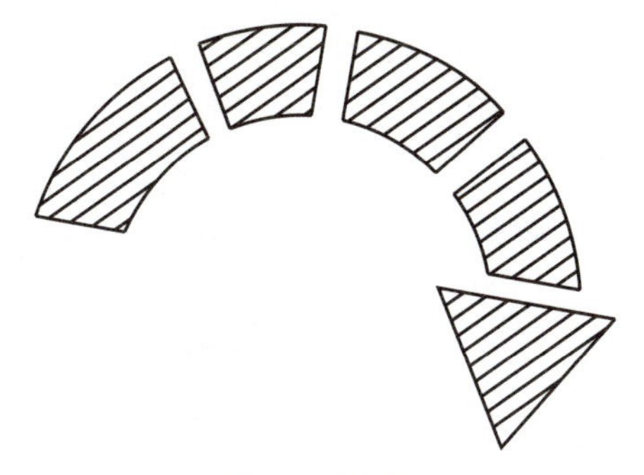

Figure 14-40

Motor Shaft Speed

Speed at which the output shaft of a hydraulic motor rotates is determined by how quickly the motor rotating group is filled with liquid. This is illustrated by the expression:

$$\text{Motor shaft speed (rev/min)} = \frac{\text{flow (gal/min)} \times 231}{\text{motor displacement (in}^3\text{/rev)}}$$

$$= \frac{\text{flow (l/min} \times 1000)}{\text{motor displacement (cm}^3\text{/rev)}}$$

The expression shows the relationship between shaft speed, flow rate, and motor displacement. The larger the flow rate, or the smaller the motor displacement, the more quickly the motor will be filled. Consequently, the more shaft revolutions will be made in one minute.

Consider an 8 in³ (131.1 cm³) displacement motor which has 15 gpm (56.85 lpm) flowing to its inlet. Motor shaft speed can be calculated by plugging the information into the shaft speed formula:

$$\text{Motor shaft speed (rev/min)} = \frac{\text{flow (gal/min)} \times 231}{\text{motor displacement (in}^3\text{/rev)}}$$

$$= \frac{\text{flow (l/min} \times 1000)}{\text{motor displacement (cm}^3\text{/rev)}}$$

$$= \frac{15 \times 231}{8 \text{ in}^3} \qquad = \frac{56.85\ 1000}{131.1 \text{ cm}^3}$$

$$= \frac{3465}{8} \qquad = \frac{56850}{131.1}$$

$$= 433 \text{ rpm} \qquad = 433 \text{ rpm}$$

With 15 gpm (56.85 lpm) flowing onto an 8 in³ (131.1 cm²) displacement motor, the formula indicates that the motor shaft speed will be 433 rpm.

The equation for motor shaft speed indicates theoretical values. It does not take into account volumetric inefficiencies which are present just as in a pump. For this reason, manufacturer's catalog must be consulted for actual values.

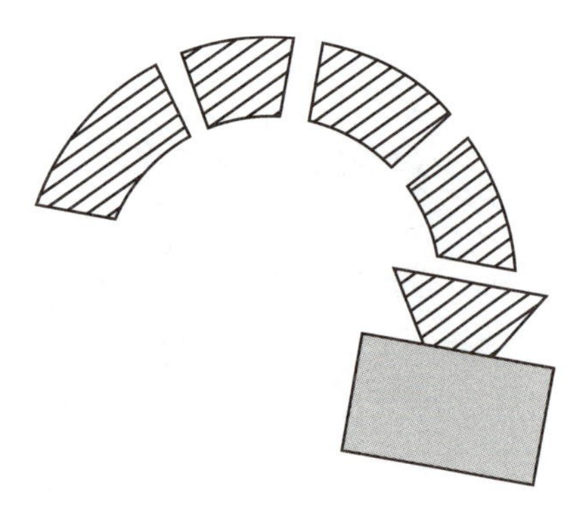

Figure 14-41

Motor Horsepower

Hydraulic horsepower is gpm (lpm) and psi (bar) flowing through a system. As hydraulic horsepower enters a motor inlet, it is converted into rotary mechanical horsepower at motor output shaft. The expression which describes this action is:

$$\text{Motor hp} = \frac{\text{shaft speed (rev/min}^2\text{) x torque. (lb. in.)}}{63025}$$

$$\text{kW} = \frac{\text{Torque T (N-m) x shaft speed (rpm)}}{9543}$$

Assume that a motor develops 620 lb-in (70 N-m) of torque at 800 rpm. To calculate motor horsepower, the equation can be used:

$$\text{Motor hp} = \frac{\text{shaft speed (rev/min}^2\text{) x torque. (lb. in.)}}{63025}$$

$$\text{kW} = \frac{\text{Torque T (N-m) x shaft speed (rpm)}}{9543}$$

$$= \frac{800 \times 620}{63025} \qquad = \frac{70 \times 800}{9543}$$

$$= \frac{496000}{63025} \qquad = \frac{56000}{9543}$$

$$= 7.9 \text{ hp} \qquad = 5.87 \text{ kW}$$

The equation indicates that a motor developing 620 lb-in (70 N-m) of torque at 800 rpm is generating 7.9 hp (5.86 kW).

Since motor power consists of values of shaft speed and torque, it might be assumed that ordinary hydraulic motors can develop power consisting of high torque at low speed. This is not the case.

The common gear, vane, or piston motor is not capable of developing smooth, low rpm shaft speeds at full torque. Because of the increased force required to start a load moving, and the increased internal motor leakage and static friction at high pressures, smooth shaft rotation at low rpm and high pressure is difficult, if not impossible to achieve. To receive smooth shaft speed as well as high torque, these motors must be operated at a minimum of 200-400 rpm.

The high operating speed of these motors generally ranges from 2400-3600 rpm.

If low shaft speed and low torque are required for an application, the ordinary hydraulic motor may be capable of satisfying the need. But, if high torque is required at a low speed, a special class of motors is used. These are identified as hi-torque low speed motors. Shaft speed for these motors ranges from 1 rpm to approximately 1000 rpm.

With basic motor types defined and with their relationships between flow and pressure indicated, we find in the next section how hydraulic motors operate in a system. We begin by seeing how motor torque is affected by backpressure.

Back Pressure Affects Motor Torque

As shown in the formula, motor torque depends on system pressure acting on the imbalance within a motor rotating group. The formula does not take into account backpres-sure at motor outlet.

In the illustrated circuit, assume a load attached to a motor shaft requires a pressure of 800 psi (55.17 bar) within the motor rotating group to overcome load resistances. This assumes that no backpressure is present at motor outlet.

With 800 psi (55.17 bar) at motor inlet and 100 psi (6.89 bar) backpressure, a force is generated on the rotating elements to turn the shaft in the opposite direction. This subtracts from the torque at the motor shaft. The motor, in effect, develops a torque equal to 700 psi (48.28 bar) acting on the rotating group.

If a motor requires 800 psi (55.17 bar) to equal load resistances, an 800 psi (55.17 bar) differential must exist from motor inlet to motor outlet. Therfore, pressure at the inlet must be at least 900 psi (62.1 bar).

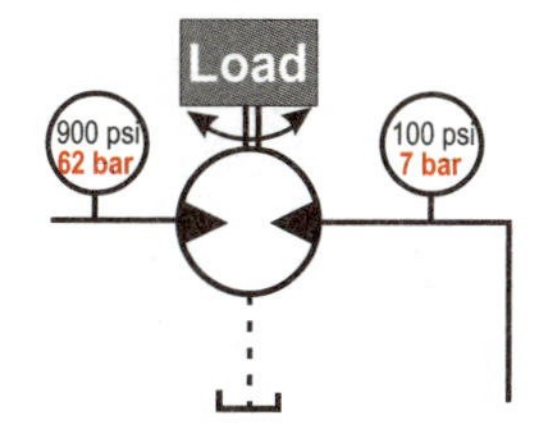

Figure 14-42

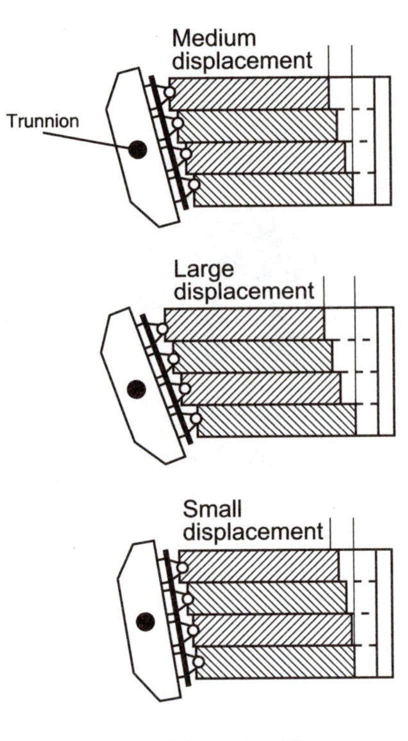

Figure 14-43

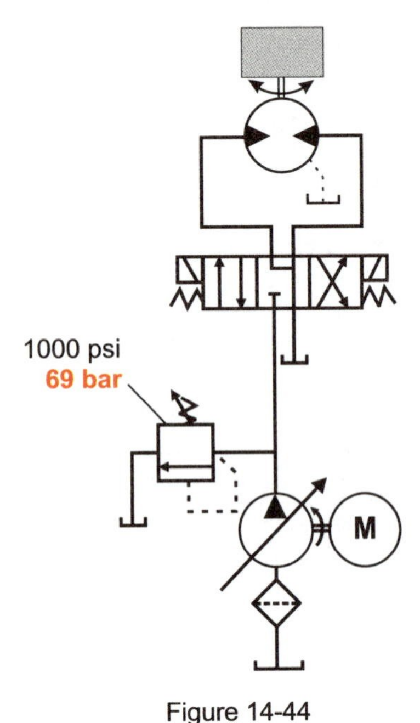

Figure 14-44

Variable Displacement Motor Affects Torque

An axial piston motor can be designed to be variable displacement. With this arrangement, output torque can be varied as required.

Maximum pressure developed by pump/electric motor is usually limited by a relief valve or a pressure compensator. With motor torque being a function of pressure and motor displacement, the maximum torque of a fixed displacement motor is reached once the relief valve or compensator setting has been reached.

In the circuit illustrated, a hydraulic motor requires a pressure of 800 psi (55.17 bar) to equal load resistances and backpressure. System relief valve is set at 1000 psi (68.97 bar).

Operating under normal conditions, pressure at motor inlet is 800 psi (55.17 bar). Now assume that motor load increases so that pressure required at motor inlet is 1100 psi (75.86 bar). With the relief valve set for a maximum pump/electric motor pressure of 1000 psi (68.97 bar), the motor stalls. Since motor torque is a function of psi (bar) and motor displacement, torque has therefore reached its maximum. Motor displacement is fixed and system pressure is maximum. With a variable displacement motor, motor displacement can be changed when this condition arises generating more of an imbalance inside of its rotating group. Consequently, more torque is developed with the same system pressure. This assumes, of course, that the motor had not been at maximum displacement initially.

Controlling Motor Speed

Shaft speed of a hydraulic motor is determined by how quickly the motor rotating group is filled with liquid. A motor receiving 5 gpm (18.95 lpm) at its inlet develops a certain shaft speed. If the same motor received 10 gpm (37.9 lpm), it would fill twice as fast; therefore, its shaft speed would be nearly twice as much.

In some cases, motor shaft speed is required to be periodically adjustable and at the same time accurate. For these reasons, a flow control valve is used.

Three types of flow control circuits were illustrated previously — meter-in, meter-out, and bleed off. Bleed off circuits generate the least amount of heat, but they are the least accurate since they only indirectly control actuator speed and cannot compensate for system leakage. Bleed off circuits are normally used in applications where control of motor speed is not critical. In illustrating motor flow control circuits, we will consider meter-in and meter-out circuits, and compare their accuracies.

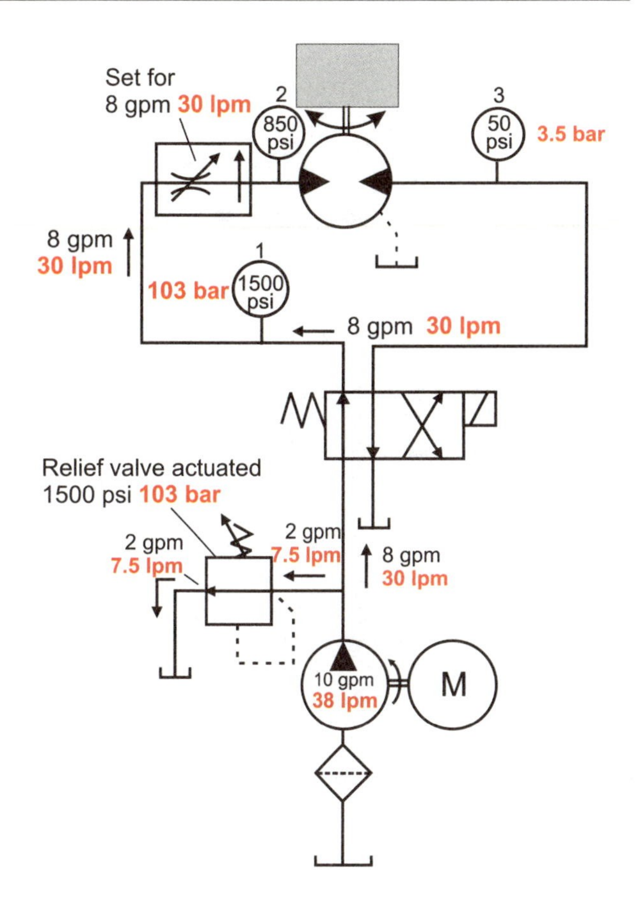

Figure 14-45

Motor Speed with a Meter-In Circuit

With a meter-in circuit, a flow control valve is positioned upstream or ahead of motor inlet; flow is controlled as it passes into the motor. In the example circuit, a pressure compensated flow control valve is used.

In the circuit, assume that the motor load requires 800 psi (55.17 bar), backpressure is 50 psi (3.45 bar), and system relief valve is set at 1500 psi (103.4 bar). Pump flow is 10 gpm (37.9 lpm), but the pressure compensated flow control is set for 8 gpm (30.32 lpm).

With the system operating, gage 1 indicates the relief valve setting of 1500 psi (103.4 bar); gage 2 indicates 850 psi (58.62 bar) ahead of the motor; and gage 3 reads a 50 psi (3.45 bar) backpressure. The pressure compensated flow control is measuring or metering 8 gpm into the motor.

The 8 gpm (30.32 lpm) flowing into the motor is not all used to fill the motor rotating group; some of it is wasted. Hydraulic motors have internal leakage just as pumps. As pressure differential across a motor increases, leakage also increases. To avoid fluid accumulation in the housing, hydraulic motors are many times externally drained, but may be internally drained as well.

As fluid enters a hydraulic motor, leakage occurs between clearances of stationary parts or across rotating elements. In vane and gear motors, fluid in an external drain is only leakage between clearances of stationary parts. Leakage fluid across vanes or gear teeth pass into other vane or gear chambers and out the outlet. With an axial piston motor, leakage across the pistons ends up in the housing along with leakage through stationary parts. An externally drained axial piston motor separates all internal leakage from motor discharge flow.

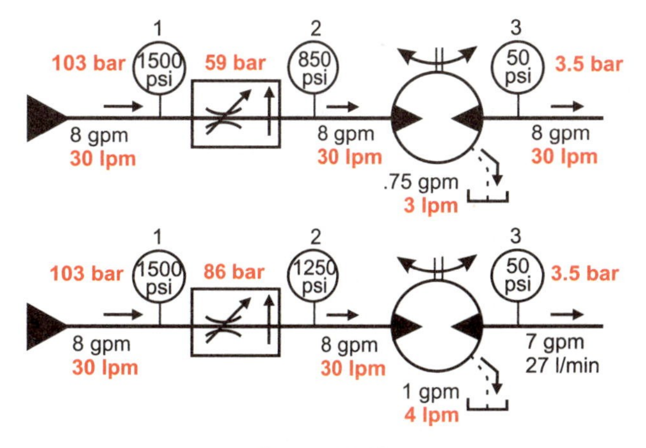

Figure 14-46

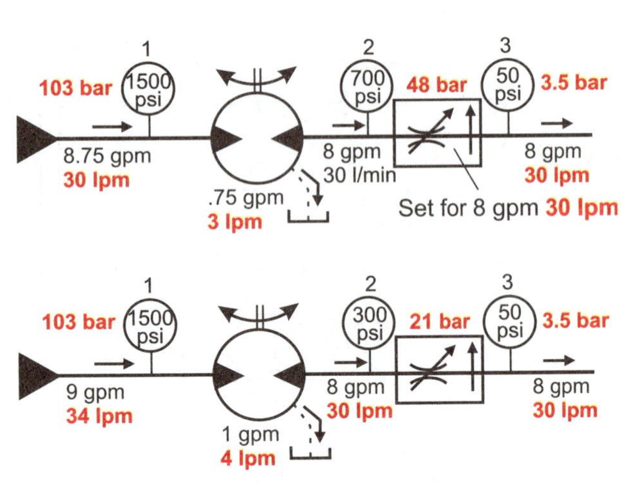

Figure 14-47

With an axial piston motor in our example, it can be seen that as 8 gpm (30.32 lpm) entered motor inlet, .75 gpm (2.8 lpm) leaked out the drain at a pressure differential of 800 psi (55.17 bar). 7.25 gpm (27.48 lpm) discharged from the motor outlet indicating that 7.25 gpm (27.48 lpm) determined motor shaft speed.

In the next illustration, motor load has increased to 1200 psi (82.76 bar). Gage 1 indicates the 1500 psi (103.4 bar) relief valve setting once again; gage 2 indicates 1250 psi (86.21 bar); gage 3 reads a backpressure of approximately 50 psi (3.45 bar). The pressure compensated flow control is still metering 8 gpm (30.32 lpm) into the piston motor. However, with a pressure differential of 1200 psi (82.76 bar) across the motor, leakage increases to 1 gpm (3.73 bar). Now 7 gpm (26.53 lpm) discharges from the motor oulet indicating that 7 gpm (26.53 lpm) is causing shaft rotation. Shaft speed reduces with an increased load.

Motor Speed with a Meter-Out Circuit

Metering flow into a motor does not provide accurate speed control. As load resistance and pressure differential increase, motor shaft speed slows. To receive a more accurate speed control of hydraulic motors, a meter-out circuit is generally used.

In the illustration, a pressure compensated flow control valve is set to measure or meter flow out of an externally drained axial piston motor. The valve is adjusted for 8 gpm (30.32 lpm).

With the system operating, gage 1 indicates the 1500 psi (103.4 bar) relief valve setting; gage 2 indicates a 700 psi (48.3 bar) backpressure at motor outlet; gage 3 reads a tank pressure of 50 psi (3.45 bar). A pressure differential of 800 psi (55.17 bar) across the motor overcomes load resistances, but also generates a leakage of .75 gpm (2.8 lpm). 8.75 gpm (33.2 lpm) enters the motor. Leakage out the drain is .75 gpm (2.8 lpm). Eight gpm (30.32 lpm) discharges from motor outlet indicating that 8 gpm (30.32 lpm) is used to develop shaft speed.

In the next illustration, motor load has increased. The motor now requires a pressure differential of 1200 psi (82.8 bar) to equal load resistances. Gage 1 indicates the relief valve setting of 1500 psi (103.4 bar); gage 2 indicates 300 psi (20.7 bar); and gage 3 reads a tank pressure of 50 psi (3.45 bar). With a pressure differential of 1200 psi (82.8 bar), 1 gpm leaks out the drain. Nine gpm (34.11 lpm) enters motor inlet; 8 gpm (30.32 lpm) discharges motor outlet through the flow control valve back to tank. Eight gpm (30.32 lpm), the discharge flow, indicates the amount of fluid turning the piston motor shaft. Since this is the same flow rate as when pressure differential was 800 psi (55.17 bar), motor shaft speed remains the same.

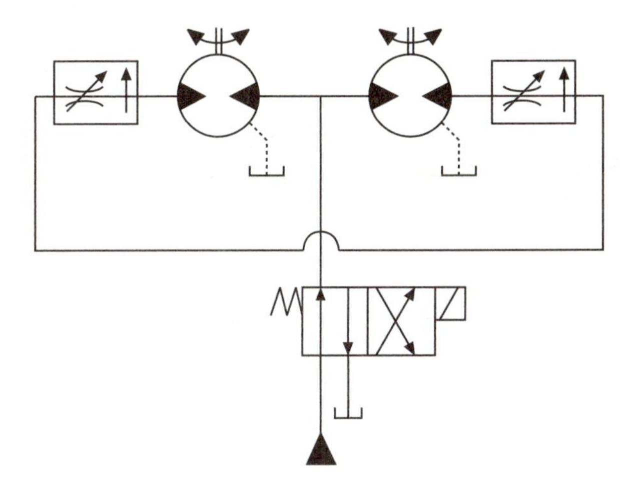

Figure 14-48

To achieve accurate speed control with an externally drained motor, a meter-out circuit is used. Since internally drained motors do not separate any leakage from motor discharge flow, speed cannot be accurately controlled as load resistance changes.

Synchronizing Two Motors

A very difficult thing to accomplish in hydraulics is synchronizing the shaft speed of two hydraulic motors.

In the illustrated circuit, a pressure compensated flow control is positioned at the outlet of each hydraulic motor; they are both adjusted for 8 gpm (30.32 lpm). Assume that the motors are of the axial piston variety. Even though the meter-out arrangement compensates for motor leakage, flow control valves will react slightly different to the same set of conditions which affects shaft speed. This brings the motor speeds out of synchronization.

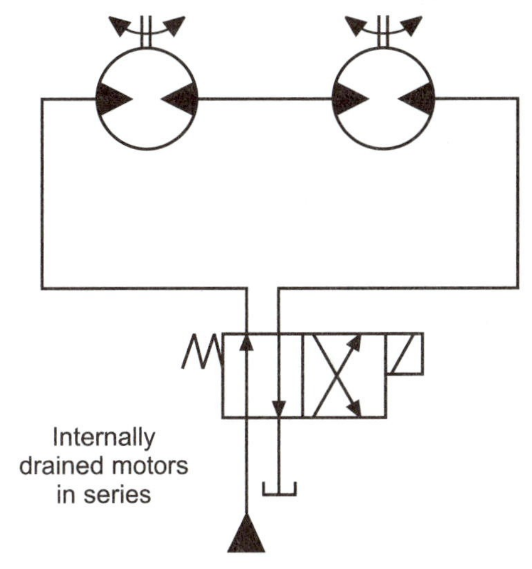

Figure 14-49

In some cases, two motors are used in series. The outlet of one motor feeds the inlet of the other. If they are externally drained, the second motor in the series will always run slower than the first motor because of leakage. Even if the motors are internally drained, because of differences in machining tolerances, leakage rates will probably be different and consequently shaft speeds vary.

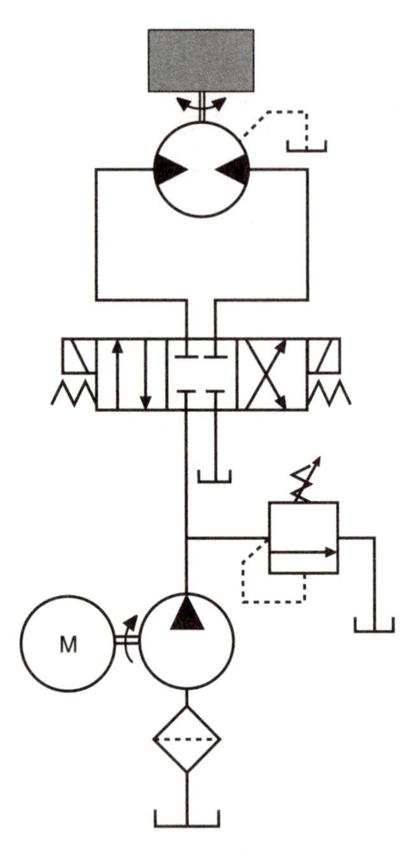

Figure 14-50

Achieving Position with a Motor

When the load attached to a motor shaft is allowed to coast to a stop, it is referred to as freewheeling.

In some situations, the load is required to achieve some sort of mechanical position as it comes to a stop. This can be accomplished to a degree with a directional valve.

In the illustrated circuit, a closed center directional valve is used to control the reversing motion of a hydraulic motor. With the valve shifted so that parallel arrows are in the circuit, motor shaft and load rotate in a clockwise direction. When the load is required to stop, the directional valve is centered. If the load in this application is light, it will appear to stop instantly and a slight system shock will be generated. A heavy load presents another problem.

With a heavy load attached to a motor shaft, an excessive shock will be generated as the directional valve is centered. This results from load inertia which attempts to keep pushing fluid from the motor outlet.

With a heavy load developing a considerable push on the fluid and no place for it to go, a high shock pressure is generated at motor outlet. With a heavy load, the load will not appear to stop immediately. Depending upon its inertia, it will continue to rotate as long as enough pressure is developed to cause motor leakage. This may continue for one, two, or more seconds.

A hydraulic motor is not used to hold a load rigidly in place. Because of the leakage it can generate within a motor and associated valving, a heavy load will continue to drift once it has been stopped. If the load must be held in place, a mechanical device such as a brake must be used.

Hydraulic Motor Wear

Hydraulic motors wear just as pumps or any other rotating elements. Wear in a hydraulic motor shows up as increased leakage and reduced shaft speed.

As illustrated earlier, while motors operate, internal leakage is generated by pressure differential across clearances of stationary parts or rotating elements. In a gear motor, this occurs between gear teeth, side plates, and housing. In a vane motor, leakage occurs between rotor, vanes, cam ring, and port plate.

In a piston motor, leakage occurs between cylinder barrel, pistons, and port plate. As wear increases, leakage increases especially across rotating elements. With the same flow entering a worn motor, shaft speed decreases. Approximately the same pressure is required to equal load resistances; but, since leakage paths have increased in size, more flow can be pushed through the clearance with the same pressure.

Finally, if wear is allowed to continue, a point is reached where all inlet flow goes into leakage at a pressure less than work pressure. This results in no movement at the motor shaft.

Assume a load requires 800 psi (55.17 bar) differential across a motor to equal resistances. As wear causes leakage to increase, motor speed decreases. Finally, when the motor is excessively worn, a pressure differential of less than 800 psi (55.17 bar) can push all incoming flow back to tank. The motor stalls.

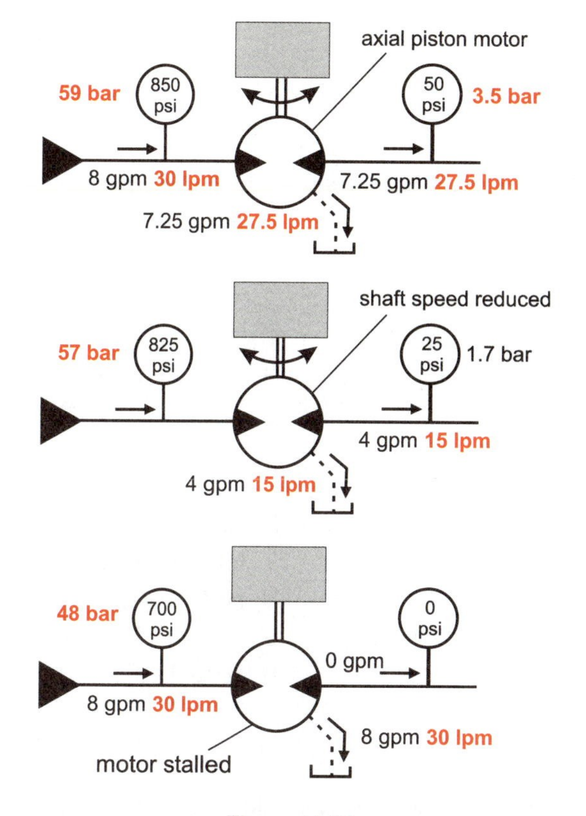

Figure 14-51

Check for Hydraulic Motor Wear

A check for hydraulic motor wear is performed by comparing drain flows and/or shaft speeds.

With an externally drained axial piston motor, all internal leakage discharges through the drain. If drain flow is excessive as compared to original conditions, motor wear is excessive. This can be checked with a flow meter or by replacing the normal external drain line with a hose. With the motor loaded, the open end of the hose can be held over a container and its discharge flow estimated or timed as it fills the container.

Since the external drain of a gear or vane motor only indicates leakage through clearances of stationary parts, drain flow could have increased relatively little between original and worn conditions. However, fluid bypassing gears and vanes could have increased significantly. Motor wear in these motors can be only indirectly determined by comparing present motor shaft speeds with original shaft speed. The same is true with an internally drained motor since leakage is not separated from motor discharge at all.

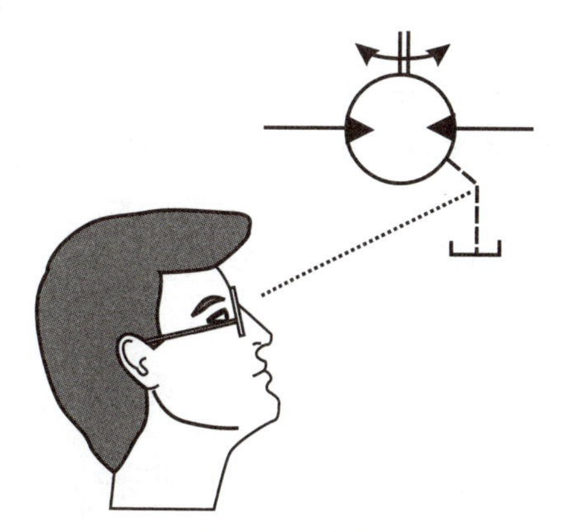

Figure 14-52

Exercise
Hydraulic Motors
50 Points

INSTRUCTIONS: In the maze of letters are hidden words which are the answers to the questions below. Circle the words.

```
M E G A I P L T P P O S R C A Q R P J
O S V I G I S R O R D T P A B I C E W
T H I C K S M L X I M P G V M F L X L
H T N Q S T B F N M Y S E I X R M T S
V S W Y U O C A M B J N R T L E T E P
P C E M F N H M C R X T O A C E S R A
L V J Q K C L O S E D K T Y V W P N N
Z N A L U F X P V A C G O I R H A A R
T H O R K E G A X K A E R O I E J L I
R B L P I D N J G A A R X N P E Q F C
J W F X E A Y C E W E W T T C L K C F
B B O P E N B V E A D U P A E I T M H
S B W Y U S R L N Y L R S C T N V E R
M T V D R I V E E B N C F U F G R J I
```

1. The internal check valve in a vane motor which uses fluid pressure to extend its vanes, performs the function of a(n) ____________ valve.
2. A motor with its inlet connected to pump outlet and its outlet conected to the reservoir is a(n) ____________ loop system.
3. A replenishing pump is generally used in a(n) ____________ loop system.
4. A ____________ type motor can be variable displacement.
5. A hydrostatic transmission with variable horsepower, speed, and torque consists of a variable displacement pump and a ____________ displacement motor.
6. Make-up check valves are used to avoid hydraulic motor ____________.
7. ____________ indicates that a load attached to a motor's shaft is allowed to coast to a stop.
8. Hydraulic motors used in combination with various pumps is known as a hydrostatic ____________ .
9. ____________ torque is the torque required of a hydraulic motor to get its load moving.
10. A bi-directional motor generally has a(n) ____________ drain.

CHAPTER 15

Reservoirs, Coolers and Filters Hydraulic Reservoirs

The obvious function of a hydraulic reservoir is to contain or store a system's hydraulic fluid.

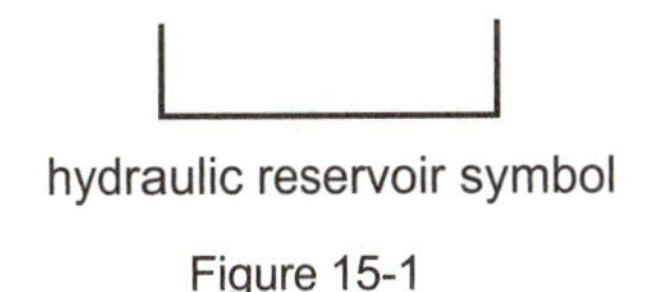

Figure 15-1

What a Hydraulic Reservoir Consists Of

In an industrial hydraulic system, where space is not a problem and consideration can be given to good design, hydraulic reservoirs consists of four walls (usually steel); a dished bottom; a flat top with mounting plate; four legs, suction, return, and drain lines; drain plug; oil level gage; filler/breather cap; cleanout covers; and baffle plate.

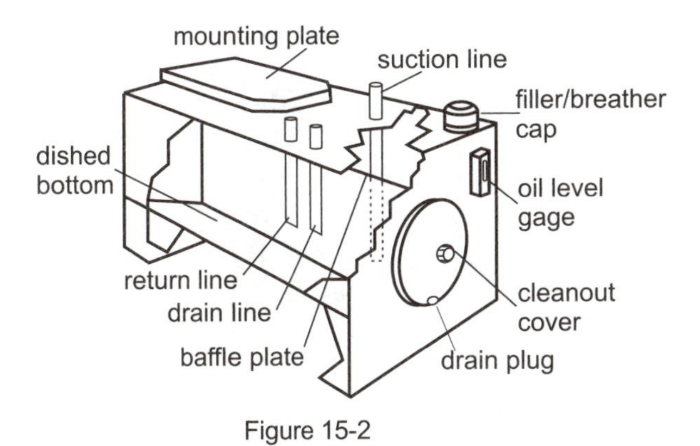

Figure 15-2

How a Reservoir Works

Besides acting as a fluid container, a reservoir serves also to cool the fluid, to allow contamination to settle out, and to allow entrained air to escape.

With fluid returning to a reservoir, a baffle plate blocks the returning fluid from going directly to the suction line. This creates a quiet zone which allows large dirt to settle out, air to rise to the fluid surface, and gives a chance for the heat in the fluid to be dissipated to the reservoir walls.

Fluid baffling is a very important part of proper reservoir operation. For this reason, all lines which return fluid to the reservoir should be located below fluid level and at the baffle side opposite the suction line.

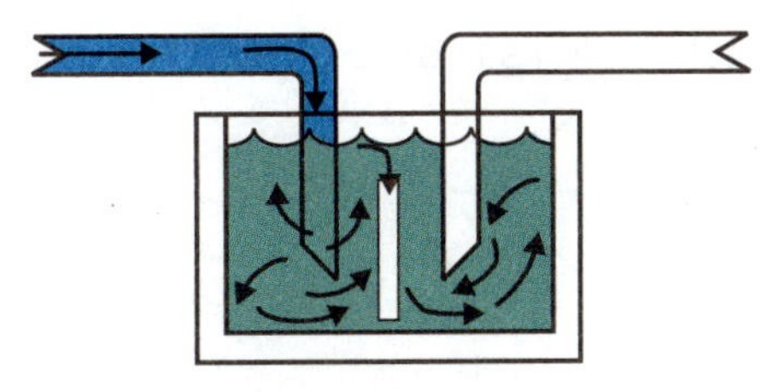

Figure 15-3

Reservoir Types

Industrial reservoirs come in a variety of styles among which are the L-shaped reservoir, the overhead reservoir, and the conventional reservoir.

The conventional reservoir is the most commonly used industrial hydraulic reservoir.

Overhead and L-shaped reservoirs afford the pump a positive head of fluid.

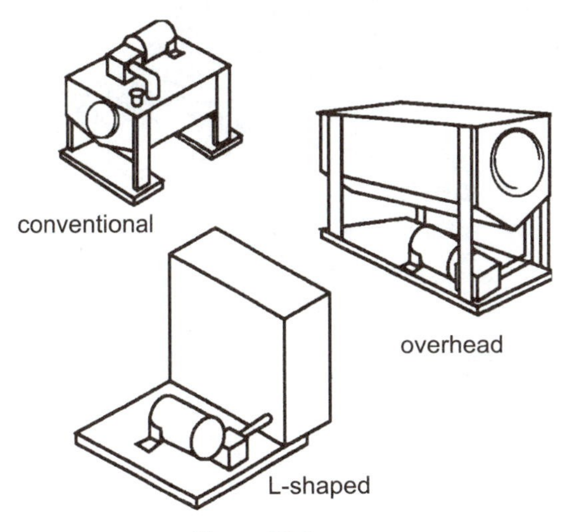

Figure 15-4

Terms and Idioms Associated with Reservoirs

SUMP - hydraulic reservoir

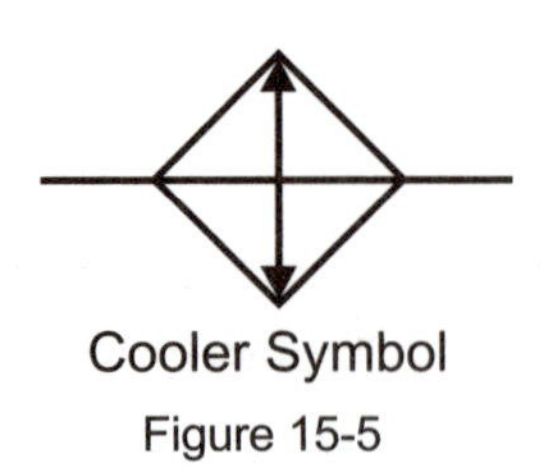

Cooler Symbol

Figure 15-5

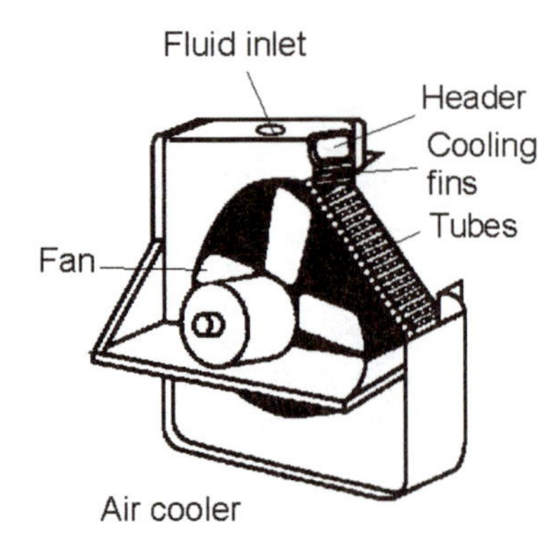

Figure 15-6

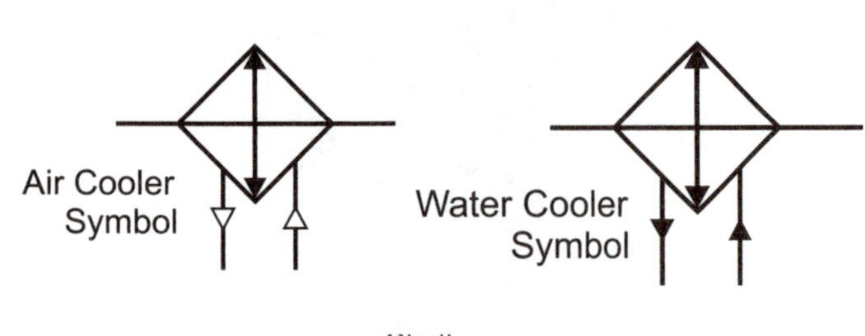

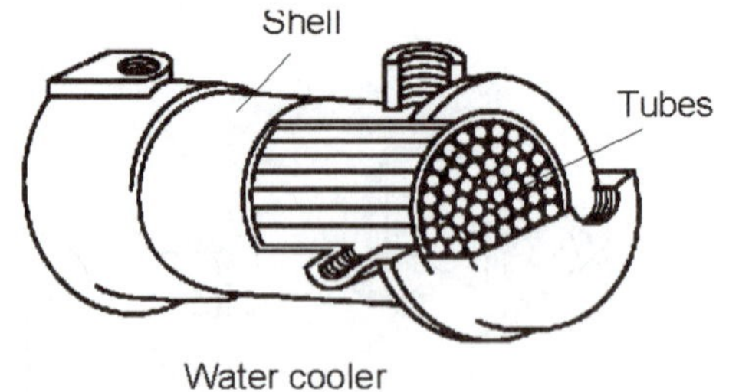

Figure 15-7

Coolers

Inefficiency in the form of heat can be expected in all hydraulic systems. Even well designed hydraulic systems can be expected to turn some portion of its input horsepower into heat. Hydraulic reservoirs are sometimes incapable of dissipating all this heat. In these cases a cooler is used.

Coolers are divided into air coolers and water coolers.

Air Cooler

In an air cooler, fluid is pumped through tubes to which fins are attached. To dissipate heat, air is blown over the tubes and fins by a fan. The operation is exactly like an automobile radiator.

Air coolers are generally used where water is not readily available or too expensive.

Water Cooler

A water cooler basically consists of a bundle of tubes encased in a metal shell. In this cooler, a system's hydraulic fluid is usually pumped through the shell and over the tubes which are circulated with cooling water.

This cooler is also known as a shell-and-tube type heater exchanger. It is a true heat exchanger since hydraulic fluid can also be heated with this device by simply running hot water through the tubes.

Coolers in a Circuit

Coolers are usually rated at a relatively low operating pressure (150 psi, 10.34 bar). This requires that they be positioned in a low pressure part of a system. If this is not possible, the cooler may be installed in its own separate circulating system.

To insure that a pressure surge in a line does not damage a shell-and-tube type cooler, they are generally piped into a system in parallel with a 65 psi check valve.

Coolers can be located in a system's return line, after a relief valve, or in a case drain line of a variable volume, pressure compensated pump.

Hydraulic Filters

The components which have been discussed and the circuits which have been illustrated to this point will function as described and perform the job they were intended to do as long as the fluid is clean. The best designed components and the most carefully thought out circuits require clean fluid to achieve optimum performance.

All hydraulic fluids contain dirt to some degree. But, the need of a filter in a system is many times not recognized. After all, the addition of this particular component does not increase a machine's apparent actions. But, this text would be sorely lacking if it did not clearly point out that dirt in hydraulic fluid is the downfall of even the best designed hydraulic systems. As a matter of fact, experienced maintenance men agree that the great majority of component and system malfunctions is caused by particles of dirt. Dirt particles can bring huge and expensive machinery to its knees.

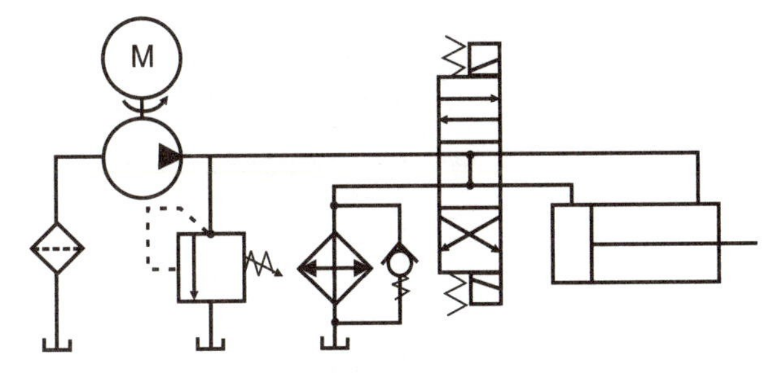

Figure 15-8

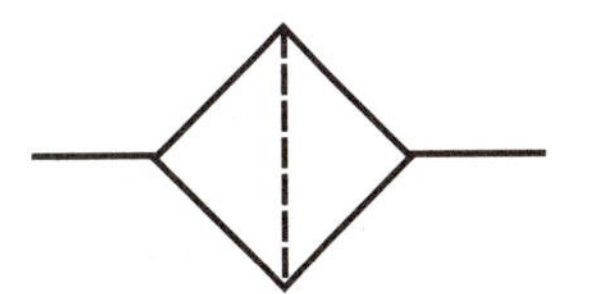

Filter Symbol

Figure 15-9

Dirt Interferes with Hydraulic Fluid

Dirt causes trouble in a hydraulic system because it interferes with the fluid which has four functions:

1. to act as a medium for energy transmission
2. to lubricate internal moving parts of hydraulic components
3. to act as a heat transfer medium
4. to seal clearances between close fitting moving parts

Dirt interferes with three of these functions. Dirt interferes with the transmission of energy by plugging small orifices in hydraulic components like pressure valves and flow control valves. In this condition pressure has a difficult time passing to the other side of the spool. The valve's action is not only unpredictable and non-productive, but unsafe.

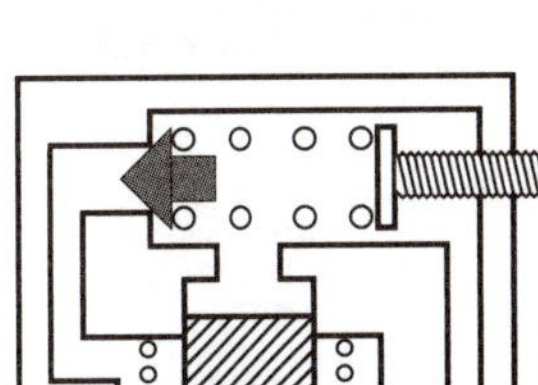

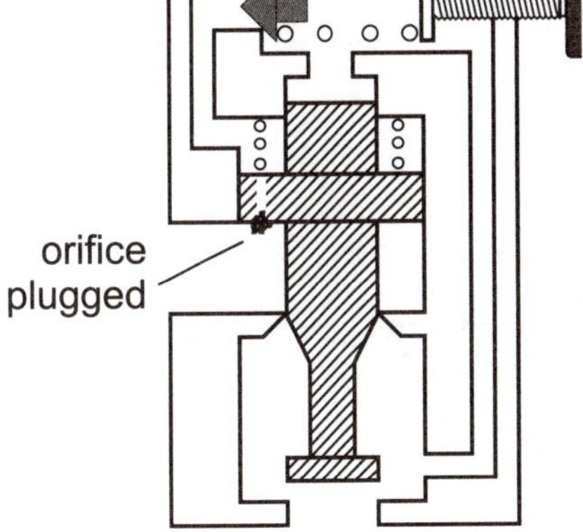

Figure 15-10

Because of viscosity, friction, and changing direction, hydraulic fluid generates heat during system operation. When the liquid returns to the reservoir, it gives the heat up to the reservoir walls. Dirt particles interfere with liquid cooling by forming a sludge which makes heat transfer to reservoir walls difficult.

Clean hydraulic systems run cooler than dirty systems. Probably the greatest problem with dirt in a hydraulic system is that it interferes with lubrication.

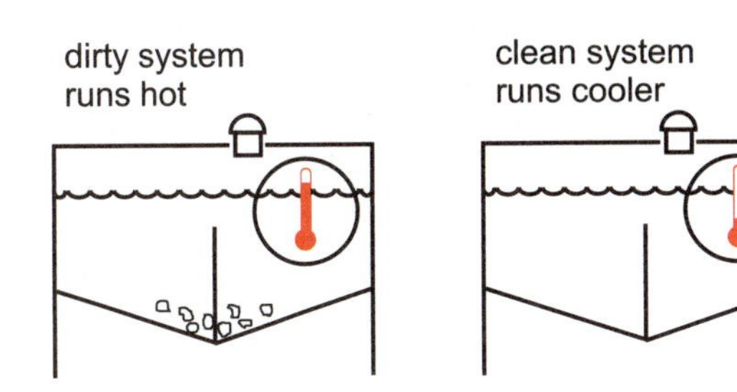

Figure 15-11

Dirt can be divided into three sizes with respect to a particular component's clearances; that is, dirt which is smaller than a clearance, dirt which is the same size, and dirt which is larger than a clearance.

Extremely fine dirt, which is smaller than a component's clearances, can collect in clearances especially if there are excessive amounts and the valve is not operated frequently. This blocks or obstructs lubricative flow through the passage. An accumulation of extremely fine dirt particles in a hydraulic system is known as silting.

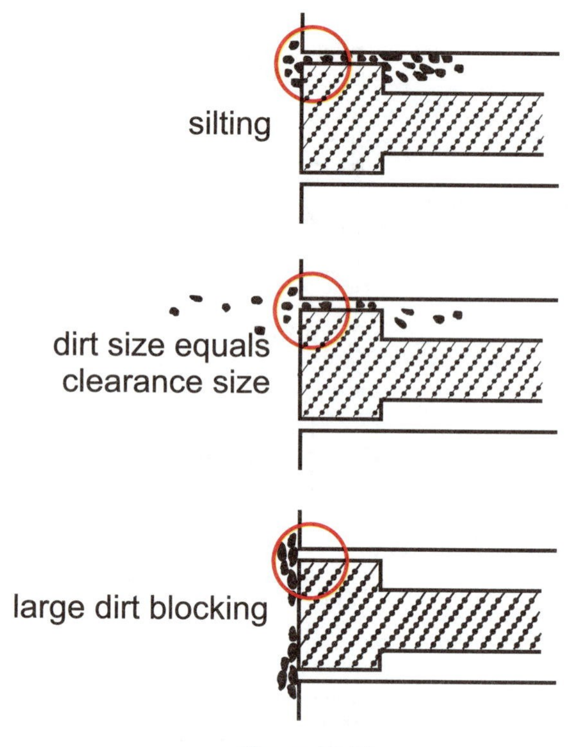

Figure 15-12

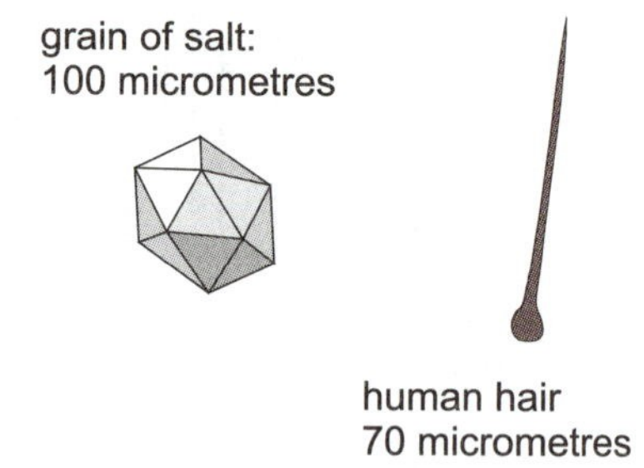

Figure 15-13

Dirt which is about the same size as a clearance rubs against moving parts breaking down a fluid's lubricative film.

Large dirt can also interfere with lubrication by collecting at the entrance to a clearance and blocking fluid flow between moving parts.

A lack of lubrication causes excessive wear, slow response, erratic operation, solenoid burn out, and early component failure.

Dirt is Pollution

Dirt in a hydraulic system is pollution. It is very similar to bottles, cans, paper and old tires floating in your favorite river or stream. The difference is that hydraulic system pollution is measured using a very small scale. The micrometre scale is used to measure dirt in hydraulic systems.

The Micrometre Scale

One micrometre (micron) is equal to one millionth of a meter or thirty-nine millionths of an inch. A single micrometre is invisible to the naked eye and is so small that to imagine it is extremely difficult. To bring the size more down to earth, some everyday objects will be measured using the micrometre scale.

- An ordinary grain of table salt measures 100 micrometres (μm).
- The average diameter of human hair measures 70 micrometres (μm).
- Twenty-five micrometres is approximately one thousandth of an inch.

Limit of Visibility

The lower limit of visibility for a human eye is 40 micrometres. In other words, the average person can see dirt which measures 40 micrometres and larger. This means that just because a sample of hydraulic fluid looks clean, doesn't necessarily mean that it is clean. Much of the harmful dirt in a hydraulic system is below 40 micrometres.

Determination of Fluid Cleanliness

Since human vision is not a proper judge, cleanliness of hydraulic fluid is determined by examining a sample of a system's fluid by visual inspection with a microscope or with the use of an automatic particle counter.

In each of these methods, the number of particles in a micrometre size range is the determining factor for cleanliness.

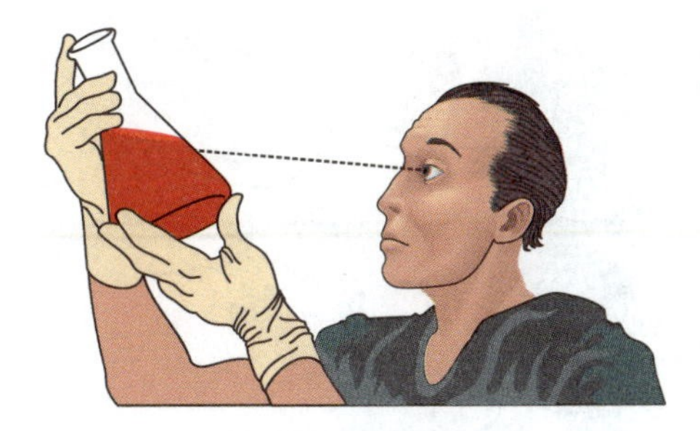

40 Micrometres

Figure 15-14

Filter Elements

The function of a mechanical filter is to remove dirt from hydraulic fluid. This is done by forcing the fluid stream to pass through a porous filter element which catches the dirt.

Filter elements are divided into depth and surface types.

Depth Type Elements

Depth type elements force the fluid to pass through an appreciable thickness of many layers of material. The dirt is trapped because of the intertwining path the fluid must take.

Treated paper and synthetic materials are commonly used porous media for depth elements.

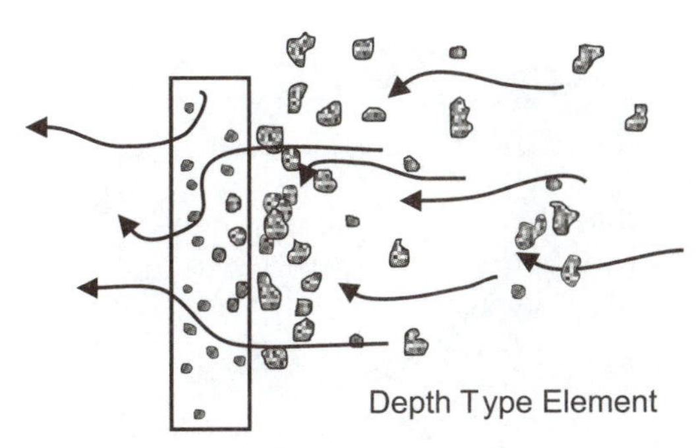

Figure 15-15

Pore Size in Depth Type Elements

Because of its construction, a depth type filter element has many pores of various size. This fact is shown by the pore size distribution curve.

A point on the curve is the number of pores per unit area of a given size in a typical depth type element.

The shape of the curve shows that there are a great deal more pores of small size than of relatively large size. This means that a large percentage of flow passes through relatively small holes.

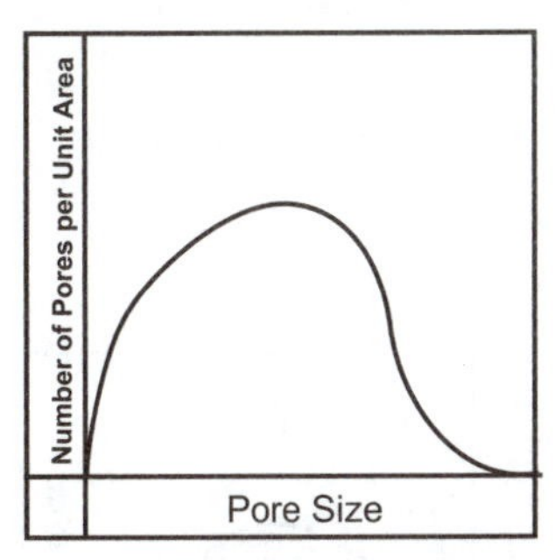

Port Size Distribution for Depth Type Element

Figure 15-16

Nominal Rating

Since there is no one, consistent hole or pore size in a depth type element, it is given a nominal rating which is based on its average pore size.

For example, a depth element with a nominal rating of 40 micrometres means that the element's average pore size is at least 40 micrometres and that initially it will remove dirt of 40 micrometres and larger and will not remove some dirt which is smaller than 40 micrometres.

NOTE: Some manufacturers do not use an element's average pore size as a basis for a nominal rating. In these cases, the nominal rating is usually an arbitrary value which indicates little.

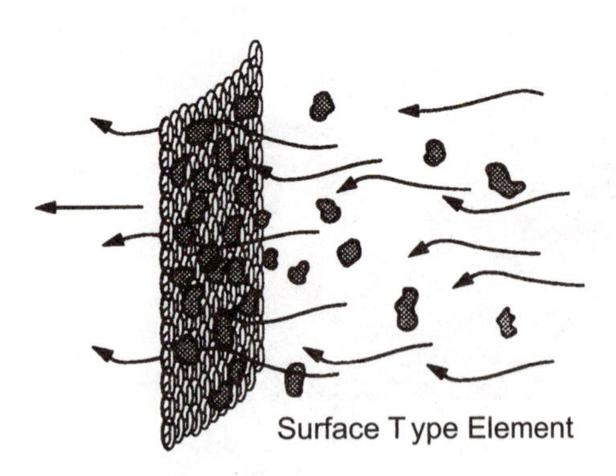

Figure 15-17

Surface Type Elements

In a surface type filter element a fluid stream has a straight flow path through one layer of material. The dirt is caught on the surface of the element which faces the fluid flow.

Wire cloth and perforated metal are common types of materials used in surface elements.

Pore Size in Surface Type Elements

Since the process used in manufacturing the cloth material and the perforated metal can be very accurately controlled, surface type elements have a consistent pore size. Because of this fact, surface type elements are usually identified by their absolute rating.

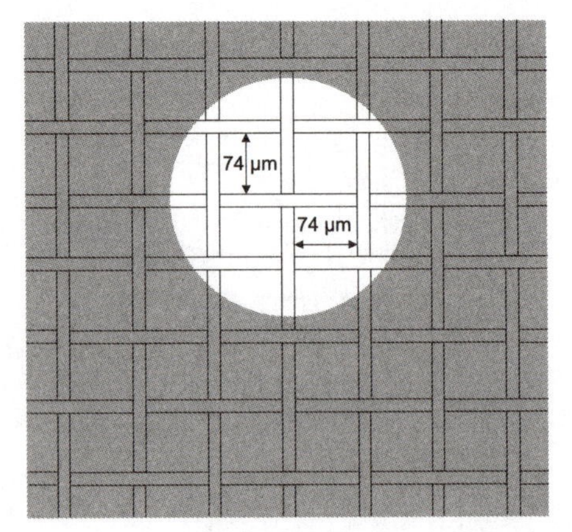

Absolute Rating
square wire mesh

Figure 15-18

Absolute Rating

An absolute rating is an indication of the largest opening in a filter element. This rating indicates the diameter of the largest hard spherical particle which can pass through an element.

Since the pore size can be accurately controlled in this type element, basically all the holes in a 200 wire mesh element are 74 micrometres square.

A 200 mesh wire element with an absolute rating of 74 micrometres means that for every square inch of material there are 200 wires running vertically, 200 wires running horizontally, and the perpendicular distance between the wires is 74 micrometres. Therefore, the largest, hard spherical particle to pass through the element would have a diameter of 74 micrometres.

The absolute rating for a depth type element would be the last point on the pore size axis of the pore size distribution curve. There may only be one hole of that size in the element, but that would still be its absolute rating.

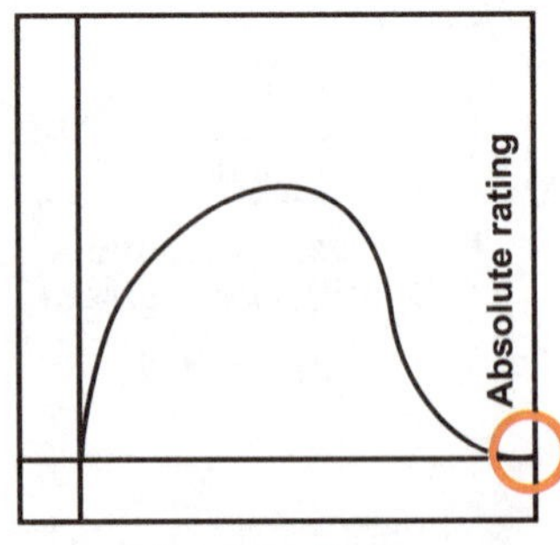

Absolute Rating for Depth Type Element

Figure 15-19

Filter Ratings in Practice

As was pointed out above, an absolute rating indicates the largest hole size in an element. Sometimes, it is deduced that an absolute rating indicates the largest particle which will pass through the element while operating in a system. Since the dirt in the common hydraulic system is not spherical particles, this deduction is erroneous.

Dirt in hydraulics fluids is any insoluble material. It comes in all sizes, shapes, and materials.

An element with an absolute rating of 74 micrometres may have a difficult time catching a long, thin particle. A sliver with an overall length of 150 micrometres and a diameter of 3 micrometres, travels like a rocket in a fluid stream. The sliver penetrates the element with its 3 micrometre dimension and probes its way through the relatively large holes. Consequently, using a 74 micrometre absolute element is not a guarantee that all dirt larger than 74 micrometres will be removed from a fluid stream.

An absolute or nominal rating only indicates a filter element's largest or average hole size. Filter ratings do not guarantee what size particles will be removed or how clean a fluid will become.

To determine how fine and what type of a filter element should be used to protect a specific hydraulic component, consult the component manufacturer or a reputable filter dealer.

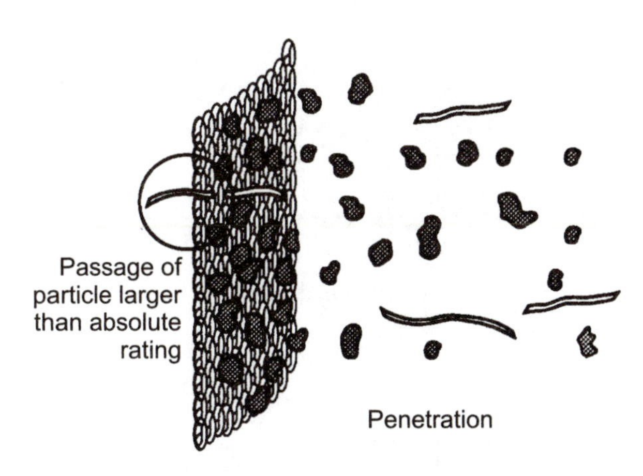

Figure 15-20

Sources of Dirt

Dirt was previously defined as any type of insoluble material in a hydraulic fluid. Dirt comes in all sizes, shapes, and materials and has several sources by which it enters a system.

Dirt Built Into a System

Hydraulic systems which are newly fabricated are often extremely dirty. As a machine is being assembled, the reservoir becomes a collection point for rust, paint chips, dust, cigarette butts, grit and paper cups. Even though the reservoir is "cleaned" before it is filled with fluid, a sample of the fluid taken shortly after start up time will show dirt particles of many foreign materials. Many dirt particles which are harmful to a system are invisible to the naked eye and cannot be removed by wiping with a rag or blowing off with an air hose.

The components which make up a new system may also be sources of system contamination. Because of storage, handling, and installation practices, new directional valves, cylinders, relief valves, and pumps may be equipped with dirt particles which enter a fluid stream after a very short time.

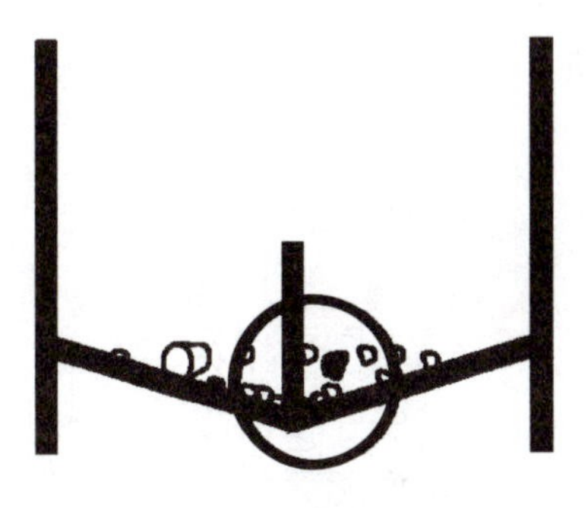

Figure 15-21

Dirt Generated Within a System

Another source of dirt is the dirt generated within the system itself. As a machine continues to operate, moving parts naturally begin to wear and generate dirt. Every internal moving part in a system can be considered a source of contamination for the entire system.

Figure 15-22

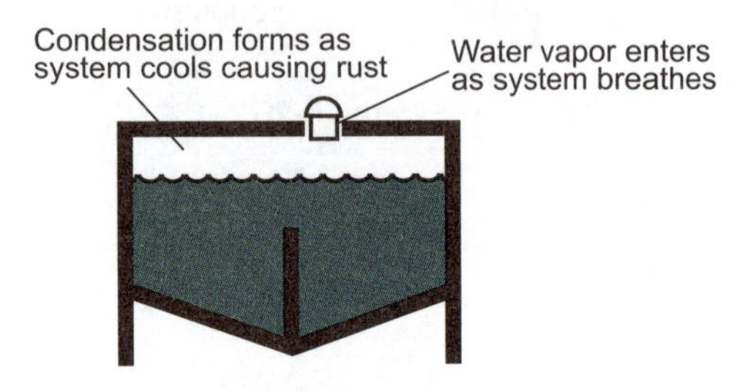

Figure 15-23

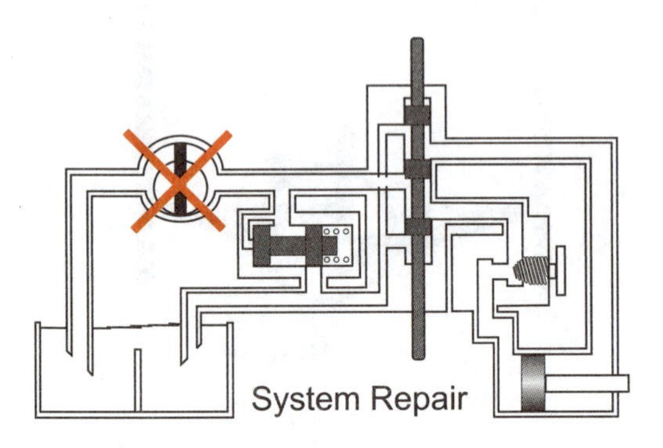

Figure 15-24

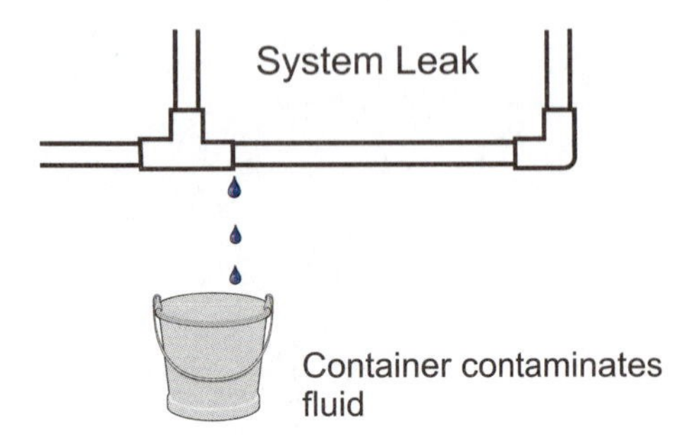

Figure 15-25

Component housings are continuously flexing from normal stresses and are constantly subjected to hydraulic shock pressures. These actions cause metal and casting sand to break loose from the housing and enter the fluid stream.

Air entering the reservoir during operation contains water vapor. When the machine is shut down, the air inside the tank cools and water vapor condenses on the walls. This causes the formation of rust which eventually is washed into the fluid.

Dirt Added to a System

An additional source of dirt is the dirt added to a system. This is many times the result of ordinary maintenance practices or a lack of maintenance.

If a pump happens to break down, the maintenance man replaces the component or repairs it right on the spot. In either case he will be working in a dirty environment, which will contaminate the system as soon as a line is cracked.

If the pump is disassembled, more than likely it will be done on a dirty workbench and the parts will be wiped off with a "not-too-dirty" rag before reassembly. When the pump is replaced, it will spread dirt throughout the system.

The usual first aid measure for a line which has sprung a leak is to put the nearest "clean" bucket under the leak. After it is repaired, the maintenance man must dump the bucket of fluid. It usually happens if no convenient way is seen to dispose of the contaminated fluid, it is poured back into the reservoir. It was one bucket short any way.

The atmosphere in which the machine lives contaminates a system's hydraulic fluid. Operation of a hydraulic system depends upon air entering the reservoir through the breather cap and pushing the fluid up to the pump. As actuators in the system are filled and discharged with fluid, the tank actually inhales and exhales dirt-laden air which is filtered by a coarse screen in the breather cap. The dirt which is not filtered out settles into the fluid.

Breather caps are very rarely, if ever, cleaned. Because of this lack of maintenance, the screen becomes plugged. As a result, the breather cap is removed from the reservoir or the atmosphere finds a path through an old, cracked seal or stripped bolt hole. Now a clear path to the fluid is open for almost anything.

Dirt can also be added to a system by means of a cylinder. After a time, a cylinder's rod wiper seal wears. In this condition, dirt is drawn into the system each time the cylinder rod retracts.

Type of Filtration by Flow

In the early days of hydraulics, filtration was not considered necessary, and this was more or less a correct assumption. Hydraulic components at that time were crude when compared to modern standards. Clearances between moving parts were large, therefore dirty fluid would not affect operation that much. Components were dirt tolerant to a large extent.

As manufacturing processes naturally improved, tolerances were improved. Component operation became much more efficient, but components were less dirt tolerant. The need for filtration was recognized.

Proportional Flow Filtration

The first measure was to place a filter in a system in such a way that a part of the total volume of a pump was filtered. This can be done by placing a filter in a system so that a portion of a pump's flow is bled off through the filter.

Full Flow Filtration

Proportional flow filtration was found to be inadequate after a time, especially as system components became more and more efficient.

The next step was to place a filter in a system so that all flow from a pump was filtered.

This full flow filtration is the type of filtration used in most modern hydraulic systems.

Type of Filtration by Position in a System

A filter is protection for a hydraulic component. Ideally each system component should be equipped with its own filter, but this is economically impractical in most cases. To obtain best results, the usual practice is to strategically place filters in a system.

In the majority of applications the fluid reservoir is a large source of dirt for a system. Since a pump is the heart of a system, one of the most expensive components in a system, and one of the fastest moving system components, it seems logical that a good place to put a filter is between a reservoir and pump.

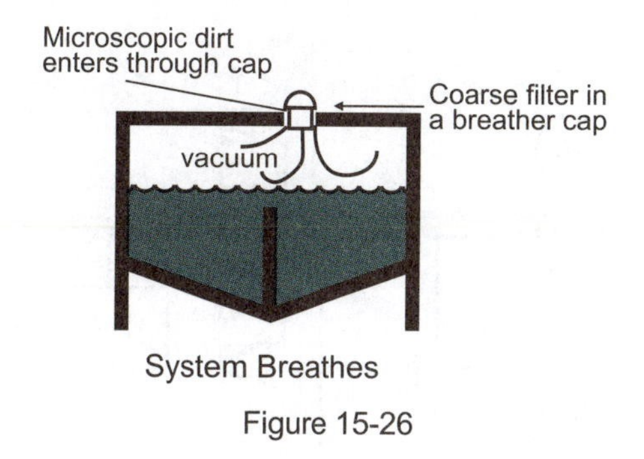

Figure 15-26

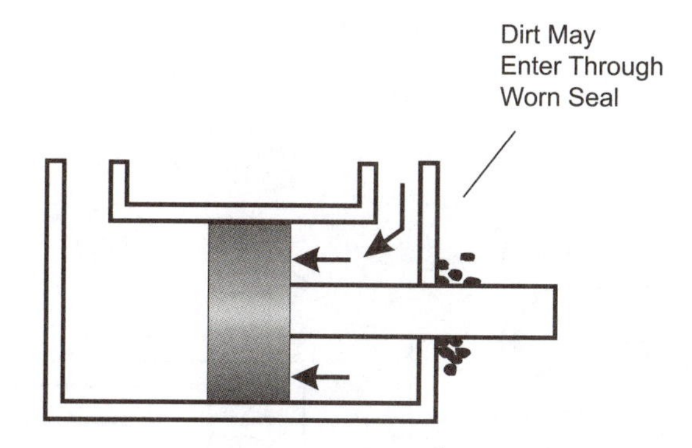

Figure 15-27

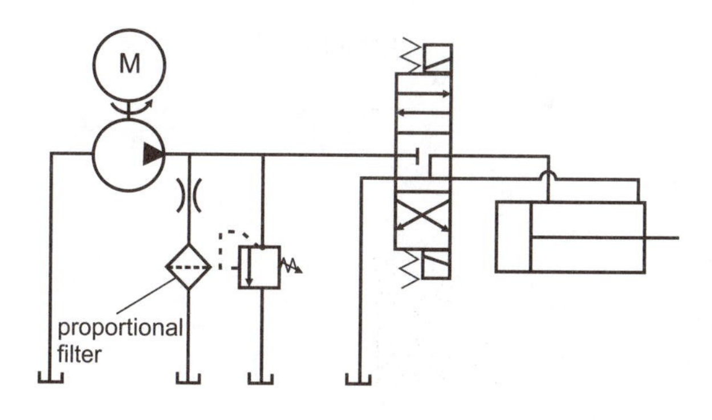

Figure 15-28

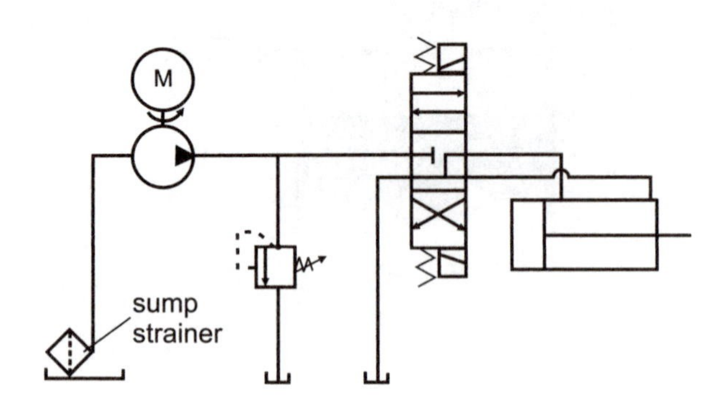

Figure 15-29

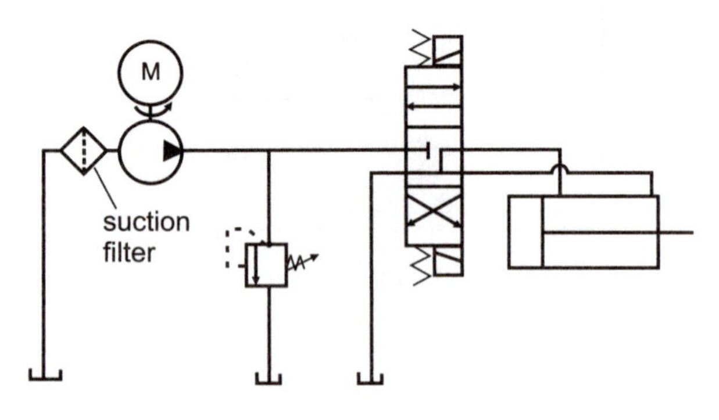

Figure 15-30

Sump Strainer

A sump strainer is usually a coarse filter element screwed onto the end of a pump's suction line.

The range of filtration for sump strainers is a perforated metal cylinder with large drilled holes to 74 micrometre wire mesh.

Advantages:

1. Sump strainers protect the pump from dirt in the reservoir.
2. Because they have no filter housing, sump strainers are very inexpensive.

Disadvantages:

1. Being below fluid level, sump strainers are very difficult to service when cleaning is necessary, especially if the fluid is hot.
2. A sump strainer does not have an indicator to tell when it is dirty.
3. A sump strainer may block fluid flow and starve the pump if not sized correctly and properly maintained.
4. A sump strainer does not protect components downstream from the particles generated by the pump.

Suction Filter

A suction filter is placed in the suction line outside of the reservoir. Range of filtration usually found in suction line filters is from 238-25 micrometres.

Advantages:

1. A suction filter protects the pump from dirt in the reservoir.
2. Since the suction filter is outside the reservoir, an indicator telling when the element is dirty can be used.
3. The filter element can be serviced without disassembling the suction line or reservoir.

Disadvantages:

1. A suction filter may starve the pump if not sized or engineered into the system properly.
2. A suction filter does not protect components downstream from the particles generated by the pump.

Pressure Filter

A pressure filter is positioned in the circuit between the pump and a system component.

Range of filtration usually found in pressure line filters is from 40-3 micrometres.

A pressure filter can also be positioned between system components. If the flow between the components can flow in two directions (as between directional valve and cylinder), the filter must be capable of handling bidirectional flow. Bidirectional pressure filters are used on the downstream side of servo valves and in closed looped hydrostatic transmissions.

Advantages:

1. A pressure filter can filter very fine since system pressure is available to push the fluid through the element.
2. A pressure filter can protect a specific component from the harm of deteriorating particles generated from an upstream component.

Disadvantages:

1. The housing of a pressure filter must be designed for high pressure because it is operating at full system pressure. This makes the filter expensive.
2. If pressure differential and fluid velocity are high enough, dirt can be pushed through the element or the element may tear or collapse.

Return Line Filter

A return line filter is positioned in the circuit just before the reservoir. Range of filtration usually found in return line filters is from 40-5 micrometres.

Advantages:

1. A return line filter catches the dirt in the system before it enters the reservoir.
2. The filter housing does not operate under full system pressure and is therefore less expensive than a pressure filter.
3. Fluid can be filtered fine since system pressure is available to push the fluid through the element.

Disadvantages:

1. There is no direct protection for circuit components.
2. In return line full flow filters, flow surges from discharging cylinders, actuators, and accumulators must be considered when sizing.
3. Some system components may be affected by the back pressure generated by a return line filter.

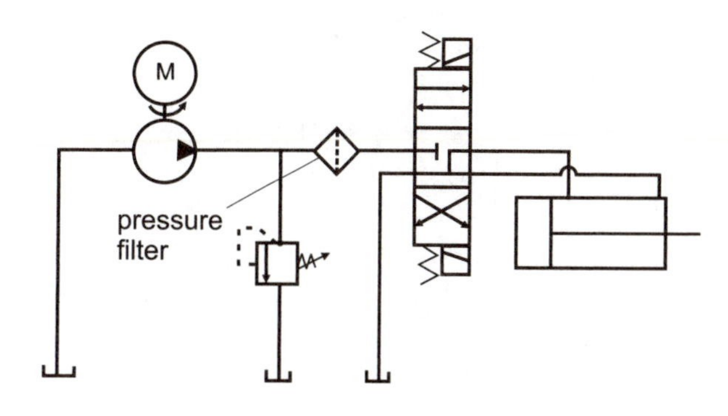

Figure 15-31

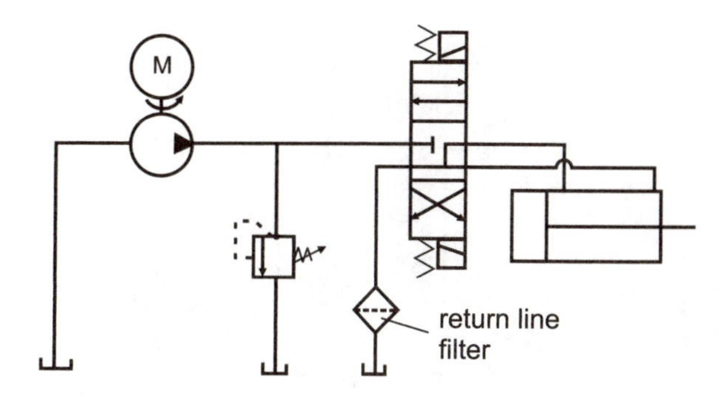

Figure 15-32

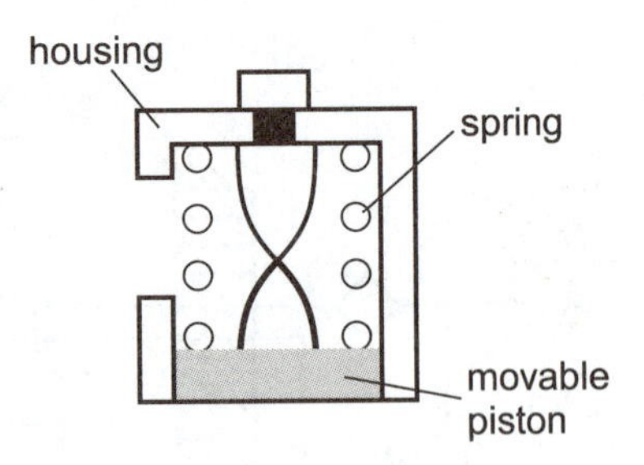

Figure 15-33

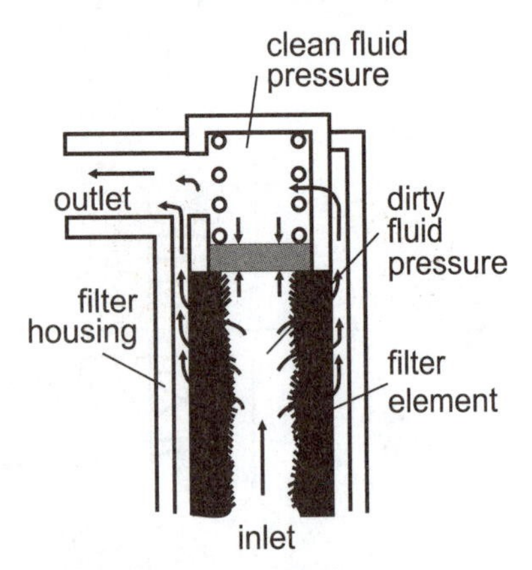

Figure 15-34

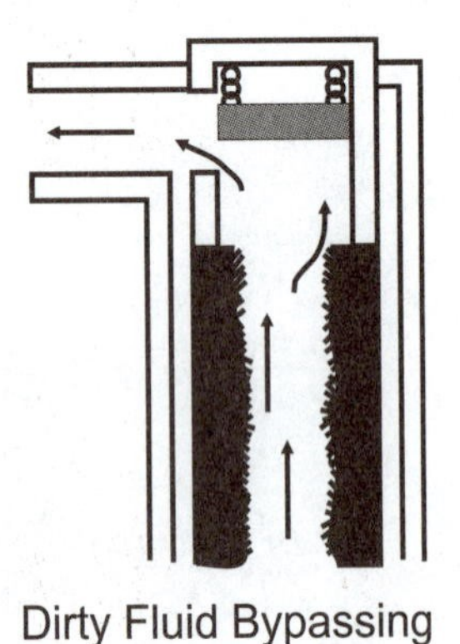

Dirty Fluid Bypassing

Figure 15-35

Filter Bypass Valve

If filter maintenance is not performed, pressure differential across a filter element will increase.

An unlimited increase in pressure differential across a filter on the suction side of a system means that a pump will eventually cavitate.

An unlimited increase in pressure differential across a filter on the pressure side means that the filter element will eventually collapse.

To avoid this situation a simple or direct acting relief valve is used to limit the pressure differential across a full flow filter. This type relief valve is generally called a bypass valve.

What a Bypass Valve Consists Of

A bypass valve basically consists of a movable piston, housing and a spring which biases the piston.

How a Bypass Valve Works

There are several types of bypass valves, but they all operate by sensing the difference in pressure between dirty and clean fluid.

In our illustration, pressure of dirty fluid coming into the filter is sensed at the bottom of the piston. Pressure of the fluid after it has passed through the filter element is sensed at the other side of the piston on which the spring is acting.

As the filter element collects dirt, the pressure required to push the dirty fluid through the element increases. Fluid pressure after it passes through the element remains the same. When the pressure differential across the filter element, as well as across the piston, is large enough to overcome the force of the spring, the piston will move up and offer the fluid a path around the element.

A bypass valve is a fail safe device. In a suction filter, a bypass limits the maximum pressure differential across the filter if it is not cleaned. This protects the pump. If a pressure or return line filter is not cleaned, a bypass will limit the maximum pressure differential so that dirt is not pushed through the element or that the element is not collapsed. In this way, the bypass protects the filter.

The whole key, then, to filter performance centers around cleaning the filter when it needs cleaning. To help in this regard, a filter is equipped with an indicator.

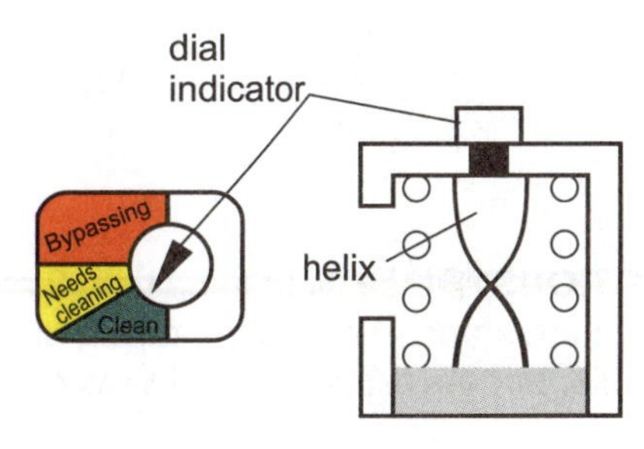

Filter Indicator

Figure 15-36

Filter Indicator

A filter indicator shows the condition of a filter element. It indicates when the element is clean, needs cleaning, or in the bypassing condition.

What a Filter Indicator Consists Of

One common type of filter indicator consists of a helix and a dial indicator which is attached to the helix.

How a Filter Indicator Works

The operation of a filter indicator is dependent upon the movement of the bypass piston. When the element is clean, the bypass piston is fully seated and the indicator shows "clean."

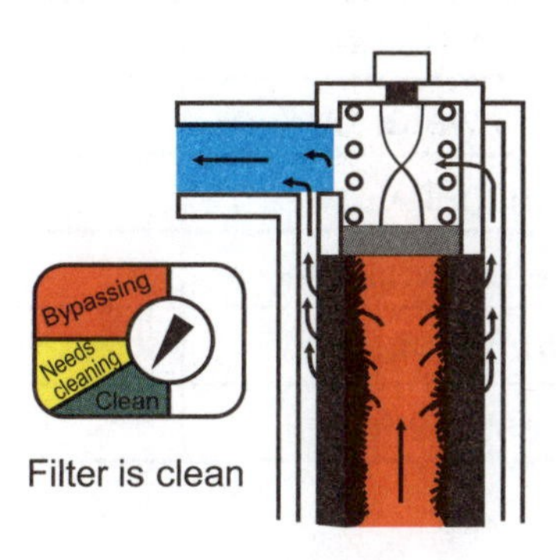

Figure 15-37

While pressure differential across the piston and element increases to the point where the filter needs cleaning, the piston moves up. During its movement, the piston twists the helix which positions the indicator dial at "needs cleaning."

If the filter element is not cleaned when indicated, pressure differential will continue to increase. The piston will continue to move up and bypass the fluid. At this time the indicator will show a bypassing condition.

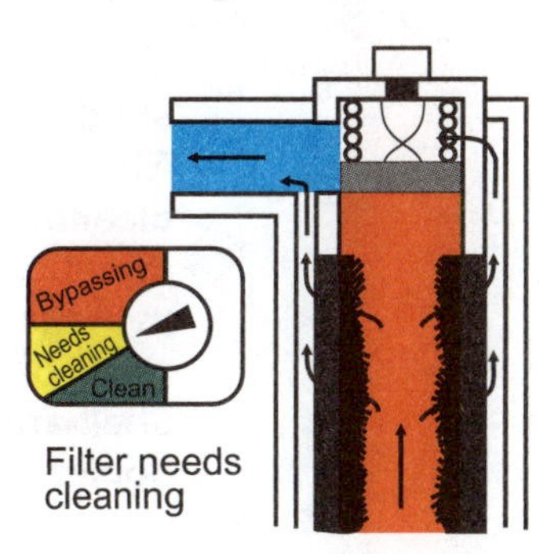

Figure 15-38

Filters Must be Maintained

A machine may be equipped with the best filters available and they may be positioned in the system where they do the most good; but, if the filters are not taken care of and cleaned when dirty, the money spent for the filters and their installation has been wasted. A filter which gets dirty after one day of service and is cleaned 29 days later gives 29 days of non-filtered fluid. A filter can be no better than the maintenance afforded it.

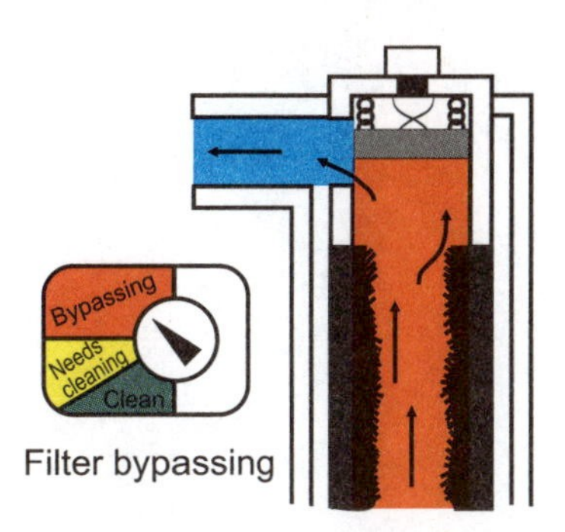

Figure 15-39

Exercise
Reservoirs, Coolers, Filters
48 Points

INSTRUCTIONS: Find a word in column 2 related to a word in column 1. Then, pair up two more words which have the same relationship, taking one from column 3 and one from column 4.
For example: bird-air & fish-water. All words are used once.

1	2	3	4
filler/breather			
VI improver			
depth element			
40% water 60% oil			
nominal			
water available			
dirt built-in			
pressure filter			
bird	air	fish	water

invert emulsion	**5% oil-95% water**	**AW**
synthetic	**absolute**	**air cooler**
EP	**water scarce**	**reservoir**
cleanout cover	**R & O**	**pump**
directional valve	**suction filter**	**wire mesh**
average	**oil-level gage**	**baffle**
new components	**surface element**	**largest**
shell-and-tube	**dirt added**	**soluble**
cooler		

Appendix A
Metric Coversions

SI Prefixes		
Prefix	SI Symbol	Multiplication Factor
tera	T	10^{12}
giga	G	10^{9}
mega	M	10^{6}
kilo	k	10^{3}
hecto	h	10^{2}
deka	da	10^{1}
deci	d	10^{-1}
centi	c	10^{-2}
milli	m	10^{-3}
micro	µ	10^{-6}
nano	n	10^{-9}
pico	p	10^{-12}
femto	f	10^{-15}
atto	a	10^{-18}

Table P1 — SI Prefixes

Derived Units			
Quantity	Unit	SI Symbol	Formula
Acceleration	Meter per Second Squared	—	m/s^2
Angular Velocity	Radian per Second	—	rad/s
Area	Square Meter	—	m^2
Density	Kilogram per Cubic Meter	—	kg/m^3
Electric Resistance	Ohm	Ω	V/A
Energy & Work	Joule	J	N.m
Force	Newton	N	$kg.m/s^2$
Frequency	Hertz	Hz	cycles/s
Power	Watt	W	J/s
Pressure & Stress	Pascal	Pa	N/m^2
Quantity of Heat	Joule	J	N.m
Specific Heat	Joule per Kilogram-Kelvin	—	J/kg.K
Thermal Conductivity	Watt per Meter-Kelvin	—	W/m.K
Velocity	Meter per second	—	m/s
Viscosity, Dynamic	Pascal Second	—	Pa.s
Viscosity, Kinematic	Square Meter per Second	—	m^2/s
Voltage	Volt	V	W/A
Volume	Cubic Meter	—	m^3

Table P2 — Derived Units

Basic Units		
Quantity	Unit	SI Symbol
Length	Meter	m
Mass	Kilogram	kg
Time	Second	s
Electric Current	Ampere	A
Thermodynamic Temperature	Kelvin	K
Amount of Substance	Mole	mol
Luminous Intensity	Canela	cd

Table P3 — Basic Units

Supplementary Units		
Quantity	Unit	SI Symbol
Plane Angle	Radian	rad
Solid Angle	Sterodian	sr

Table P4 — Supplementary Units

Table P5 — English to Metric and Metric to English Conversions

	English to Metric			Metric to English		
	To Convert From	**To**	**Multiply By**	**To Convert From**	**To**	**Multiply By**
Area	sq. in. (in^2)	sq. mm (mm^2)	645.16	square millimeters	square inches	0.00155
	sq. in. (in^2)	sq. cm (cm^2)	6.4516	(mm^2)	(in^2)	
	sq. ft. (ft^2)	sq. meters (m^2)	0.0929			
Density	pounds/cubic ft (lb/ft^3)	Kilograms/cubic meter (kg/m^3)	16.02	kilograms/cubic meter (kg/m^3)	pounds/cubic ft (lb/ft^3)	0.0624
Energy	British thermal units (Btu) (1 J = Ws = 0.2388 cal)	joules (J)	1055	joules (J)	British thermal units (Btu)	0.000947
Force	pounds - force (lbf) (1N = 0.102 kgf)	newtons (N)	4.448	newtons (N)	pounds - force (lbf)	0.2248
Length	inches (in)	millimeters (mm)	25.4	millimeters (mm)	inches (in)	0.03937
	feet (ft)	meters (m)	0.3048	meters (m)	feet (ft)	3.281
	miles (mi)	kilometers (km)	1.609	kilometers (km)	miles (mi)	0.621
Mass (Weight)	ounces (oz)	grams (g)	28.35	grams (g)	ounces (oz)	0.035
	pounds-mass (lb)	kilograms (kg)	0.4536	kilograms (kg)	pounds-mass (lb)	2.205
	short tons (2000 lb) (tn)	metric tons (1000 kg) (t)	0.9072	metric tons (1000 kg) (t)	short tons (2000 lb) (tn)	1.102
Power Pressure	horsepower (550 ft. lb/s) (hp)	kilowatts (kW)	0.7457	kolowatts (kW)	horsepower (550 ft. lb/s) (hp)	1.341
	pounds/square inch (psi)	kilograms (f)/square cm (kg (f)/cm^2)	0.0703	kilograms (f)/square cm (kg (f)/cm^2)	pounds/square inch (psi)	14.22
	pounds/square inch (psi)	kilopascals (kPa)	6.8948	kolopascals (kPa)	pounds/square inch (psi)	0.145
	pounds/square inch (psi)	bars (100 kPa)	0.06895	bars (100 kPa)	pounds/square inch (psi)	14.503
Stress	pounds/square inch (psi) (1 N/mm^2 = 1 MPa)	megapascals (MPa)	0.006895	megapascals (MPa) (1 N/mm^2 = 1 MPa)	pounds/square inch (psi)	145.039
Temperature	degrees fahrenheit (°F)	degrees celsius (°C)	5/9 (after subtracting 32)	degrees celsius (°C)	degrees fahrenheit (°F)	9/5 (then add 32)
Torque or Bending	pounds-force-foot (lb-ft)	Newtons-meter (Nm)	1.3567	Newtons-meter (Nm)	pounds-force-foot (lb-ft)	0.737
Moment	pounds-force-inch (lb-in)	Newtons-meter (Nm)	0.113	Newtons-meter (Nm)	pounds-force-inch (lb-in)	8.85
Velocity	feet/second (ft/s)	meters/second (m/s)	0.3048	meters/second (m/s)	feet/second (ft/s)	3.2808
Viscosity	dynamic (centipoise)	pascal-second (Pas)	0.001	pascal-second)Pas)	dynamic (centipoise)	1000
	kenematic-foot2/sec (ft^2/s)	meter^2/sec (m^2/s)	0.0929	meter^2/sec (m^2/s)	foot2/sec (ft2/s)	10.7643
Volume	cubic inch (in^3)	cubic centimeter (cm^3) (milliliter)	16.3871	cubic centimeter (cm^3)	cubic inch (in3) (milliliter)	0.061
	quarts (qt)	liters (1000 cm^3)	0.9464	liters (1000 cm^3)	quarts (qt)	1.057
	gallons (gal)	liters	3.7854	liters	gallons (gal)	0.2642

Units of Measurement – Unit Names

Quantity	Name of Unit	Value	Symbol
Length	**Meter**	**Base Unit**	**M**
	centimeter	0.01m	cm
	millimeter	0.001m	mm
	micrometer	0.000001m	µm
	kilometer	1,000m	km
Mass	**Kilogram**	**Base Unit**	**Kg**
	milligram	0.000 001kg	mg
	gram	0.001 kg	g
	tonne	1,000kg	t
Time	**Second**	**Base Unit**	**s**
	minute	60s	min
	hour	60 min	h
	day	24h	d
Area	**Square Meter**	**SI Unit**	**m sq**
	square millimeter	0.000 001m sq	mm sq
	square centimeter	0.000 1m sq	cm sq
	hectare	10,000 m sq	ha
	square kilometer	1,000,000 m sq	km sq
Volume	**Cubic Meter**	**SI Unit**	**m cu**
	cubic centimeter	0,000 001 m cu	cm cu
Volume (Fluids)	**Liter**	**0.001m cu**	**1**
	milliliter	0.001 l	ml
	kiloliter	1,000 l (1m cu)	kl
Velocity	**Meter per Second**	**SI Unit**	**m/s**
	kilometer per sec	0.27 m/s	km/h
	knot	1 n mile/h	kn or 0.514 m/s
Force	**Newton**	**SI Unit**	**N**
	kilonewton	1,000 N	kN
	meganewton	1,000 000 N	MN
Energy	**Joule**	**SI Unit**	**J**
	kilojoule	1,000 J	kJ
	megajoule	1,000 000 J	MJ
Power	**Watt**	**SI Unit**	**W**
	kilowatt	1,000W	kW
	megawatt	1,000 000W	MW

Metric Units of Length

Metric Unit	Equivalent Value
1 metre	39.37 inches 3.28083 feet 1.09361 feet 1000 millimetres 100 centimetres 10 decimetres 0.01 kilometres
1 centimetre	0.3937 inch 0.0328083 foot 10 millimetres 0.01 metres
1 millimetre	39.370 mils 0.03937 inch 0.001 metre
kilometre	3280.83 feet 1093.61 yards 0.62137 mile 1000 metres

US Units of Length

US Unit	Equivalent Value
1 inch	1000 mils 0.0833 foot 0.0277 yard 25.40 millimeters 2.540 centimeters
1 foot	12 inches 1.33333 yard 0.0001893 miles 0.30480 meter 30.480 centimeters
1 yard	36 inches 3 feet 0.0005681 mile 0.914402 meter
1 mile	63360 inches 5280 feet 1760 yards 320 rods 8 furlongs 1609.35 meters 1.60935 kilometers

Metric Units of Volume

Metric Unit	Equivalent Value
1 cubic metre	61023.4 cubic inch 35.3145 cubic feet 1.30794 cubic yard 1000 litres
1 cubic decimetre	61.0234 cubic inch 0.035145 cubic foot 1000 cubic centimetres 1 litre
1 cubic centimetre	0.0000353 cubic foot 0.0610234 cubic inch 1000 cubic millimetres 0.001 litre (1ml)
1 cubic millimetre	0.000061023 cubic inch 0.0000000353 cubic foot 0.001 cubic centimetre
1 litre	1 cubic decimetre 61.0234 cubic inches 0.351345 cubic foot 1000 cubic centimetres 0.001 cubic metre 2.202 lbs of water

US Units of Volume

US Unit	Equivalent Value
1 cubic yard	46656 cubic inches 27 cubic feet 0.76456 cubic metre
1 cubic foot	1278 cubic inches 0.03703703 cubic yard 28.317 cubic decimetres 0.028317 cubic meter 7.4805 gallons
1 cubic inch	16.3872 cubic centimetres (16.3872 ml)
1 gallon (US)	3.78543 litres
1 gallon (British)	4.54374 litres

Metric Units of Weight

Metric Unit	Equivalent Value
1 gram	15.432 grains 0.022046 lb 0.3527 oz.
1 kilogram	1000 grams 2.20462 lb 35.2739 oz.
1 metric ton	2204.62 pounds 0.984206 ton of 2240 pounds 22.0462 cwt 1.10231 ton of 2000 pounds 1000 kilograms

US Units of Weight

US Unit	Equivalent Value
1 ounce	437.5 grains 0.0625 pounds 28.3496 grams
1 pound	7000 grains 16 ounces 453.593 grams
1 ton (2000 pounds)	1.01605 metric tons 1016.05 kilograms

SI Prefixes

Prefix	Symbol	Value (Words)	Value (Numerical)
Giga	G	One thousand million	1,000,000,000
Mega	M	One million	1,000,000
Kilo	k	One thousand	1,000
Milli	m	One thousandth	0.001
Micro	μ	One millionth	0.000,001
Nano	n	One thousand millionth	0.000,000,001

Fraction to Decimal

Fraction	Decimal (mm)	Decimal (in)
1/16	1.58750	0.0625
1/8	3.17501	0.125
3/16	4.76251	0.1875
1/4	6.35001	0.2500
5/16	7.93752	0.3125
3/8	9.52502	0.3750
7/16	11.11252	0.4375
1/2	12.70003	0.5000
9/16	14.28753	0.5625
5/8	15.87503	0.625
11/16	17.46253	0.6875
3/4	19.05004	0.7500
13/16	20.63754	0.8125
7/8	22.22504	0.8750
15/16	23.81255	0.9375

Celsius to Fahrenheit

Celsius (°C)	Fahrenheit (°F)
-9	15.8
-8	17.6
-7	19.4
-6	21.2
-5	23
-4	24.8
-3	26.6
-2	28.4
-1	30.2
0	32
5	41
10	50
15	59
20	68
25	77
30	86
35	95
40	104
45	113
50	122

Metric to US Units

Quantity	Metric Unit	US Equivalent Value
Length	1 cm	0.394 in
	1 m	3.28 ft
	1 m	1.09 yd
	1 km	0.621 mile
Mass	1 g	0.0353 oz
	1 kg	2.20 lb
	1 tonne	0.984 ton
Area	1 cm sq	0.155 in sq
	1 m sq	10.8 ft sq
	1 m sq	1.20 yd sq
	1 ha	2.47 ac
	1 km sq	247 ac
Volume	1 cm cu	0.0610 in cu
	1 m cu	35.3 ft cu
	1 m cu	1.31 yd cu
	1 m cu	27.5 bushels
Volume (fluids)	1 ml	0.0352 fl oz
	1 litre	1.76 pint
	1 m cu	220 gallons
Force	1N (newton)	0.225 lbf
Pressure	1 kPa	0.145 psi
Velocity	1 km/h	0.621 mph
Temperature	°C	$\frac{9 \times {}^{\circ}C}{5} + 32^{\circ}F$
Energy	1 kJ	0.948 Btu joule
Power	1 kW	1.34 hp

US to Metric Units

Quantity	US Unit	Metric Equivalent Value
Length	1 in	25.4 mm
	1 ft	30.5 cm
	1 yd	0.914 m
	1 mile	1.61 km
Mass	1 oz	28.3 g
	1 lb	454 g
	1 ton	1.02 tonne
Area	1 in sq	6.45 cm sq
	1 ft sq	929 cm sq
	1 yd sq	0.836 m sq
	1 ac sq	0.405 ha
	1 sq mile	259 ha
Volume	1 in cu	16.4 cm cu
	1 ft cu	0.02383 m cu
	1 yd cu	0.765 m cu
	1 bushel	0.0364 m cu
Volume (fluids)	1 fl oz	28.4 ml
	1 pint	568 ml
	1 gallon (US)	3.79 litre
Force	1 lbf (pound force)	4.45 N
Pressure	1 psi (lb/sq in)	6.89 kPa
Velocity	1 mph	1.61 km/h
Temperature	°F	$\frac{5}{9}(f - 32)^{\circ}C$
Energy	1 Btu	1.06 kJ thermal unit
Power	1 hp	0.746 kW

Appendix B
Basic Hydraulic Circuitry

The following legend will be used for the color coding of the drawings. It is as follows:

Red - operating or system pressure

Yellow - reduced flow or reduced pressure

Orange - intermediate pressure inside flow control

Green - intake or drain

Blue - return line flow

No color - inactive fluid

Solenoid Operated Relief Valve

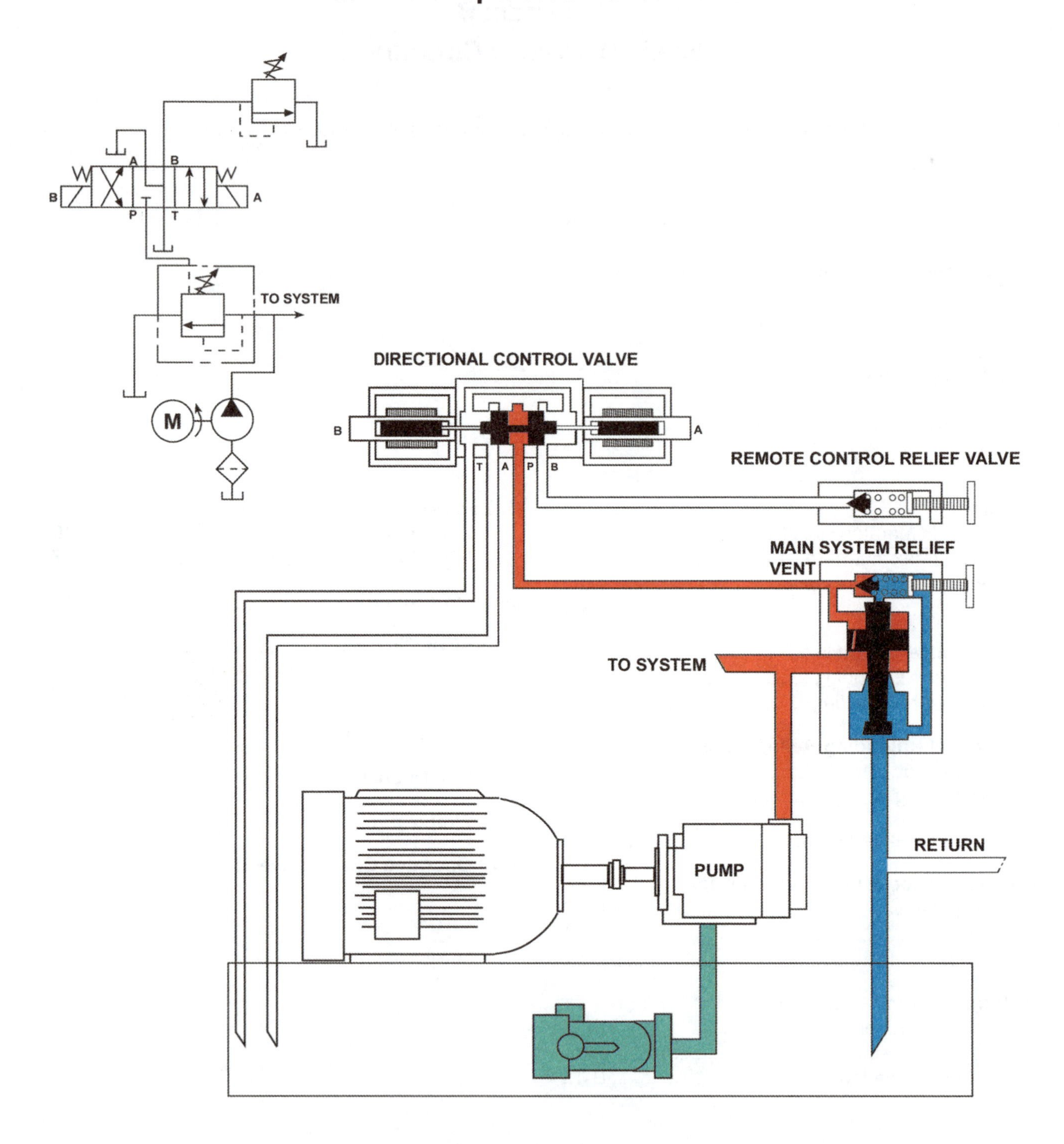

High Maximum Pressure

Directional control valve is not energized causing vent line to be plugged. Pressure at pump is determined by relief valve setting.

Solenoid Operated Relief Valve

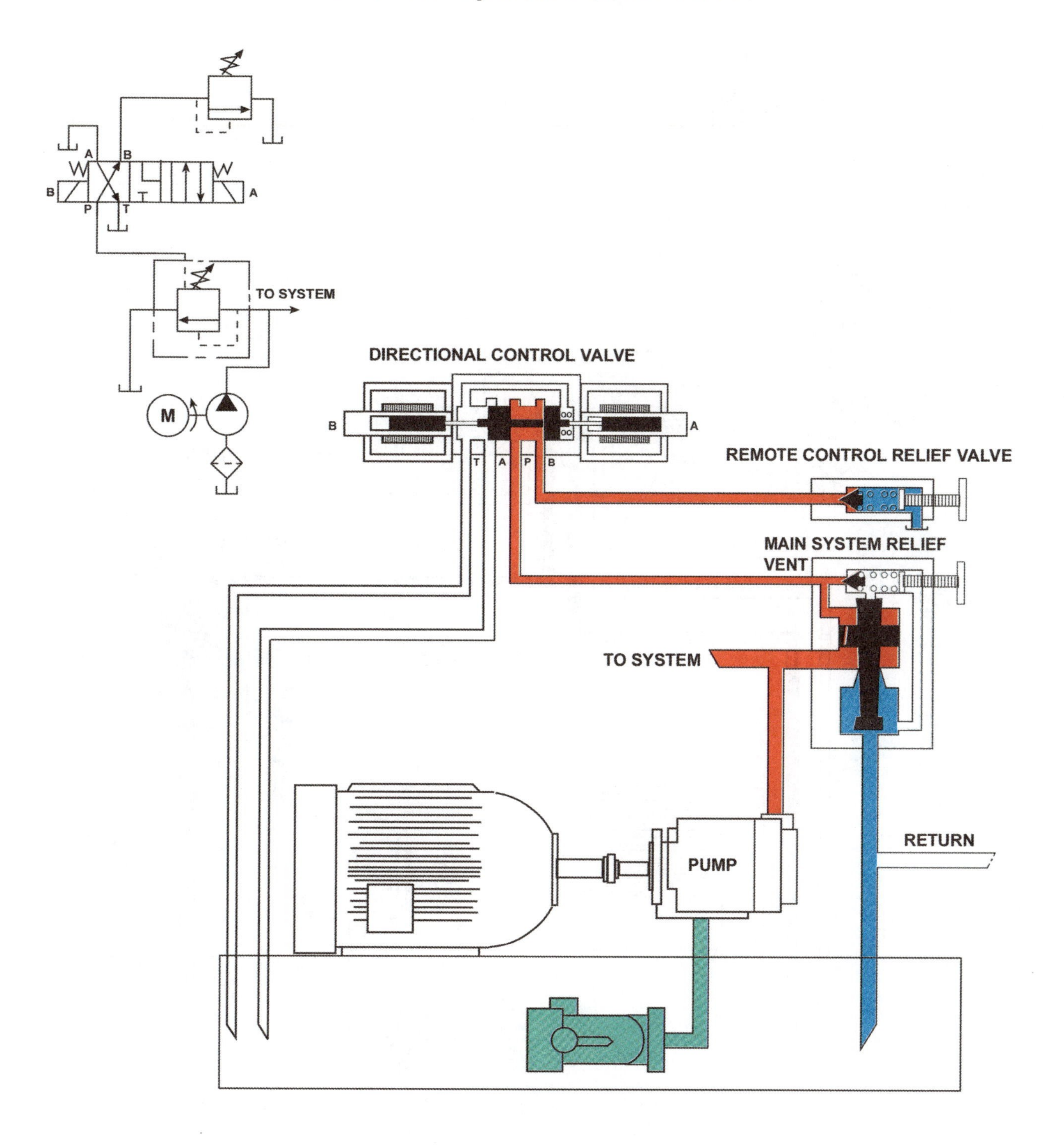

Intermediate Pressure

Solenoid "B" of directional valve is held energized. The spool shift thereby connecting the vent of relief valve to pressure port of remote control relief valve. System pressure is limited by the remote control relief valve which remotely controls the main relief.

Solenoid Operated Relief Valve

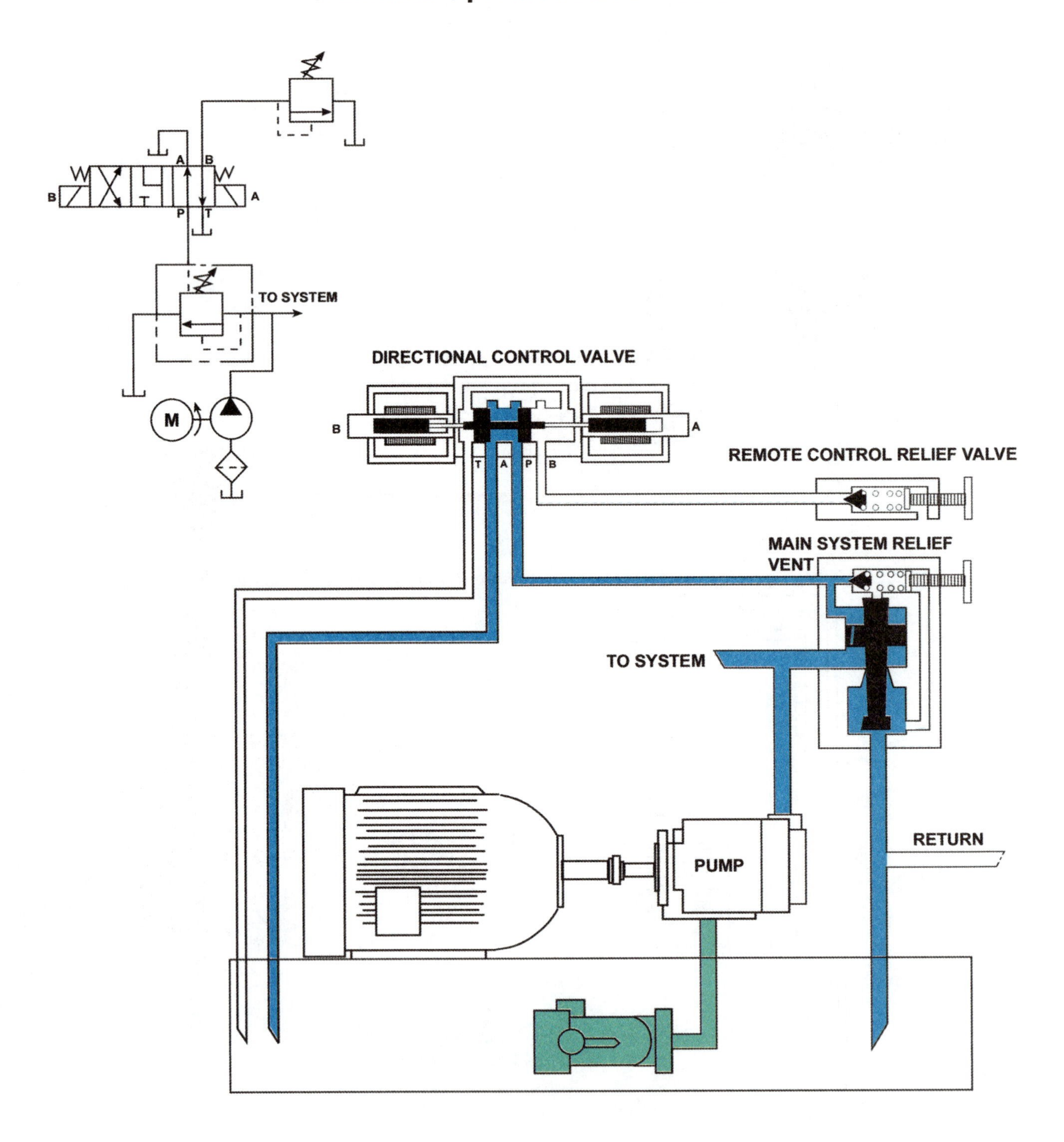

Venting

Solenoid "A" is energized connecting the vent port of the main relief to tank. By releasing this pilot pressure, the only pressure holding the spool closed is the relatively light pressure of the spring. This results in the pump applying a relatively low pressure to return its flow to tank.

Regenerative Circuit - Cylinder Extending

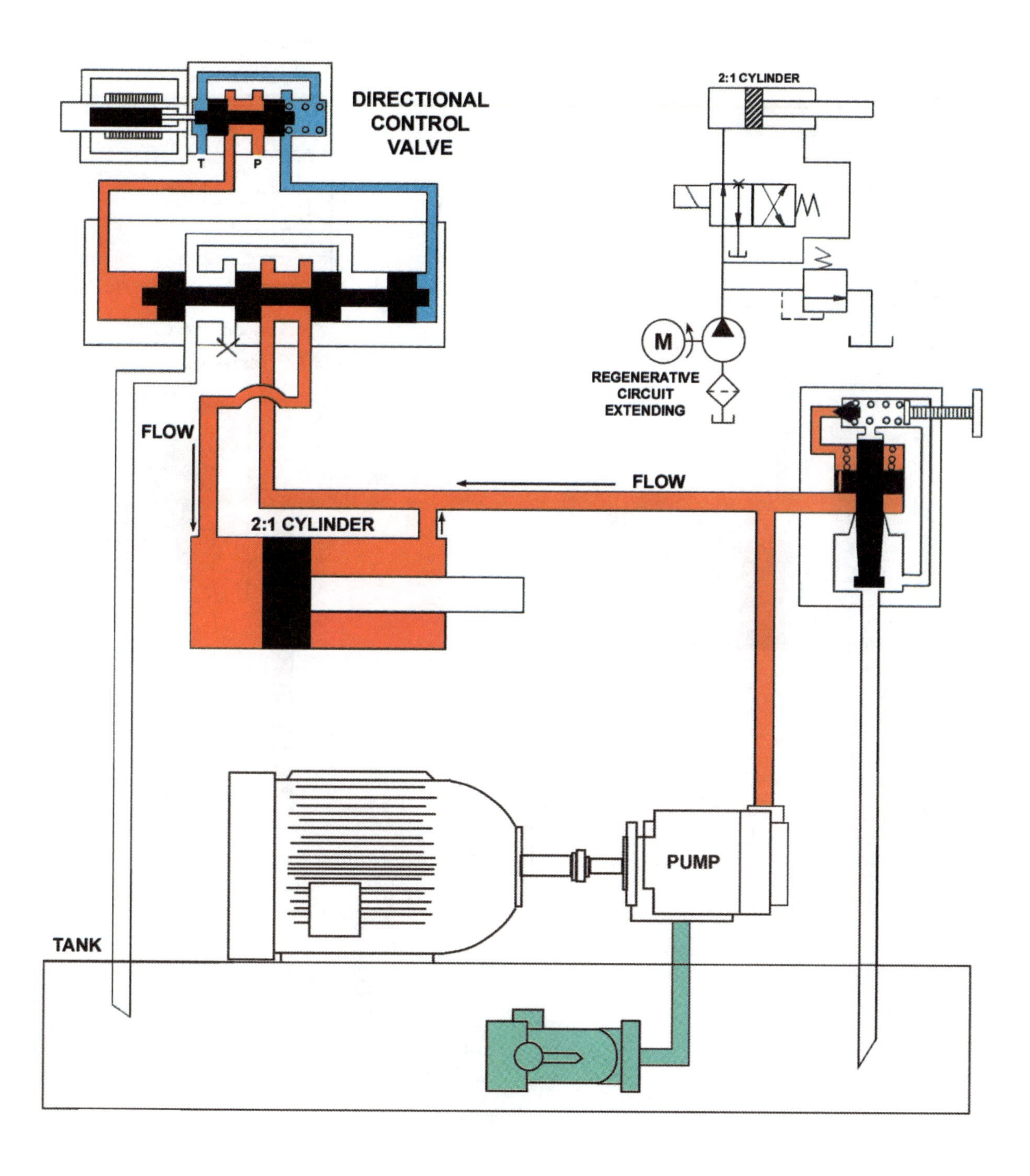

The regenerative circuit which is illustrated consists of a pump, relief valve, a directional valve with a blocked port, and a 2:1 cylinder. With the directional valve in the position shown, both sides of the cylinder piston are subjected to the same pressure. The unbalanced force which results extends the rod. Fluid discharging from the rod end is added to the pump's flow. Since in a 2:1 cylinder the discharge fluid from the rod end is always half the volume entering the cap end, the only volume which is filled by pump flow is the other half of the cap end volume. To calculate the rod speed of a 2:1 cylinder when it is regenerating, the cross-sectional area of the rod is used in the calculation.

Regenerative Circuit - Cylinder Retracting

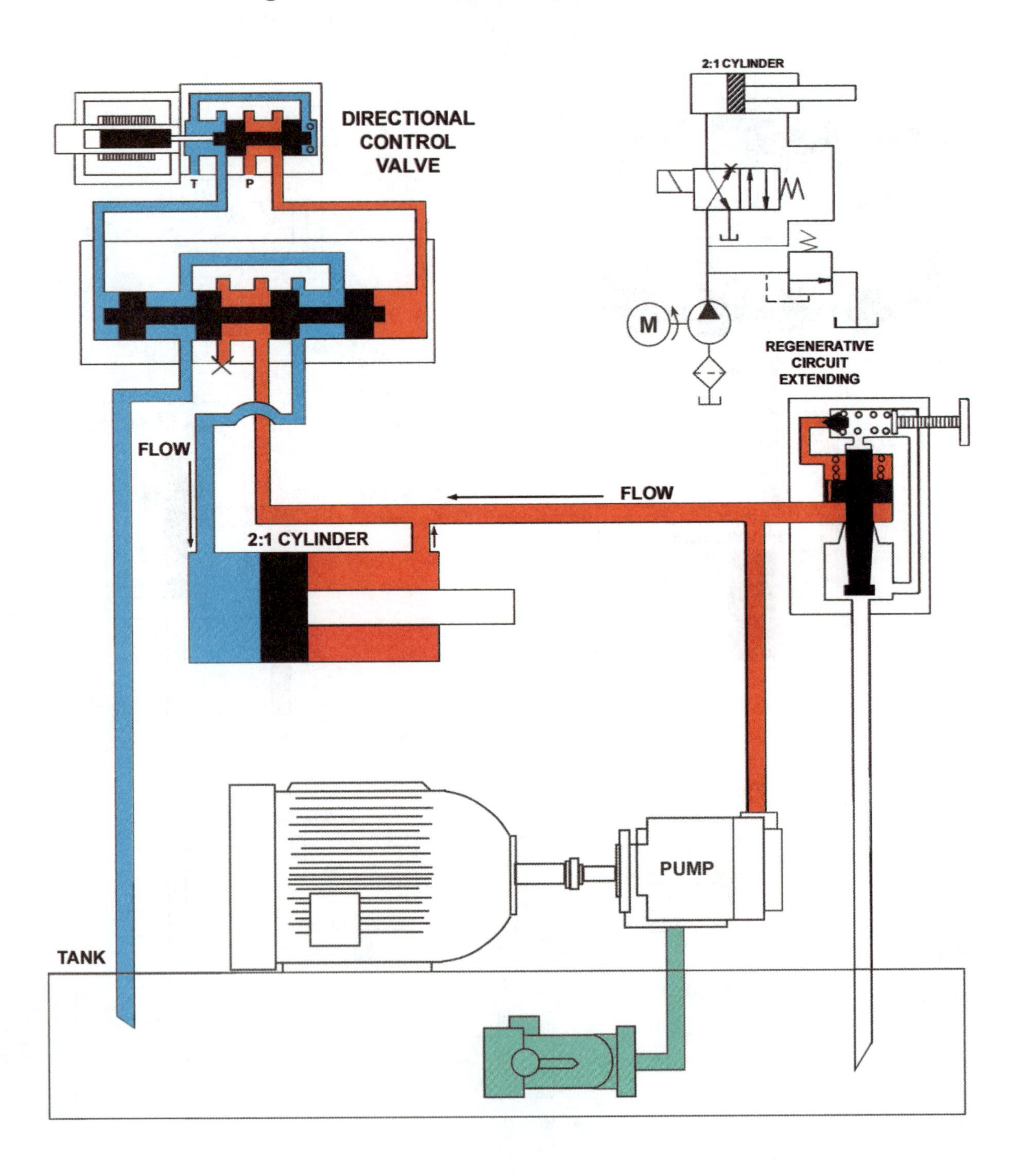

To retract the cylinder rod, the directional valve is shifted. The cap end of the cylinder is drained to tank. All pump flow and pressure is directed to the rod end side. Since the pump is filling the same volume as at the cap end side (half cap end volume) the rod retracts at the same speed.

Differential Unloading Relief Valve

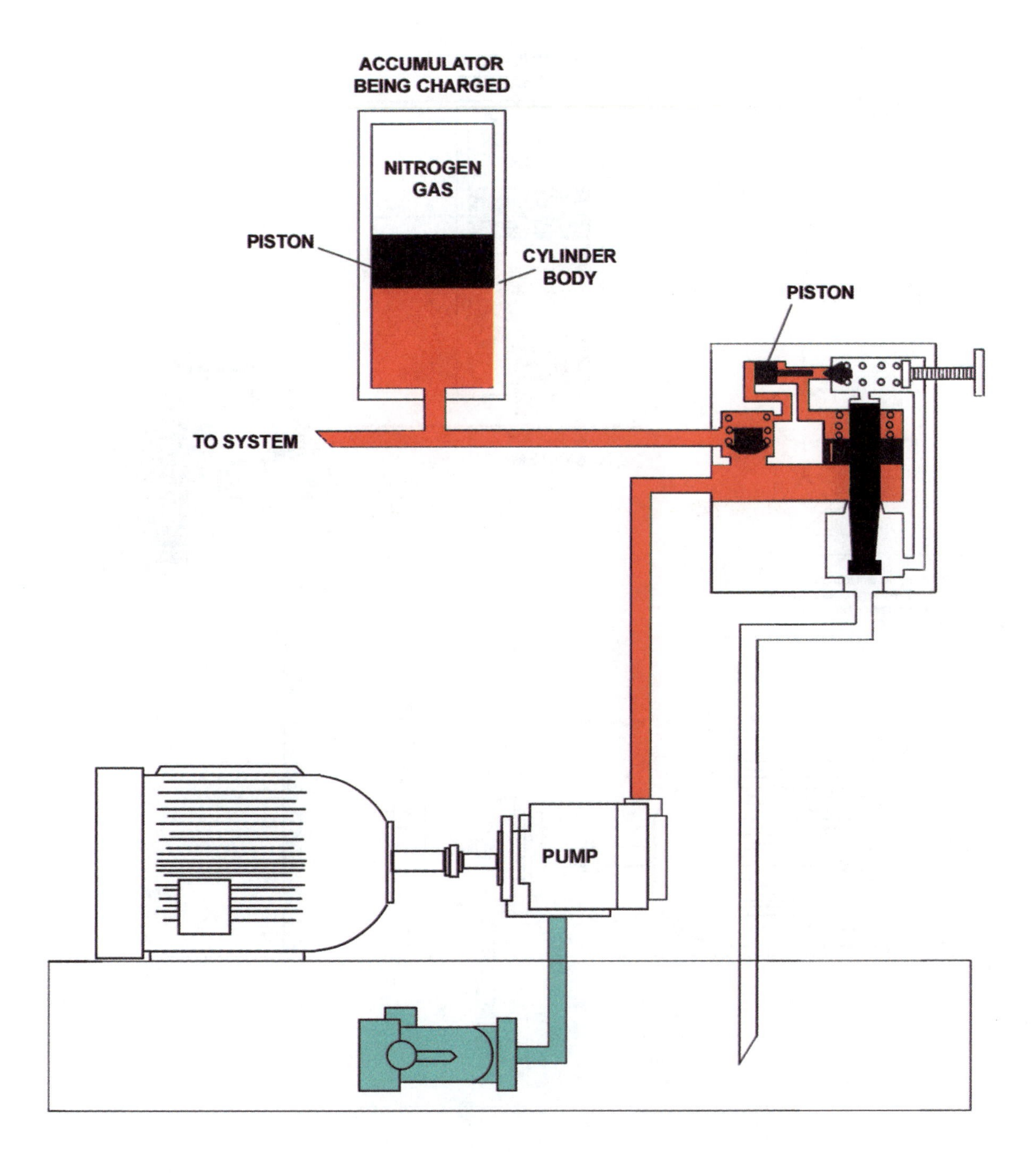

The differential piston fits in a bore opposite the pilot valve dart. At each end of the piston, the areas exposed to pressure are equal. During the time the accumulator is being charged, pressure at each end of the piston is equal.

Differential Unloading Relief Valve

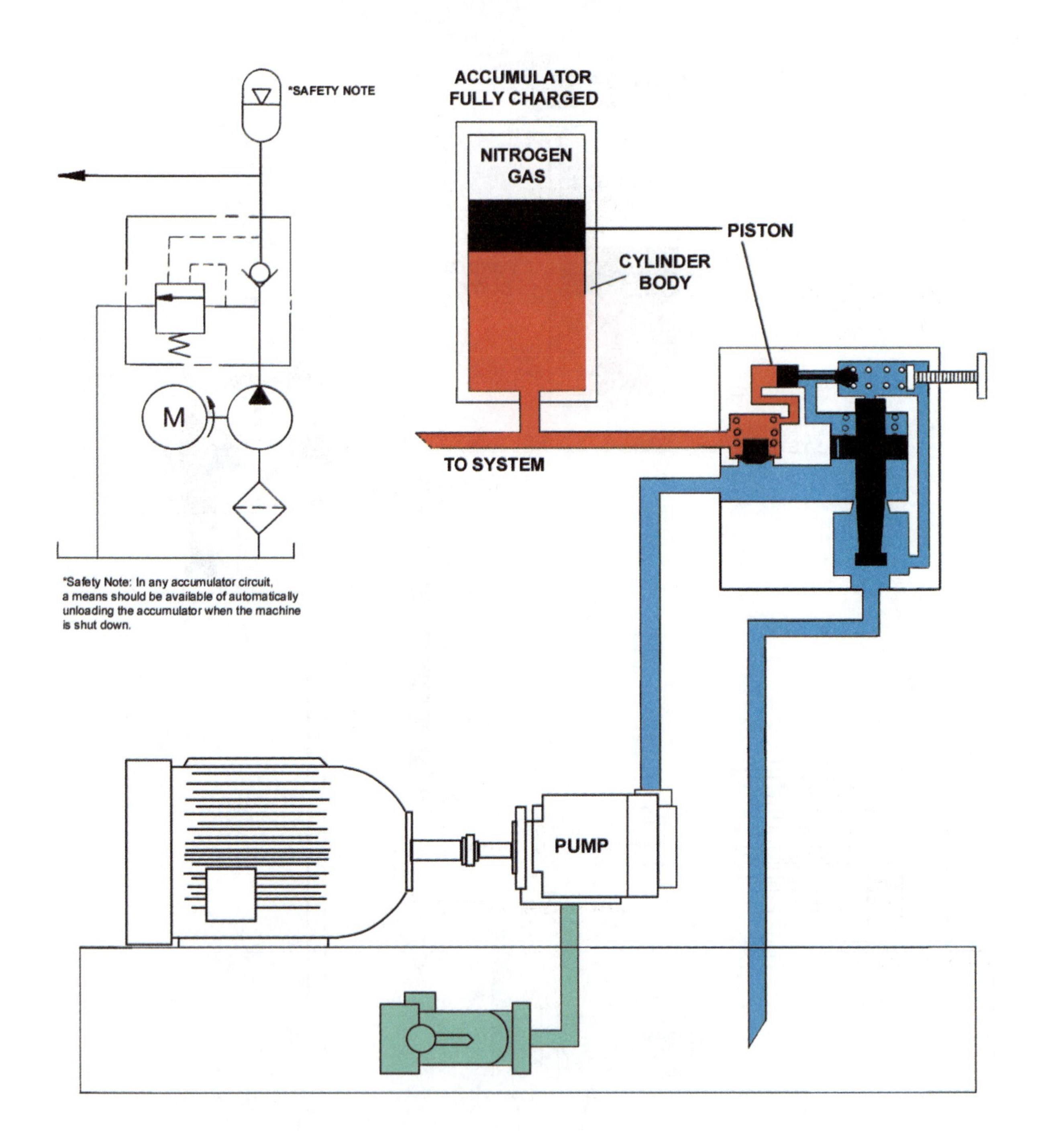

When the accumulator is charged, the piston is forced toward the pilot dart and forces the pilot dart completely off its seat. This in effect releases the main spool spring chamber of pilot pressure. The relief valve is vented. The spool moves up and allows flow to go to tank at a low pressure. At the same time the check valve closes so that the accumulator cannot discharge through the relief valve. At this point one pressure has been achieved - the accumulator's maximum pressure.

Accumulator Bleed Down Circuit

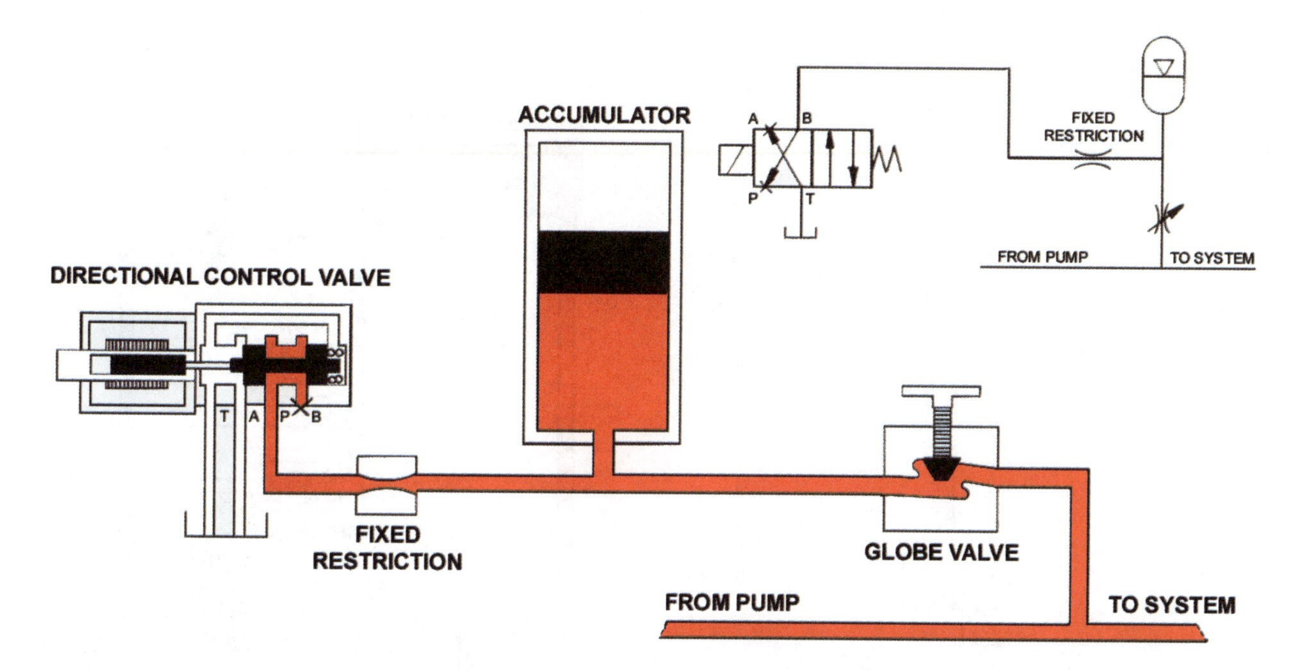

When the system is shut down, the solenoid is de-energized and the spring pushes the valve to its normally open position The accumulator bleeds down safety through the needle valve. Therefore, any-time the electric motor is shut down, the accumulator automatically bleeds.

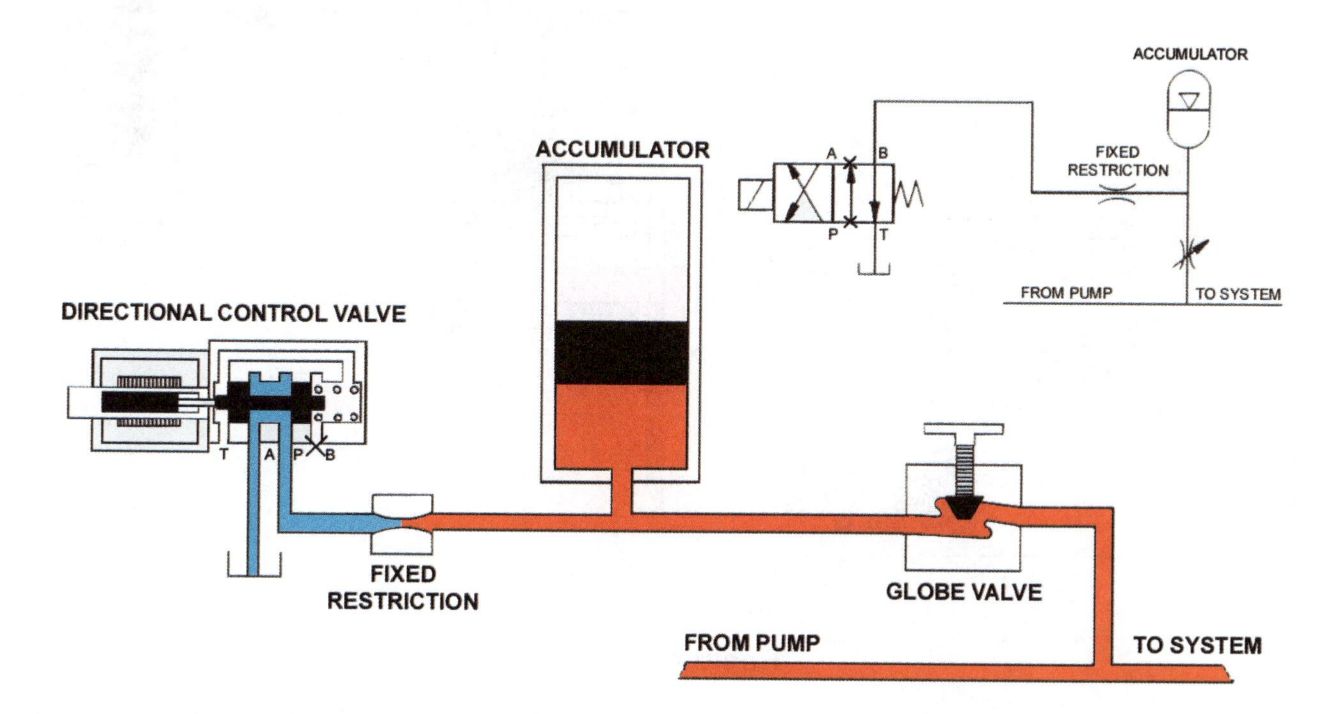

In any accumulator circuit, a means should be available of automatically unloading the accumulator when the system is shut down. This can be accomplished by using a spring offset, solenoid operated, 4-way valve that has been converted into a normally open 2-way.

In the example circuit, the solenoid of the converted 2-way valve can be energized when the electric motor is started. This blocks the flow through the valve and allows the accumulator to charge.

Rapid Advance & Feed Type Circuit

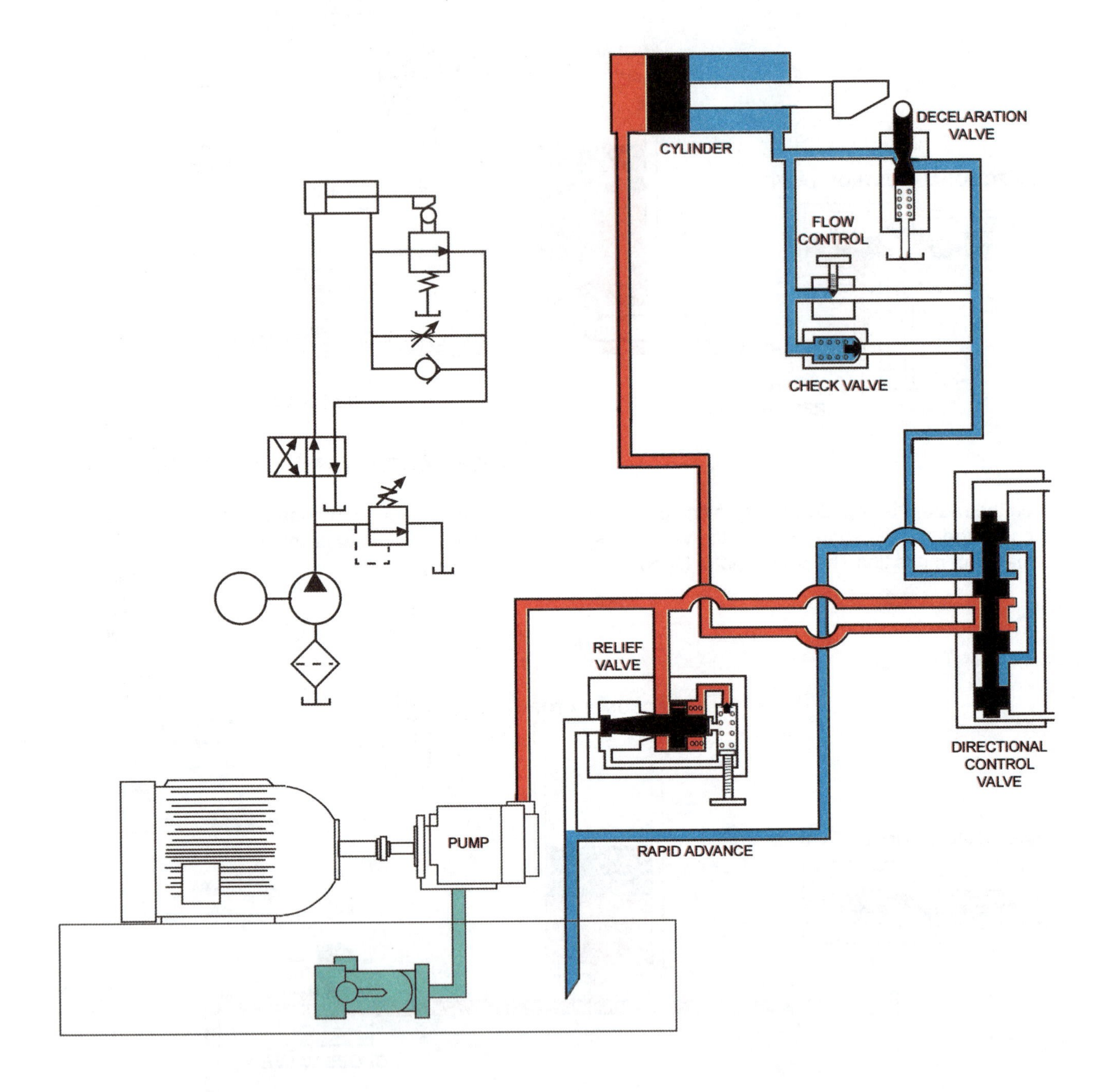

Rapid Advance

In many hydraulic circuits, a rapid traverse is needed until the machining portion of the stroke is reached. This is known as rapid advance and feed type circuitry.

For this portion of the circuitry, the directional valve has been shifted and pump flow is directed to the cap end of the cylinder. Oil flow from the head is free to flow through the deceleration valve. The fluid will move through the directional control valve and back to tank.

Rapid Advance & Feed Type Circuit

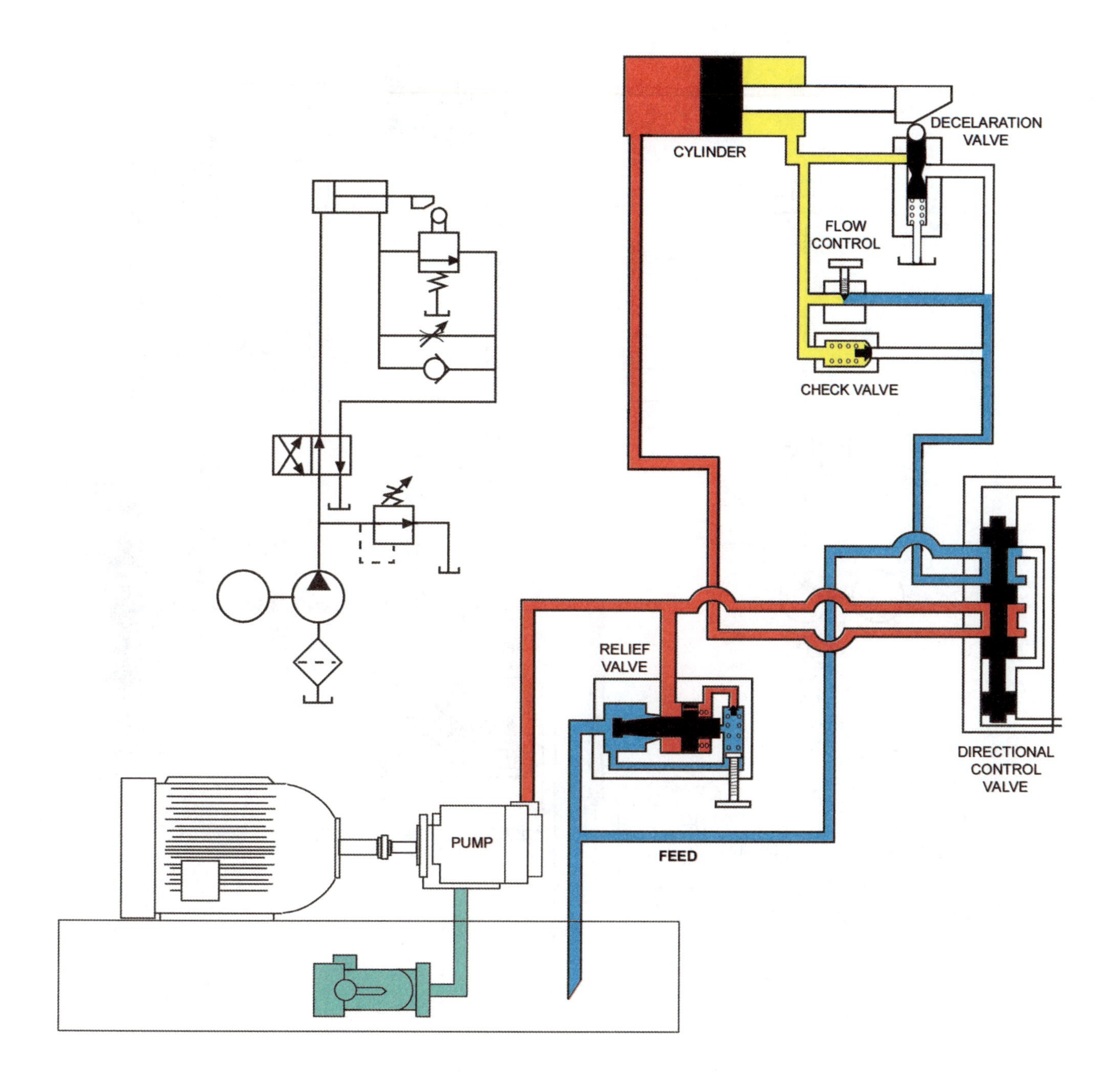

Feed

It is at this point in the circuitry that the cam connected to the cylinder rod actuates the deceleration valve. As the cam depresses the plunger, flow through the valve is gradually cut off. This valve allows a load attached to the cylinder rod to be slowed down at any point of its stroke, where cylinder cushions are not in effect. For the remainder of the stroke, oil leaving the head end will pass through the flow control valve (set at the necessary feed rate) through the directional control and back to tank. It should be noted that the relief valve has opened because flow control has now become apparent to the circuitry.

Rapid Advance & Feed Type Circuit

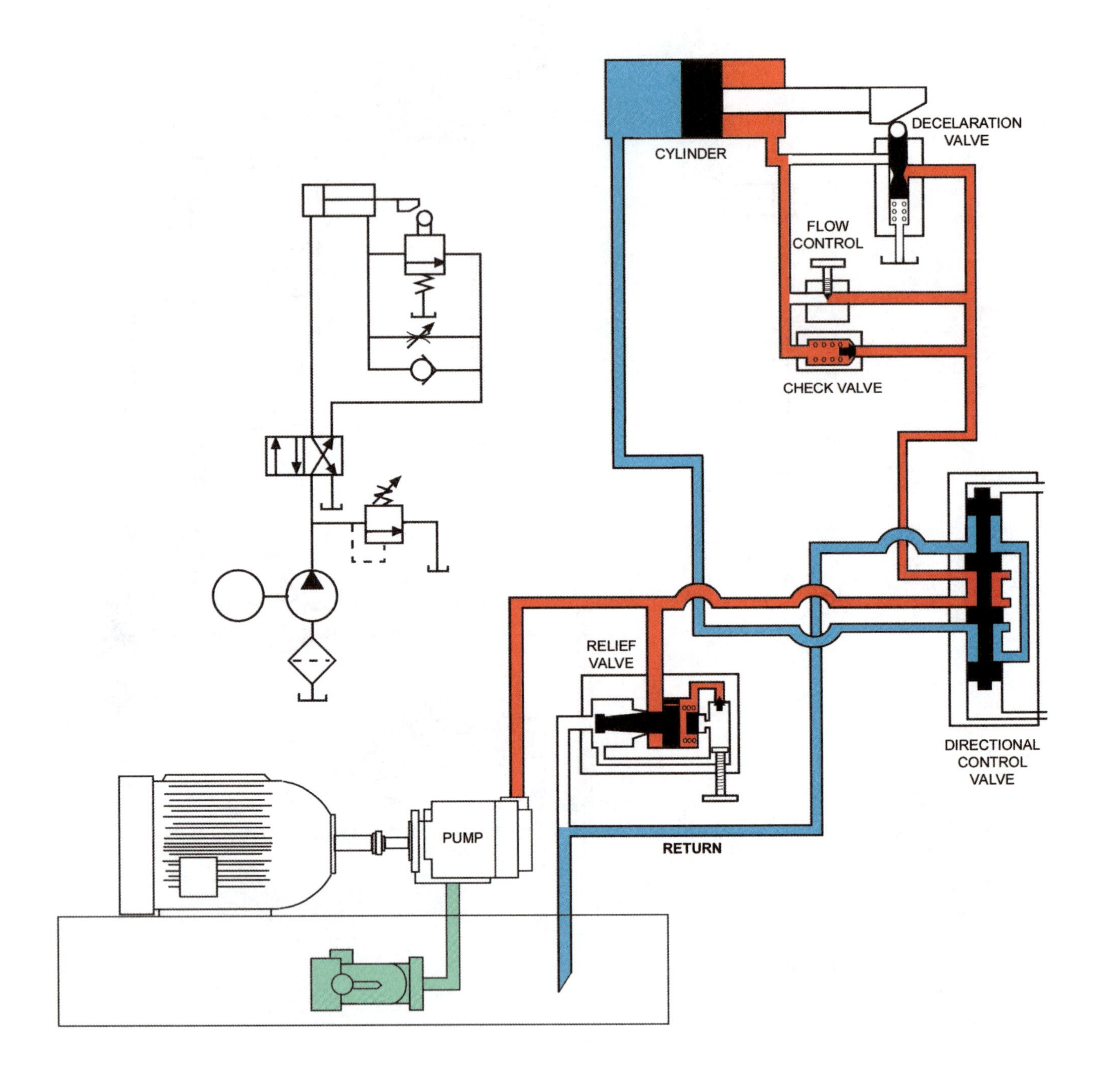

Retract

In this sketch, pump flow is directed through the directional control valve to the check valve, flow control and deceleration valve. Since the check valve has the least resistance, most of the flow will pass through its path. Fluid exiting the cap end of the cylinder is directed through the directional control and back to tank.

Automatic Pump Unloading

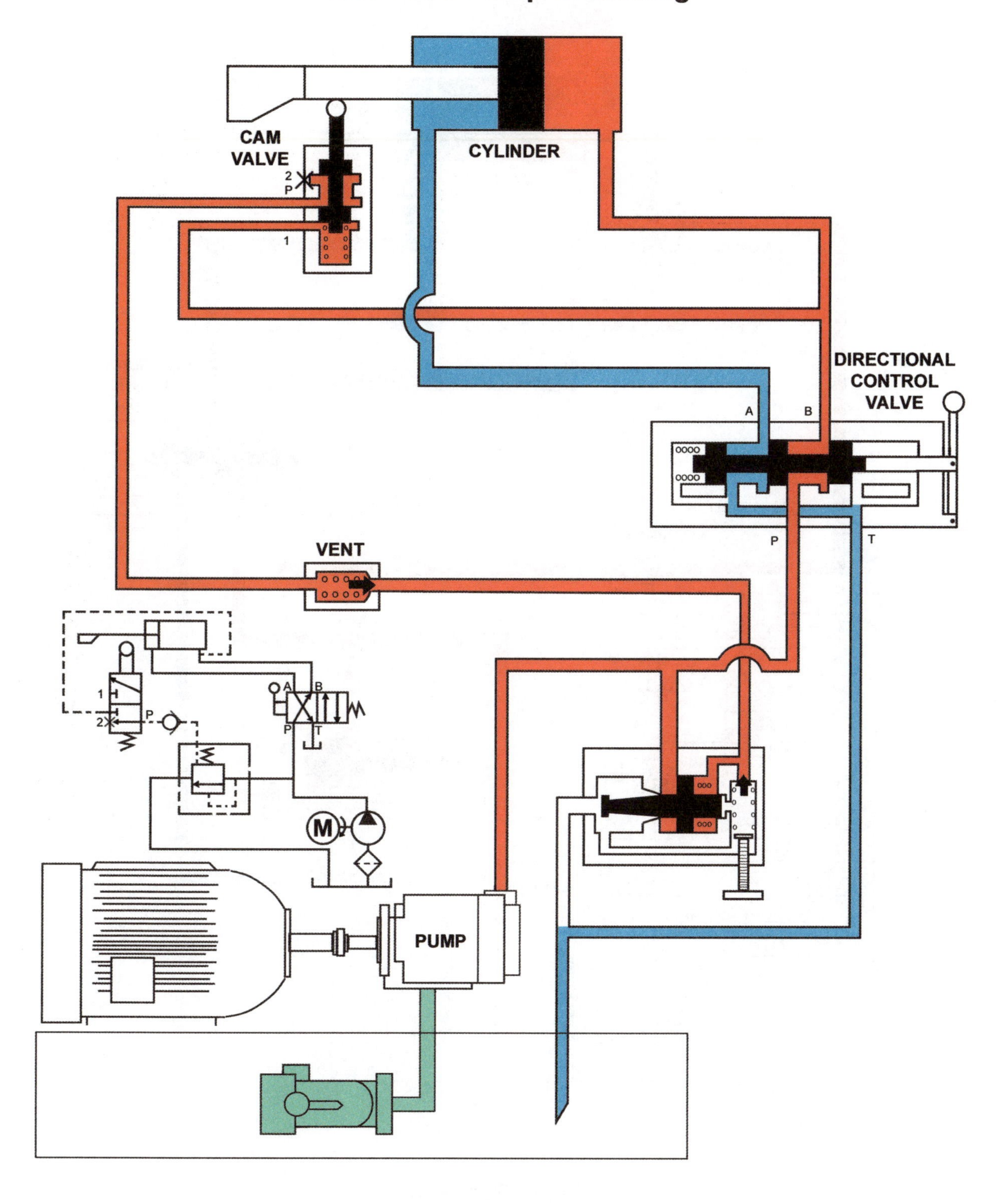

Cylinder Extending

To make the cylinder extend, the directional control valve is pushed. This directs the flow from the pump to the cap end of the cylinder, as well as closes the .3 bar (5 psi) check valve. By closing the check valve, pilot flow from vent port stops, and working pressure is obtainable.

Automatic Pump Unloading

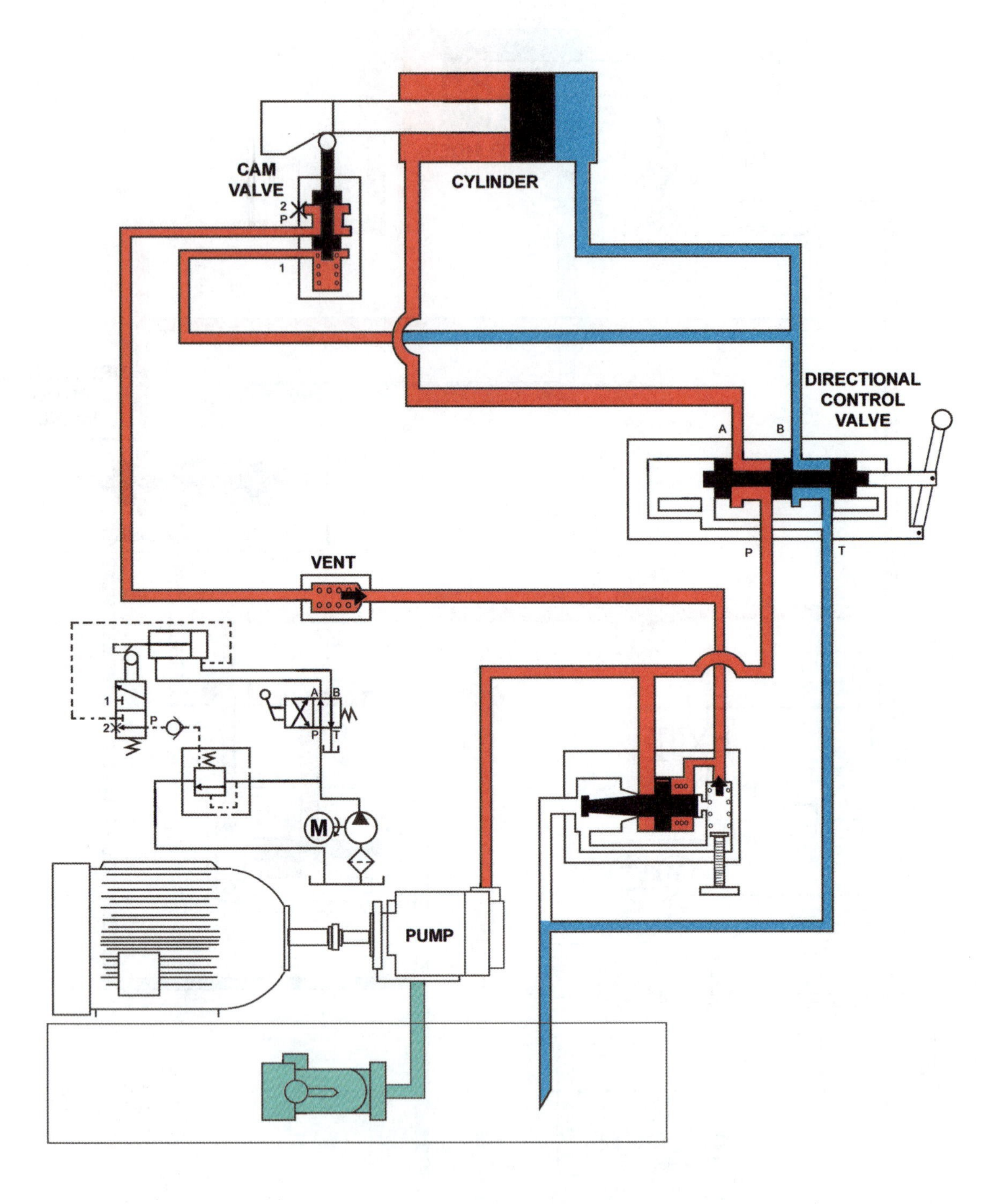

Cylinder Retracting

For retraction, the directional control valve handle is released. This directs pump flow to the head end of the cylinder. The vent line of the relief valve remains closed until the cylinder is fully retracted.

Automatic Pump Unloading

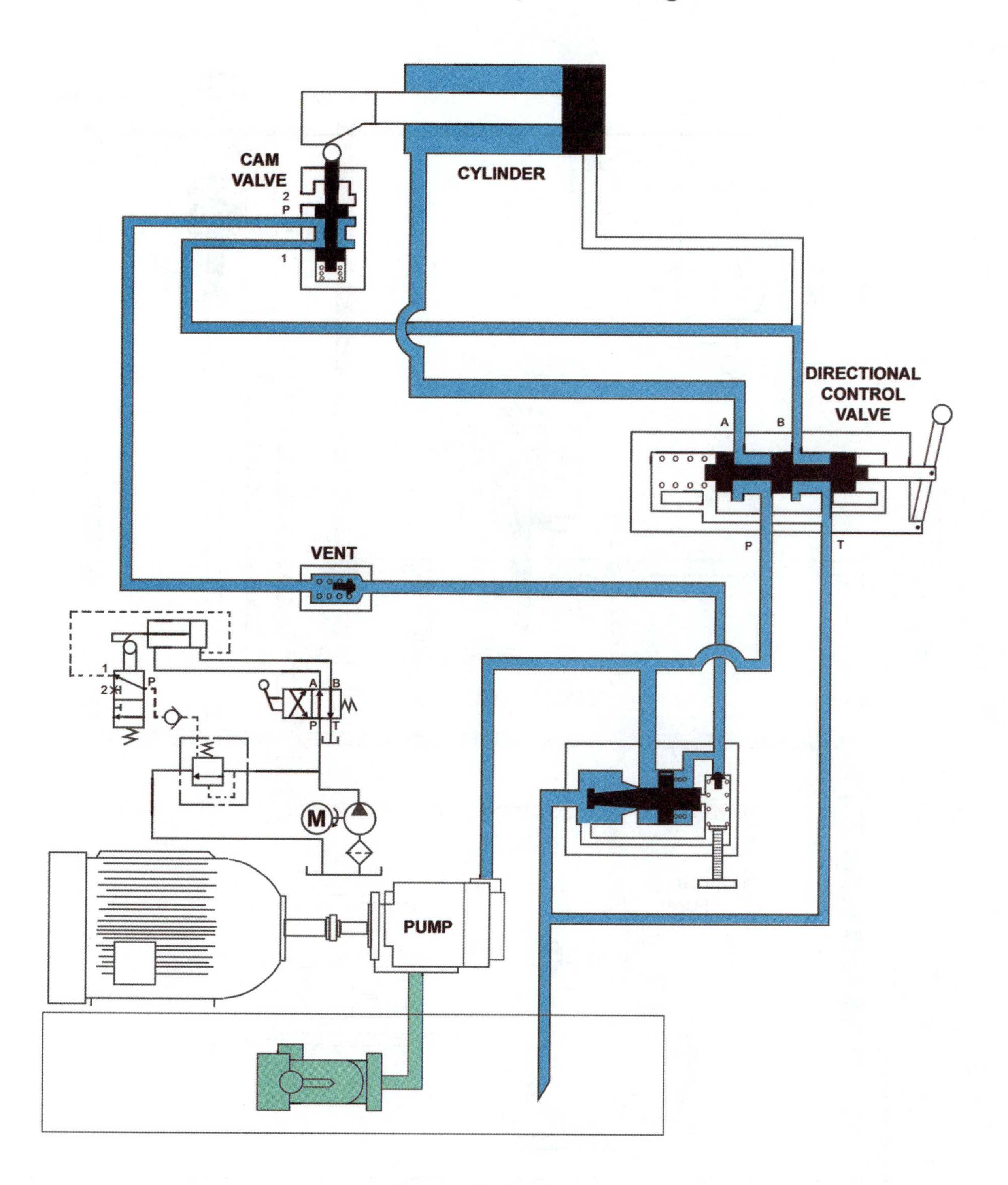

Pump Unloaded

At the end of the retraction stroke, the cam on the cylinder will depress the cam valve. This will allow the fluid from the vent port of the relief to return to tank. This, in turn, will allow the main relief to open causing the pump to run at low pressure for efficient idle operation.

Hi-Lo System

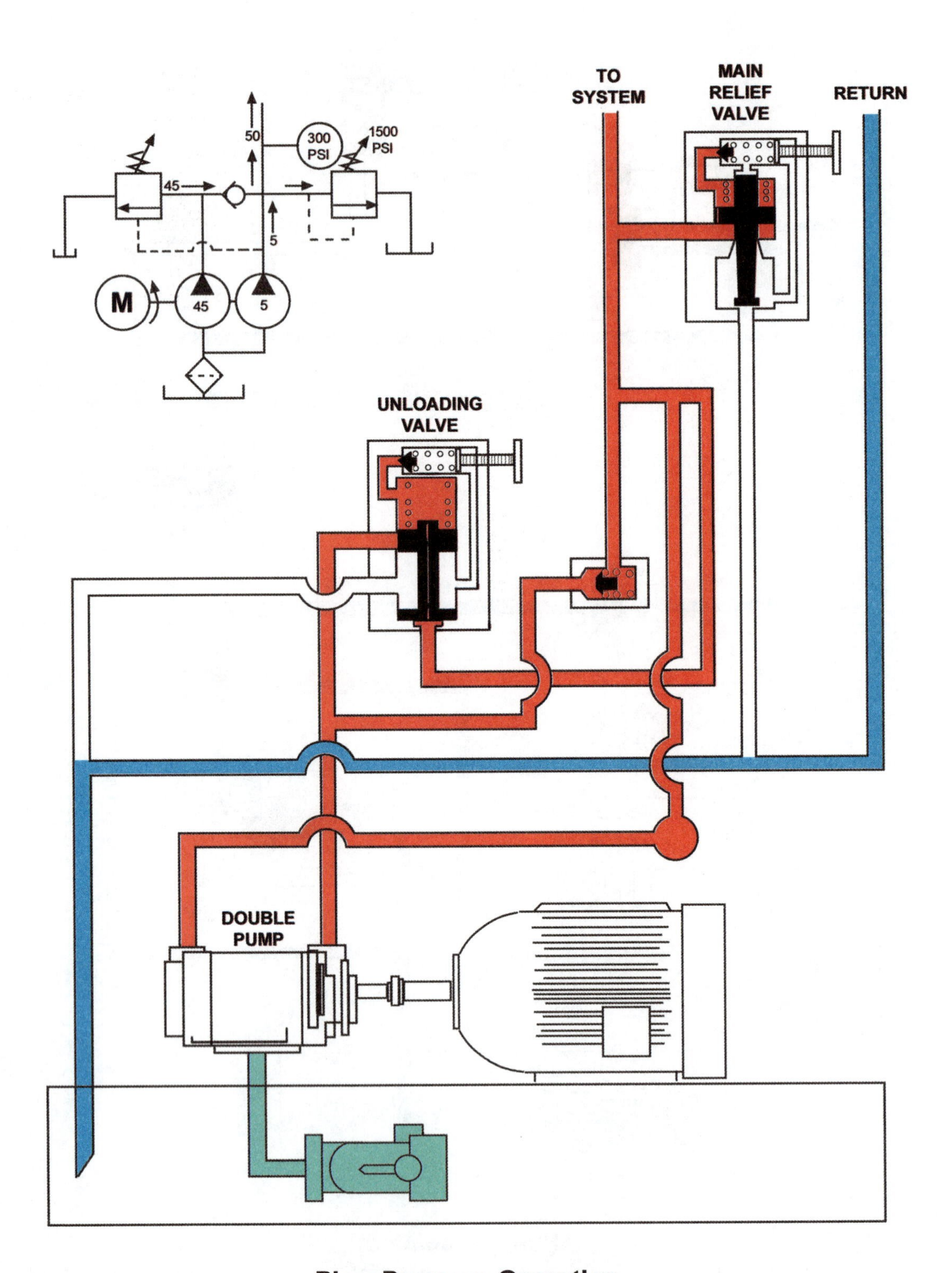

Plow Pressure Operation

A hi-lo system satisfies a system demand by combining a 171 l/min (45 gpm) and 19 l/min (5 gpm) pump flow. When the electric motor is turned on, the 171 l/min (45 gpm) passes through the check valve adding to the 19 l/min (5 gpm flow); 190 l/min (50 gpm) passes out into the system, possibly extending a cylinder at a relatively low pressure.

Hi-Lo System

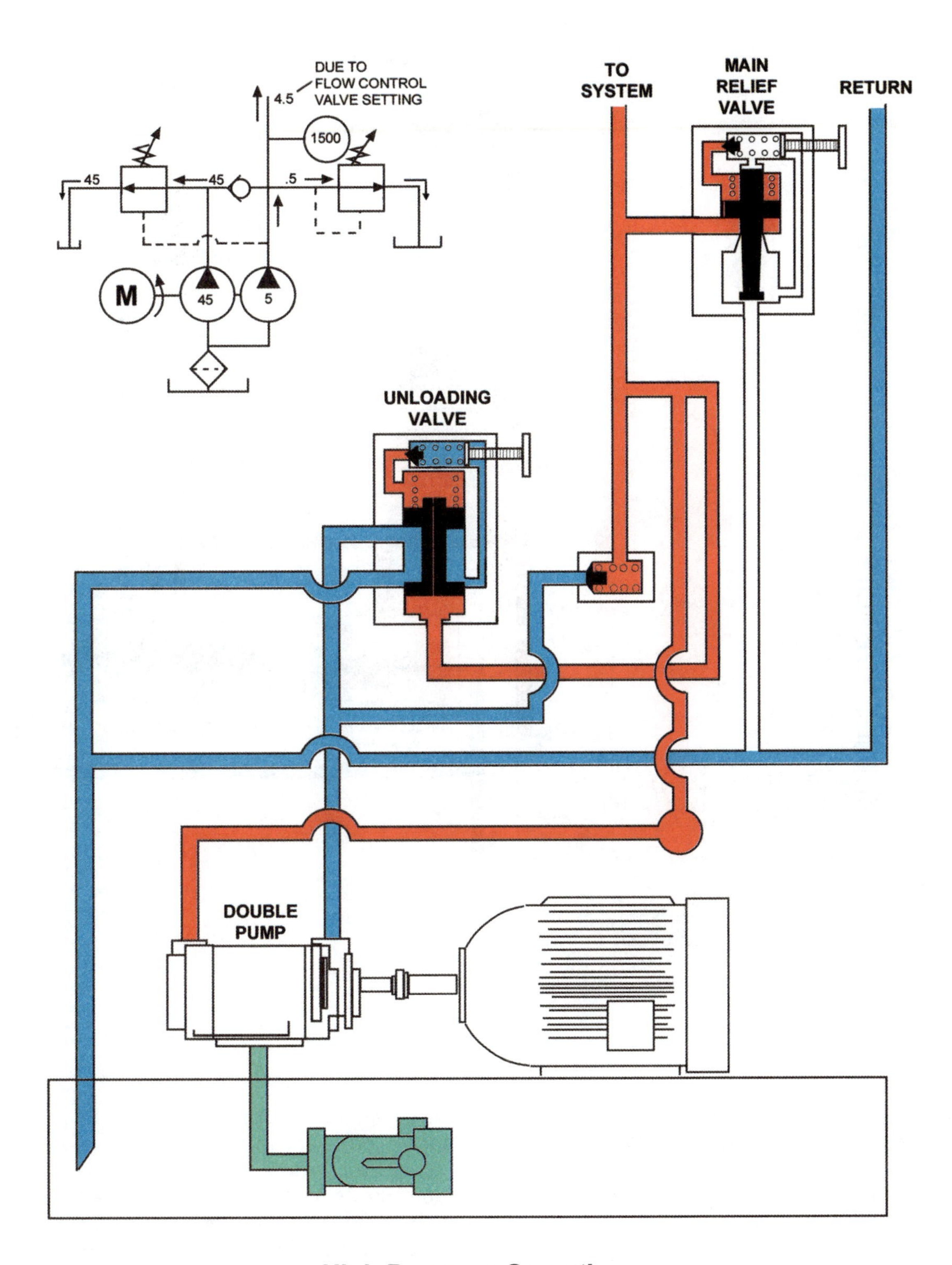

High Pressure Operation

When the work load is contacted and the work pressure is desired, pump/electric motor pressure starts climbing toward the relief valve setting of 104 bar (1500 psi). As it passes through the 34.5 bar (500 psi) pressure level, the normally closed unloading valve opens allowing the 171 l/min (45 gpm) pump to unload while the 19 l/min (5 gpm) pump continues to work. This action eliminates unnecessary power generation by the 171 l/min (45 gpm) pump when it is not needed.

Meter-in Circuitry

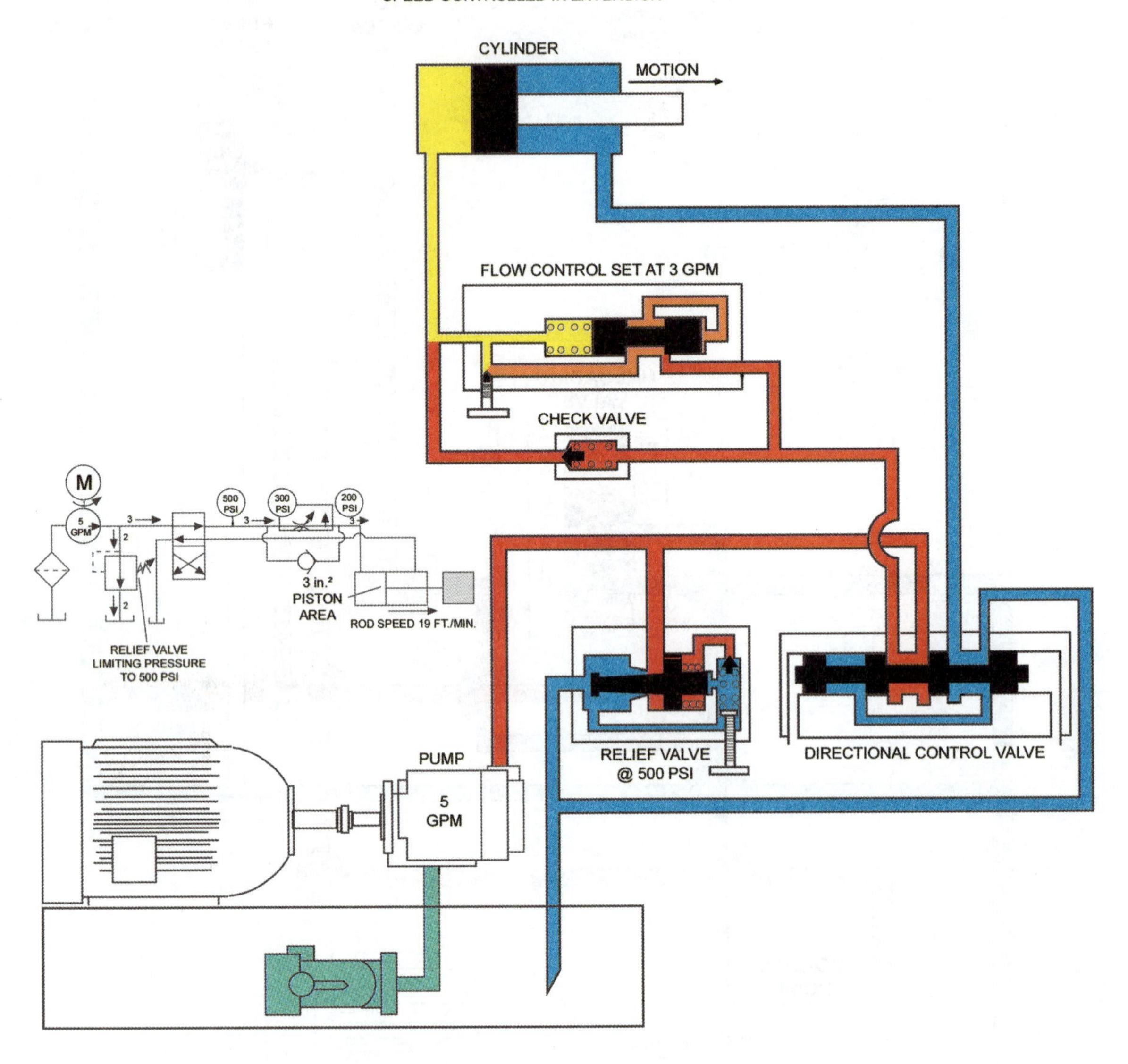

In the circuit illustrated, the restrictor type pressure compensated flow control valve is set for 11 l/min (3 gpm). Relief valve setting is 34.5 bar (500 psi). Work load pressure is 14 bar (200 psi). The spring biasing the compensator spool has a value of 7 bar (100 psi).

During system operation, the work load pressure of 14 bar (200 psi) plus the 7 bar (100 psi) spring, bias the compensator spool.

The pump attempts to push its total flow of 19 l/min (5 gpm) through the needle valve orifice. When pressure ahead of the needle valve reaches 21 bar (300 psi), the compensator spool moves and causes a restriction for the incoming fluid. The pressure at the flow control inlet rises to the relief valve setting of 34.5 bar (500 psi). As the fluid passes over the resttriction made by the compensator spool, 14 bar (200 psi) of the 34.5 bar (500 psi) is transformed into heat. The pressure ahead of the needle valve is limited to 21 bar (300 psi). Of this 21 bar (300 psi), 14 bar (200 psi) is used to overcome the resistance of the load; 34.5 bar (100 psi) is used to develope a flow rate through the needle valve orifce. The flow rate in this case is 11 l/min (3 gpm). The remaining 7.5 l/min (2 gpm) is dumped over the relief valve.

Meter-out Circuitry

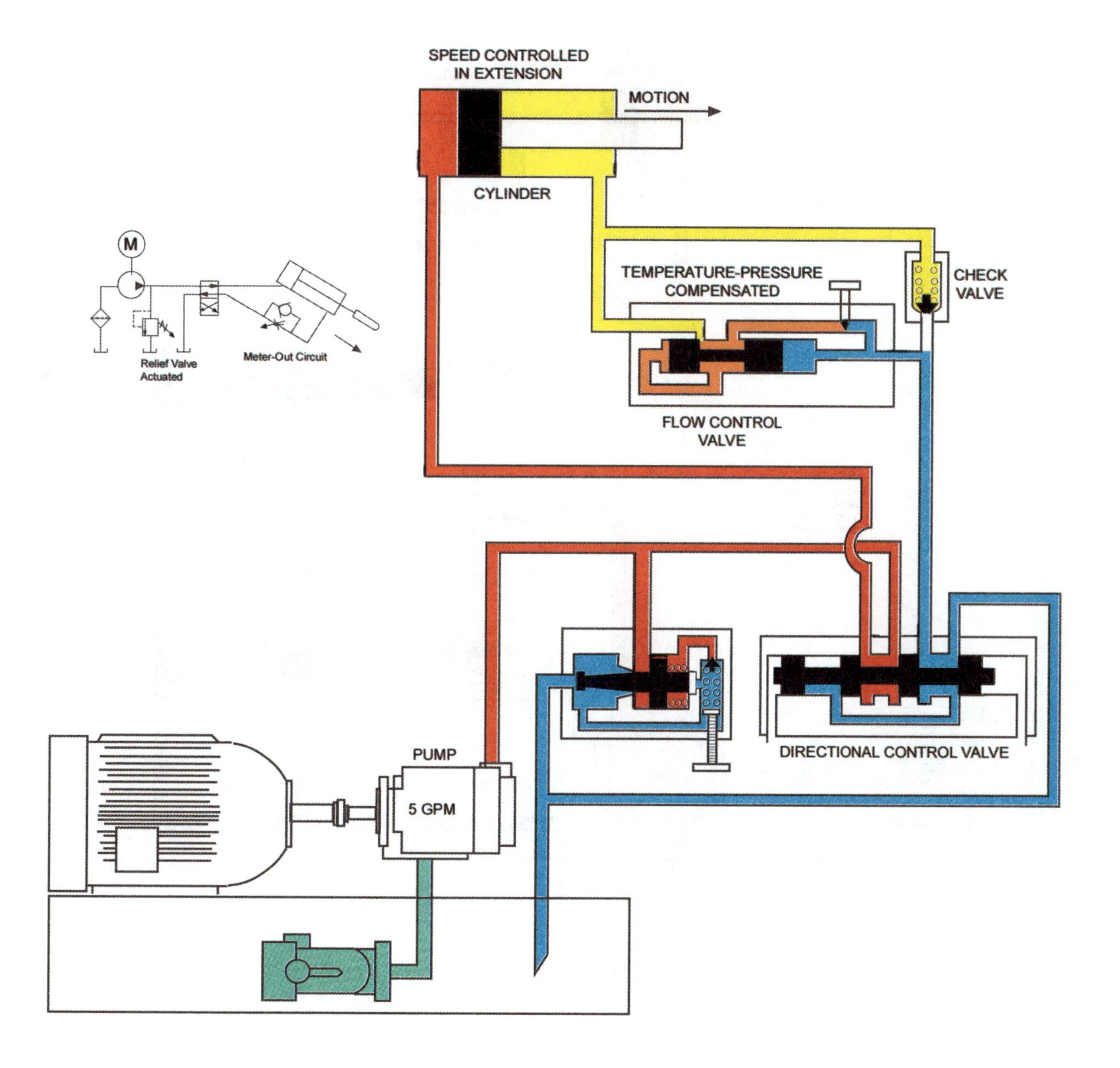

If the speed of an actuator must be precise through the workday, a temperature-pressure compensated flow control could be used.

In some cases, the work load changes direction (load passing over the center point of an arc) or the work load pressure suddenly changes from full to zero pressure (drill breaking through stock). This causes the load to run away.

A flow control valve placed at the outlet port of an actuator controls the rate of flow exiting the actuator. This is a meter-out circuit and gives positive speed control to actuators used in drilling, sawing, boring and dumping operations. A meter-out circuit is a very popular industrial hydraulic flow control unit.

Bleed-Off Circuitry

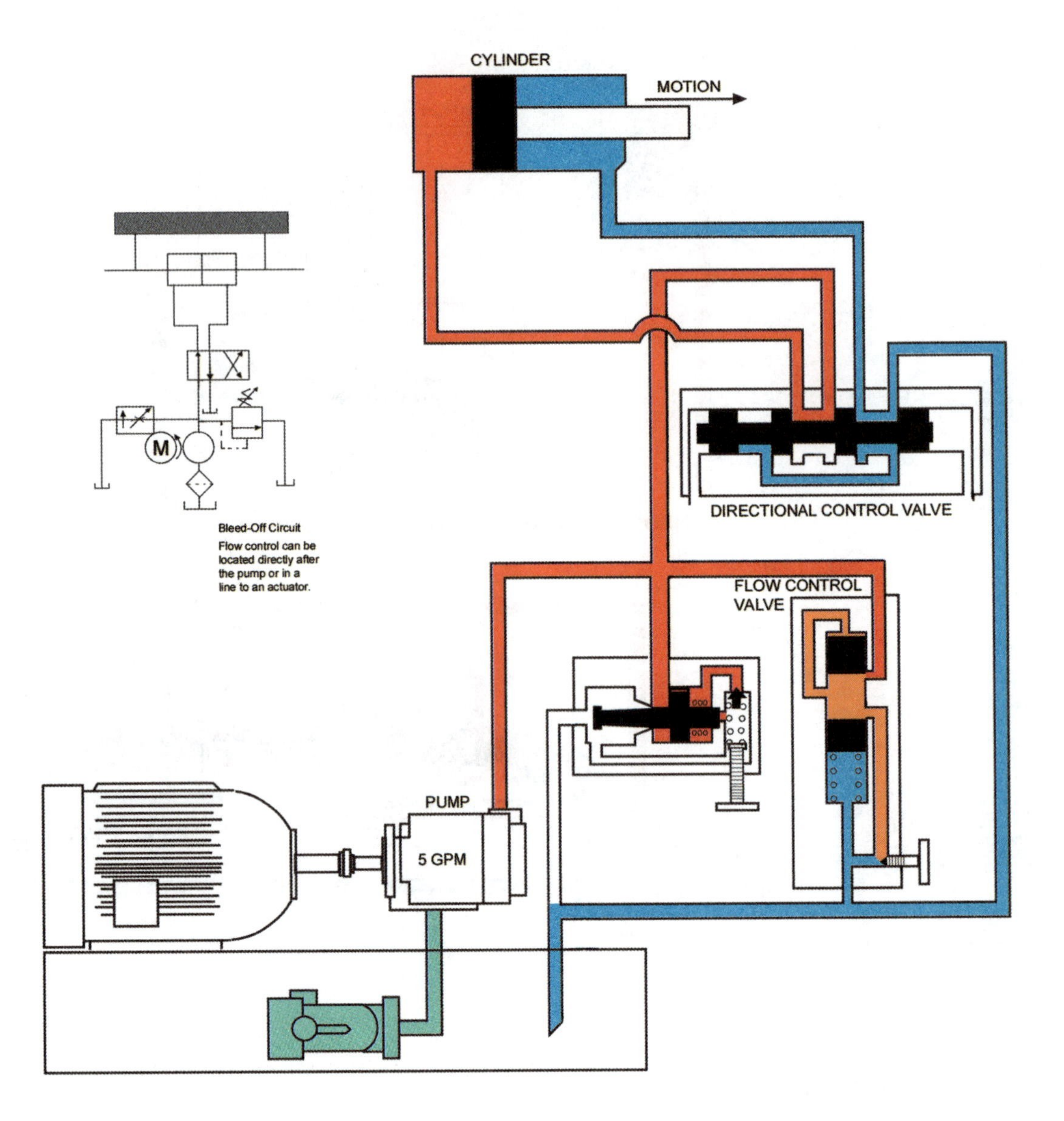

Another type of flow control circuit is the bleed-off circuit. In this circuit, the flow control valve does not cause an additional resistance for the pump. It operates by bleeding off to tank a portion of the pump's flow at the existing system pressure.

Besides generating less heat, a bleed-off circuit can also be more economical than a meter-in or meter-out circuit. For instance, if a flow rate of 379 l/min (100 gpm) had to be reduced to 341 l/min (90 gpm), a 341 l/min (90 gpm) flow control valve would be needed in a meter-in circuit and, depending upon the size of the cylinder, approximately a 256 l/min (70 gpm) flow control in a meter-out circuit. Whereas in a bleed-off circuit, a 38 l/min (10 gpm) flow control could be used.

Even with these apparent advantages, a bleed-off circuit is not a very popular flow control circuit. This is because a flow control in a bleed-off arrangement only indirectly controls the speed of an actuator. It can precisely meter flow to the tank, but if leakage through various system components increases, actuator speed will decrease.

A bleed-off circuit can be used in any application where precision flow regulation is not required; and where the load offers a constant resistance as in reciprocating grinding tables, honing operations and vertically lifting a load.

Counterbalance Valve

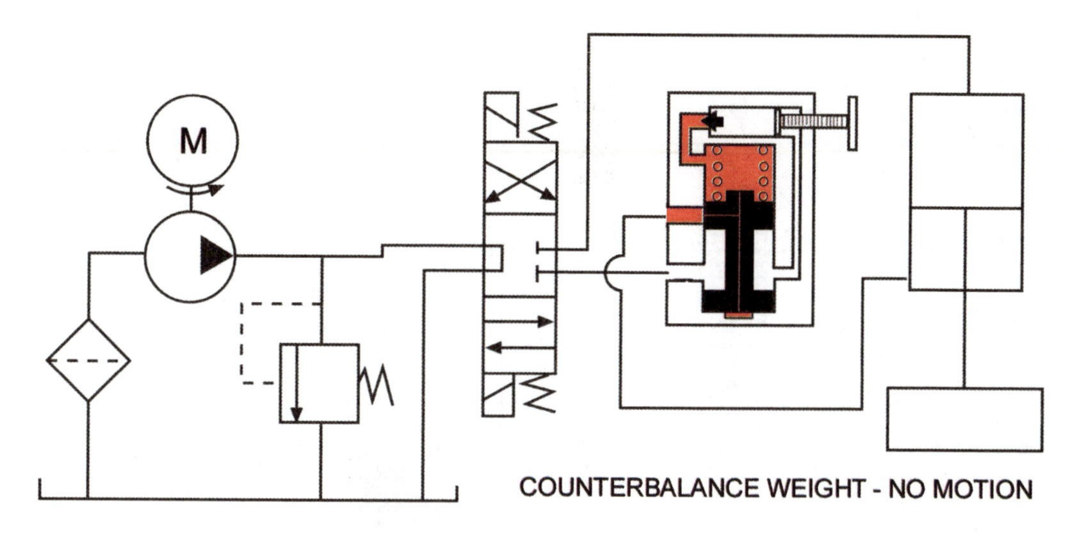

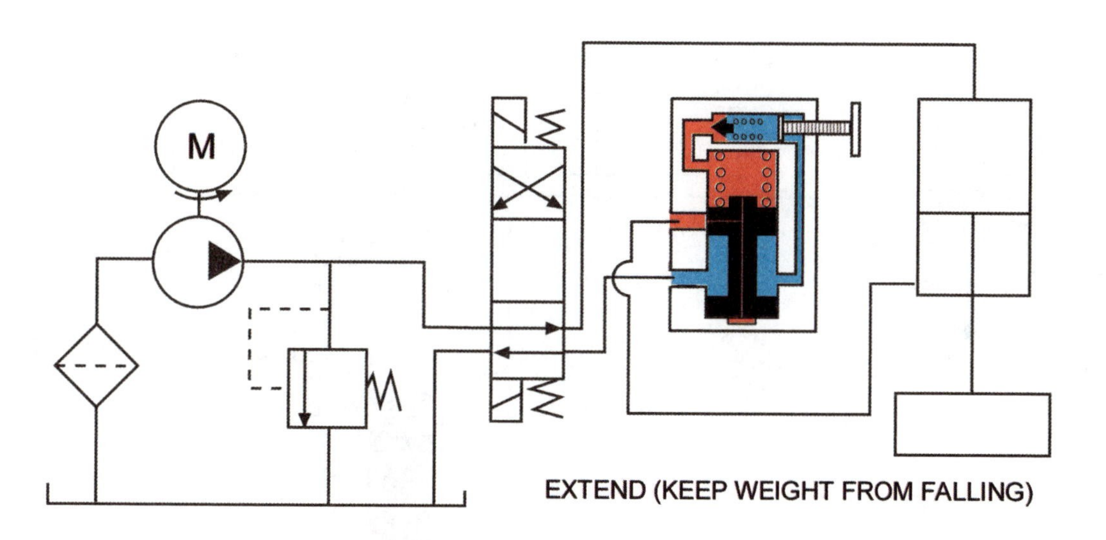

In a press circuit, when the directional valve directs flow to the cap end of the cylinder, the weight of the platen attached to the cylinder rod will fall uncontrollably. Pump flow will not be able to keep up.

To avoid this situation, a normally closed pressure valve is located downstream from the press cylinder. The spool in the valve will not connect primary and secondary passages until a pressure, which is sensed at the bottom of the spool, is greater than the pressure developed by the weight of the platen. (In other words, when fluid pressure is present at the cap end of the piston.) In this way the weight of the platen is counterbalanced throughout its downward stroke.

A normal requirement of the counterbalance valve is that reverse flow must be able to pass through the valve.

Since normally closed pressure valves sense pressure form the primary passage, as soon as flow is reversed, pressure in the primary passage falls off. The spool is de-actuated. Primary and secondary passages are disconnected. Flow through the valve is blocked. Since we cannot go through the valve, we go around the valve by using a check valve. This check is typically built right in the valve. (It has not been shown in this case.)

Pressure Reducing Valve

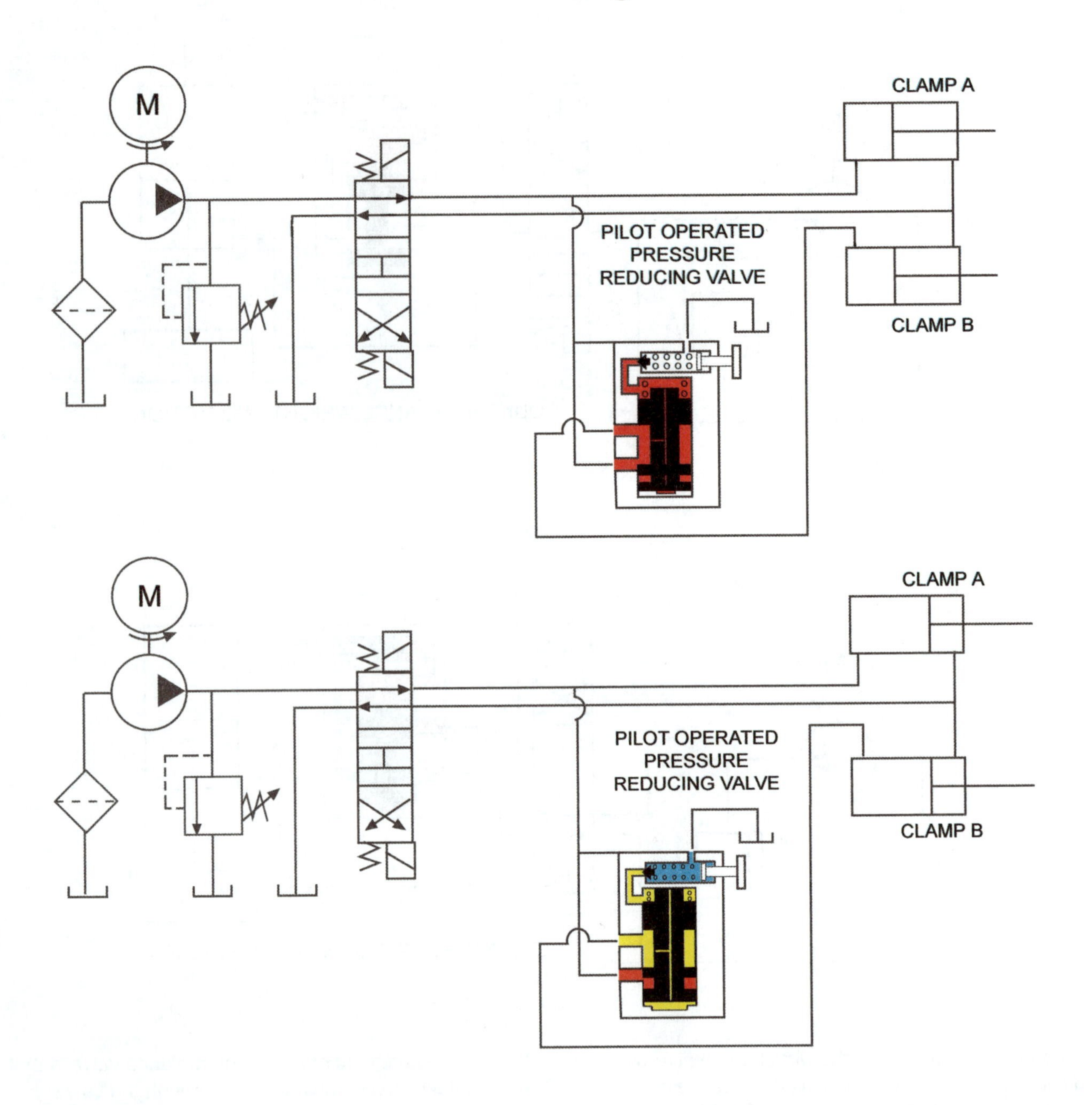

A pressure reducing valve is a normally open pressure control valve.

A pressure reducing valve operates by sensing fluid pressure after it has passed through the valve. As pressure downstream equals the setting of the valve, the spool is partially closed causing a restricted flow path. This restriction turns any excess pressure energy ahead of the valve into heat.

If pressure after the valve drops off, the spool will open and allow pressure to build once again.

The illustrated clamp circuit requires that clamp cylinder B apply a lesser force than clamp A. A pressure reducing valve placed just ahead of clamp cylinder B will allow flow to go the cylinder until pressure reaches the setting of the valve. At this point, the valve spool is actuated causing a restriction to that leg of the circuit. Excess pressure ahead of the valve is turned into heat. Cylinder B clamps at a reduced pressure.

Brake Valve

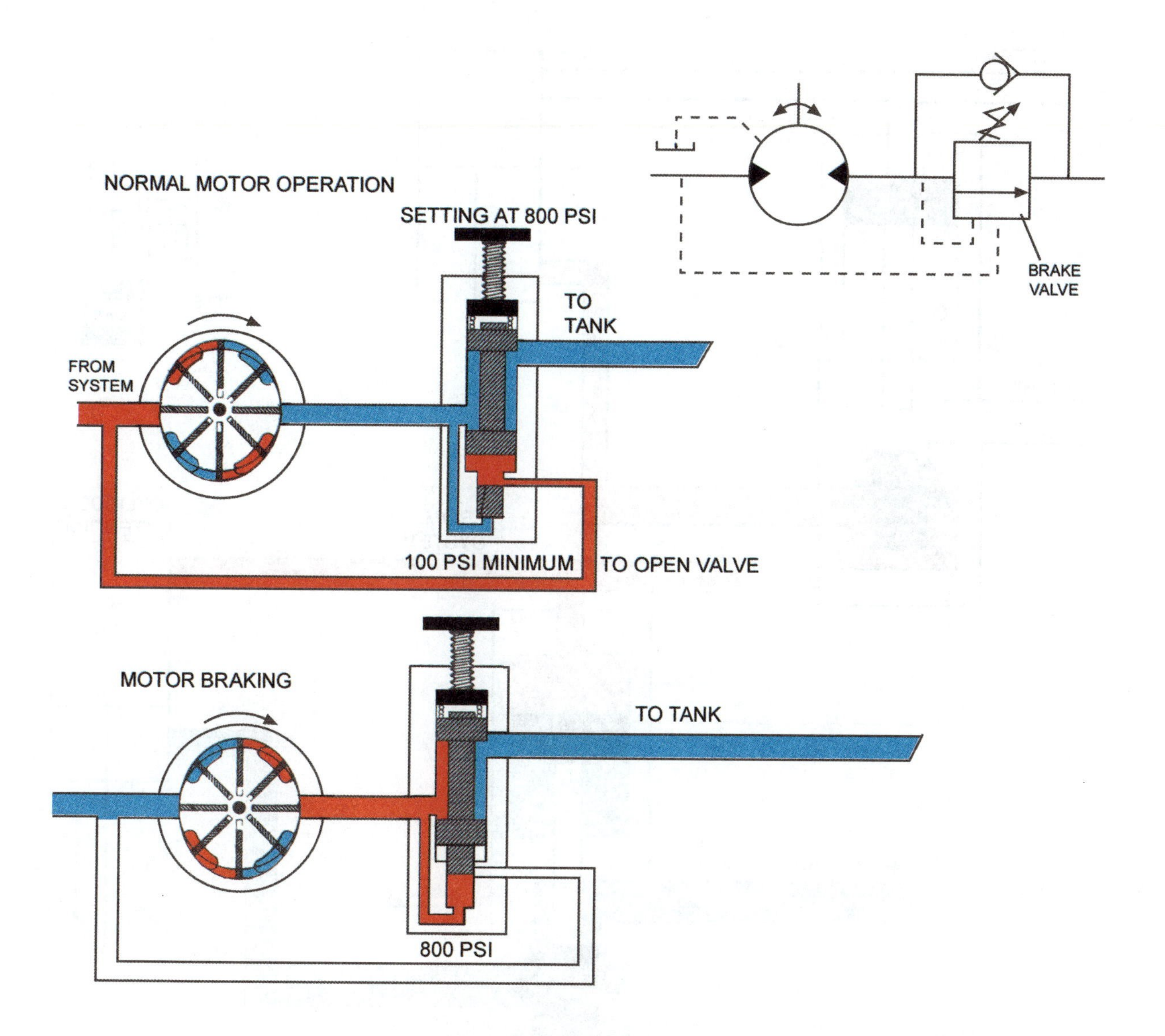

A brake valve consists of a valve body with primary and secondary passages, internal and remote pilot passages, spool, piston, bias spring and spring adjustment.

A brake valve is a normally closed valve. Assume that the spring biasing the spool is adjusted for 55 bar (800 psi) direct operation. When pressure in the internal pilot passage reaches 55 bar (800 psi), the piston moves up pushing the spool and opening a passage through the valve. If pressure falls below 55 bar (800 psi) the valve closes. This operates as the directly operated counterbalance valve which we saw earlier.

The piston on which the internal pilot pressure acts, has much less cross sectional area than the spool. The area ratio is frequently 8:1. With the remote pilot connected to the opposite motor line, a pressure of only 7 bar (100 psi) is needed to open the valve since it acts upon the bottom of the spool with eight times more area than the piston.

With a brake valve set for 55 bar (800 psi), the valve will be open when 7 bar (100 psi) is present in the motor inlet line. Pressure at motor inlet will be whatever it takes to turn the load only [assuming that his pressure is above 7 bar (100 psi)]. If the load attempts to run away, pressure drops off in the motor inlet line. The brake valve closes and does not reopen until a backpressure of 55 bar (800 psi) is generated to slow down the load.

A brake valve is a normally closed pressure control valve whose operation is directly tied to the needs of a motor load.

Pilot Operated Check Valve

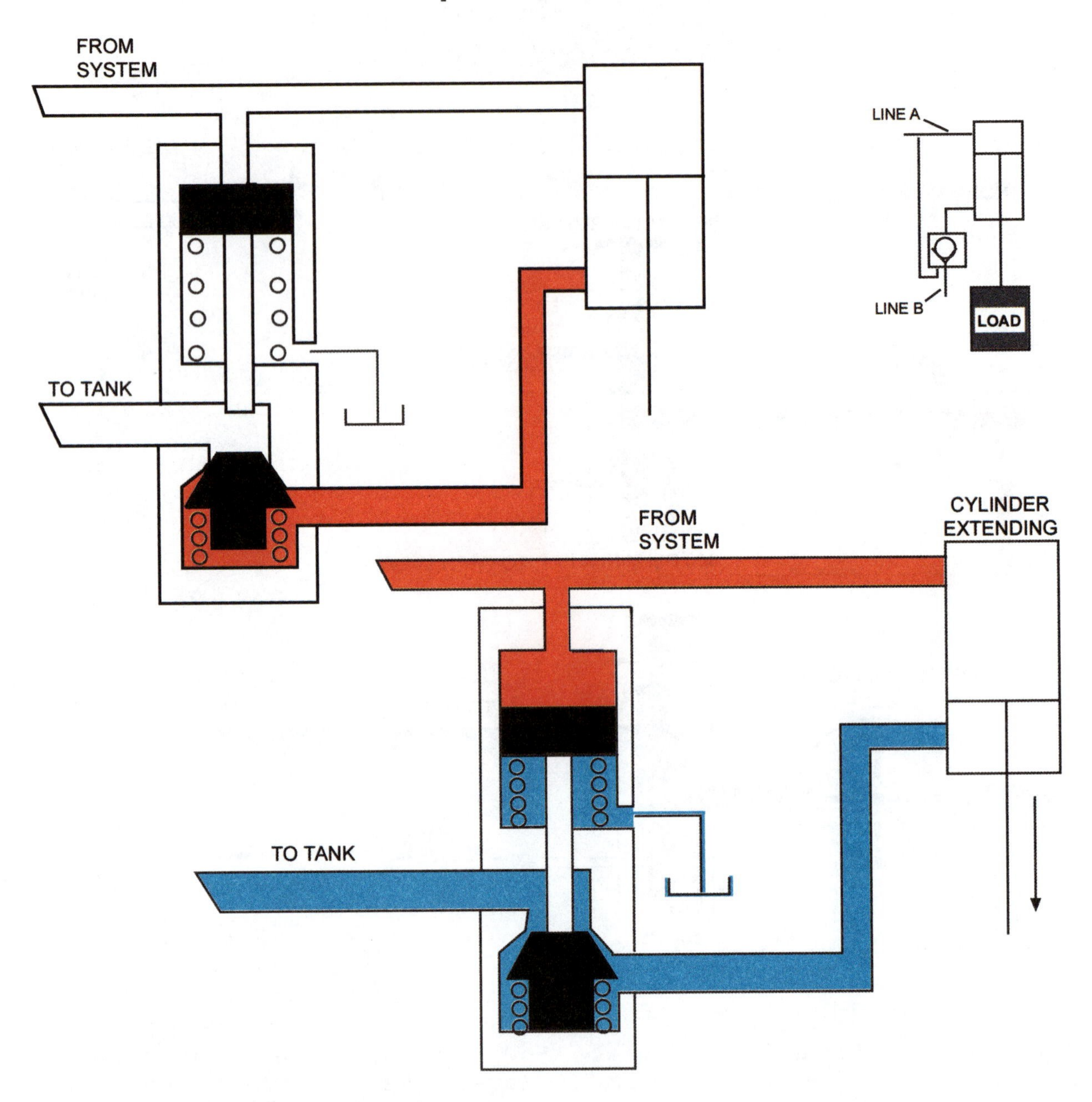

A pilot operated check valve allows free flow from its inlet port to its outlet port just as an ordinary check valve.

Fluid flow attempting to pass through the valve from outlet to inlet port will force the poppet on its seat. Flow through the valve is blocked.

When enough pilot pressure is sensed at the plunger piston, the plunger is moved and unseats the check valve poppet. Flow can pass through the valve from outlet to inlet as long as sufficient pilot pressure is acting on the plunger piston.

With a pilot operated check valve blocking flow out of cylinder line B, the load will stay suspended as long as the cylinder seals remain effective. When it is time to lower the load, system pressure is applied to the cylinder piston through line A.

Pilot pressure to operate the check valve is taken from this cylinder line. The check valve will remain open as long as enough pressure is available in line A.

To raise the load, fluid flow can easily pass through the valve since this is the valve's free flow direction.

Appendix C
Hydraulic Graphic Symbology

Excerpted from Parker Hannifin Bul. 0240-B1, Hydraulic Maintenance Technology

One of the first steps in diagnosis and testing is to get a schematic of the hydraulic, electric and any other system on the machine to be analyzed. The schematic is the "road map" to the hydraulic system. What if no schematic exists? One line of thought to remedy this problem is for you to draw the schematic or at least have someone do this. If you draw the schematic, it will help you better understand the system and greatly simplify your troubleshooting. Therefore, a thorough understanding of the graphic symbols for fluid power systems and components is necessary.

ANSI (American National Standards Institute) and ISO (International Organization for Standardization) are the two recognized standards bodies for fluid power system and component graphic symbols.

Many companies create their own symbols for their components; however, this practice should be discontinued and avoided because of the confusion it can and does cause.

The ISO symbols have become the standard for international usage; therefore, we'll look at these symbols in greater detail. There are six categories that comprise all of the symbols. It is the intent of this section to thoroughly acquaint you with these symbols and then utilize them in developing simple systems as well as diagnose problems using some actual system schematics.

"Graphic symbols shown in accordance with ISO 1219, Fluid Power Systems and Components - Graphic Symbols."

When this phrase is seen on test reports, schematics, catalogs and sales literature it means that all symbols used comply to the ISO Standard, ISO 1219-1976.

General Symbols

The symbols commonly used in drawing hydraulic equipment and any accessory equipment are either pictorial, cutaway or graphic symbols. There is no prescribed layout or scale size for the symbols. Generally in drawing a circuit diagram, the draftsperson will represent the equipment in the non-actuated (at rest) condition. This does not preclude other conditions from being represented, but it should be clearly stated that some other condition exists when such a situation occurs.

Pictorial symbols are very useful for showing the interconnection of components. They are not standardized from a functional basis.

Cutaway symbols emphasize construction. These symbols are complex to draw and the functions are not readily apparent.

Graphic symbols emphasize the function and methods of operation of components. These symbols are simple to draw. Graphic symbols are capable of crossing language barriers, and can promote a universal understanding of fluid power systems.

When the draftsperson draws the diagram, they have the choice of using either complete graphic symbols, simplified graphic symbols, or composite graphic symbols.

Complete graphic symbols are those which give symbolic representation of the component and all of its features pertinent to the circuit diagram.

Simplified graphic symbols are stylized versions of the complete symbols

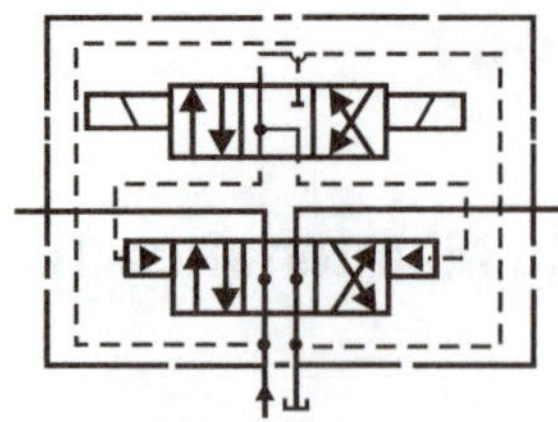

Detailed directional valve

Simplified directional valve

Composite graphic symbols are an organization of simplified or complete symbols. Composite symbols usually represent a complex component.

General Characteristics of Symbols

All graphic symbols will take on one of the basic elementary forms, such as:

Circles
Rectangles
Arcs
Dots
Squares
Triangles
Arrows
Crosses
Lines - solid or dashed

When drawing these symbols, there are some clear rules to follow in order to make the graphic symbols recognizable on the international level.

1. Symbols show connections, flow paths, and functions of components represented.
2. They can indicate conditions occurring during transition from one flow path arrangement to another.
3. Symbols do not indicate construction, nor do they indicate values, such as pressure, flow rate, and other component settings.
4. Symbols do not indicate the location of actual valve ports, direction of shifting of spools, or the actual mounting of actuators in machines.
5. Symbols may be rotated or reversed without altering their meaning except in cases of lines to reservoirs, vented manifolds and accumulators.
6. The means of operating fluid power components are shown as part of the symbol (where applicable).
7. Symbols using words or their abbreviations are avoided.
8. Simplified symbols are shown for commonly used components.

Lines

The basic symbols of lines consists of continuous lines for main flow lines; long dashes for pilot pressure lines; short dashes for drain lines.

Lines that are connected by tees and crosses are denoted by a dot at the point of intersection. Lines that cross over one another but do not intersect, can be denoted by not putting a dot at the crossover point or by putting a loop in one line.

Double lines are mechanical connections such as motor shafts, hand levers and cylinder rods. Several components that are assembled as a complete unit are represented by a series of long and short lines that enclose them.

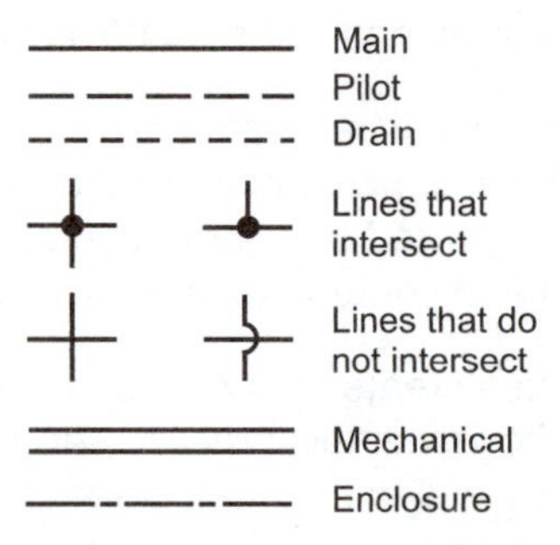

Circles, Semi-Circles Squares, Rectangles, and Diamonds

The basic symbol of a circle or semi-circle represents energy conversion components such as a pump and/or motor. Smaller diameter circles are used as measuring instruments (pressure and vacuum gauges), non-return valve, rotary connection, mechanical links and rollers.

The square and rectangle are generally reserved for control valves except for non-return type valves.

The single square can also be used to represent an internal combustion engine. Therefore, the basic symbol can be used to represent several different components.

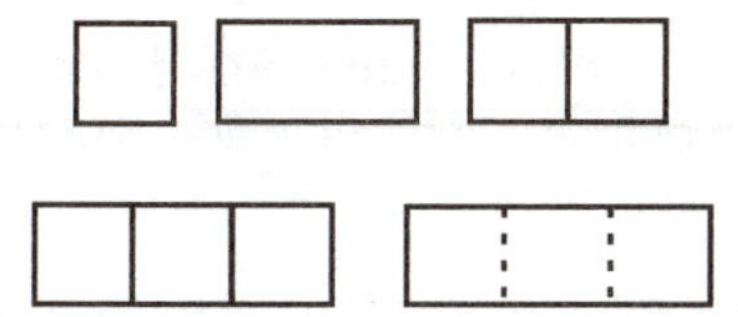

Commonly known as envelopes, the squares joined in series will tell you how many distinct operating positions a valve may assume such as two, three or four. When there are three positions indicated, but they are separated by dash lines, it means that the valve is being used as a two position valve with some types of cross over center condition.

Filtration, cooling and heating of fluids will use the basic symbol of the diamond to represent the conditioning aparatus.

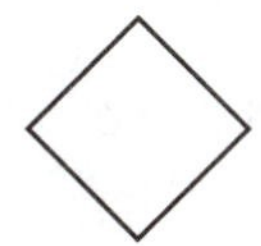

Miscellaneous Symbols

Other basic symbols would include flow line connections, springs and flow restrictions, which are orifices in the flow path; one that is affected by fluid viscosity changes and the other is unaffected by viscosit changes.

The triangle is considered a functional symbol that represents the direction of flow of the fluid and the nature of the fluid whether liquid or gaseous.

Arrows indicate direction, direction or rotation and path and direction of flow through a valve.

In the path and direction of flow, arrows note the perpendicular line to the arrow head. This indicates as a general rule that when the arrow moves, the interior path always remains connected to the corresponding exterior path as will be shown in pressure control valves.

A sloping arrow, generally a 45° slope, indicates the ability to vary or regulate in a progressive manner.

Energy Conversion Components

The next category represents the components that convert energy. Pumps convert mechanical energy into hydraulic energy; motors, cylinders and rotary actuators convert hydraulic energy into mechanical energy.

Pumps

There are two basic types of pumps, fixed and variable displacement capacity. Fixed displacement pumps are represented as shown and can be either unidirectional or bidirectional. The symbol does not, however, tell you whether the pump is a vane, piston or gear pump.

Two-stage or double pumps are represented by two individual unidirectional pump symbols joined together with a mechanical connection symbol (double horiontal lines).

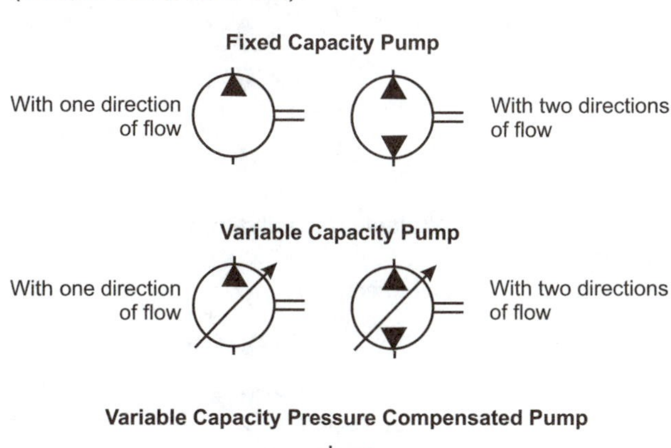

Variable volume displacement and pressure compensated pumps are also unidirectional and bidirectional. The sloping arrow through the pump represents variable volume and the square with the arrow inside represents pressure compensation.

When the pump symbol is drawn, note the two small external lines that are in series with the arrowhead(s). These are the inlet and outlet of the pump.

Arrowhead(s) inside the circle shall be solid and point outward to indicate hydraulic fluid flow is directed outward, into the circuit.

All pumps are driven by a prime mover that would be coupled to the pump using the mechanical connection symbol (double horizontal lines).

Motors

Motor symbols are quite similar to the pump symbols for fixed and variable displacement as well as unidirectional and bidirectional rotation. The major difference is in the position of the internal arrowhead(s). Note the arrowhead is solid, but points inward to indicate fluid is directed into the component.

The semi-circle, called an oscillating motor, is representative of a rotary actuator. This symbol, like the other motor symbols, does not indicate the type of motor being used.

Fixed Capacity Motor

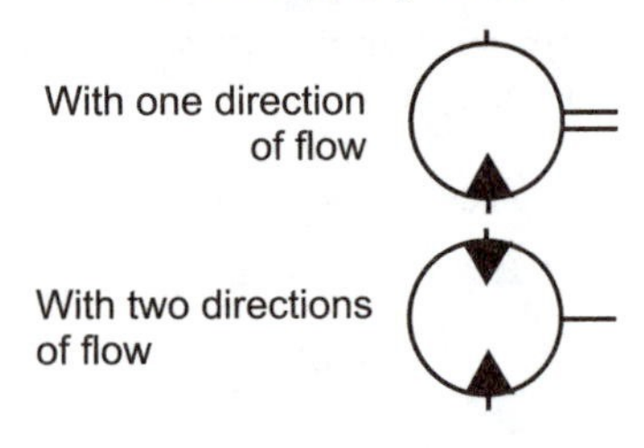

Variable CapacityMotor

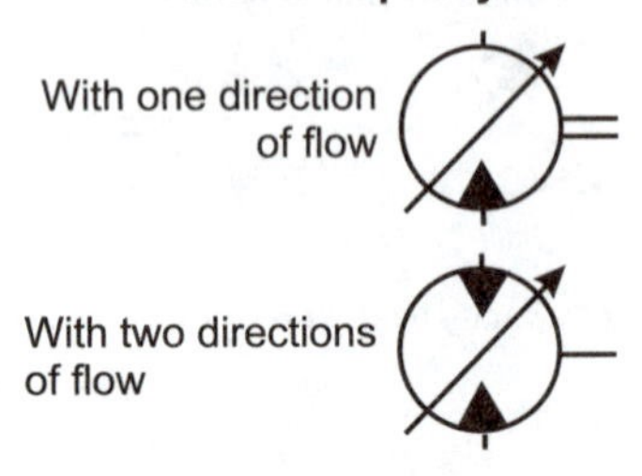

Variable Capacity Pressure Compensated Motor

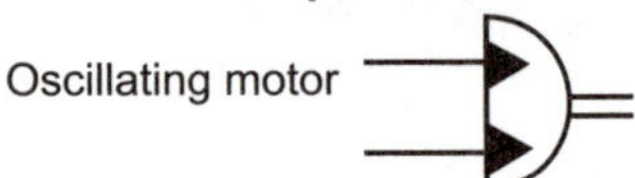

Pump-Motor Units

A more specialized motor is the pump/motor unit with two functions. It either operates as a pump or as a rotary motor.

There are fixed volume pump/motors that can reverse the direction of flow, have single flow direction or be bidirectional in flow.

The variable volume pump/motors are similar with the addition of a sloping arrow to indicate the volume may be changed.

Pump/Motor Units

Fixed capacity pump/motor unit with reversal of the direction of flow

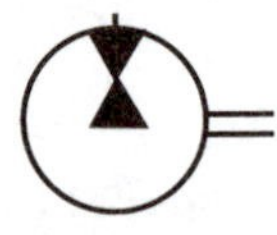

Variable capacity pump/motor unit with reversal of the direction of flow

Cylinders

Cylinders are used to convert hydraulic energy into mechanical force by the application of fluid pressure to the surface area of the piston.

Single Acting Cylinders

The simplest type of cylinder is the single acting cylinder in which the fluid pressure acts on one side of the piston and in one direction. Usually in the forward stroke.

The symbol for a cylinder that needs an external force will have the rod end of the cylinder opened to indicate it is open to atmosphere. At the base end is a small vertical line to indicate the port where fluid enters and leaves the cylinder.

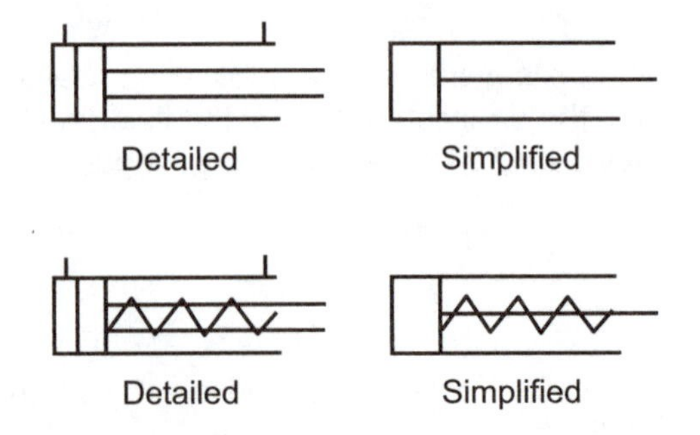

There is also the single acting spring returned cylinder. In the example, fluid pressure will stroke the cylinder forward while spring force will retract the cylinder.

Double-Acting Cylinders

Double-acting cylinders are either single or double rod type. The small vertical lines at each end represents the ports where fluid enters and leaves the cylinder.

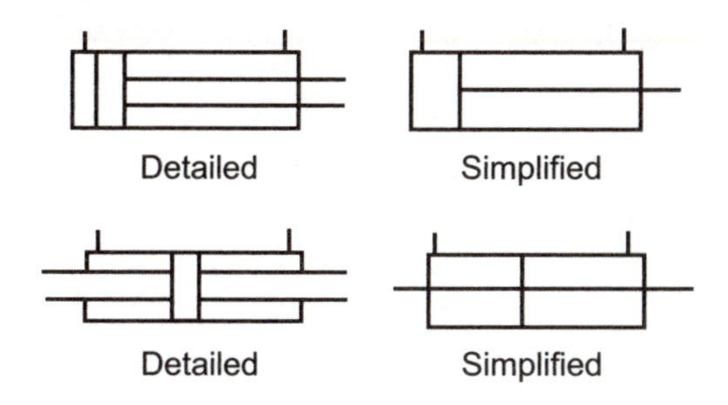

Differential Cylinders and Cylinders with Cushions

Differential cylinders better known as 2:1 cylinders are denoted by the large rod size in the symbol. This is also a double acting cylinder as represented by the two small vertical lines (ports).

Cylinders that are equipped with cushion assemblies at either or both ends of the cylinder are represented as shown. The cushion symbol does not indicate the cushion type. That is, straight, tapered, stepped, etc., but it will indicate whether it is fixed or adjustable.

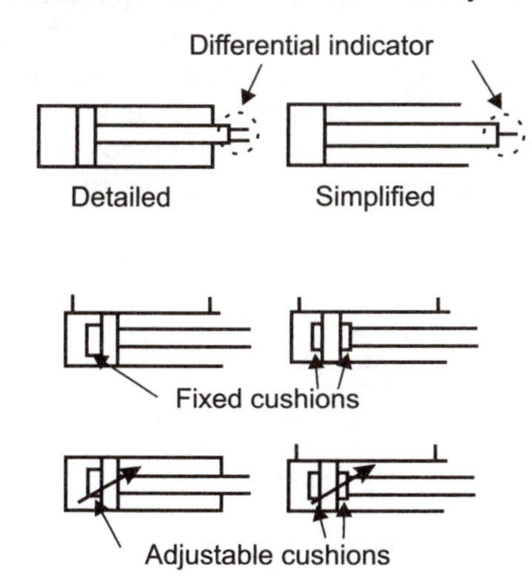

Telescopic Cylinders

When cylinders contain more than one rod and these rods collapse inside one another, it is called a telescoping cylinder. They can be either single acting or double acting.

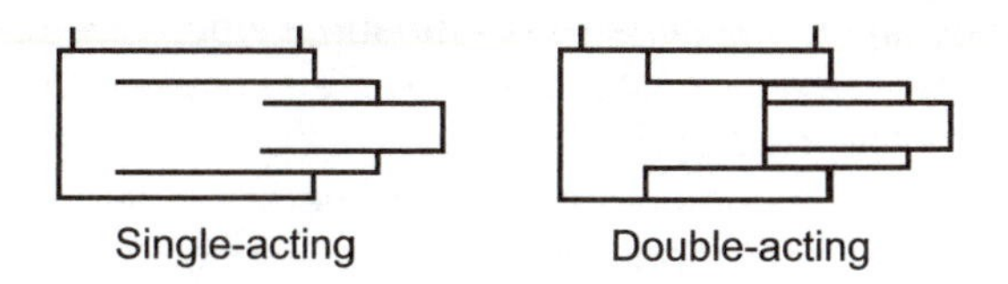

Pressure Intensifier

A specialized cylinder containing two pistons of different areas, connected with a common rod is known as a pressure intensifier. The intensifier can have similar fluids entering and leaving the ports as shown by the open triangles at each end of the symbol. There is also a center port indicated by the small vertical line that is a bleed-to-atmosphere port.

When dissimilar fluids (e.g. air and oil) are used as indicated by the open and darkened triangles, the cylinder body is divided into two parts. To the left, air pressure enters and there is a small vertical line representing an air-bleed. In the right half, hydraulic fluid exits and a drain line is shown on the bottom.

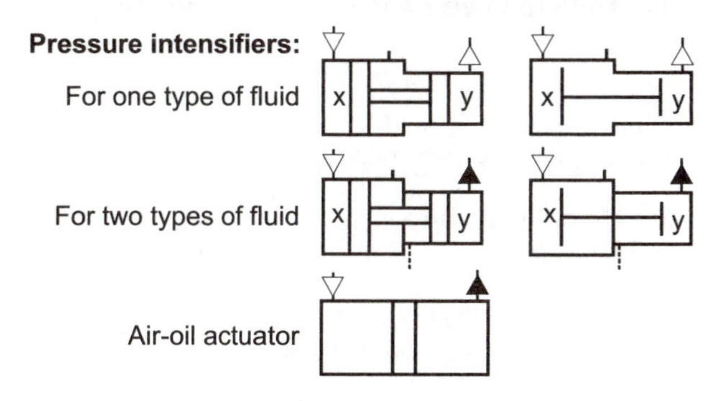

Control Valves

In control valve symbology, the square is used for housing valve functions such as controlling flow, direction of flow and/or pressure. The single square indicates a component used for controlling flow or pressure, various conditions may exists inside with regard to the operating condition of the component.

Two or more squares together indicate a directional control valve having as many positions as there are squares. The external piping as represented by the small vertical lines are normally connected to the square representing the inoperative or non-actuated condition. The operating positions are determined by imagining the squares moving right or left until the ports of the various conditions represented inside the squares, line up with the external piping.

Many times in a schematic, the same type valve can be used several times. To save time in drawing the symbol repeatedly, a simplified symbol for the valve is used. The number inside the square refers to a note on the schematic in which the symbol for the valve is drawn in detail.

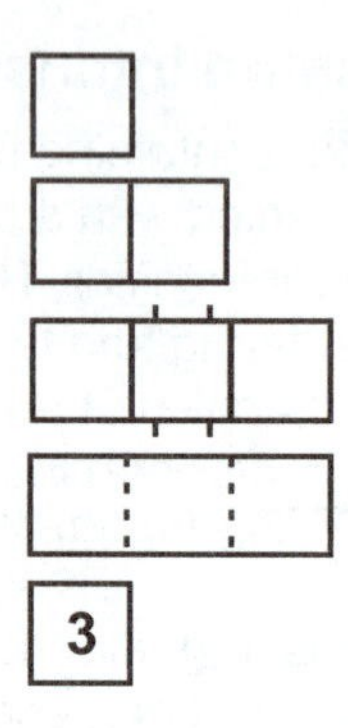

NOTE: Though this practice is used in very complex diagrams, it can be very time consuming and confusing having to refer to the referenced valve each time you encounter this symbol.

Flow Paths

Various flow paths may exist in one directional valve symbol. The most common types are:

One flow path, the arrowhead shows direction of flow and it can be drawn in either direction.

Two closed ports, represented by two "T's" in the square, indicates no flow between the ports.

Two flow paths in a single square indicates a four ported valve with flow in two directions.

Two flow paths in a single square with cross connection indicates that all four ports are connected.

One flow path in a bypass condition and two closed ports, indicate that the flow, e.g. from the pump enters the valve and exits the valve back to tank while the other ports are blocked.

The number of ports in a valve, which allows flow through them or blocks flow at the port indicated in one square are commonly referred to as the number of ways a valve has e.g., 2-way, 3-way or 4-way.

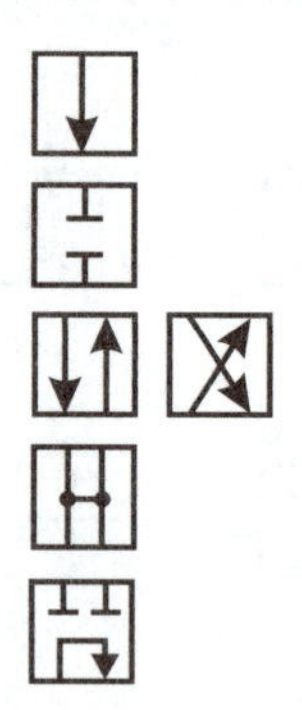

Directional Control Valve Designations 2/2 or 3/2

The 2-way, 2-position, manually controlled valve is often used as a safety or interlock or an on-off function valve. It can be operated in one direction by fluid pressure, manually or electrically and returned to its normal non-actuated position by a spring.

The symbol for a 3-way, 2-position valve is designated by two squares and three ports in each square and this valve is operated by pressure at either end. The other symbol at first glance looks like a 3-position valve controlled by a solenoid and returned by a spring. But notice the center square is separated by dotted lines. This designates the center position as a transitory intermediate condition or crossover condition. Therefore, this too is a 2-position valve.

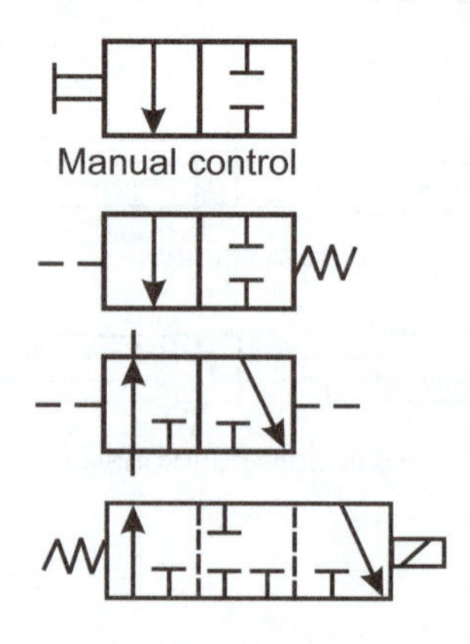

4/2 Directional Valve

A 4-way, 2-position directional valve is designated by two squares each containing four ports. Commonly used to give a reciprocating action to double-acting cylinders or to reverse the rotation of a motor, these valves can be operated directly by either solenoids or fluid pressure. Some are operated by the use of a second 4-way, 2-position valve called a pilot valve.

The pilot valve can be operated by two solenoids or a solenoid on one side and a return spring on the other to return the pilot valve to its normally non-actuated position.

The detail symbol shows an enclosure line around the two valves to indicate that these valves are one unit. The simplified symbol indicates the method of main directional valve controls, e.g. solenoid control, pilot operated, spring offset (return); and the type of main valve section only, a 4-way, 2-position directional control valve.

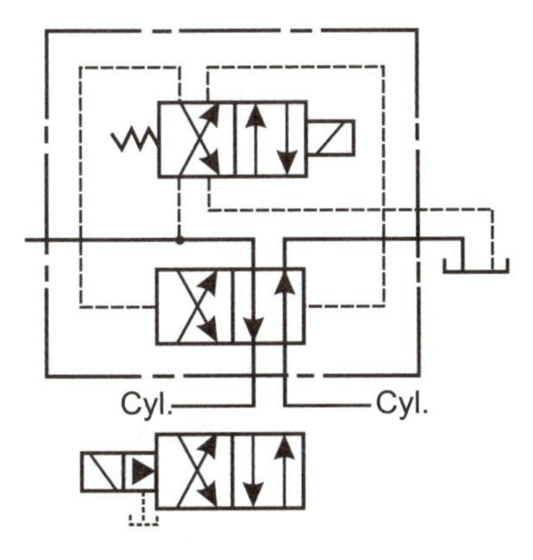

4/3 Directional Valve

4-way, 3-position directional valves are distinguished by a neutral or intermediate center (at rest) position. This center condition can consist of various flow paths. The four most common are: open center, closed center, tandem center and float center.

Open center

Closed center

Tandem center

Float center

These valves are commonly operated by pilot pressure in both directions by means of a second 4-way, 3-position valve called a pilot valve. The center position, for a spring centered, is a float center. Another center position used is a regenerative center for directional valves that have pressure centered main valves.

The pilot valve is controlled by two solenoids and centering springs. The main valve is controlled by pilot pressure and centering springs as shown in the detailed symbol. The simplified symbol shows the main valve positions and the control mechanism. This valve nomenclature is: solenoid controlled, pilot operated, spring centered, 4-way, 3-position directional valve.

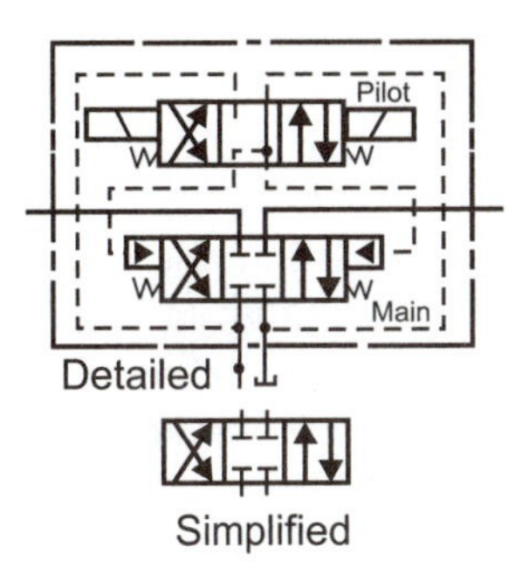

4/3 or 4/4 Mobile Directional Valves

4-way, 3-position or 4-position valves commonly used in mobile hydraulic systems are distinguishable by the open center flow path indicated in the center position.

These valves are commonly operated by levers into the various positions and spring returned to the center position. In some cases an electrical or mechanical device is added to the valve to hold the spool in a desired position. This device is known as a detent.

Throttling Directional Control

Throttling directional control valves or proportional valves, have 2 or 3 positions. The center position has an infinite number of intermediate conditions with varying degrees of throttling. Note that all the symbols have parallel lines at top and bottom along the length of the envelopes. These parallel lines indicate an infinite positioning capability, for example: a 2-way, 2-positioned tracer valve with one throttling orifice is shown operated by a plunger against a return spring or a 4-way, 3-position type with four throttling orifices.

Another example of a throttling valve's control is also controlled by pressure against a return spring.

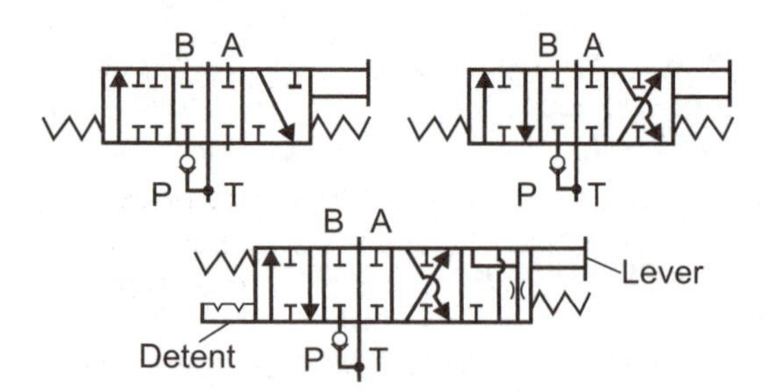

Electrohydraulic Servo Valve

Electrohydraulic servo valves accept an analogue or digital electrical signal and provide a controlled flow (via set spool position) in response to varied current. There are three basic types: the single stage valve operated electrically (direct), the two stage with mechanical or electrical feedback and pilot pressure operation of the main valve unit; the third type uses hydraulic feedback to indicate valve position and is operated by pilot pressure.

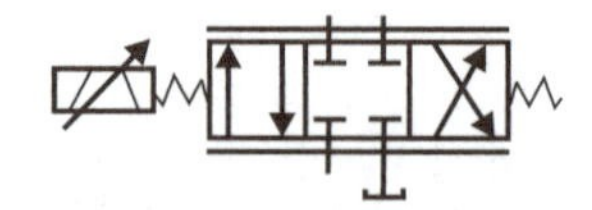

single stage with direct operation

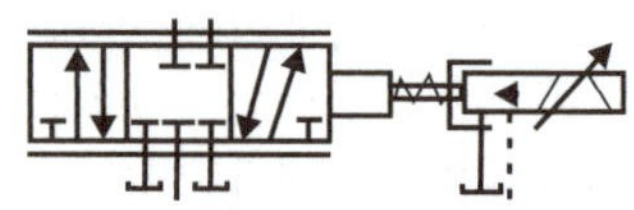

2-stage with mechanical feedback with indirect pilot operation

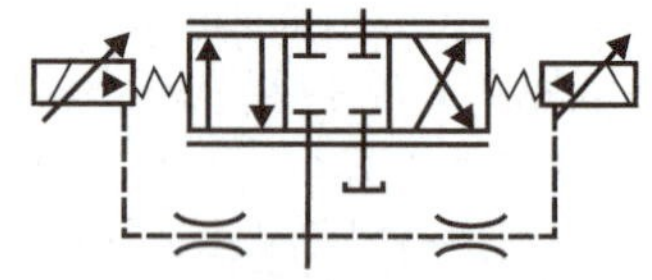

2-stage with hydraulic feedback with indirect pilot operation

Non-return Valves Check Valves

Non-return valves are a form of directional control valves. However, they only allow free flow in one direction. Normally known as check valves; there are four common types.

A free type check valve contains no spring biasing the moveable member (poppet or ball). It allows flow through it when the incoming pressure is higher than the outlet or back pressure.

A spring loaded check is held closed by spring force bias and opens only if inlet pressure is higher than spring pressure plug outlet back pressure.

A common use for this type check valve is shown in conjunction with an orifice symbol. These two symbols are contained in a rectangular box to indicate it is one unit.

Pilot operated checks can be either opened by or closed by an external pilot pressure.

Often these symbols are drawn incorrectly because people view the arrowhead as indicating direction of free flow. However, the arrowhead represented the seat for the moveable member (ball or poppet). Free flow is in the opposite direction. If any doubt exists, contact the manufacturer.

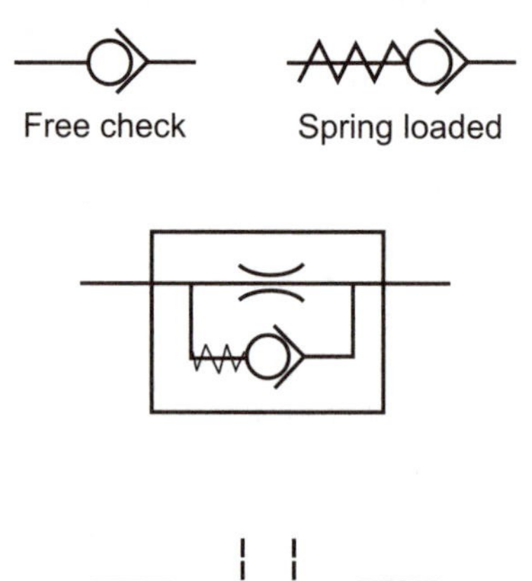

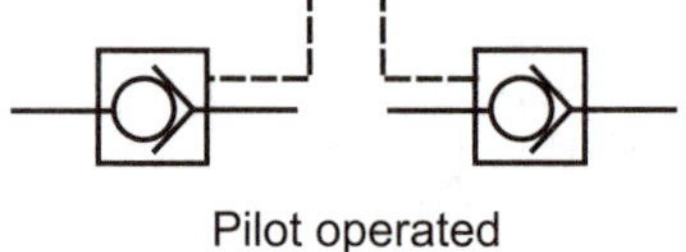

Pilot operated

Shuttle Valves

Another type of directional control valve is a shuttle valve. The ball will move to one side or the other depending upon which inlet port has the higher pressure, thus connecting that pressure to the outlet port and blocking the opposite inlet port.

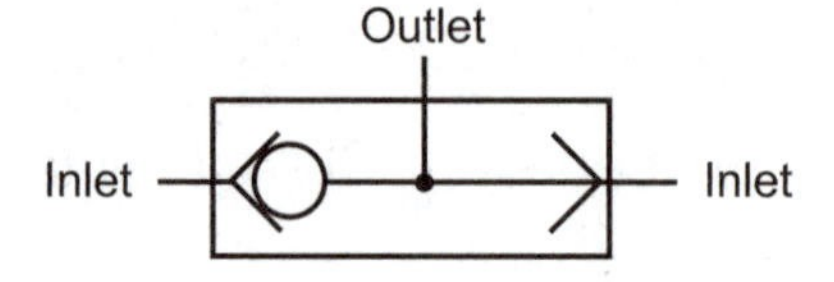

Pressure Control Valves

These symbols in general are represented by a single square with an arrow indicating direction of flow and flow path condition, e.g. normally passing (open) or normally non-passing (closed).

Pressure Relief Valves

Pressure relief or safety valves come in two styles: direct acting fixed or variable and a pilot operated.

Direct operated reliefs are represented by internal pilot pressure (dotted line) working against a fixed or variable spring pressure.

With pilot control, inlet pressure is limited to the combination forces of spring and pilot pressure. These symbols are presented in a normally non-passing (closed) condition.

In some diagrams a pilot operated relief valve is represented in a detailed drawing with an orifice between the main valve section and the pilot valve. The two combined symbols are represented inside an enclosure symbol to show that they are contained in one unit.

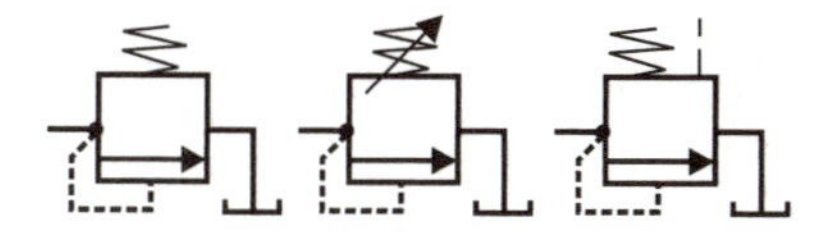

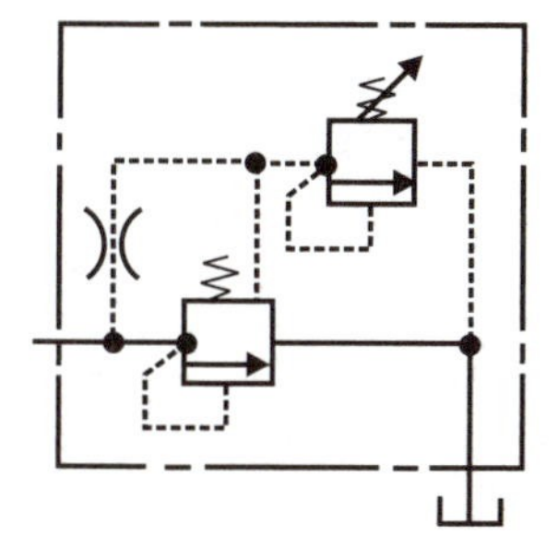

Proportional Relief Valve

Another type of pressure relief valve is a proportional relief valve. The inlet pressure is limited by a fixed ratio as determined by the size ratio between the large rectangle and small square.

The valve can be presented in two ways. One is to show internal pilot pressure having to develop a proportionally higher pressure than the remote pilot pressure. The other representation is the reverse action, internal pilot pressure has to be proportionally lower than remote pilot pressure.

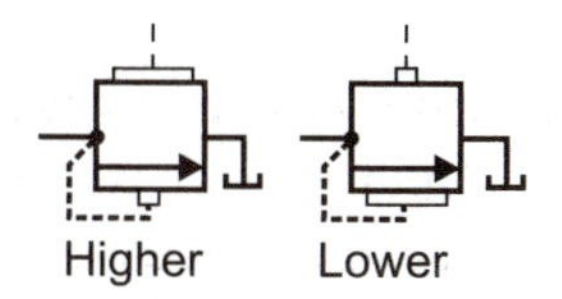

Sequence Valve

Sometimes in a system, one operating leg of the system must complete an operation before another operating leg can begin its cycle. This is often controlled by a sequence valve.

This valve will remain in its normally non-passing (closed) condition until a specific pressure has been reached in the upstream portion of the system.

Sequence valves require a check valve for reverse flow, can be directly or remotely piloted and are always externally drained.

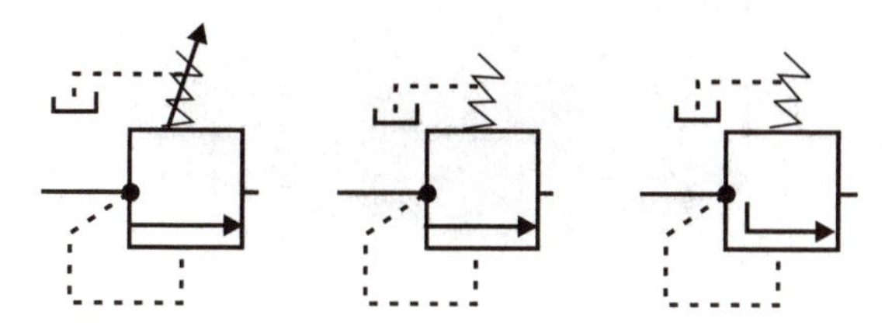

Pressure Reducing Valve

In a circuit where one leg may require a lower pressure than the rest of the system, a pressure reducing valve may be installed.

This valve is a normally passing (open) valve as indicated by the interior arrow flow path being aligned with the inlet and outlet ports. The internal pilot is sensing outlet or secondary pressure of the valve. The outlet pressure setting is determined by either spring pressure or external pilot pressure.

This valve is also represented with or without a relief port. This relief port allows secondary pressure that exceeds the pressure setting to be vented back to tank. The double arrowhead interior arrow indicates this flow capability. The perpendicular leg on the interior arrow indicates that the interior flow path always remains connected to the secondary or outlet path and therefore should pressure at the outlet become excessive, the movable member will align with the tank port to relieve.

To achieve reverse flow, a check valve is needed, and this valve requires external draining of the spring cavity.

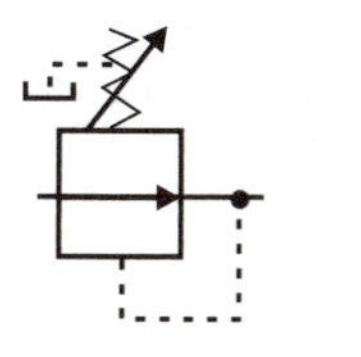

without relief port

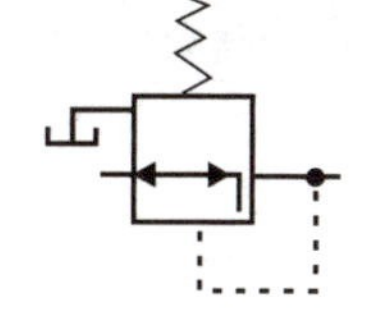

with relief port

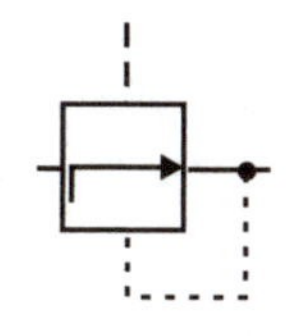

without relief port with remote control

Differential and Proportional Regulator Valve

These two pressure regulator valves both reduce outlet pressure to a lower pressure than inlet pressure.

A differential pressure valve is represented by a square that has a normally non-passing (closed) interior. As inlet pressure enters the valve, it is moved to a passing position. Pressure in the secondary or outlet port is also directed to the spring side of the valve, thus regulating outlet pressure to a fixed amount determined by the spring valve. This valve can also be represented with an adjustable spring symbol.

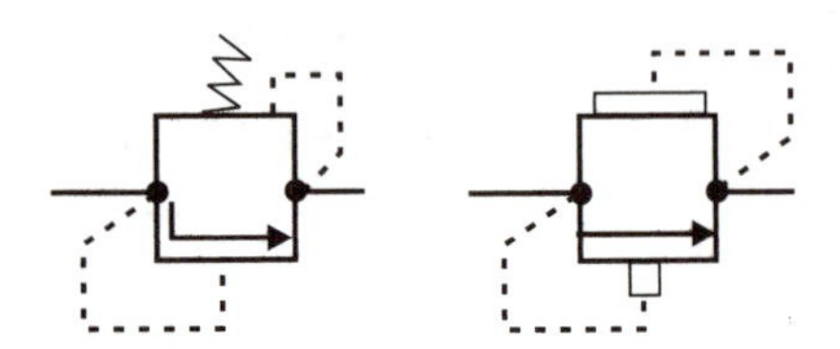

A proportional pressure valve is represented by a square that has a normally non-passing interior. The larger rectangle represents the larger control area. The outlet pressure is reduced by a fixed ratio as determined by the size ratio between the large rectangle and the small square.

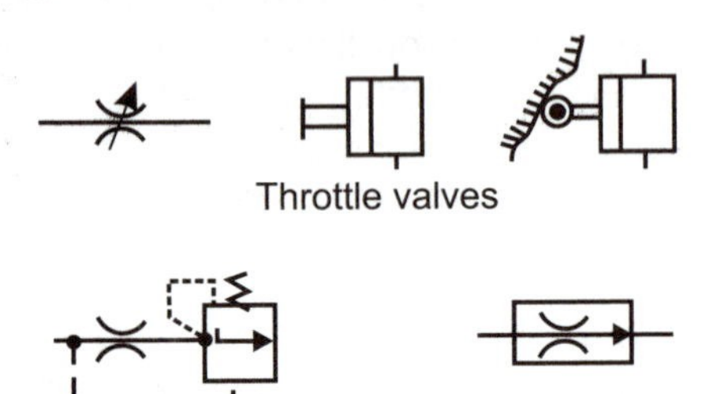

Throttle valves

Fixed flow pressure compensated restrictor

Flow Control Valves

Flow control valves affect the fluid flow rate. One type is known as a throttle valve. The simplified symbol represents an adjustable restriction commonly called a needle valve. However, a throttle valve detailed symbol shows that the valve can be manually controlled and the state of the valve can be fully non-passing (closed) or fully passing (open).

It can also be mechanically controlled against a return spring. These valves are commonly known as throttling valves.

The throttle valves would be affected by variations in the pressure and fluid temperature. There are flow valves that compensate for these variations in pressure and temperature.

Shown are a fixed flow rate, pressure restrictor type; another is fixed flow rate, pressure compensated bypass type with a relief port to reservoir.

The flow rate can also be adjustable as shown with pressure compensation. To show that the valve also compensates for fluid viscosity variation due to temperature, note the different restriction symbol.

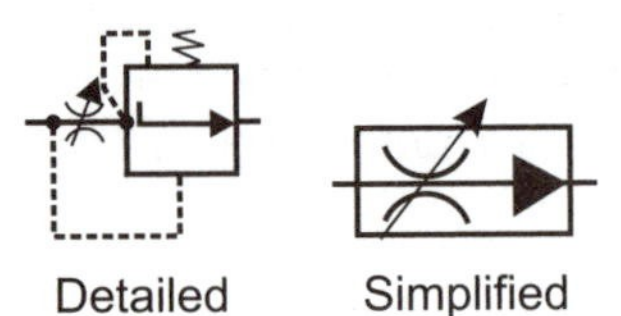

Adjustable flow pressure compensated restrictor

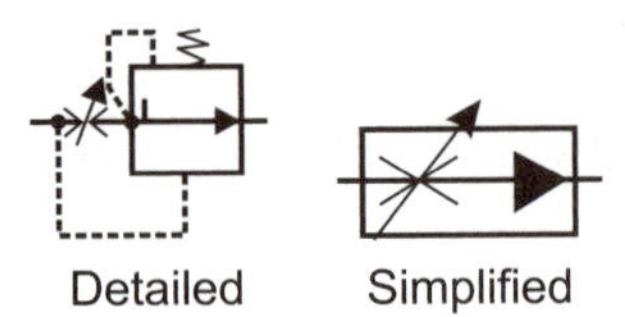

Adjustable flow pressure-temperature compensated restrictor

Flow Divider Valve

The output flow of a pump can be divided into flows of fixed ratios using a flow divider valve. The resulting flows are substantially independent of one another and of pressure variations in the lines. The symbol shown represents a fixed ration; however, variable outputs can be presented by arrows drawn on a 45° slant through the restriction. This symbol does not differentiate between the types of flow dividers available.

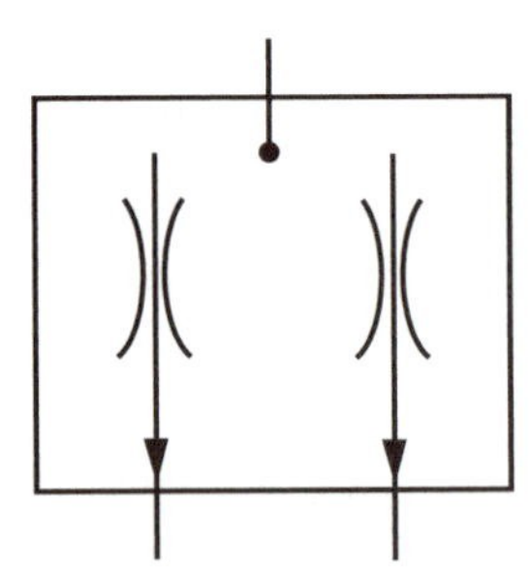

Shut-off Valve

The shut-off valve is primarily used in a closed condition but can also be represented in an open condition by separating the two arrowheads. The symbol can be used to represent a valve or a pipe plug in a valve passage.

Source of Pressure

Energy transmission and conditioning devices consists of prime movers, pressure sources, flow conduits, energy storage devices, filtration and temperature control devices.

When there is some pressure source present in a circuit, a general simplified symbol is used. But when the pressure source must be specified as hydraulic or pneumatic, a closed triangle indicates hydraulics and an open triangle indicates pneumatics.

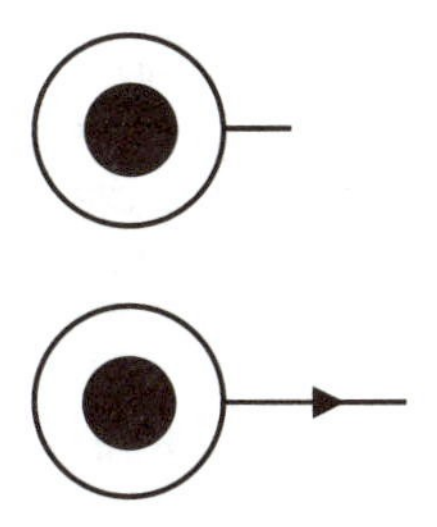

Prime Movers and Power Take-offs

The prime mover of a hydraulic pump is typically either an electric motor or an internal combustion engine. The two horizontal lines represent a mechanical connection.

In some equipment, such as in mobile applications, a PTO is used and the point where an auxiliary piece of equipment can be connected is represented by a line with an X. The X means that it is plugged, if there is an arrow pointing into the X, the power take-off is connected to some device.

Flow Lines

Long solid continuous lines are the symbols for main working lines, return lines and feedlines; long dashed lines represent pilot control pressure lines; short dashed lines represent drains or bleed lines.

A line that is flexible, such as a rubber hose, that connects two moving parts is represented by a concave line with dots at each end. The dots are points of connection.

Sometimes electrical lines are drawn in a hydraulic circuit diagram and they are represented as a solid line with an electrical symbol pointing to the line.

Lines that are connected, such as with tees or crosses, are denoted by a dot at the point of junction. Lines that cross but are not junctured are represented without a dot or by having a slight hump in one of the crossing lines at the crossing point.

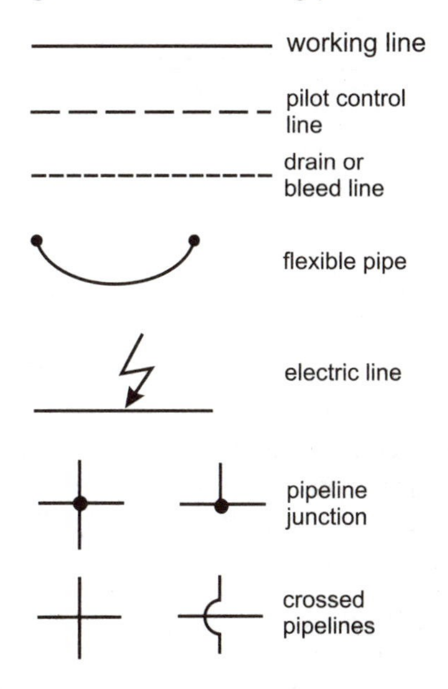

Quick Couplings

Quick couplings allow fast installation and removal of components without having to use wrenches. Basically there are two types available and represented.

The quick couplings can be equipped with a mechanically opened non-return valve (check valve) or without a check valve. They can be drawn in pairs to represent a connected or unconnected condition; or singularly in the port of valve, as an example, to represent the method of junction to the circuit.

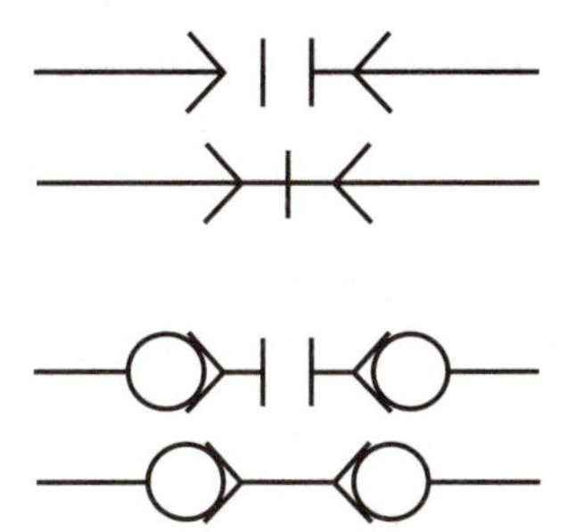

Rotary Connection

When a component(s) must rotate, the line(s) connecting is surrounded by a circular arrow. The number of lines passing through the circular arrow represents the number of flow paths, e.g. one way or three way.

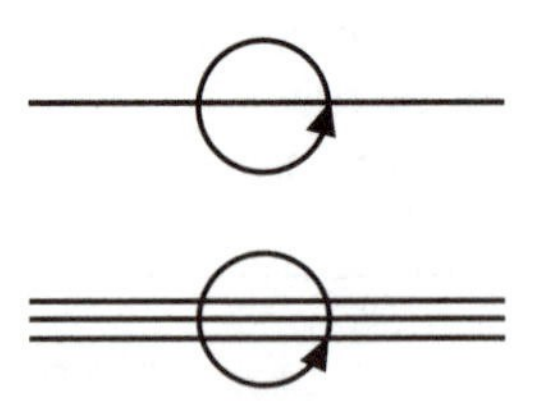

Reservoirs

There are three styles of reservoirs: open, closed or sealed, and pressurized. The lines entering and leaving the reservoir are important in their positioning. If the line terminates above the fluid level, it is considered a drain or bleed line with low pressure and a small volume returning to tank. When the line is drawn below the fluid level to the bottom of the tank, this represents a suction or return line. A line that is drawn from the bottom and below the tank represents an overhead tank.

Pressurized tanks are completely closed to atmosphere, the lines entering and leaving the tank are shown the same as the open-to-atmosphere tank.

Accumulators

When an accumulator is used in a circuit, it can be a weight loaded, spring loaded, or compressed gas type. The accumulator stores fluid under pressure. The oval shape represents the vessel for storage. If the accumulator elements, such as inert gas and fluid are separated, there is a line drawn across the center of the oval.

The element that develops the pressure will be represented by a square for a weight loaded accumulator; a spring or a spring loaded accumulator; an open triangle for a compressed gas (nitrogen) accumulator.

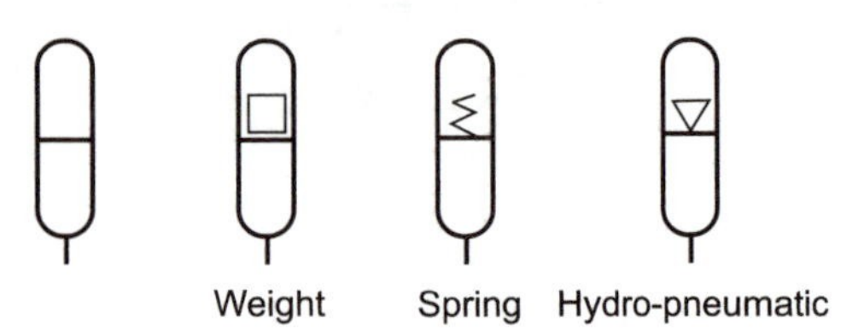

Fluid Conditioners, Filters and Heat Exchangers

No matter what type of filter is used, the symbol is always the same. What determines if the filter is actually a sump strainer, suction filter, pressure, or a return filter is its location in the hydraulic circuit diagram.

A temperature controller tries to maintain the fluid temperature between two predetermined limits. The arrows in the diamond indicate that heat may be induced or extracted as required.

Heat exchangers can either heat up, cool down, or maintain the temperature of the fluid. In the heater symbol, the arrows inside the diamond indicate the induction of heat to the circulating fluid. The cooler symbol can be represented in two ways. The arrows inside the diamond indicate the extraction of heat, but there is no representation of the type of coolant and coolant flow. Coolant can be either water or air as represented by the arrows for water flow outside the diamond.

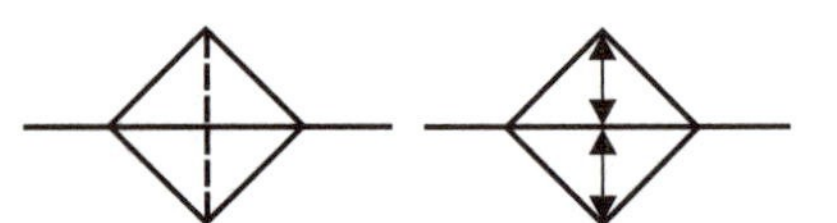

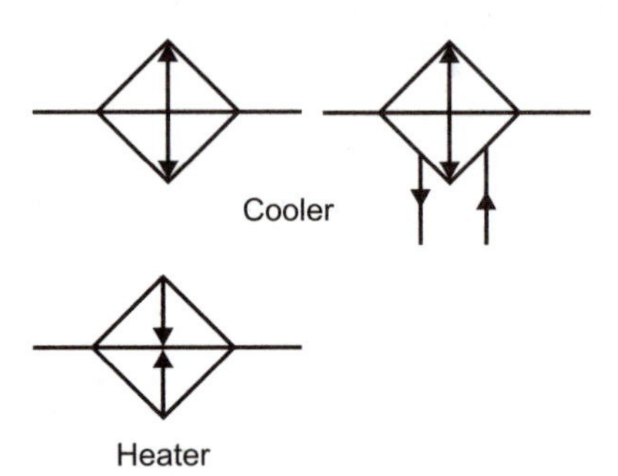

Mechanical Components

There are a variety of control mechanisms used in controlling actuation of valves, cylinders, motor/pump, etc. These mechanisms can be of the mechanical, electrical, hydraulic or pneumatic type.

Rotating shafts, such as a pump shaft are represented by double parallel lines. The arrows indicate unidirectional or bidirectional rotation.

In some applications of directional valves it is desirable to mechanically hold the valve in one position. This is represented by the detent symbol. Detents can be spring loaded balls or spring loaded electromagnets; the symbol does not indicate the type being used.

Mechanical locking devices that can stop, as an example a conveyor, at various positions are controlled by some type of actuator that would be drawn inside the square. The actuator could be a solenoid.

Overcentering devices such as stroke limiter are used to prevent a valve spool from stopping on dead center position.

When a device is designed to pivot on an axis, the simplified symbol can represent this. However, when the devices have to traverse and/or have a fixed fulcrum point, the complete symbols are used.

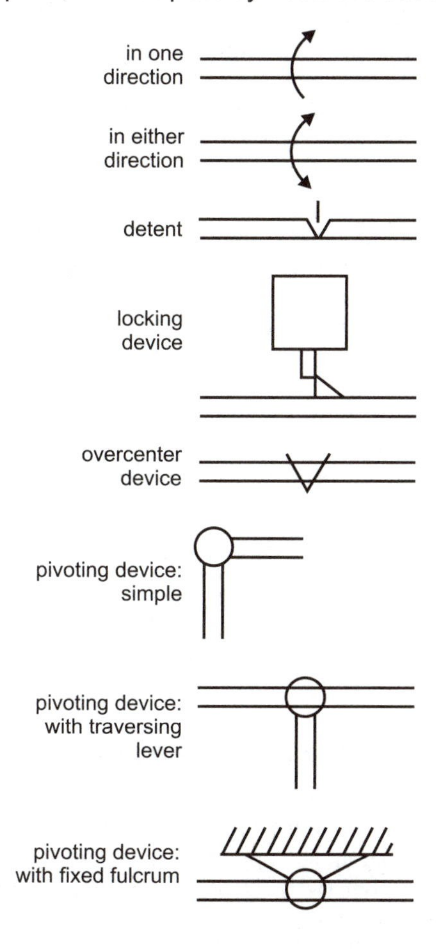

Actuator Methods

The actuator symbols are incorporated into the component symbol that is being controlled. Depending upon which side of the component symbol the control symbol is adjacent to, the square is affected by actuation of the control.

Actuator methods can be manual (human) controls, mechanical controls, electrical controls, pressure controls, or a combination of two or more controls.

Manual Controls

The general symbol for manual (human effort) control doesn't indicate any specific type of control. There are pushbuttons or palm buttons, hand lever and foot pedal controls.

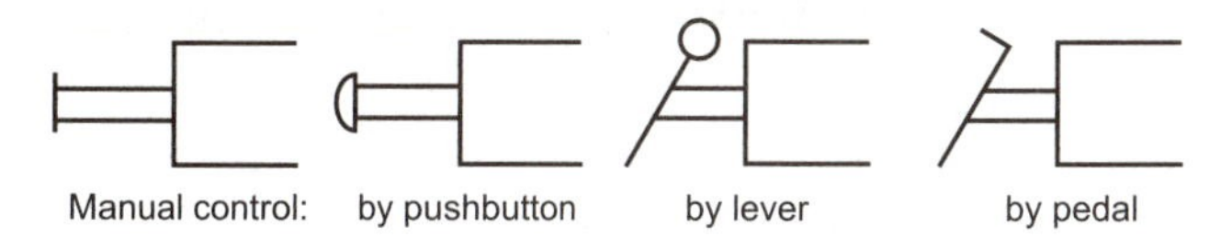

Mechanical Controls

Mechanical controls can be plunger or tracer, spring, roller for bidirectional control, or a roller control that operates in only one direction.

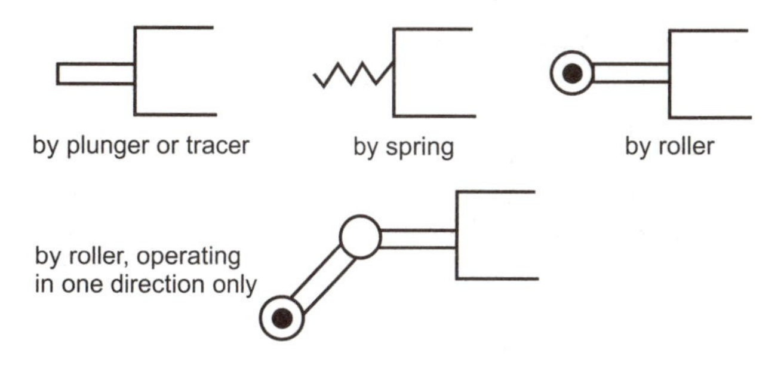

Electrical Controls

Electrical controls most commonly used are solenoids. The diagonal line in the rectangle indicates a winding. A single diagonal line indicates the solenoid can operate in one direction and is directly connected to the device being controlled. When the symbol is represented with 2 diagonal lines, it means the solenoid is capable of operating in both directions.

If the double lines have an arrow drawn on a 45° slant through them, it means that the solenoid operates in a variable way progressively in both directions.

The other electrical control is the electric motor. A circle with an M inside indicates the motor, rotation is bidirectional as indicated by the double headed arrow, or unidirectional if arrow has only one head.

Another electrical symbol is the torque motor represented by two triangles, an arrow and a small circle.

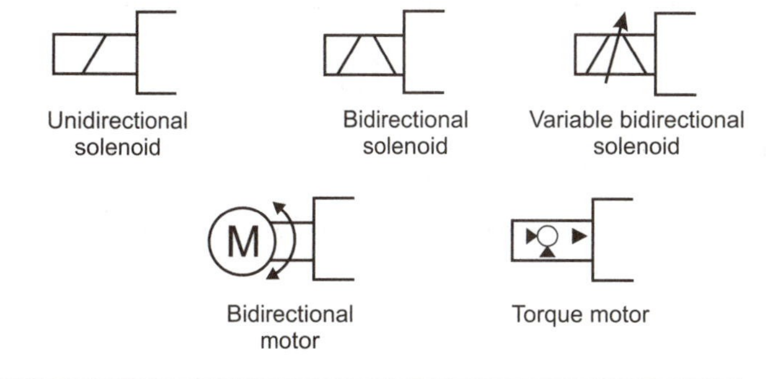

Pressure Controls

Control of a component can be done by either applying pressure or releasing pressure. Direct acting control shows either hydraulic or pneumatic pressure being applied directly to the square. Also, hydraulic or pneumatic pressure could be released directly from the square.

Control can be accomplished by different size control areas. The larger rectangle on the left represents the priority phase of the control.

Indirect control symbols indicate pilot operated directonal control valves. Pressure comes into the rectangle in the form of hydraulic or pneumatic pressure and can be released from the rectangle.

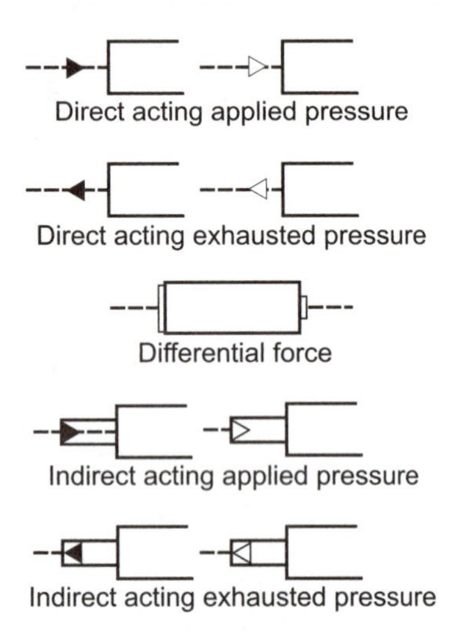

Direct acting applied pressure

Direct acting exhausted pressure

Differential force

Indirect acting applied pressure

Indirect acting exhausted pressure

Combined Controls

By combining two or more control symbols we can simplify, for example, the symbol for a solenoid controlled, pilot operated, spring centered directional valve. The pilot valve is controlled by the solenoid, the main valve would be controlled by pressure and springs.

If a valve can be controlled by either a solenoid or pressure, the control symbols are stacked.

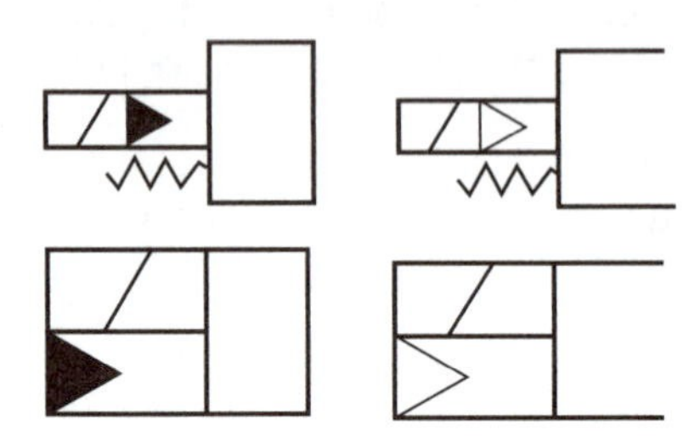

Miscellaneous Equipment

In a hydraulic circuit pressure, temperature and flow must be measured.

To measure either pressure or vacuum, a pressure gauge is installed. To represent the point of installation, a line is drawn to the circle. The arrow in the inside of the circle indicates pressure or vacuum. The location of this symbol in the circuit diagram will determine if it is pressure or vacuum that is being measured.

Temperature of the hydraulic fluid can be measured with a thermometer. The symbol is similar to a pressure gauge except for the thermometer symbol inside the circle.

The flow rate out of the pump or into and out of any component or leg of a hydraulic circuit is measured with a flow meter or an integrating flow meter.

Another often used symbol is the electric pressure switch. The dashed line is hydraulic pressure that actuates the switch, the spring holds the switch in either a normally passing or normally non-passing condition. Over the contact points you may find NO or NC to represent whether the switch is normally open or closed.

Some springs are shown as variable by placing an arrow on a 45° slant through the spring.

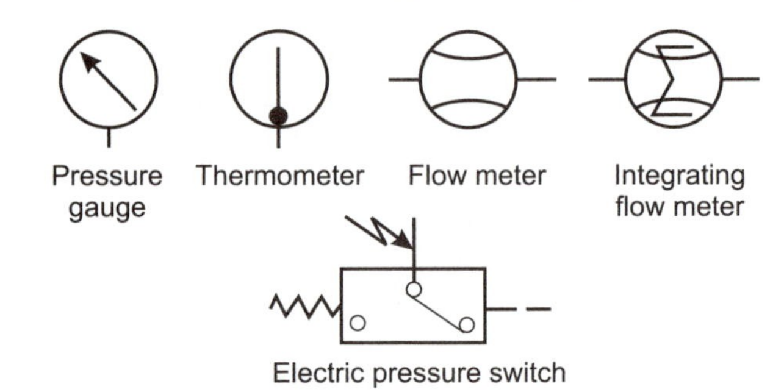

Pressure gauge | Thermometer | Flow meter | Integrating flow meter

Electric pressure switch

Mechanical Feedback Control

In applications such as a copying control, the mechanical connection between the controlling apparatus' moving part and the controlled apparatus is represented by this symbol: 1=controlled apparatus, 2=control apparatus.

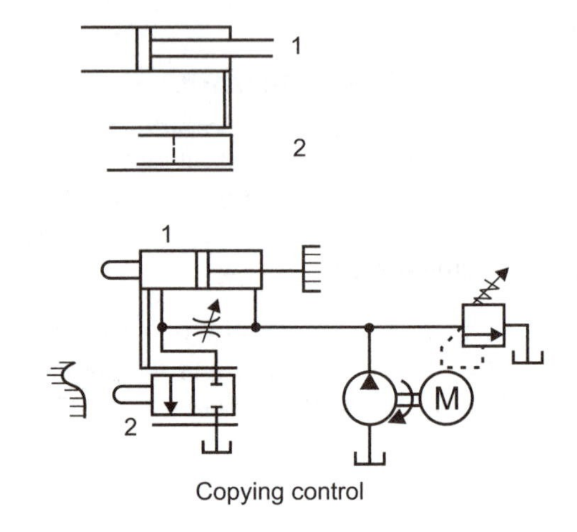

Copying control

Index

D

E

S